HABITAT SELECTION IN BIRDS

PHYSIOLOGICAL ECOLOGY

A Series of Monographs, Texts, and Treatises

A complete list of titles in this series is available from the Publisher upon request.

HABITAT SELECTION IN BIRDS

Edited by

MARTIN L. CODY

Department of Biology
University of California at Los Angeles
Los Angeles, California

ACADEMIC PRESS, INC.
Harcourt Brace Jovanovich, Publishers
San Diego New York Berkeley Boston
London Sydney Tokyo Toronto

ACADEMIC PRESS, INC.
1250 Sixth Avenue, San Diego, California 92101

United Kingdom Edition published by
ACADEMIC PRESS INC. (LONDON) LTD.
24–28 Oval Road, London NW1 7DX

Library of Congress Cataloging in Publication Data

Main entry under title:

Habitat selection in birds.

(Physiological ecology)
Includes index.
1. Birds—Habitat. 2. Birds—Behavior. 1. Cody, Martin L., Date.
QL673.H2 1985 598.2'522 84-20380
ISBN 0–12–178080–5 (alk. paper)
ISBN 0–12–178081–3 (paperback) (alk. paper)

PRINTED IN THE UNITED STATES OF AMERICA

89 90 91 92 10 9 8 7 6 5 4

Contents

Part III Habitat Selection in Specific Habitat Types

Chapter 6 Habitat Selection in Grassland and Open-Country Birds
Martin L. Cody

Chapter 7 Habitat Selection in Variable Environments: Shrub–Steppe Birds
John A. Wiens

Chapter 8 Habitat Selection in Temperate Marsh-Nesting Birds
Joanna Burger

Chapter 9 Dispersion Patterns and Habitat Responses of Birds in Northern Hardwoods Forests
Thomas W. Sherry and Richard T. Holmes

Chapter 13 Physiological Consequences of Microhabitat Selection
Glenn E. Walsberg

Chapter 14 Morphological Aspects of Habitat Selection in Birds
Hans Winkler and Bernd Leisler

Chapter 15 Habitat Selection: Behavioral Aspects
Peter H. Klopfer and Jörg U. Ganzhorn

Chapter 16 Habitat Selection by Nonbreeding, Migratory Land Birds
Richard L. Hutto

Chapter 17 Habitat Selection in Island versus Mainland Birds
Jacques Blondel

Contributors

Numbers in parentheses indicate the pages on which the authors' contributions begin.

Rauno V. Alatalo (59), Department of Zoology, Uppsala University, S-751 22 Uppsala, Sweden

Jacques Blondel (477), Centre National de la Recherche Scientifique, Centre d'Etudes Phytosociologiques et Ecologiques, 34033 Montpellier, France

Joanna Burger (253), Department of Biological Sciences, Bureau of Biological Research, Rutgers University, Piscataway, New Jersey 08854

Martin L. Cody (3, 85, 191), Department of Biology, University of California at Los Angeles, Los Angeles, California 90024

Hugh A. Ford (367), Department of Zoology, University of New England, Armidale, New South Wales 2351, Australia

Jörg U. Ganzhorn* (435), Abteilung Verhaltenphysiologie, Universität Tübingen, Tübingen, Federal Republic of Germany

Carlos M. Herrera (341), Estación Biológica de Doñana, 41013 Seville, Spain

Richard T. Holmes (283), Department of Biological Sciences, Dartmouth College, Hanover, New Hampshire 03755

Richard L. Hutto (455), Department of Zoology, University of Montana, Missoula, Montana 59812

Stewart W. Janes (159), Department of Biology, University of California at Los Angeles, Los Angeles, California 90024

Peter H. Klopfer (435), Department of Zoology, Duke University, Durham, North Carolina 27706

Bernd Leisler (415), Max-Planck-Institut für Verhaltenphysiologie, Vogelwarte Rodolfzell, D-7760 Radolfzell-Möggingen, Federal Republic of Germany

*Present address: Gluckstrasse 1, D-7032 Sindelfingen, Federal Republic of Germany.

Arne Lundberg (59), Department of Zoology, Uppsala University, S-751 22 Uppsala, Sweden

Douglass H. Morse (131), Division of Biology and Medicine, Brown University, Providence, Rhode Island 02912

David C. Paton (367), Department of Zoology, University of Adelaide, Adelaide, South Australia 5000, Australia

Michael L. Rosenzweig (517), Department of Ecology and Evolutionary Biology, University of Arizona, Tucson, Arizona 85721

Thomas W. Sherry (283), Department of Biological Sciences, Dartmouth College, Hanover, New Hampshire 03755

John Terborgh (311), Department of Biology, Princeton University, Princeton, New Jersey 08544

Staffan Ulfstrand (59), Department of Zoology, Uppsala University, S-751 22 Uppsala, Sweden

Glenn E. Walsberg (389), Department of Zoology, Arizona State University, Tempe, Arizona 85287

John A. Wiens (227), Department of Biology, University of New Mexico, Albuquerque, New Mexico 87131

Hans Winkler (415), Institut für Limnologie der Österreichischen, Akademie der Wissenschaften, A-5310 Mondsee, Austria

Preface

The topic of habitat selection has been central in behavioral ecology for several decades. Perhaps because of their exceptional mobility and the consequent implications of habitat choice or because of their overt and observable daytime behavior or perhaps just because they are so pleasant to watch, birds have contributed to habitat selection studies far beyond their numerical importance or their significance to biomass or trophic levels. For whatever reason, habitat selection has become almost synonymous with avian habitat selection, although certainly important work has been done with other higher vertebrates, such as Wecker's mouse studies [*Sci. Am.* (1964) **3**] and that by Kiester *et al.* on perch and substrate choice in arboreal *Anolis* lizards [*Ecology* (1975) **220**]. Even intertidal invertebrates, insects, or plants might make choices about which habitats to settle in or on, but here the sense is progressively further from a cognizant weighing of alternatives and closer to that of merely a better survival for the fortunate few that happen to land in the more favorable sites. And clearly, birds have the option of making *post facto* adjustments to their choice of habitat, an option beyond the repertoire of organisms with roots, strong attachment organs, or strictly temporary winged or swimming juvenile phases.

Habitat selection in birds, along with its evolutionary and ecological implications, has a long tradition in the literature, from Charles Darwin's time through the subsequent century and a half of natural history observations to recent and more quantitative work on distributional ecology. Key figures in the development of the theme and its pivotal questions are Joseph Grinnell, who discussed in the early 1900s why congeneric bird species so often occupy different habitats, and David Lack, whose earlier work was similarly centered around the role of divergence into different habitats in the adaptive radiation of bird species. Early reviews of the subject were written by Svärdson [*Oikos* (1949)] with an emphasis on the effects of interspecific competition on habitat use, and by Hildén [*Ann. Zool. Fenn.* (1964)], whose multifaceted approach guided many of the studies of the last two decades. The relevance of habitat selection for ecological studies on diversity and distribution was clearly stated by Peter Klopfer in his 1969 book "Habitats and Territories" (Basic Books, New York) and again in 1973 ("Behavioral Aspects of Ecology," Prentice-Hall, Englewood Cliffs, New Jersey). The subject constituted a large part of Lack's 1966 book "Ecological Isolation in Birds" (Blackwell, Oxford) and of Robert MacArthur's "Geographical Ecology" in 1972 (Harper and Row, New York), the classic statement of how habitat selection at the local level is related to biogeographical patterns. In my own monograph on bird communities (1974, Princeton University Press, Princeton, New Jersey), I set differences in habitat preference relative to differences in foraging ecology and behavior in the context of alternative aspects of resource partitioning.

The present book is divided into several parts. An introductory chapter serves to make the reader aware of the diversity of the subject of habitat selection in birds. Many of the various aspects of habitat selection introduced in the first chapter are developed in subsequent chapters, and thus it serves to some extent as an overview of the subject and as a "lead-in" to subsequent work.

A group of chapters treats specific taxonomic groups in Part II. Chapter 2 deals with the factors that affect habitat selection in a single species, the Pied Flycatcher (*Ficedula hypoleuca*), a species in which, because of sequential polygamy, habitat selection factors differ between males and females. After this are chapters that treat habitat selection in Old World sylviine warblers from the Sahara to the Arctic (Chapter 3) and, in considerable contrast, the New World parulid warblers (Chapter 4), which by and large are more foraging height and site specific, less flexible in habitat choice, and less interactive among species. Chapter 5 examines habitat selection in raptorial birds and the extent to which good habitat means high prey density or vegetation structurally suitable to a hawk's foraging behavior and morphology.

The chapters in Part III cover habitat selection from the point of view of certain habitat types. Perhaps the simplest of these is grassland (Chapter 6), where vegetation structure is easily quantified; but food resources are less easily measured, and their variability between sites and years must strongly affect habitat selection patterns in the birds. A species with a certain morphology and foraging behavior must find both suitable vegetation structure and minimum levels of food availability. Year-to-year variation in food resources also affects habitat use in shrub–steppe birds, whose different species react apparently independently to variable production by different floristic components in the vegetation (Chapter 7). Productivity may be much higher in marshes than in grasslands, but marsh-nesting birds are easily located by predators, and this plus strong intra- and interspecific aggression for food and nest sites modify species' use of marsh habitats (Chapter 8).

In Chapters 9 and 10 habitat selection in temperate and tropical forest, respectively, is discussed at necessarily a lower resolution in the latter habitat, where the challenges of dealing with the several hundred bird species that together occupy Amazonian floodplain forest must count among the greatest to face the field ornithologist.

Part IV contains chapters by authors with quite different perspectives on the questions of avian habitat selection. Chapters 11 and 12 deal with frugivorous and nectarivorous birds. Both are groups in which there is likely to be strong coevolutionary feedback between the birds as consumers and their food resources and habitats. Subsequent chapters (13–15) treat physiological, morphological, and behavioral aspects of habitat selection, and each brings a fresh perspective to the topic. In Chapter 16 the problems and choices faced by birds in transit between breeding and wintering areas are discussed; migrants may pass through habitats very different from those at either end of the migration, and these alien habitats may be stocked with resident competitors that make their use by the transients no easier. On the other hand, birds on islands may face relaxed competitive regimes compared to those on mainlands and expand in habitat use, although a reduced number of coexisting species does not necessarily mean reduced competition (Chapter 17).

Finally, the theoretical aspects of the subject are developed in the last chapter (18). Habitat selection theory was begun some fifteen years ago and has proceeded along several divergent lines: Lotka–Volterra models, optimization models on patch use, and stochastic models of regional coexistence in which immigration and extinction among and within habitat patches are the dominant processes. These models have a good deal to offer the field ornithologist from, at least, a conceptual framework in which to cast habitat selection studies to, at best, concrete, testable hypotheses.

It is hoped that this book will appeal to a wide range of ecologists, including those with behavioral and evolutionary bents as well as those in the field facing real-life situations that demand description and interpretation. It might even appeal to researchers who study organisms other than birds; indeed, many of the basic questions concerning habitat selection are taxon independent and may be asked of certainly most vertebrates and of many mobile nonvertebrates, such as butterflies and robberflies, with the behavioral machinery to search for and evaluate habitat quality. The book is essentially a current statement on how a wide variety of field ornithologists view a broad but fascinating topic, and I hope that some of its fascination will be passed on to the reader.

I am grateful to several of my colleagues for advice and comment on various chapters, including Roy Siegfried, Tom Howell, Carl Bielh, and Stewart Janes. I should especially like to thank John Emlen for the perspective he kindly lent to the introductory chapter; habitat selection in birds is a subject very dear to his heart, and perhaps no other field ornithologist has done more to explore its various dimensions. My own fieldwork, reported in Chapters 1, 3, and 6, has been supported by the National Science Foundation and Resources for the Future (research in Northwestern Mexico) and by the Guggenheim Foundation and the National Geographic Society (research in North and South Africa). I received assistance in fieldwork from Hartmut Walter (Sardinia) and Ted Case (Canary Islands); to all of these persons and agencies I express my thanks.

Martin L. Cody

A view of the riparian vegetation of *Tessaria–Cecropia* on a meander loop of the Rio Manu, Peruvian Amazonia. A succession of vegetation occupies these meander loops as rivers constantly change course over the level floodplain. The bewildering diversity of birds in these habitats presents the ornithologists with great challenges, from identification and field censuses to elucidation of habitat selection patterns (see Chapter 10, this volume). Photo by John Terborgh.

HABITAT SELECTION IN BIRDS

Part I

Introduction

Chapter 1

An Introduction to Habitat Selection in Birds

MARTIN L. CODY

Department of Biology
University of California at Los Angeles
Los Angeles, California

HABITAT SELECTION IN BIRDS

ISBN 0-12-178080-5

I. ROLE OF HABITAT SELECTION IN BIRD BIOLOGY

A. Historical Overview

Presumably since man first started foraging for animals and plants in the bush he has realized that different organisms live in different places and that different birds are found in different habitats. Such observations became commonplace in the literature of the Victorian naturalists and travelers and were then recognized to have some bearing on the more subtle differences between related species. For example, it was the association of different types of song with the different habitats in which he heard them that brought Gilbert White (1906) to the realization that there was not just one but three *Phylloscopus* warblers in his parish. The bird singing with a "sibilous, grasshopper-like" trill was usually found in "only the tops of beechen woods," *P. sibilatrix,* and contrasts to the "joyous and easy laugh" of *P. trochilus* and the "harsh chirp of two notes" that distinguishes the third species, *P. collybita*. Darwin, a supreme observer of nature, recorded numerous examples in his travelogue of the Beagle expedition (1897). He noted that the two species of geese on the Falkland Islands were segregated into an upland species, *Anas magellanica* (*Chloeophaga picta*), and a rocky-shore species, *A. antarctica* (*C. antarctica*). Later he found the mockingbird *Mimus orpheus* (*M. triurus* Viellot, 1818) common around habitation in Patagonia, and first realized that the mockingbird of the scrubby plains might belong to a different species [*M. patagonica* (*M. patagonicus* LaFresnaye and D'Orbigny, 1837)] because of its "slightly different tone of voice," the taxonomic difference being later confirmed in London by Mr. Gould. The fact that later, in the Galapagos Islands, he encountered three species of *Mimus* (*Nesomimus*), all obviously closely related but all resident on different islands (Charles: *M. trifasciatus;* Albemarle: *M. parvulus;* James and Chatham: *M. melanotis*), was cause for further and significant speculation!

More modern studies of habitat selection were also closely tied to questions about the taxonomy of congeneric species and indeed were first approached from that point of view. Joseph Grinnell's papers from early in the present century give a clear idea of the differences in habitats occupied between chickadees *Parus* spp.) in the boreal forests of the northwestern United States and between thrashers (*Toxostoma*) of southern California chaparral and desert edge (Grinnell, 1904, 1917). A more ecological approach was developed by David Lack (1933), who watched the colonization of pine plantations on Breckland Heath, southern England, by species apparently drawn to appropriate ''ancestral'' habitat through ''psychological'' factors. He noted, however, that, where more birds were available than their preferred habitat could accommodate, some species would expand into other habitat types. A later analysis of woodland birds (Lack and Venables, 1939) confirmed that, while some species had strong preferences for habitats with particular features, such as conifers, broadleaf forests, an abundance of nest holes, or taller trees, others in certain circumstances were quite flexible. Hole-nesting species, for example, are more widely distributed in the winter than in the breeding season, Siskins (*Carduelis pinus*) breed in spruce and fir but winter in pines, and Bramblings (*Fringilla montifringilla*) breed in Scandinavian birch and pines but winter in English beech woods. Beyond these seasonal shifts were noted differences in habitat selection between different regions, such as the strong preference of Lesser Whitethroats (*Sylvia curruca*) for conifers in continental Europe but not in Britain.

The role of competition from other species in habitat selection was made explicit by Svärdson (1949), who emphasized that, while intraspecific population pressure tended to broaden habitat use, interspecific competition tended to limit it. His study of Wood Warblers (*Phylloscopus sibilatrix*) showed that habitat choice was innate in young males but was perhaps reinforced by the singing of older males in the habitat; later arrivals in the breeding areas were often the younger birds, which were attracted to the vicinity of singing conspecifics.

B. What Habitat Selection Means to a Small Bird

1. Various Determinants of Habitat Selection

Perhaps no other taxonomic group has, and presumably exercises, the potential for habitat selection that birds do. Birds are extremely mobile and wide ranging, and of the range of habitats they pass through or over, only specific ones are used for breeding or foraging or wintering. The uniqueness of birds with respect to habitat choice was discussed by Hildén (1965), who distinguished between and summarized the ultimate and proximate factors involved in the choice. The evolution of habitat preferences is determined by, and determines, the bird's morphological structure and behavioral functions, its ability to obtain food and shelter successfully in the habitat. The proximate stimuli for the choice of habitat might be structural features of the landscape, foraging or nesting opportunities, or the presence of other species. Such factors might operate independently, hierarchically as a system of sequential decisions or overrides, or synergistically in a complex fashion or "gestalt."

For the average small temperate passerine, such as a Song Sparrow (*Melospiza melodia*) or a Garden Warbler (*Sylvia borin*), the choice of suitable habitat must be made repeatedly, often in both wintering quarters and in breeding territory, and perhaps continuously during migration. These choices affect the potential for both survival and reproduction and must be the product of many generations of natural selection. Besides providing food and shelter from predators and the weather, breeding habitats must ensure the best chances of securing mates, nesting, and raising young. But not all of these factors will be important for all birds at all times: some species, for example, may be essentially immune from predation by virtue of their speed or size, others, from inclement weather by virtue of size or a benign climate; and some birds may have already mated when they set up breeding territories. For migrants, the selection of wintering grounds may be independent of reproductive considerations, but not if they spend the nonbreeding season in flocks in which social status is defined and later bears on securing prime territory and attracting mates (e.g., Fretwell, 1972). Thus, at certain times particular features of the habitat may predominate, and at other times they may be overridden by yet other features. It is against the background of variations of critical resources in time and space that habitat selection studies must be conducted.

2. *Some Examples of Important Habitat Features*

Many papers have pointed out the special significance of certain habitat features to particular birds. For Belted Kingfishers (*Megaceryle alcyon*), for example, stream quality is related to the riffles in which they preferentially feed (Davis, 1982), and perch sites located in open ground with low cover are critical for Ferruginous Hawks (*Buteo regalis;* Wakeley, 1978). The topographical differences between plains and mountains differentially affect flying conditions and hence foraging success for the two vultures, White-backed Griffon (*Gyps fulviceps*) and Rüppell's Griffon (*G. ruppellii*) in East Africa (Houston, 1975), and in northeastern Rwanda, six species of Lapwings (*Vanellus* spp.) select different foraging substrates, such as lakeshores, floating vegetation, and dry grassland (van de Weghe and Monfort-Braham, 1975), which would be inconsequential to foraging vultures. Pairs of Japanese Wagtails (*Motacilla grandis*) on winter territories forage along rivers, but White Wagtails (*M. alba*) forage further from running water, and mostly singly (Higuchi and Hirano, 1983). Three sympatric species of grebes in North Dakota choose breeding ponds on the basis of differences in pond size and the extent of the emergent vegetation (Faaborg, 1976).

Most small birds apparently distinguish habitats on the basis of structural characteristics. Thus, in Patagonian mockingbirds (see Section I,A), Gochfeld (1978) found that *Mimus triurus* avoided low desert scrub and increased in density with vegetation height (correlation $r = 0.798$). *Mimus patagonica* showed no such preference but occurred throughout taller as well as low scrub habitats. With reference to chickadees and thrashers (see Section I,A), species of the former select largely nonoverlapping breeding habitats in western Washington (Sturman, 1968), with Chestnut-backed Chickadees (*Parus rufescens*) in the taller coniferous vegetation and Black-capped Chickadees (*P. atricapillus*) in habitats where bushes and middle-story trees are common. Sympatric thrashers in southern Texas segregate according to vegetation density, with Curve-billed Thrashers (*Toxostoma curvirostre*) in more open brushlands and Long-billed Thrashers (*T. longirostre*) in denser habitat (Fischer, 1980).

C. Habitat Selection in Bird Ecology and Behavior

1. Adaptations to Habitat

The role of habitat selection in bird studies, whether their emphasis is on ecology, behavior, or physiology, is all-pervasive. The tautology that each bird is adapted to its particular environment or habitat gains respectability if it can be shown that, for example, specific features of habitat are responsible for the physiological differences between related species or that a particular species has evolved a certain foraging behavior in relation to certain structural features of habitat. In the hardwood forests of the northeastern United States, for example, Robinson and Holmes (1982) showed that the breeding passerines could be sorted into five major foraging categories, which were related to habitat structure in fairly obvious ways. But if food resources change dramatically, as they do during caterpillar outbreaks, these categories have to be modified. Observations such as that by Ulfstrand *et al.* (1981) that bill lengths of resident Great Tits (*Parus major*) are greater in deciduous forests than in conifers point up the importance of local morphological adaptations. These may be well developed in species in which different populations occupy very different sorts of habitat, such as in Song Sparrows, which in the western United States live in a wide variety of habitats from oak woodlands to coastal marshes and from coastal sage scrub to alpine meadows. This must be balanced by the realization that in some species, especially migrants which often must tolerate a great variety of habitats year-round, such local fine tuning of morphology may not be possible or advantageous.

2. Role in Community Structure and Dynamics

Habitat selection, by whatever process and of whatever specific aspects, results in certain species sets co-occupying a particular habitat type. Some species may occur there because they can forage successfully in the canopy; others, because the lack of ground cover is conducive to their litter-scratching proclivities. Thus, within a habitat, there are still ample opportunities for species to select different parts of the vegetation or different structural niches in which to center their activities. Among others, segregation by tree species or foliage type, or by foraging height in the vegetation, are obvious possibilities.

Given that similar parts of the same habitats may be utilized by bird species, there is still the potential for ecological segregation among species by dint of different foraging methods or use of different prey-capture tactics or equipment. The subject of resource partitioning in bird communities cannot be fully expounded here, but Lack (1971) and Cody (1974a) give a more complete treatment. Suffice it to say that, of the methods of resource subdivision widely recognized by researchers, habitat selection among different vegetation types and foraging site selection within habitats have been accorded prominent positions.

D. Importance to Speciation and Reproductive Isolation

The role of habitat selection in speciation and adaptive radiation was developed by Lack (1940, 1944), who emphasized the importance of habitat segregation in closely related species and of habitat diversity in the multiplication of congeneric species. Studies that have considered the history, evolution, and distribution of the habitats themselves, as well as of the bird species that occupy them, are able to shed most light on the process of adaptive radiation. Mengel's (1964) study of the evolution of parulid warbers in relation to successive glaciations in North America and their effects on the distributions of forest types and Hubbard's (1973) work on the evolution of arid-zone quail, finches, and thrashers in the southwestern United States and northwestern Mexico in relation to the expansion and contraction of arid areas with glaciation events are excellent examples of this sort of work.

Adaptive radiations within genera are common in topographically diverse regions with a small pool of colonists, such as on isolated large islands or archipelagos. The vanga-shrikes (Vangidae; see Rand, 1936) of Madagascar and the drepaninine honeycreepers of the Hawaiian Islands (Amadon, 1950) are examples; segregation of incipient species by habitat within islands, such as Madagascar or Australia (Keast 1961, 1980), or between islands within archipelagos, such as Hawaii or the Galapagos, has clearly played a major role in the radiation. Within topographically isolated continental regions, such as Chile, cut off by high mountains in the east, the Pacific Ocean to the west, and extremely dry deserts to the north, there are relatively few genera but many species per genus, the great majority of which segregate by habitat (Cody, 1970). Repeated invasions from without the country, as well as divergence in isolated habitats within, may have contributed to the adaptive radiations, among which species of ground-tyrants (*Muscisaxicola*), of dipper-like furnariids (*Cinclodes*), and of finches (*Phrygilus*), conceivably ancestral to the Galapagos Geospizinae, are examples.

E. Habitat Selection and Species Diversity

Studies on bird species diversity became widespread after Robert MacArthur and associates (MacArthur and MacArthur, 1961; MacArthur *et al.*, 1962, 1966) showed that, within a given region such as the northeastern United States, information theory measures of bird species diversity and of habitat structure are linearly related, Subsequent work has extended the application of MacArthur's techniques to birds and habitats worldwide, and to organisms besides birds, but has also revealed limitations not only to the generality of the parameters but also to the circumstances in which the basic relation itself, of higher diversity in structurally more complex habitats, may be upheld (see Section IV,E).

Within broad limits, these diversity studies show that structural aspects of habitats can be used to predict diversity and are at least correlated to features of the habitat that the birds themselves use. In general, the numbers of species that pack into a habitat, defined as α-diversity, are directly related to structural diversity, and in turn structural diversity is related to either resource diversity or the numbers of ways in which resources can be partitioned.

A second component of species diversity is the rate at which species composition changes with changing habitat type, or β-diversity, and is directly related to habitat selection. If bird species are narrowly restricted in the range of habitats they occupy along a habitat gradient, then for a given level of α-diversity, β-diversity will be higher than if species are more broadly distributed over habitats. In areas of comparable habitat diversity but different-sized bird faunas, it is the β-diversity component that increases far more with faunal size than does α-diversity within habitats (Cody, 1975, 1983a). Regions of overall low bird diversity, such as Chile, exhibit low β-diversity and are characterized by bird species with extremely wide habitat tolerances (Cody, 1970); the Rufous-naped Sparrow (*Zonotrichia capensis*), with a distribution from Mexico to Tierra del Fuego and a habitat range from grasslands and scrub to *Nothofagus* forest, is an extreme example.

II. A FIELD EXAMPLE: BIRDS ON THE NEVADO DE COLIMA, MEXICO

To illustrate some of the various aspects of bird habitat selection, I shall use bird census data collected in March through April 1982 and April through May 1983 in the state of Jalisco in western Mexico. The censuses were made at a series of 20 study sites located within a limited geographic area and covering the range of natural habitats from sea level on the Pacific coast to nearly 4000 m at tree line on the Nevado de Colima, the prominent mountain near the state line in the south.

A. Geographical Setting

The major topographical feature in northwestern Mexico is the Sierra Madre Occidental, a mountain range running approximately 1200 km south from near the United States border to near Guadalajara. Further south, across Laguna Chapala, a series of more isolated mountains, the Eje Volcánico Transversal, runs from the Pacific coast east to the south of the Valley of Mexico. The western section of this area covers the states of Colima and Jalisco and is dominated by the twin summits of the Nevado de Colima, Jalisco, 4330 m, and the smoking Volcan de Colima on the Jalisco–Colima boundary. I selected study sites and censused birds in all of the major vegetation types in the area bordered by the coast from Puerto Vallarta in the north and Manzanillo in the south inland to Colima and Ciudad Guzman in the south and to Cocula in the north.

B. Vegetation Zonation with Elevation and Distance Inland

Jalisco is the western part of the region referred to as Nueva Galicia by Rzedowski and McVaugh (1966), who have described the vegetation of the area. At lower elevations toward the coast, the three most important types of vegetation are, with increasing precipitation, thorn forest, tropical deciduous forest, and tropical subdeciduous forest. These habitats are of increasing stature, respectively, and leguminous trees and shrubs are important in all of them. In the drier and shorter thorn forest, *Acacia, Ruprechtia,* and *Achatocarpus* are common; in the deciduous forest, *Bursera* spp., *Ceiba,* and *Lysiloma* dominate; and in the taller and more mesic subdeciduous forest, *Brosimum, Ficus,* and the palm *Orbignya* are characteristic. At higher elevations, one finds a subtropical scrub vegetation and in the driest inland valleys, a crassicaulescent scrub with many cacti, especially *Opuntia streptacantha,* and other succulents. Much of the region's interior is covered with pine–oak woodlands, with oaks (e.g., *Quercus castanea, Q. acutifolia, Q. macrophylla*) commoner in lower and drier sites and pines (e.g., *Pinus oocarpa, P. hartwegii, P. ayacahuite*) at higher and wetter sites. Between 2800 and 3400 m on the Nevado de Colima, fir (*Abies religiosa*) joins the pines and oaks, and *P. hartwegii* gives way to alpine vegetation at around 4000 m.

C. Bird Diversity, Density, and Turnover

The results of earlier bird collections on the Nevado de Colima and in the state of Colima to the south, where habitats are generally similar to those I studied in

Jalisco, are summarized in Schaldach (1963). The 20 study sites I used are shown and labeled in Fig. 1a. Each site is characterized by its elevation and by a measure of habitat structure, log(HhA), which is derived from the foliage profile, a plot of vegetation density against height. This measure increases in direct proportion to the height H of the vegetation, the total amount of foliage A, and the ''half height'' h at which half the foliage A is above and half below. Approximate isohyets of annual precipitation, which show at least a threefold variation over this region, are also included in the figure.

A total of 202 bird species was censused at the sites, using the spot-mapping technique (see, e.g., Chapter 10, this volume). This species number excludes raptorial and nocturnal species, aerial feeders such as swallows and swifts, and brood parasites, which are species either broadly ranging and not habitat specific or not accurately censused by the spot-mapping techniques (M. L. Cody, in preparation). Of these 202 species, 25 were wintering or migrating, and another 5 species were wintering at some sites and breeding at others. Contours of bird species numbers are given in Fig. 1b, and total bird densities are included also.

Total bird species (63) and breeding species (53) were highest in the pine–oak vegetation at 2660 m on the Nevado; breeding species were fewest (12) in the timberline pines at 3610 m. Migrant or wintering species varied from 2 to 11 among sites and were most diverse in the low-elevation tropical forest. Both total bird species and breeding species were correlated with my measure of habitat structure but not spectacularly so ($r = 0.61$, $r^2 = 38\%$); the residuals were positively related to elevation ($r = 0.33$) and to precipitation. Migrant species numbers are less strongly related to habitat structure ($r = 0.36$) and to numbers of breeding species ($r = 0.33$).

Breeding bird density varied from 2.58 to 13.09 pairs/ha and was strongly associated with total annual precipitation ($r = 0.80$); migrant densities were generally around 1 individual/ha but again were more weakly correlated with environmental factors ($r = 0.33$ with precipitation) than were breeding species.

An initial indication of habitat use by these birds in spring is given by the numbers of sites occupied by species. Breeding species averaged 3.6 sites/species, and wintering species, just fractionally higher at 3.9. Thus both breeding and wintering species are quite restricted in the habitat range they utilize and a more precise indication of this is given in Fig. 1c. Here sites are classified or grouped according to the numbers of species they support in common. Differences in species occurrence, or species turnover, are measured between and among sites by crossing the contours in the figure. In this way species sets typical of certain habitat types are identified. Thorn scrub species, for example, are typical of the lowest-structure habitats to the left, and different species sets are characteristic of the tropical deciduous forest (bottom center) and tropical subdeciduous forest (lower right). Lower-elevation oaks (center) support generally different bird species than the pine–oak and pine–oak–fir vegetation at higher elevations, and a further subset of species is found at the highest elevations in pine–fir forest. The challenge of habitat selection studies is in part to interpret patterns of species turnover such as that in Fig. 1c. These patterns are presumably the result of many interacting factors: some species may be relatively restricted in the range of habitats they occupy, and others, restricted in elevation despite a wider range of their apparently preferred habitat structure. How do factors such as habitat structure, competitor species (both resident and wintering), resource productivity and predictability, and the areal extent of habitats and identity of adjacent habitat all interact to produce the observed bird distributions?

Fig. 1. Bird census data from a variety of habitats in Jalisco, western Mexico. (a) Above and inset is a map of northern Mexico showing the area in Jalisco, on the Pacific coast of western Mexico (shaded), from within which 20 study sites over a wide variety of habitat types were selected and censused. Each study site is numbered and named on the figure, and characterized by elevation above sea level (ordinate) and an index of vegetation structure (abscissa). The index of structure is log (HhA), where H is vegetation height, h is the vegatation half height, at which half the vegetation (leaf area) lies above and half below, and A is the area under the foliage profile, a plot of vegetation density against height above ground. Thus short and open habitats rank to the left, and tall, dense, and high-canopied habitats rank to the right. The census habitats range from short thorn scrub, at left, to tropical deciduous woodland at lower elevation and oak woodland at higher elevation, in the center of the gradient, to tropical subdeciduous forest (low elevation) to mixed pine–oak and pine–oak–fir forests at higher elevations on the Nevado de Colima, to the right of the gradient. Isohyets of precipitation are included on this habitat plane. (b) Bird species diversity, in number of species (- - -), increases generally from left to right, with the index of habitat structure, but peaks in site 12 at 2600 m in mixed pine–oak–fir forest with over 60 species recorded. Total bird density, in pairs/hectare (···), follows a similar pattern but falls off the most structurally diverse forests, to the extreme right. (c) Each bird census is compared to each other, and the bird species turnover, or β-diversity, is assessed as the average number of species gained and number of species lost between census pairs. These turnover figures show among which habitats bird species identities remain relatively constant and among which habitats there is a relatively large or rapid change in bird identities. Thus, sites 16 and 17, low and tall thorn scrub, are most similar in bird species, with a $7\frac{1}{2}$ species turnover, site 1 (thorn mátorral) is similar to 16 + 17, with an 11 species turnover, and site 5, succulent scrub, joins this group next with a 12 species turnover. The turnover contours allow bird faunal groups to be identified within the set of habitats. The strong faunal groups are (a) the thorn-scrub birds to the right, (b) the tropical deciduous woodland birds in the lower center, (c) the oak-woodland birds in mid-elevation habitats, (d) the tropical subdeciduous forest birds in tall, low-elevation habitats in the lower right, (e) the birds of the pine, pine–oak, and pine–oak–fir forests of high elevation, within which a subgroup of typically temperate forest birds segregates the highest forests. (d) The distribution of five species of jays, largely habitat specific but with two species in many taller low-elevation habitats. Steller's jay may still be extant at the highest elevations on the Nevado de Colima but was not found by the author. These jays are all similar in size. (e) Six species of orioles (*Icterus*) and one of cacique occupy largely different habitats, except that the wide-ranging, Streak-backed Oriole cohabits several habitats with other species. Oriole body size decreases somewhat with the elevation of habitats. (f) Wren species (Troglodytidae) differ considerably in body size, and here the distributions over habitat of the ten species censused are distinguished according to four body size ranges (1) large wrens, >17 cm (····), (2) medium wrens, 13–14 cm (- - - -), (3) small to medium wrens, 12 cm (–––), (4) tiny wrens, 9–10 cm (——). (*Continued on pp. 14–15.*)

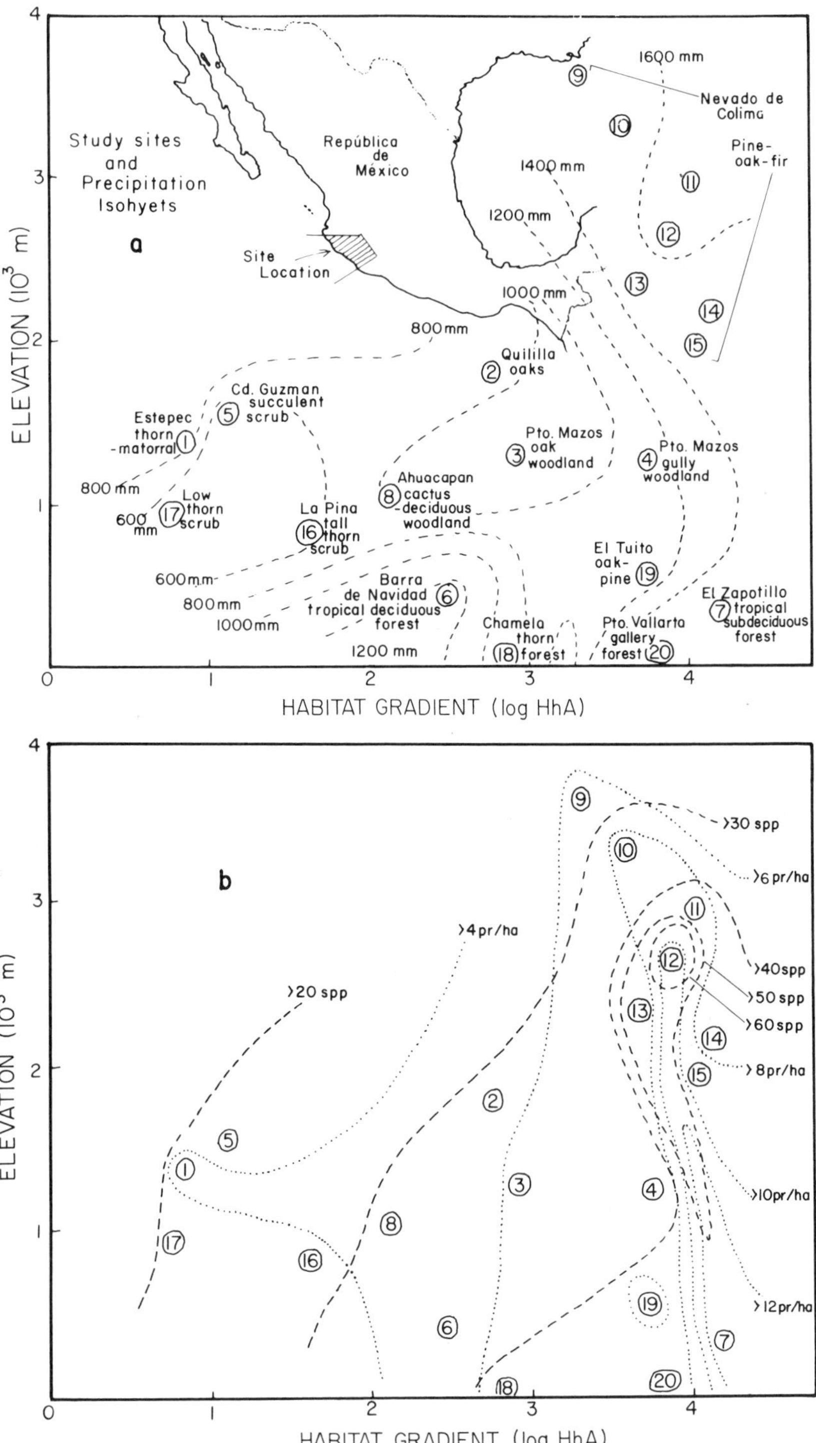

Study sites and Precipitation Isohyets
a
República de México
Site Location
1600 mm
Nevado de Colima
Pine-oak-fir
1400 mm
1200 mm
1000 mm
800 mm
Quililla oaks
Cd. Guzman succulent scrub
Estepec thorn -matorral
Pto. Mazos oak woodland
Pto. Mazos gully woodland
Ahuacapan cactus -deciduous woodland
Low thorn scrub
600 mm
La Pina tall thorn scrub
El Tuito oak-pine
Barra de Navidad tropical deciduous forest
El Zapotillo tropical subdeciduous forest
Chamela thorn forest
Pto. Vallarta gallery forest
ELEVATION (10³ m)
HABITAT GRADIENT (log HhA)
b
>30 spp
>6 pr/ha
>4 pr/ha
>20 spp
>40 spp
>50 spp
>60 spp
>8 pr/ha
>10 pr/ha
>12 pr/ha

Fig. 1a and b

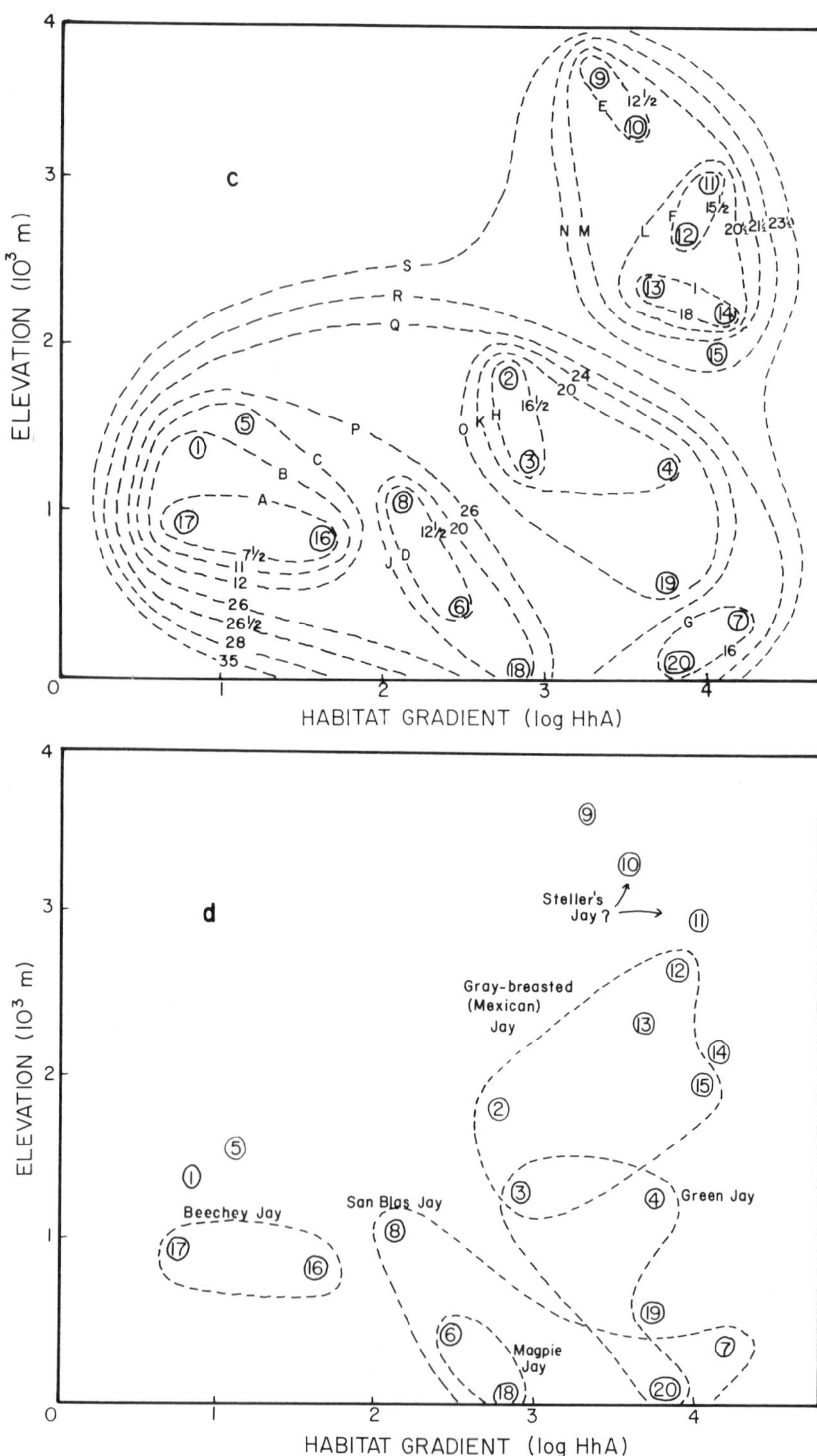

Fig. 1c and d *(Continued)*

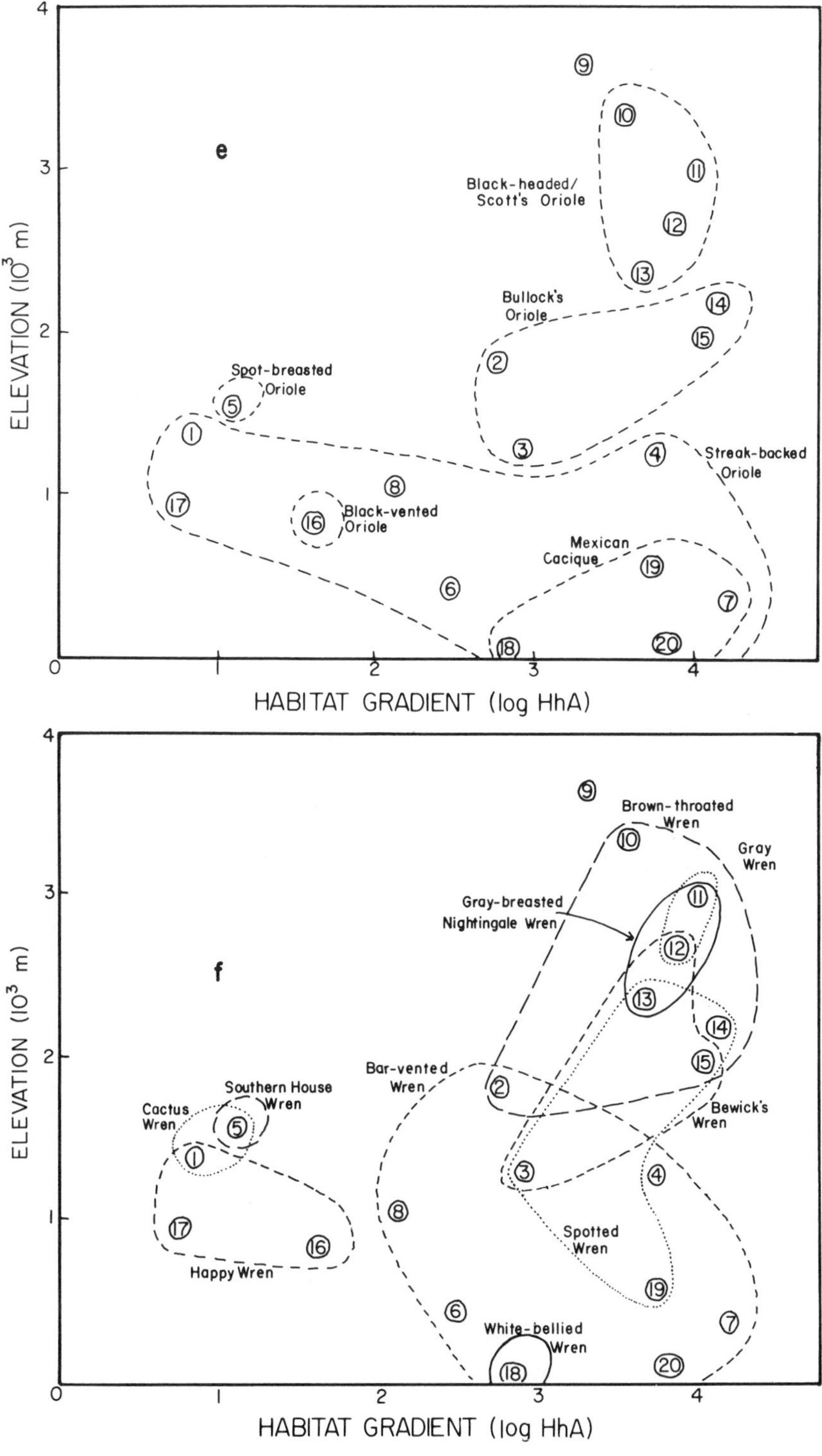

Fig. 1e and f (*Continued*)

Many of these factors can be observed in operation in data sets such as these; a few of them will be indicated next.

D. Variation in Habitat Occupancy: Breeding Species

Some breeding species occupy very distinct habitats at the census sites, distinctions not apparent in the crude measure of habitat structure, log(*HhA*), used as abscissa in the figures. For example, the cactus scrub of site 5 near Ciudad Guzman supports a breeding bird community very similar to that of the Sonoran Desert, which reaches its southern limits hundreds of kilometers to the north. At this site such typical desert species as Cactus Wren (*Campylorhynchus brunneicapillus*), Roadrunner (*Geococcyx californianus*), Pyrrhuloxia (*Pyrrhuloxia sinuata*), House Finch (*Carpodacus mexicanus*), Curve-billed Thrasher (*Toxostoma curvirostre*) and Ash-throated Flycatcher (*Myiarchus cinerascens*) are common, and many are found at no other of my census sites. Similarly, the two highest-elevations sites are distinguished by such species as Pygmy Nuthatch (*Sitta pygmaea*), Pine Siskin (*Carduelis pinus*), and Golden-crowned Kinglet (*Regulus satrapa*), species typical of high-elevation coniferous forest far to the north.

Distributions of species in fairly distinct ecological and taxonomic groups are shown in Figs. 1d–f. The five species of jays in Fig. 1d are classified in four genera, with Beechey Jay (*Cissilopha beecheii*) and San Blas Jay (*C. sanblasiana*) congeneric and segregated by habitat at lower elevations. San Blas Jay overlaps Magpie Jay (*Calocitta formosa*) in tropical deciduous woodland and Green Jay (*Cyanocorax yncas*) in wetter forest, and the latter marginally overlaps Gray-breasted Jays (*Aphelocoma ultramarina*) in the lowest oaks. Thus the jays appear to be habitat segregated, with two species present only in some low-elevation and high-structure habitats. The gap in higher-elevation pine–fir habitats should be occupied by Steller's jay (*Cyanositta stelleri*), which was reported from the Nevado by Schaldack (1963) but not collected; I failed to see this species during the 3 weeks I spent in that habitat and so did J. Terborgh (personal communication).

The same picture is presented by orioles (*Icterus* spp.) and the Mexican Cacique (*Cassiculus melanicterus*) in Fig. 1e. Note that two species, Spot-breasted Oriole (*I. pectoralis*) and Black-vented Oriole (*I. wagleri*), were found at just a single site, whereas Streak-backed Oriole (*I. sclateri*) is very widely distributed from forest to low scrub. Also of interest is the status of Northern Oriole (*I. galbula bullocki*) in mid-elevation oak and pine–oak; it alone of the six species is supposedly nonbreeding here, and I have no data to the contrary. The orioles resident in the higher pine–oak–fir I recorded in my notebook as Scott's oriole (*I. parisorum*) in a habitat that contrasts sharply with the low and open shrublands of the Mojave and Sonoran deserts the species occupies in the southern United States. This species was not found by Schaldach, but he recorded the similar *I. graduacauda* in mountains to the northwest of the Nevado. My identification may be questionable.

Unlike jays and orioles, among which size differences are small, the neotropical wrens come in many different sizes, and some sites support as many as four species (Fig. 1f). The three large *Campylorhynchus* species (Gray, Spotted, and Cactus Wrens) are all habitat segregated, as are the two *Troglodytes* species (Southern House Wren and Brown-throated Wren) and the two *Thryothorus* species (Happy and Bar-vented Wrens). Thus congeners co-occur at no sites, and maximum wren diversity occurs with a small species (Gray-breasted Woodwren, 9 cm), two medium-sized species in different genera (Brown-throated and Bewick Wrens, 12–14 cm), and the large Grey Wren (18 cm).

Different taxonomic groups reach their highest diversity in different sorts of habitat. Warblers (Parulidae) are most diverse in the pine–oak forest around 2500 m, but flycatchers (Tyrannidae) reach maximum diversity in the tropical subdeciduous forest (Fig. 2a). The speculation is this: Warblers are generally similarly sized and subdivide resources within habitats largely via foraging site differences (MacArthur, 1958; see also Chapter 3, this volume); this may be facilitated in the higher-elevation forests by the more continuous distribution of foliage vertically and from the inside to the outside of the trees. In the low-elevation tropical forests the foliage is more similar among tree species, and leaves are larger and perhaps more suitable for hover-gleaning than search-and-probe tactics. The foliage is also more discontinuously distributed, with many large gaps located vertically and between tree insides and outsides. In addition, the range of insect sizes is larger, and this might help explain why the tropical forests, with diverse gaps and a larger size range of insects, as well as backup fruit resources, support many sizes of sit-and-wait flycatchers but fewer warblers.

In Fig. 2b the ranges of some breeding warblers are shown; interspecific overlaps are considerable in the high-structure forests, but all except two of the species shown are in different genera, indicating considerable morphological differentiation (although not in body sizes). The two congeners are Slate-throated Redstart (*Myioborus miniatus*) and Painted Redstart (*M. picta*), with the former common throughout the pine–oak–fir, and the latter present only at the lower and more open pine–oak site.

The two largest genera of foliage gleaners in the region are the warblers (*Basileuterus*) and vireos (*Vireo* spp.; see Fig. 2c), with three species of the former and no interspecific habitat overlap. There are six species of vireos, with a good deal of overlap between species in taller habitats, just as in North American species (see Section IV,G,4). Unlike *Basileuterus* species, however, there are size differences among species, with Yellow-green Vireo (*V. flavoviridis*) about 30% heavier than Golden Vireo (*V. hypochryseus*) and Warbling Vireo (*V.

gilvus) intermediate in size between Hutton's Vireo (*V. huttoni*) and Solitary Vireo (*V. solitarius*). Bell's Vireo (*V. bellii*) is the only scrub species, and although singing and territorial, it is supposedly a wintering species here.

Species in two genera of flycatchers, *Myiarchus* and *Contopus*, are shown in Fig. 2d. Three *Myiarchus* species are generally habitat segregated at lower elevations [Ash-throated Flycatcher (*M. cinerascens*), Nutting's Flycatcher (*M. nuttingii*), and Brown-crested Flycatcher (*M. tyrannulus*)], but the fourth, Dusky-capped Flycatcher (*M. tuberculifer*) is ubiquitous at all elevations in taller forests. The two pewees overlap with very unequal ranges (like the *Myioborus* redstarts), with Greater Pewee (*C. pertinax*) broadly distributed and Western Wood-pewee (*C. sordidulus*) restricted to high elevations.

The ranges of thrushes and other thrushlike birds are shown in Fig. 2e. In low, open, and dry habitats are Northern Mockingbird (*Mimus polyglottus*) and Curve-billed Thrasher (*Toxostoma curvirostre*), both Mimidae, with Rosy Thrush-tanager [*Rhodinocichla rosea* (Thraupidae)] in deciduous and thorn forest. In taller vegetation Blue Mockingbird [*Melanotis caerulescens* (Mimidae)]

Fig. 2. (a) In different species groups diversity contours over habitats are different. The diversity of warblers (Parulidae, in number of species (·····), peaks with a dozen species in the pine–oak–fir forests of high elevations, whereas the diversity of flycatchers (Tyrannidae), in numbers of species (- - -), peaks with eight species in the tropical lowland forest. (b) Distributions of ten species of breeding warblers. All of these species are in different genera except the two redstarts, one of which (Slate-Throated Redstart) is widely distributed and the other (Painted Redstart) is confined to lower and more open pine–oak woodland. (c) Three species in the parulid genus *Basileuterus* (- - -) occur in taller habitats, but all are of similar body size, and the species are elevationally segregated with no habitat overlap. Six species of *Vireo* [Vireonidae (- - - -)] are present, with larger species (Yellow-green Vireo, Solitary Vireo), medium-sized species (Golden Vireo, Warbling Vireo), and smaller species (Hutton's Vireo, Bell's Vireo) habitat segregated, with the single exception of the two large species in site 4. (d) The distribution of breeding species in the two tyrannid genera *Myiarchus* (- - - -) and *Contopus* (–––). Three *Myiarchus* species are smaller in size than the fourth, Brown-crested Flycatcher, which overlaps smaller species in lowland habitats. The two *Contopus* species differ in size, and coexist in the high-elevation pine–fir forests. (e) The distributions of 11 thrushes and thrushlike birds are shown over the habitat range. The shorter and more open habitats are occupied by mockingbirds and thrashers (Mimidae) and a thrushlike tanager (Thraupidae) (——). In taller habitats the large ground-foraging *Turdus* thruses (- - - -) are largely elevation segregated and are joined in high-elevation forests by smaller ground-foraging thrushes [two nightingale-thrushes (*Catharus*) (–––)] and by pouncing and sallying flycatching thrushes [solitaire and bluebird (·······)]. (f) Numerous migrant and overwintering birds are found over the habitat range, and the distributions of 16 species are shown here. Four wintering sparrow species occur (——), three of them in the lowest and most open habitats, and the fourth, Lincoln's Sparrow, at high elevations. Six species of Parulid Warblers (–––) are shown, with some species, such as Yellow-rumped Warbler, very widely ranging and others, such as American Redstart and Northern Waterthrush, appearing very habitat specific. Some species, such as Cedar Waxwing (- - - -), are habitat specific because their wintering food supply, in this case the fruit of *Myrtillocactus* cacti, are restricted to certain habitat types. [(- - -) also indicates all other species shown.] Some species, such as Ruby-crowned Kinglet (····), are extremely broadly distributed over habitats, but the density contours included show that this species prefers the high-elevation pine–oak–fir forests. (*Continued on pp. 19–21.*)

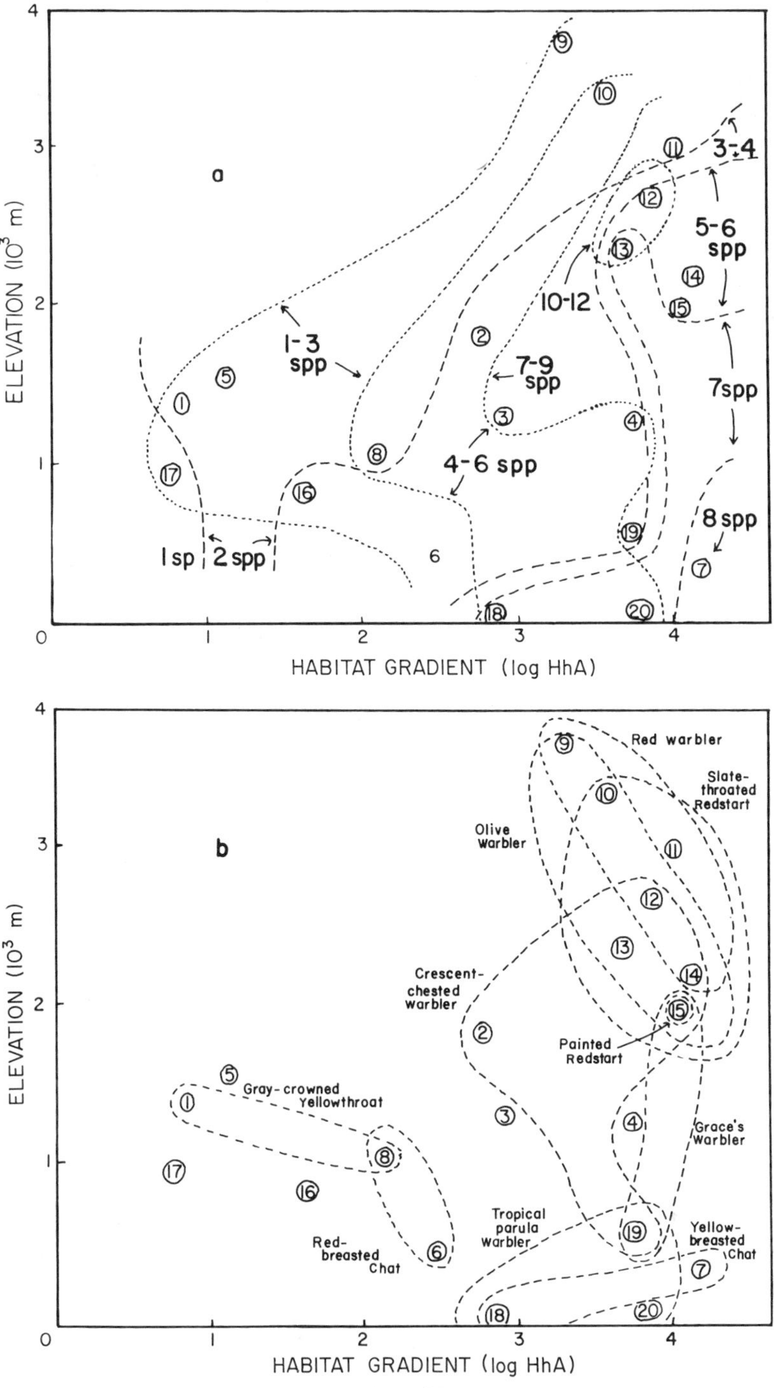
a
ELEVATION (10^3 m)
HABITAT GRADIENT (log HhA)
1-3 spp
4-6 spp
7-9 spp
10-12
3-4
5-6 spp
7spp
8 spp
1 sp
2 spp
b
ELEVATION (10^3 m)
HABITAT GRADIENT (log HhA)
Red warbler
Slate-throated Redstart
Olive Warbler
Crescent-chested warbler
Painted Redstart
Gray-crowned Yellowthroat
Grace's Warbler
Red-breasted Chat
Tropical parula Warbler
Yellow-breasted Chat

Fig. 2a and b

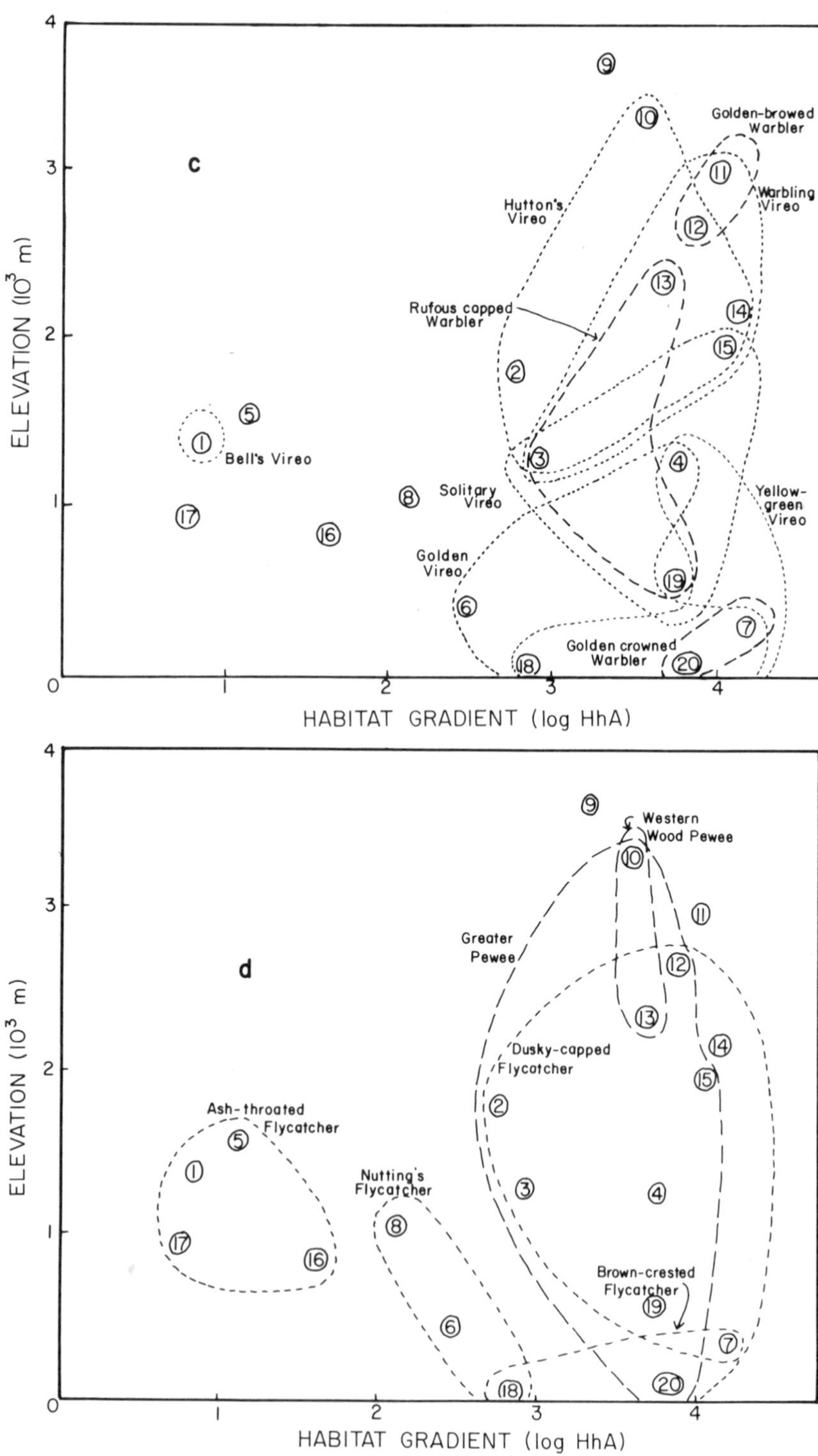

Fig. 2c and d (*Continued*)

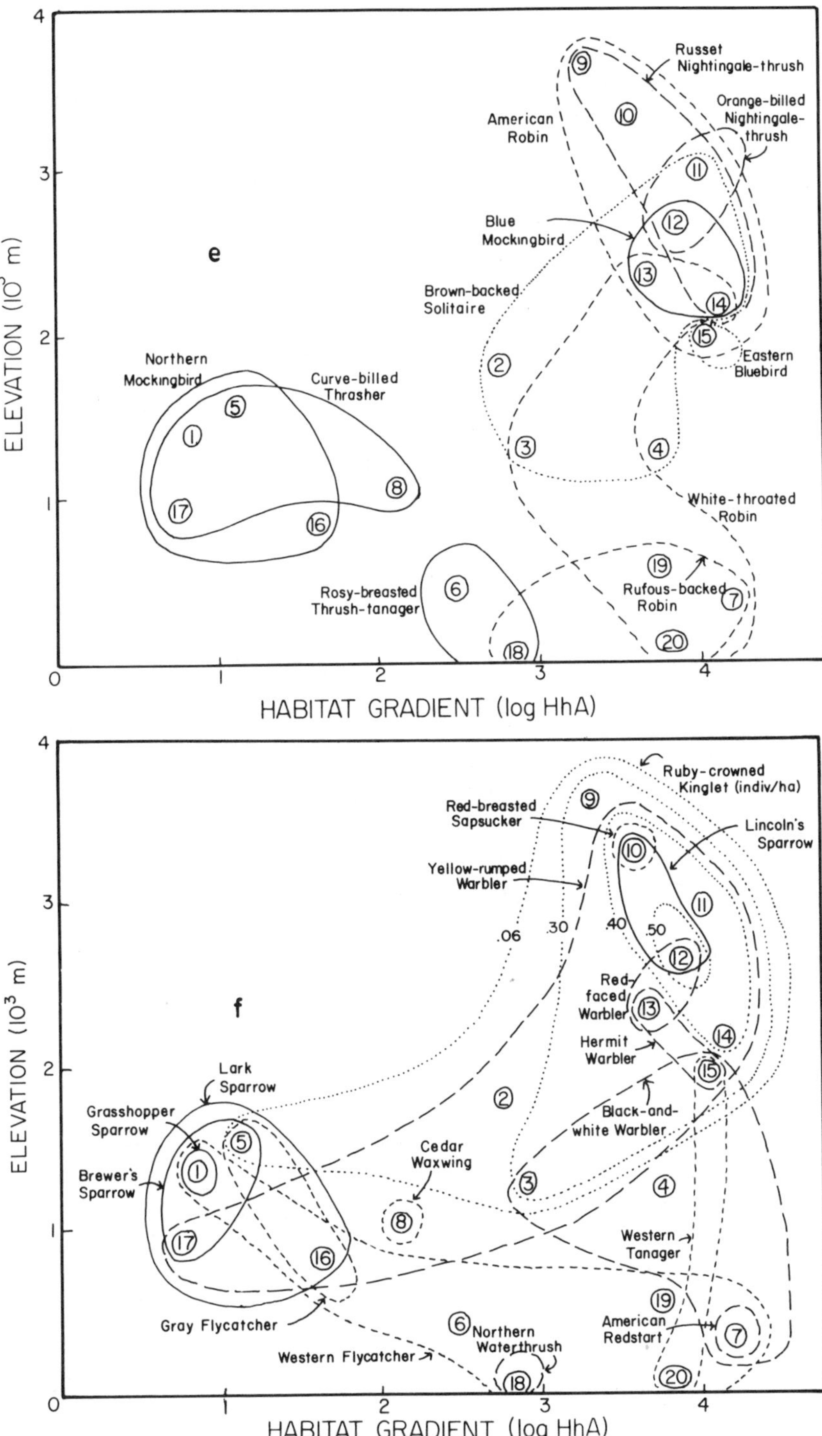

Fig. 2e and f (*Continued*)

forages in the shrub layer of forests, and two other species, Brown-backed Solitaire (*Myadestes obscurus*) and Eastern Bluebird (*Sialia sialis*), flycatch and pounce near ground level. These last two are true thrushes (Turdidae), as are all other species in the figure, and forage on the ground in the typical thrush fashion. Three larger species are similarly sized (around 24 cm), with partially overlapping habitat ranges [American Robin (*Turdus migratorius*), White-throated Robin (*T. assimilis*), and Rufous-backed Robin (*T. rufopalliatus*)], and similarly two smaller species (around 16 cm) overlap at higher elevations [Russet Nightingale-thrush (*Catharus occidentalis*) and Orange-billed Nightingale-thrush (*C. aurantiirostris*)].

E. Variation in Habitat Occupancy: Visitors

Besides the breeding warblers, flycatchers, and thrushes of Fig. 2, there are many wintering species. Just two of these are flycatchers, both *Empidonax* species, Gray Flycatcher (*E. wrightii*) winters in low scrub (sites 5 and 16), and Western Flycatcher (*E. difficilis*), in a wide range of lowland habitats (Fig. 2f). The wintering birds avoid overlap with resident congeners and conspecifics, Western Flycatcher, Pine Flycatcher (*E. affinis*), and White-throated Flycatcher (*E. albigularis*), which breed above 1500 m.

Wintering thrushes are also limited to two species, Swainson's Thrush (*Catharus ustulatus*) and Hermit Thrush (*C. guttatus*). The former occurs in tall forest from 350 to 3500 m, but the latter was found only at 2000 m (site 15) in open pine–oak.

Wintering warblers are diverse and common, and include 12 parulids and the Ruby-crowned Kinglet [*Regulus calendula* (Sylviidae)]. This kinglet is very widely distributed (see Fig. 2f), and judging by its density in sites 9, 10, and 11, it is not much affected by overlap with resident Golden-crowned Kinglets (*R. satrapa*) in those sites. Other wintering warblers, such as Yellow-rumped Warbler (*Dendroica coronata*), Nashville Warbler (*Vermivora ruficapilla*), Orange-crowned Warbler (*V. celata*), and MacGillivray's Warbler (*Oporornis tolmiei*) are nearly as ubiquitous (averaging 9.5 sites/species from sea level to 3600 m). In contrast, there are several warbler species which are restricted to a single habitat, and examples are Northern Waterthrush (*Seiurus noveboracensis*) in lowland thorn forest (site 18), American Redstart (*Setophaga ruticilla*) in tallest tropical forest (site 7), Virginia's Warbler (*V. virginiae*) in tropical scrub (site 5), and Red-faced Warbler (*Cardellina rubifrons*) in pine–oak (sites 12 and 13).

These interspecific differences in habitat use exist independently of variations in abundance, for some widely distributed species, such as MacGillivray's Warbler, are not common, and some restricted species, such as the Northern Waterthrush, are quite common in their specific, preferred habitat. Of course, different species behave differently and are differentially conspicuous, but such differences cannot explain the observed variations in habitat breadth.

F. Strategies of Habitat Segregation

From the foregoing it appears that ecologically similar species (e.g., jays) commonly segregate by habitat, especially if they are congeners (e.g., orioles, wrens, *Myiarchus* flycatchers, and *Basileuterus* warblers). But wrens, thrushes, flycatchers, and warblers in different genera frequently overlap in habitat. Some species have apparently wide habitat ranges, and others, quite restricted; often, within a genus, one member is widely distributed, and the other, restricted (e.g., *Myioborus* redstarts, wintering *Catharus* thrushes, and *Vermivora* warblers). In some genera species differ in body size (*Vireo, Contopus*) or in other morphological aspects (various genera of flycatchers and warblers), and among these species more overlap in habitat is apparent.

Some wintering species may be habitat specific [some warblers, thrushes, and flycatchers, Western Tanager (*Piranga ludoviciana*)], but others are apparently habitat generalists (Ruby-crowned Kinglet, some warblers and thrushes), and others are intermediate. In some cases wintering species seem to avoid habitats occupied by breeding congeners (*Empidonax* flycatchers), but in other instances they do not (*Catharus* thrushes, *Dendroica* and *Vermivora* warblers). In yet other instances, the migrants appear to track specific habitat resources, such as the Cedar Waxwings (*Bombycilla cedrorum*), which feed on the fruits of the columnar cactus *Myrtillocactus* in site 8, and the Red-breasted Sapsucker (*Sphyrapicus varius*), which requires the alder (*Alnus*) trunks in site 10. Many of the more widely distributed wintering warblers seem to be gregarious and feed together in what are apparently high-productivity hotspots, but the others that are habitat specific forage singly and sing from and apparently defend winter territories.

These data illustrate some of the diveristy of factors that can affect habitat selection in breeding and wintering species. Some of these factors are elaborated below, but many of them remain speculative and most appropriate for further research.

III. HABITAT SELECTION IN A GEOGRAPHIC PERSPECTIVE

A. Changes in Species Ranges and Habitats Occupied

1. Recent Range Changes

Changes in a species' geographic range might be the result of at least two quite different processes: first, a species might be simply extending its occupancy of the same habitat type into areas from which it was hitherto excluded by constraints such as colonization time, a topographical barrier, or a shortage of recruits; second, a species might be extending into new or different habitats to which it has lately become adapted or for which it was preadapted and has previously lacked appropriate cues for a broader habitat choice or recognition.

Such a species might be said to be constrained by a behavioral or a genetic barrier. Often it is not easy to distinguish between these alternatives. For example, many bird species have recently expanded their breeding ranges north in western Eruope. Järvinen and Ulfstrand (1980) estimate that Scandinavia has gained 2.8 species/decade and lost 0.6 species/decade since 1850, but the colonization rate has been higher since 1950. Some of the new birds are taking advantage of habitat changes, but others expand into habitats that have been unchanged for long periods of time. Further, since 38% of the successful colonists since 1900 are migrants, changes in their wintering habitats in Africa might account for greater overwintering survival and the push for more breeding habitat. Changing forest structure appears to be the main cause for bird population changes in northern Finnish forests (Järvinen and Väisänen, 1978), with some species such as the Siberian Jay (*Perisoreus infaustus*) and Redstart (*Phoenicurus phoenicurus*) declining as the older forests are cleared, and others such as the Willow Warbler (*Phylloscopus trochilus*) and the Spotted Flycatcher (*Muscicapa striata*) increasing as the forests open up and edge habitat becomes commoner. Other species have expanded west into the older extant forests, and the Red-flanked Bluetail (*Tarsiger cyanurus*) is an example (Mikkola, 1973).

In the northeastern United States there have been similar range expansions to the north, with Tufted Titmouse (*Parus bicolor*) and Northern Cardinal (*Cardinalis cardinalis*) being examples. Range shifts are also recorded from the southwestern United States, with records of Mexican Crow (*Corvus imparatus*) (Arvin and Arvin, 1975; Tyler, 1976) and Brown Jay (*Psilorhinus morio*) (Shifflet, 1975) in Texas, and several interior pine–oak species now breeding in southern California (Johnson and Garrett, 1974). The latter shifts, at least, are thought not to be due to habitat changes, but are perhaps due to subtle shifts in climate over the last decades.

2. *Habitat Tracking*

Some habitats have been altered in structure and in extent by man's activities, and bird species that follow these changes can be regarded as simply tracking habitat changes. The expansion of the grassland Brewer's Blackbird (*Euphagus cyanocephalus*) more than 1000 km east-northeast from the native prairies of Minnesota into recently deforested areas of Ontario (Stepney and Power, 1973) and similar range extension in prairie Horned Larks (*Eremophila alpestris;* Hurley and Franks, 1976) are examples. The northern extensions of typical Saharan birds following increasing desertification in North Africa (François, 1975) provide other examples. More subtle are the population increases of Blue Grouse (*Dendragapus obscurus*) on Vancouver Island after logging activities provide more of their preferred open habitat (Bendell and Elliot, 1966), and the competitive shift in favor of Blue-winged Warblers (*Vermivora pinus*) over Golden-winged Warblers (*V. chrysoptera*) in the northeastern United States as the earlier successional scrub favored by the latter becomes scarcer and the former's wider habitat tolerance gives it an advantage (Confer and Knapp, 1981). The advantage of Song Thrushes (*Turdus philomelos*) over Blackbirds (*T. merula*) where habitats have been man modified in Poland provides a similar case history (Dyrcz, 1969).

B. Breeding versus Wintering Habitats: A Compromise

1. The Compromises

Most high-latitude passerines spend about half their time wintering in very different habitats than those in which they breed, and this must impose morphological and perhaps behavioral and physiological constraints on their abilities to efficiently use either habitat extreme. Fretwell (1972) has argued convincingly that selection for survival in wintering habitats may affect choice of breeding habitats and may even dominate adaptive morphology and that adaptations to breeding habitat may therefore be far from optimal; this theme has been re-echoed by many researchers whose studies of breeding birds show apparent divergence from what might be considered more optimal community structure. Finch bills seem more strongly influenced by their granivorous winter diet than their mostly insectivorous diet in the breeding season, for example. Townsend's Solitaires are flycatching insectivores during the breeding season but are frugivorous in winter, having diets of up to 97% juniper berries in one instance (Salomonson and Balda, 1977); it seems unlikely that they could possess a morphology equally well adapted to both sorts of regimen.

Wing morphology and body size are greatly affected by the migration to winter quarters in warblers. The largest *Dendroica* warblers breed in conifers or woodland and are mobile and gregarious opportunists on their wintering grounds (Greenberg, 1979). Sylviine warblers with the longest migrations can be identified by their wing morphology (Gaston, 1974), and this might well compromise their habitat- and foraging-site selection in summer.

2. Habitat Selection in Winter Quarters

In some habitats wintering species appear to "fit into" niches vacated by similar species of summer visitors which are wintering elsewhere. In Californian oak woodlands, for example, wintering Yellow-rumped Warblers and Ruby-crowned Kinglets are reasonable morphological counterparts of summer-visiting Orange-crowned Warblers and Blue-gray Gnatcatchers (*Polioptila caerulea*), respectively, though there are necessarily dietary and behavioral differences (Wagner, 1981). Even year-round residents may shift their habitat occupancy during the nonbreeding season, as in Swedish crows (Loman, 1980) and many riparian birds in the Colorado River valley (Rice *et al.*, 1980). In south Texas the two resident thrashers (above) are joined by Brown Thrashers (*Toxostoma rufum*) in winter (Fischer, 1981); habitats are subdivided among species, and the habitat differences are fine-tuned by interspecific territoriality.

Some wintering bird communities seem just as well organized with respect to differential habitat selection as are breeding communities; aggregations of wintering sea-ducks (Stott and Olson, 1973; Nilsson, 1972) give this impression. Wintering waders in southern Florida are thought to exhibit narrower niche breadths there during a time of stronger resource limitation than on arctic breeding grounds (Baker and Baker, 1973), a trend supported by Rice *et al.* (1980) in Arizona; Alatalo (1981a), however, showed that breeding forest birds in Finland have narrower habitat ranges in the breeding season. The data on wintering species in Jalisco (Section II,E) indicate no generalizations either way.

The question of how migrant birds ''fit into'' their wintering habitats has always intrigued ecologists. Karr (1976) pointed out differences between the neotropics and Africa, with grasslands and savannas used less by migrants in the former region and lowland forests used less in the latter. In Jalisco, numbers of migrant species varied with habitat, just as did numbers of breeding species, with a tendency to avoid deciduous vegetation and to reach higher densities, but not higher diversities, in higher-elevation pine–oak than lower-elevation subdeciduous forest. In southern Texas, Emlen (1972) found that woodlands supported more wintering birds than did grasslands, which in turn were preferred over brushlands; a wide variety of patterns of habitat use occurred in both residents and visitors.

3. *Birds in Transit*

Many migrants in transit make temporary use of habitats quite different from either their breeding or their wintering quarters; in the southwestern United States we see many woodland and forest warblers crossing, resting, and feeding in the Mojave and Sonoran deserts in fall and especially in spring, where they seem quite out of place. Rappole and Warner (1976) have related habitat choice in transient Northern Waterthrushes in south Texas to their physiological state, with some individuals in Zugstimming classified as ''flyers,'' gregarious with broad habitat usage, and others in Zugdisposition as ''feeders,'' aggressively defending individual territories in specific habitat. Questions on the habitat use of migrants are discussed in detail later, in Chapter 16, this volume.

C. Colonization of New and Exotic Habitats

1. *Pine Plantations*

Pine trees are physiognomically distinct, and pine plantations are an example of a distinct habitat now found in many regions of the world to which they are not native. In nineteenth–century plantations in northern France, bird species numbers increase with vegetation age differently than in native deciduous forests; the conifers are, in addition, comparatively underutilized by summer visitors (Constant *et al.*, 1973). Pines have been planted in Australia since before the turn of the century and support generally fewer bird species than do native woodlands (Disney and Stokes, 1976); some native birds, such as the White-tailed Black Cockatoo (*Calyptorhynchus boudinii*), find in pine cones fortuitous resemblances to native food resources (Saunders, 1974). Earlier I contrasted the use of pines (*Pinus radiata*) by birds in South Africa and Chile to that by birds in California, where these pines are native. In South Africa the plantations are virtual ornithological deserts, but in Chile both bird densities and species numbers are comparable to those in California. The difference appears related to patterns of β-diversity, which is very high in South Africa, with habitat-specific birds, and very low in Chile, with habitat-generalist birds (Cody, 1975).

2. *Eucalyptus Plantations*

Some Australian eucalypts, tall and stately trees, are widely planted in Mediterranean-climate regions, including California, Sardinia, Chile, and South Africa. In Australia they support a diverse bird fauna, but their use by birds elsewhere is minimal, even in Chile, perhaps because the trees often lack their native insect fauna and therefore have little in the way of bird food resources (personal observation). Some use of eucalypt nectar is made by hummingbirds in both Chilean and Californian plantations, and in Africa the trees primarily serve as nest sites (Smith, 1974).

3. *Birds in Cities*

An extreme example of man-modified environments are cities, where usually little of the native habitat remains. This may suit some birds well; American goldfinches (*Carduelis tristis*) have higher nesting success in the city of Guelph, Ontario, than in natural habitats (Middleton, 1979), perhaps because of lower nest predation. Emlen (1974) analyzed the composition of the urban bird community of Tucson, Arizona, and found 65% of the birds belonged to three nonnative species [House Sparrow (*Passer domesticus*), Starling (*Sturnus vulgaris*), Inca Dove (*Scardafella inca*), this last species colonizing the area about a century ago] preadapted to the urban scene, the first two in Europe and the third in Mexico. Of the remainder, 30% belonged to species ranging widely across North America with generalized habitat requirements and only 5% to birds characteristic of the Sonoran Desert which surrounds Tucson.

D. Shifts in Habitat Selection on Islands

Since Crowell (1962) described differences between habitats occupied in Bermuda and in Florida by Northern Cardinals (*Cardinalis cardinalis*), Gray Catbirds (*Dumetella carolinensis*) and White-eyed Vireos (*Vireo griseus*), many studies have emphasized that birds have different and generally more extensive habitat ranges on islands than they do on mainlands. These studies illustrate both that habitat selection is a flexible behavior and that a signficant part of the process is owed to the competitive environment. On islands in the Caribbean, for example, Lack *et al.* (1973) found that over two-fifths of the bird species on St. Vincent and A. W. Diamond (1973) found over one-half of the species on St. Lucia occupied both lowland and montane forest, unlike the usual restriction to either high- or low-elevation forest in tropical birds, as on Greater Antillean islands or on the mainland. Diamond (1973) also noted that the Caribbean endemic *Dendroica adelaidae* was widespread over all habitat types on St. Lucia but restricted to arid scrub on all other islands. Cox and Ricklefs (1977) compared bird occupancy in nine habitat types in mainland Panama, land-bridge Trinidad, and both Greater and Lesser Antillean islands in the Caribbean and found that, as species numbers declined on smaller and more isolated islands, the numbers of habitats occupied by species and their densities within habitats both increased. Habitat range appeared to be strongly influenced by interspecific effects on the mainland but largely independent of them on the small island of St. Kitts. There is also a substantial amount of information on habitat range expansions of birds on the Soloman, Fiji, and New Hebridean Islands in the southwest Pacific (J. M. Diamond, 1975; Gorman, 1975; Diamond and Marshall, 1977).

The phenomenon of increased densities per species in species-poor situations, such as islands, is called ''density compensation'' and has been reviewed recently by Wright (1980). MacArthur *et al.* (1972) first used the term to describe density changes of birds in the Pearl Islands, off the Pacific coast of Panama, in their study, which also described many species shifts in habitat between mainland and the islands. Terborgh and Faaborg (1973) described the same phenomenon in the Caribbean between the large island of Puerto Rico and its small satellite, Mona; species were up to 11 times commoner on species-poor Mona.

One of the most detailed studies of mainland–island habitat shifts and density compensation is that of Yeaton (1974), who compared birds in chaparral and pine forest on the southern California mainland with those on Santa Cruz island 50 km offshore. The chaparral had five fewer species on the island, absentees that include Wrentits (*Chamaea fasciata*), the commonest species in mainland chaparral, Anna Hummingbird (*Calypte anna*), and Nuttall's Woodpecker (*Dendrocopus nuttallii*). Island chaparral supported species such as Allen's Hummingbird (*Selasphorus sasin*), Acorn woodpecker (*Melanerpes formicivorus*), and Blue-gray Gnatcatchers, which are typical of oak woodlands on the mainland. The island pines lacked nine species that breed in mainland pines but gained six new species; most of the additions are typical of either chaparral or oak woodland habitats. Density compensation in these island communities was to a large extent predictable from mainland community structure and niche overlaps there.

Density compensation is also observed in isolated mainland habitats with reduced species diversity, such as the patches of Afromontane forest in the southwestern Cape of South Africa (Cody, 1983b). But in that habitat relatively few species from other habitats invade the species-poor patches, a finding again in accordance with the high β-diversity of the region. These studies point up the role of other species in restricting the range of habitats occupied by birds, a theme developed further in Section IV,D.

IV. LOCAL FACTORS IN HABITAT SELECTION

A. Philopatry, Site Fidelity, and Choice

One behavioral aspect of habitat selection concerns the degree to which free choices of habitat are made as opposed to individuals imprinting on and returning to natal habitats. Certainly the extensive data on site fidelity, where adult individuals return to exactly or approximately the same territories year after year, indicate that there may be a very limited choice in some species (perhaps exercised just once, as a yearling). Svärdson (1949) reported how young male Wood Warblers were influenced in habitat and territory choice by older males which had returned earlier from the wintering grounds. In the thrashers studied by Fischer (1981), individuals returned to the same wintering territories the following year, and the same was true for wintering Greenish Warblers (*Phylloscopus trochilioides*) in southern India (Price, 1981).

The role of previous nesting experiences is shown by studies of bluebirds. Pinkowski (1979) found that adult Eastern Bluebirds (*Sialia sialis*) more often select nest boxes than do yearlings, but birds will switch nest sites following an unsuccessful breeding attempt. In a nest-box study of Mountain Bluebirds (*S. currucoides*), Herlugson (1981) used two types of nest boxes, and found that females chose the same territory and the same sort of nest box the year after a successful breeding season. Unsuccessful females switched territories *and* nest-box types, while first-time breeders tended to choose their natal type of nest box. Clearly both experience and the natal circumstances may play a role in habitat selection, but this is worked out in very few species.

B. Sex Differences in Habitat Choice and Use

The normal sequence of events in species which do not overwinter on breeding territories is that males return from migration first, select territories, and sing in territory defense and to attract the later-arriving females. Thus, whereas males may select habitats alone, females may choose either males, habitats, or both. In the Yellow Warblers (*Dendroica petechia*) studied by Ficken and Ficken (1966) in New York state, males defended territories both with and without lower shrubbery; females, however, preferentially mated with males whose territories had plenty of low shrubbery.

In polygynous species, the dominant males may defend superior territories that attract several females, while males with inferior territories go unmated. Thus male quality and territory quality may be closely correlated, and the females' choice therefore may be unambiguous. The polygyny model of Orians (1969) uses territory quality as its independent variable and receives a good deal of support from studies of grassland birds (see Chapter 6, this volume). Polygynously mated males have territories that are larger or are of better quality than those of monogamous males in Bobolinks (*Dolichonyx oryzivorus;* Martin, 1971; Wittinberger, 1980), Dickcissels (*Spiza americana;* Harmeson, 1974; Zimmerman, 1982), Savannah sparrows (*Passerculus sandwichensis;* Welsh, 1975), and Field Sparrows (*Spizella pusilla;* Walkinshaw, 1968).

Females may also use different habitats than males, as in wintering Sparrow Hawks (*Accipiter nisus*). The larger females occupy territories where larger prey, such as thrushes and starlings, are common, and the smaller males habitat with smaller prey, such as finches (Opdam, 1975). Males and females use breeding territories differently in the grassland Henslow's Sparrow (*Ammodramus henslowii*), in which females forage closer to the nest than males (Robins, 1971). Female Red-eyed Vireos (*Vireo olivaceus*) forage closer to nest height and males closer to song perch height in woodlands (Williamson, 1971), and in four species of spruce-woods warblers (*Dendroica*) studied by Morse (1968), the males also foraged higher than the females.

C. Census Repeatability

If the same breeding bird species use the same habitat year after year, between seasonal changes, this can be taken as evidence that consistent decisions on habitat selection are made repeatedly. This depends, however, on the same availability of birds between years to make the choice, on constancy in the habitat and its salient features, and on adequate sampling methods. Holmes and Sturgis (1975) report on the remarkable consistency of species composition in a New England hardwood forest over five consecutive summers, but abundances were more variable. In montane ponderosa pine forest in the Rocky Mountains, breeding birds numbered from 19 to 23 species over 5 years (Winternitz, 1976); 10 species bred every year, but each of 5 species bred just once. In the 5 commonest species (all residents or short-distance migrants which together comprise about 50% of the total bird density of 2.33 pairs/ha), however, densities were very stable year to year. These 5 species totaled a minimum of 1.05 pairs/ha, and a maximum of 1.11 pairs/ha over the 5 years. (See also data on grassland birds, Chapter 6, this volume.)

Census repeatability may be highest in habitats with mostly resident species in less seasonal habitats. Järvinen (1979) compared 15-year censuses from central to northern Europe and found greater consistency at lower latitudes; he attributed this to lower environmental predictability in the north.

The same site in the same habitat may produce very different resources from one year to the next, in which case, censuses may also differ year to year. In pine–oak woodlands in southeastern Arizona, for example, insect production varies with rainfall, and year-to-year turnover in bird species composition is as high as 50% as different species track years of high or low productivity (Cody, 1981).

In contrast to repeated censuses at different times in the same place, censuses of similar habitats in different places may yield very similar or completely different results. Alder (*Alnus*) forests are similar in vegetation profile between Costa Rica and Washington state, but although there are some parallels in their bird communities, different species and species numbers characterize the two sites censused by Stiles (1978). Not surprisingly, biogeographic considerations have a great deal to do with which species actually occupy a given habitat in a given region. This point is emphasized in a study of bird occupancy of two habitat types, renosterveld and fynbos, in different parts of Cape Province, South Africa, but these sites were much more similar in latitude and climate than were the *Alnus* forests (Cody, 1983a). While "core" species (those commoner species which together comprise around two-thirds of the total bird density) are predictable in identity and relative abundance across all sites, there are rarer fringe species whose identities differ among sites and appear to be a function of whatever habitat other than renosterveld or fynbos is common near the census sites.

D. Habitat Selection and Population Density

Svärdson (1949) noticed that some species occupy a broader range of habitats in years of higher population density and concluded that intraspecific population pressure forced some individuals to breed in less-suitable habitats. Glas (1960) showed such a relationship for Chaffinches (*Fringilla coelebs*), Alatalo (1981a) for Goldcrests (*Regulus regulus*), and Zang (1982) for Blue Tits (*Parus coeruleus*).

Habitats which are intrinsically less suitable, for example, in terms of their food resources, may be in fact quite tolerable if population densities there are lower than in prime habitats. This is the basis of the model of Fretwell and Lucas (1969), in which habitats are ranked according to their intrinsic suitability to a species; *de facto* habitat quality declines, as bird density increases, to a threshold at which the next best habitat begins to fill up. Their observations on similar breeding success in Field Sparrow populations at different densities in different habitats led to the model, which has since received much attention. The inherent feature of the model is the trade-off between low density of conspecific competitors, or large territory size, and low territory quality, i.e., low food density. The fact that many species hold larger territories where food resources are lower supports the trade-off concept (see Section IV,E). A similar trade-off is seen in the nesting territories of Herring Gulls (*Larus argentatus*), among which numbers of chicks fledged per nest in two different habitats, the preferable but crowded rocky slopes and the less-desirable but less-crowded puffin rookery, were equal (Pierotti, 1982). Of course, high nesting density may be advantageous as a protection from predation, especially in colonially breeding alcids (Grant and Nettleship, 1971; Birkhead, 1977; Harris, 1980).

E. Habitat Structure

1. Importance of Structural Characteristics

The importance of habitat structure or physiognomy in bird habitat selection was apparent from MacArthur's work on species diversity. His measure of habitat structure was derived from plots of vegetation density against height, called foliage profiles [see Karr (1971), Willson (1974), and Terborgh (1977), for a similar approach]. But in fact, individual bird species may cue on specific structural features, such as cactus in Sonoran Desert (Tomoff, 1974) or certain tree species in hardwood forests (Holmes and Robinson, 1981), where insectivorous birds prefer yellow birch and avoid beech and sugar maple.

The way in which bird species change with habitat structure can be seen readily in studies of succession (e.g., Shugart and James, 1973). Typically both species diversity and bird density increase with seral stage and with increasing habitat complexity, with more resource specialists in the later stages and more census variability in the earlier, successional stages (May, 1982).

2. Habitat Structure: Measurement and Statistics

Apart from measuring foliage profiles, there have been several suggestions on how habitat structure might be measured (e.g., James and Shugart, 1970; Fox, 1979), with the emphasis on measures of vegetation density in both vertical and horizontal directions. With the ready availability of multivariate statistical packages on computers, there is a tendency to collect measurements on a great variety of structural features and let the computer generate new axes of habitat structure, in the form of linear combinations of the original variables, using principal components analysis (PCA) or discriminant function analysis (DFA). These two techniques are quite different in principle, as are ordination and classification. PCA is an ordination technique; if many variables (vegetation height, density at various heights, percentage cover of different layers, etc.) are measured in many different habitats, PCA successively generates new axes of variation which are linear combinations of the original variables. The new axes are ordered such that the first accounts for maximum variation in the original data, the second for most of the remaining variation, and so on. DFA, on the other hand, is a classification technique. The habitat data are grouped, e.g., one group for each bird species, and new axes are constructed to maximize the differences along them of the groups or species. James (1971) and those who have followed her methodology [Anderson and Shugart (1974), Whitmore (1975), Smith (1977), Collins *et al.* (1982)] measured up to two dozen or more habitat variables, among which vegetation height and percent canopy cover often recurred as those most significantly related to bird distributions. Other recent studies using PCA are those of Nudds (1983) on breeding waterfowl and Sabo (1980) on subalpine bird communities. Cody (1981) reviewed some of the advantages and disadvantages of multivariate statistical techniques, and also observations on species that select habitats in the early spring before the foliage has developed. If species were to select habitats after fledging in the late summer and then return to these habitats in the spring (Brewer and Harrison, 1975), this complication might be circumvented.

3. Manipulation of Habitat Structure

Some insights into the significance of habitat structure to birds are gained from manipulations, fortuitous or otherwise, to the vegetation. In some cases apparently drastic alterations to habitats have very little effect on the birds, as in the case of herbicides on the brush layer in Sierra Nevada forests (Beaver, 1976) or fire on south Florida pines (Emlen, 1970). Burned coniferous forest in the Sierra Nevada supported a few more species and larger species than unburned forest, presumably because of higher productivity of bird food supplies (Bock and Lynch, 1970). Franzreb and Ohmart (1978) found very similar bird diversities in mixed coniferous forest from which many overstory trees had been removed versus unlogged forest, although bird densities were 15–20% higher in the virgin areas. The effects of removing the brush layer in sequoia forest were documented by Kilgore (1971); two species (Western Wood-pewee and American Robin) increased in density as a result, and four others [Rufous-sided Towhee (*Pipilo erythrophthalmus*), Mountain Quail (*Oreotyx pictus*), Hermit Thrush, and Nashville Warbler], three of which feed on the ground, declined to zero or near zero. It would appear that the decline of the three ground feeders was influenced not so much by a direct effect on their food resources, but rather by removal of cover which provides safe foraging opportunities.

4. What Habitat Structure Means to Birds

Clearly, habitat structure as we measure it can mean or translate into very different resources for different sorts of birds; foraging sites, nesting sites, and protection from the elements and from predators are just the obvious possibilities. Even though we may be able to measure structural features that correlate well to the density of a certain species, and in that sense we may be able to predict where to find the species and how common it will be, the correlation alone does not tell us what it is about the structural variable that the bird responds to. Comparatively little attention has been paid to actually interpreting habitat structure from this point of view.

Perhaps the most detailed study in this vein is that of Holmes *et al.* (1979), who attempted to interpret the physical structure of forest vegetation in terms of its influence on the foraging behavior of insectivorous birds (see Chapter 9, this volume). The structural differences between the foliage of red spruce and white spruce are believed by Morse (1976) to influence foraging efficiency in Black-throated Green Warblers (*Dendroica virens*). In simpler grassland habitats the influences of structure are more easily measured; Brownsmith (1977) found that Starlings forage at half the speed, with twice as many stops and as much stationary time, in tall grass as in shorter grass, and Maher (1973) attributed the lack of nesting success of Horned Larks in atypically dense grassland to a reduced foraging efficiency of the parents there.

5. *Habitat Heterogeneity*

Vegetation structure may vary horizontally to produce patchy or heterogeneous habitats. If the scale of this patchiness is small, bird diversity may be enhanced (Roth, 1976) as different species find different patches or mixes to their liking, a result confirmed theoretically by Levin (1974). If it occurs on a larger scale, species abundances may become very unequal, and diversity may decline (Rotenberry, 1978). A clear example of habitat patchiness is that of tree falls in forests; in Neotropical forest such gaps are preferred by the ant-vireos (*Dysithamnus puncticeps*) and the wren (*Cyphorhinus phaeocephalus*) but are avoided by other species (Schemske and Brokaw, 1981).

F. Habitat Selection and Productivity

1. *Territory Size, Bird Density, and Productivity*

Productivity can affect habitat selection and bird communities at various levels. Higher densities of Silvereyes (*Zosterops lateralis*) are correlated with higher densities of *Ficus* fruits on Heron Island, Australia (Catterall *et al.*, 1982), and increased summer rainfall produces more seeds and higher sparrow densities in the Sonoran Desert (Dunning and Brown, 1982). Higher productivity may permit smaller territory sizes, as in Ovenbirds (*Seiurus aurocapillus*) (Stenger, 1958) and Wrens (*Troglodytes troglodytes*) (Cody and Cody, 1972). Red Wattlebirds (*Anthochaera carunculata*) defend smaller territories and permit fewer intruders where nectar supplies are more abundant (Ford, 1981; see Chapter 12, this volume), and Red Grouse (*Lagopus scoticus)* territory size, which varies over an order of magnitude, is inversely correlated with the proportion of heather and heather edge in the territory (Miller and Watson, 1978).

2. *Productivity and Habitat Occupancy*

Territory size can increase with decreasing food supplies only to a limited degree, after which a habitat becomes unacceptable. This may happen during a drought as Smith (1982) recorded; aspen woodlands were affected more than other habitats in this study, and several species, especially hummingbirds, were eliminated from habitats they occupied in normal years. In Arizona pine–oak woodland, drought induced a sixfold reduction in sticky-board insect catches and a 50% reduction in total bird density (Cody, 1981). Further, species numbers declined from 20 to 16, but 5 of these 16 were species not present in the original 20-species community. They were species such as Ash-throated Flycatcher, House Finch, and Lucy's warbler (*Vermivora luciae*), typical of lower-elevation, drier, and more open vegetation. The 9 bird species absent in the drought year were typical of riparian habitats and higher-elevation coniferous forest, as well as pine–oak; Pygmy Nuthatch, Painted Redstart, and Western Wood-pewee are examples. For these species, low productivity seems to override considerations of habitat structure; suitable habitat structure may be necessary but not sufficient, in poor years, for their occupancy of the pine–oak.

Further examples are provided by American White Ibises (*Eudocimus albus*) in Florida, which shift their nesting colonies as a response to a shifting food supply (Kushlan, 1976), and Australian honeyeaters, which shift over habitats seasonally to track a seasonally variable nectar supply (Ford, 1979).

3. *Edge Effects*

Edges or ecotones between habitat types seem to be areas of higher productivity. Although species diversity increases from the edge to the interior of forested islands in the Gulf of Bothnia, bird density is highest at the edges (Helle and Helle, 1982). But edges are more easily searched by nest predators, and cowbird parasitism was found to be higher at field–forest edges than in the interior habitat by Gates and Gysel (1978); there again may be a trade-off, between nesting in the higher-productivity edges and in the poorer, but safer, interior habitat.

G. Habitat Selection and Competition

1. *The Role of Competition*

Bird species may find some habitats suitable in terms of structure and level of productivity, but the presence of other species there might significantly reduce food availability and render the habitat unacceptable. Lack (1933), Svärdson

(1949), and Kendeigh (1945) used this explanation for the many cases of allopatry and habitat segregation in congeneric species, and Terborgh and Weske (1975) showed that around two-thirds of the bird species on a Peruvian habitat gradient had their distributional limits determined by interspecific competition. Interspecific competition might operate only locally or periodically to effect habitat segregation between species (Udvardy, 1951). Besides such circumstantial evidence as morphological character displacement in sympatric species (e.g., Ficken *et al.*, 1968), there are observations on direct aggresive interactions between species [e.g., Gorton (1977) in Song Sparrow and Bewick's Wren (*Thryomanes bewickii*), Thiollay (1981) in *Buteo* hawks; see Cody (1974a, 1981)]. Although the adaptive nature of interspecific territoriality has been questioned (Murray, 1971), there exists, besides the large amount of indirect evidence, experimental corroboration for an adaptive basis to this behavior. Lyon *et al.* (1977), for example, showed that the order in which other hummingbird species were admitted to territories of Blue-throated Hummingbirds (*Lampornis clemenciae*) which were made increasingly indefensible was predictable from a knowledge of the degree of ecological overlap between the intruders and the territory owners, and the insect food supply of Bell Miners [*Manorina melanophrys* (Meliphagidae)] was rapidly cleaned up by other species when the Bell Miners were removed from territories they had previously defended against these other species (Loyn *et al.*, 1983).

There is further evidence of the role of competition from habitat expansions in island birds (see Section III,D) and from habitat contraction where related species occur in sympatry. Again the effects may be apparent at the level of local habitat use, as in Field Sparrows (*Spizella pusilla*), which make more extensive use of trees when Chipping Sparrows (*S. passerina*) are not present (Hebrard, 1978), or on a grander scale, as in Common Nighthawks (*Chordeiles minor*), which use a wide variety of habitats over the United States and Mexico but are excluded from desert habitats where they encounter Lesser Nighthawks (*C. acutipennis*) in the southwest (Caccamise, 1974). Here there is further direct experimental evidence of the role of interspecific competition, as in *Melanerpes* woodpeckers (Williams and Batzli, 1979). When Red-headed Woodpeckers (*M. erythrocephalus*) are removed, Red-bellied Woodpeckers (*M. carolinus*) expand their habitat range, and Downy Woodpeckers (*Picoides pubescens*) extend their foraging range in the vegetation. Further examples of interspecific constraints on habitat use follow, and many others are discussed later in this volume, in chapters on Parulid (Chapter 4) and Sylviid (Chapter 3) warblers, raptors (Chapter 5), and nectarivorous (Chapter 12) and grassland birds (Chapter 6).

2. *Titmice (Paridae)*

Some half dozen *Parus* species are widely distributed in western Europe, and several co-occur in a variety of habitats. Comparing foraging habitats in oak woodlands at sites with two versus three species present, Herrera (1978) found that Blue Tits (*Parus caeruleus*) feed less on the ground, tree trunks, and twigs and more on leaves, in the presence of Crested Tits (*P. cristatus*); the larger Great Tit (*P. major*) was little affected. Breeding success in Great Tits, however, is negatively correlated with Blue Tit density (Dhondt, 1977; Dhondt and Eykerman, 1980a; Minot, 1981), but Blue Tits are more seriously affected by Great Tits in the wintering season than in the breeding season (Dhondt and Eykerman, 1980b). Willow Tit (*P. montanus*) foraging sites in spruce forest are constrainted by both Great and Creasted Tits. In the presence of the former, Willow Tits shift to the upper and outer parts of the trees but, when together with the latter, make the reverse shift to lower and inner spruce sites (Alatalo, 1981b).

3. Thrushes (Turdidae)

In nothern Connecticut both Wood Thrush (*Hylocichla mustelina*) and Veery (*Catharus fuscens*) are species of wetter forest. There is some territorial overlap with Veery in 90% of the Wood Thrush territories, and discriminant function analysis of vegetation in territories shows that the species are distinct in habitat at the level of one standard deviation (Bertin, 1977). Wood Thrushes have recently extended their ranges further north into Maine, where they fit into the wet-to-dry forest gradient between Veery and Hermit Thrush (*C. guttatus*). There is considerable habitat overlap, however, and similar species pairs are interspecifically territorial (Morse, 1971). In Vermont there are four species of *Catharus* and the Wood Thrush in forest habitats; these species drop out one at a time as one progresses south in the Appalachian Mountains (Noon, 1981). Among the five species in New England, DFA produces a first axis emphasizing canopy cover, vegetation height, and different deciduous–coniferous preferences, but there is a lot of interspecific overlap, especially between Veery and Wood Thrush. These two species however, segregate on the second DFA axis, based on differing preferences for ground cover and shrub density. In the southern Smoky Mountains, where these two thrushes are the only breeding species of the original five in Vermont, they show a considerably expanded occupancy of habitats.

4. Vireos (Vireonidae)

The five breeding vireos on and around the Nevado de Colima (Section II,D) show extensive habitat co-occupancy among species, and the same is true for North American species. James (1976) studied three species in southern Ontario and found that two of them, Red-eyed Vireo (*V. olivaceus*) and Warbling Vireo (*V. gilvus*), segregate by habitat whereas the third, Yellow-throated Vireo (*V. flavifrons*), uses the same habitats as the first two species but forages in the interior of trees as opposed to the exterior. In northeastern Ontario, Red-eyed Vireo and Philadelphia Vireo (*V. philadelphicus*) show no interspecific differences in habitat use, although the former occupies a wider habitat range; the two are interspecifically territorial (Rice, 1978). Robinson (1981) studied these same two species further south where, although there was interspecific aggression, territories overlapped extensively. Overlap in foraging heights and in aggression varied seasonally, and relations between the two were most amicable in years of higher caterpillar density. Where these species are interspecifically territorial in Ontario, the vegetation is simpler, and the partial subdivision of resources by foraging height, as happens in the New England birds, may not be possible.

H. Effects of Habitat Extent and Patchiness

1. Real Islands

Small islands have fewer species than large islands because such islands are more difficult to find and colonize and also because they support smaller populations of a species, and smaller populations are more likely to go extinct (MacArthur and Wilson, 1967). Even though their range of habitats may be similar to that of larger islands or mainlands, small islands will lack a number of species which are sensitive to habitat area. This sensitivity varies with species, as is apparent from Willis (1974), who monitored bird populations on Barro Colorado Island for 11 years. During this time, the smaller Spotted Antbird (*Hylophylax naevioides*) maintained its population size, the medium-sized Bicolored Antbird (*Gymnopithys bicolor*) declined by 50%, and the large Occelated Antbird (*Phaenostictus mcleannani*) became extinct. Thus, larger birds need a larger critical area of habitat to avoid extinction. Other studies of birds on lake islands are published by Howe (1979) and Ahlen and Nilsson (1982).

2. Habitat Islands

Some habitats occur naturally as isolated islandlike patches; meadows (Eriksson, 1981) or mires (Kolmodin and Nilsson, 1982) in wooded country are examples. In the case of forest patches in New Jersey (Galli *et al.*, 1976), Britain (Moore and Hooper, 1975), Japan (Higuchi *et al.*, 1982), and Ecuador (Leck, 1979), these are relict stands, the results of extensive logging. Martin (1981) censused shelter belts, patches of trees planted as windbreaks in native grassland. These studies show how species numbers increase with patch size but, futher, that species have different threshold levels of habitat area, below which they occur only sporadically and above which they occur regularly. For example, American Robins were recorded in shelter belts one-fifth the minimum habitat area needed for Black-billed Cuckoos (*Coccyzus erythrophalmus*). In forest patches in Japan, the woodpecker (*Dendrocopos kizuki*) became scarce when patch size dropped below 100 ha, yet the bulbul (*Hypsipetes amourotis*) occurred regularly in patches down to 0.1 ha (Higuchi *et al.*, 1982). In English woods,

Blackbirds will persist in patches as small as 10 ha, but Coal Tits (*Parus ater*) need 80 km^2 or more to maintain steady populations (Moore and Hooper, 1975). For these species, minimum habitat area is as critical a necessity as are minimally satisfactory habitat structure and productivity.

3. *More General Effects of Habitat Area*

In other instances habitats occur as isolated patches or terminate in long peninsular-like projections because of topographic effects or reasons of climate associated with topography. The same conclusions appear to apply: species diversity is reduced in more isolated habitats, because of higher extinction rates and thresholds, which are bird-size specific, to minimum habitat areas necessary for population survival. Habitat use may be correspondingly increased by species that can survive in isolated habitat patches and habitat range expanded by species that occur on isolated mountains with a diversity of habitats. Recent examples are given by Abbott (1981) of birds in the Porongurup Range in southwestern Australia, with 19 fewer bird species than in larger ranges to the west, and by Thompson (1978) of pine-covered buttes in Montana grassland. Rabenold (1978) found that spruce–fir forest in North Carolina, the southern edge of *Picea rubens* range, held only half as many bird species as the equivalent Maine forests. Although there was a more generalized foraging of birds in the species-poor southern forest, there was no density compensation; the reason, he believed, was lower food levels in the south and stronger seasonal food pulses in the north.

Afromontane forest occurs as patches along South Africa's Indian Ocean coast, with the size of patches decreasing and their isolation from larger forest tracts increasing from northeast to southwest. Bird species numbers decrease threefold along a series of these patches, and species sensitive to habitat area limitations drop out sequentially. Reduced food levels are not a direct cause of fewer species in smaller patches, as density compensation in some groups is complete, and total bird density varies only 10%, in contrast with 300% changes in species numbers (Cody, 1983b). Species appear to be differentially sensitive to lower carrying capacities in smaller habitat patches, a point which can be treated theoretically.

Most habitats occur as patchworks, as vegetation maps show, with more or less discontinuity between patches in the form of corridors along mountains or valleys. Both the total area of habitat in a topographically circumscribed region and the frequency distribution of different habitat types help to determine species turnover between habitats, or β-diversity (Cody, 1983a). In particular, β-diversity increases as the structural differences between habitats increase but is higher also when the areal extents of two similar habitats are high: there are many species differences between common habitat types but fewer between rare and common habitats.

I. Special Features of Habitats

1. Feeding Sites

Many species that share habitats feed in different sites. The classical example is that of MacArthur's (1958) warblers (*Dendroica*) in spruce forests. In this case, the habitat simultaneously provides foraging opportunities for several related species. There are many other examples, such as many duck species that breed on Lake Myvatn, northern Iceland, but have rather distinct foraging sites (Bengtson, 1971).

2. Perch Sites

For some species, such as flycatchers, shrikes, and other sit-and-wait hunters like puffbirds (Bucconidae) and some raptors (see Chapter 5, this volume), perch-site requirements may be quite specific, Marion and Ryder (1975) described the perch site preferences for wintering raptors in the short-grass plains of Colorado. Ferruginous Hawks used the lower fence posts, Rough-legged Hawks (*Buteo lagopus*), trees and poles, and Golden Eagles (*Aquila chrysaetos*), only the highest perch sites. Perch sites may be important as singing posts for open-country passerines without aerial song flights; some parulid warblers in lower habitats have different nest and perch requirements, but the warblers of taller habitats do not (Collins, 1981).

3. Nest Sites

Many bird species have rather specific nest-site requirements. Colonial seabirds and freshwater birds are among them (Burger, 1974; Burger and Gochfeld, 1981; see Chapter 8, this volume), and in mixed colonies there are often distinct species-specific preferences. This applies to two species of tropic birds (*Phaeton*) on Aldabra Atoll (Prŷs–Jones, 1980) and to five heron species on the North Carolina coast (McCrimmon, 1978). Where good nesting sites are in short supply, there may be interspecific competition for them. White and Conroy (1975) described the different nest-site preferences of congeneric Gentoo Penguin (*Pygoscelis pappua*), Adélie Penguin (*P. adeliae*), and Chinstrap Penguin (*P. antarctica*) on South Orkney Island but emphasized that all species need an easy access between nest and sea, and ice-free ground. Penguin populations may have increased in response to whaling activities and a corresponding increase in penguin food, and this seems to have aggravated interspecific nest-site competition between Adélie and Chinstrap Penguins in mixed colonies (Trivelpiece and Volkman, 1979).

Nest-site preferences are often distinct in related species of land birds also. *Buteo* hawks, for example, select nesting trees of different characteristics in woodland in the Appalachians (Titus and Mosher, 1981), and Eastern Kingbirds

(*Tyrannus* spp.) nest in smaller, lower, but denser trees than congeneric Western Kingbirds in Manitoba (MacKenzie and Sealy, 1981).

Nest-site specificity might evolve because nesting success is enhanced by particular microclimatic features. This is implied in the nest orientation of Cactus Wrens (Austin, 1974), of hummingbirds at higher elevations in the Rocky Mountains (Calder, 1973), and of Warbling Vireos with nests open to the eastern, but not the western, sky (Walsberg, 1981; see Chapter 13, this volume). In this context the spectacular communal nest of the Sociable Weaver (*Philetairus socius*) is worth mentioning; the temperature range inside the nest is one-half that on the outside (Bartholomew *et al.*, 1976).

4. *Hole-Nesting Species*

Birds that use holes for breeding need to assess likely habitats for hole availability; the requirement is much more specific and less likely to be met ubiquitously than nest sites for open-nesting species. Hole breeders often contest nest sites interspecifically (e.g., mynahs and starlings in Japan, Ogasawara, 1976), and with nest-hole abundance sometimes as low as 6 holes/ha in woodlands, hole occupancy may run as high as 93% (van Balen *et al.*, 1982). Further, the composition of the breeding bird community may change quite dramatically when nestboxes are added. Great Tits and Pied Flycatchers (*Ficedula hypoleuca*) are species that may greatly increase with the addition of nestboxes to the habitat (van Balen *et al.*, 1982; Enemar *et al.*, 1972; LeClercq, 1976). Pied flycatcher density has increased 16-fold when nest boxes are available in excess of demand (Lennerstedt, 1983; see also Chapter 2, this volume), but such increases apparently do not affect the densities of other species in the habitat (Enemar and Sjöstrand, 1972). There is some evidence that non-nest-box titmice decrease when Great Tits densities are elevated through the provision of boxes (van Balen *et al.*, 1982).

5. *Nest Sites and Predation*

Predation rates on nests may vary as a function of nest site, predators selecting for specific nest sites and against others. In Wood Pigeons (*Columba palumbus*) predation rates varied up to >60% and were related to nest height in London parks (Tomialojć, 1978), and in Blackbird nests away from paths and buildings, success in avoiding nest predation varied with height and cover (Osborne and Osborne, 1980). Hooded Crow (*Corvus corone*) nests in conifers are safer from nest predation, perhaps because of better nest concealment (Loman, 1979).

Protection for the nest can also be afforded by particularly aggressive social insects, as Moreau (1942) pointed out. The estrildid (*Lonchura cuccullata*) nests in association with a wasp (Woodall, 1975), and another Nigerian ploceid nests close to nests of "a large and vicious red ant" (MacLaren, 1950). Other bird species that are particularly alert or aggressive toward potential nest predators can provide the same sort of service; examples are birds selecting nest sites near raptorial birds (Walsh and Walsh, 1976), in gull and tern colonies (Dwernychuk and Boag, 1972; Koskimies, 1957), and near aggressive waders, such as the Lapwing (*Vanellus vanellus;* Dyrcz *et al.*, 1981). Göransson *et al.* (1975) found that predation on artificial nests in Lapwing territories was only 26% that of the control rate on nests without Lapwing protection. In northern Europe, the Fieldfare (*Turdus pilaris*), a large, noisy, and aggressive thrush, can help to drive out nest predators, such as crows, from its own territories, particularly effectively when the thrushes are loosely colonial (Anderson and Wiklund, 1978; Wicklund and Anderson, 1980; see also Chapter 3, this volume). They can also reduce predation rates on the nests of other bird species, such as Merlins (*Falco columbarius;* Wicklund, 1979) and Bramblings and Redwings (*Turdus iliacus*), which appear to nest preferentially in Fieldfare territories (Slagsvold, 1980a,b).

6. *Roosting Sites*

The judicious selection of roosting sites may enhance survival in birds living through cold winters. Jackdaws (*Corvus mondeula*), for example, may roost in towns because of the more favorable microenvironment (Tast and Rassi, 1973) or because towns provide better predator protection or enhanced foraging efficiency (Gyllin *et al.*, 1977); Starlings select winter roosts where the nighttime temperatures are higher (Yom–Tov *et al.*, 1977) and thereby save the equivalent cost of a 31-km flight.

In the hot summers of the southwestern United States, in contrast, roost sites may be selected to keep birds cool; Spotted Owls (*Strix occidentalis*) roost in cool places in deep canyons, a special habitat requirement that may restrict their distribution (Barrows, 1981).

The positions of individual birds in social roosts are of interest. In Rooks (*Corvus frugeligus*), the adults perch above the juveniles except on windy nights (Swingland, 1977); the advantage to the upper roost sites is obvious, especially since, as Yom–Tov (1979) has shown, Starling feathers defecated on absorb more water and give less protection to their owner.

V. COEVOLUTIONARY ASPECTS OF HABITAT SELECTION

A. Specific Adaptations for Specific Resources

Specific resources of the habitat, on the one hand, and aspects of a bird's ecology, morphology, or behavior, on the other, may be tightly coupled as reciprocal selective influences and may justify the term "coevolution." Of course, warbler bills have evolved for insectivory, and finch bills, for granivory;

but the evidence for reciprocal effects on insects and seeds is usually wanting. In the case of oxpeckers (*Buphagus*), the Yellow-billed Oxpecker (*B. africanus*) feeds preferentially on the sparsely furred or naked hides of such larger animals as black rhinoceros or buffalo, while the Red-billed Oxpecker (*B. erythrorhynchus*) prefers the furred hides of smaller animals, such as antelopes (kudu, impala, and eland). Both pick off ectoparasites and likely enhance survival in their hosts; the difference in feeding substrate is reflected in the width and length of the oxpeckers' bills (Buskirk, 1975). Acorn Woodpeckers, an example of birds with specific food requirements, not only require habitats with acorns but also tall trees to use for acorn storage (Roberts, 1979); in turn, the density and distribution of oak trees are probably affected by the woodpeckers' activities as well as those of other acorn consumers and transporters, such as Scrub Jay (*Aphelocoma caerulescens*).

B. Nectarivorous Birds

Nectarivorous birds can be expected to have close associations with a habitat in terms of the sorts of nectar resources the habitat supplies. This is because flower corollas and birds' beaks can coevolve in shape to mutually exclude other birds and flowers from the association; the long, decurved bill of hummingbird *Lafresnaya* is adapted to the corolla of its customary food plant *Siphocampylus,* and that of the hermit *Phaethornis eurynome,* to its usual nectar source, *Sinningia polyantha* (Snow and Snow, 1980; Snow and Teixeira, 1982). But such resources may be dissociated from the gross structure of the habitat, since, for example, Rufous Hummingbirds (*Selasphorus rufus*) migrate from Alaska to Mexico and back utilizing mostly flowers with red, tubular corollas of a certain length. They winter in such places as the high pine–oak–fir on the Nevado de Colima, where a species of *Salvia* provides their nectar requirements; in passage through the Mojave Desert, they use *Castilleja* species, and elsewhere, while in migration, breeding, or wintering, many other red, tubular corolla flowers provide their food.

While tropical hummingbirds provide good examples of close coadaptation between hummingbirds and flowers, in temperate hummingbirds the associations appear to be less specific. At higher latitudes in these and in other nectarivorous bird systems, there may be a more limited selection for strong specialization in the face of lower resource predictabilities (see Chapter 12, this volume). In the Australian Meliphagidae, for example, six specis of *Meliphaga* coexist in southern Australia; four of them segregate by habitat and two via differences in feeding behavior (Ford and Patton, 1976). Just one of these species, the Yellow-throated Honeyeater (*M. flavicollis*), occurs in Tasmania, where it occupies all habitats and also forages on bark substrates, indicating no close ties to specific food sources.

Nectarivorous birds are represented by the endemic sub-family Drepanidinae in the Hawaiian Islands, where some species show a good deal of flexibility in resource use but where some of the more specialized species with more extreme bill morphologies have already become extinct. Three of the commoner nectarivores are species of *Vestiaria, Himatione,* and *Loxops,* with the former preferring high-density patches of *Metrosideros* trees, the second, *Metrosideros* trees in general, and the third, *Sophora* trees; but these specializations were only apparent during the times of low nectar availability (Pimm and Pimm, 1982).

C. Frugivorous Birds

Some birds are specialized in morphology (bill, foot, digestive system) and in behavior and ecology (habitat, food and foraging-site selection, breeding system, and breeding behavior) for particular fruit food supplies, either hard or soft. Other species are less specialized, but still ecologically committed to a frugivorous life-style (see Chapter 11, this volume). The mannikins (Pipridae) are common frugivores in the Neotropics which show various levels of specialization. The piprid (*Chiroxiphia linearis*), for example, needs several times the quantities of nonpreferred fruit when its preferred fruits are unavailable, although it can still maintain weight on the less desirable food (Foster, 1977). Waxwings feed insects to nestlings for the first few days but later feed them fruit and nest in high densities where *Prunus* fruits are common in Michigan (Rothstein, 1971). Elsewhere they use different fruits (e.g., Sage, 1973) and, while wintering in Jalisco and Oaxaca, Mexico, are common in habitats where the fruits of the columnar cactus (*Myrtillocactus*) are available and readily eaten (see Section II,F). A reciprocal interdependence of food and consumer is apparent between Pinyon Jays (*Gymnorhinus cyanocephalus*) and pinyon pines (Ligon, 1978) and between Clark's nutcrackers (*Nucifraga columbiana*) and whitebark and pinyon pines (Hutchins and Lanner, 1982; Vander Wall and Balda, 1977), in which dispersal and establishment of the trees are much affected by the nutcracker. Chapter eleven (this volume) treats coevolutionary aspects of the subject in more detail.

D. Brood Parasite Systems

One component of habitat selection in species of nest parasites, such as cuckoos, cowbirds, and honeyguides, must be the availability of host nests. Many nest parasites have large geographic ranges, such as that of the cuckoo (*Cuculus canorus*) across Eurasia, and occur in many different habitats but are very selective of certain host nests. These cuckoos are migratory; they presumably select the appropriate host nest, with eggs matched by those of the particular female cuckoo, by imprinting on hosts and/or host habitats as nestlings or fledglings (imprinting of course may apply more generally in habitat selection). In some European locations, several different genetic types or *gentes* of females coexist locally, with eggs color matching those of different host species (Southern, 1963); thus individuals of different *gentes* will have different habitat preferences, for habitats where different host species are common.

Brood parasites can dramatically reduce nesting success in their hosts and, thus, exert selection on their hosts to successfully avoid the attentions of the parasites. Such selection might alter, in potential hosts, their use of habitats. In woodland habitats in South Africa, where up to six cuckoo species are found, some commonly, the host Bleating Bush Warbler (*Camaroptera brachyura*), of the Emerald Cuckoo (*Chrysococcyx cupreus*), is extremely secretive and mouselike around its nest in dense ground-level vegetation and may cycle in density in forests with the right sort of vegetation, as a consequence of brood parasitism. Other species in these habitats, such as Forest Weaver (*Symplectes bicolor*), build elaborate hanging nests that might serve to reduce brood parasitism, and others, like the Chorister Robin (*Cossypha dichroa*), indulge in vocal mimicry of their potential parasites [in this case, of the Red-chested Cuckoo (*Cuculus solitarius*)] that might keep them out of host territories (Cody, 1983a; M. L. Cody unpublished observation; W. R. Siegfried, personal communication). These and other potential coevolutionary influences on habitat selection have not been well investigated and our knowledge of them is scanty.

VI. THEORETICAL APPROACHES TO HABITAT SELECTION

There has been a variety of theoretical approaches to habitat selection problems, beginning with studies such as those of Levins and Culver (1971) and Horn and MacArthur (1972), which showed that the outcome of coexistence between competing species is quite different where habitats occur as mosaics of partially isolated patches. Levin (1977) and Slatkin (1974) developed this theme, with the latter showing that two similar species can coexist (i.e., in the same habitat) over some range of the system's parameters, a range that is increased if predation rates on these species increase. Hanski (1980) also concludes that regional coexistence is possible where local coexistence between species is not and further develops the role of predation in promoting coexistence in systems in which two species occupy like habitat patches but the density of either species varies considerably among patches. Where mean population density is proportional to the variance in population density, the predator can facilitate coexistence (Hanski 1981).

If two species with different patch preferences occupy a gradient of two habitat patch types, they may coeixst over some parts of the gradient but are less likely to if the patches are quite different or unproductive or if population turnover rates through time are low (Cody, 1974b). Lloyd and White (1980) considered how the coexistence of species is enhanced if they occupy a mosaic of habitats in each of which their carrying capacities are different, and Levins (1980) considered questions of coexistence in temporally variable environments. Another recent and comprehensive treatment of the topic is that of Yodzis (1980). Again starting from basic Lotka–Volterra competition equations, Rosenzweig and colleagues (Rosenzweig, 1979, 1981; Pimm and Rosenzweig, 1981) derived the expected outcomes of competition in regions with either one or both of two habitat types and included considerations of whether habitat selection is density dependent or independent, costly or cost free, and affected more by intra- than interspecific competition.

Fretwell and Lucas (1969) initiated a somewhat different approach to the theory of habitat selection, by modeling the effects of intraspecific populational pressures on the sequential occupancy of habitats ranked by their intrinsic quality or suitability (see Section, IV,D). MacArthur (1972) extended this line of thinking to a geographic scale and discussed the trade-offs between more competition from both conspecifics and allospecifics in higher quality habitats and less constrained use of resources in intrinsically poorer habitats; his "principle of equal opportunity" implies that habitat use will be determined by a balance of resource production in habitats and both intra- and interpopulational pressure to utilize them. Further discussion of the conceptual and theoretical questions regarding habitat use and selection is developed in Chapter 18 (this volume).

REFERENCES

Abbott, I. (1981). The avifauna of the Porongurup Range, an isolated habitat in south-western Australia. *Emu* **81,** 91–96.

Ahlen, I., and Nilsson, S. G. (1982). Samband mellan fågelfauna och biotoporeal på öar med naturskog i Mäloren och Hjälmoren. *Vår. Fågelv.* **41,** 161–184.

Alatalo, R. V. (1981a). Habitat selection of forest birds in the seasonal environment of Finland. *Ann. Zool. Fenn.* **18,** 103–114.

Alatalo, R. V. (1981b). Interspecific competition in tits *Parus* spp. and the goldcrest *Regulus regulus:* Foraging shifts in multispecies flocks. *Oikos* **37**(3), 335–344.

Amadon, D. (1950). The Hawaiian Honeycreepers (Aves, Drepanididae). *Bull. Am. Mus. Nat. Hist.* **95,** 151–262.

Anderson, S. H., and Shugart, H. H. (1974). Habitat selection of breeding birds in an east Tennessee deciduous forest. *Ecology* **55,** 828–837.

Arvin, J., and Arvin, J. (1975). Mexican crow invades Texas. *Auk* **92,** 387–390.

Austin, G. T. (1974). Nesting success of the cactus wren in relation to nest orientation. *Condor* **76,** 216–217.

Baker, M. C., and Baker, A. E. M. (1973). Niche relationships among six species of shorebirds on their wintering and breeding ranges. *Ecol. Monogr.* **43,** 193–212.

Barrows, C. W. (1981). Roost selection by spotted owls: An adaptation to heat stress. *Condor* **83,** 302–309.

Bartholomew, G. A., White, F. N., and Howell, T. R. (1976). The thermal significance of the nest of the Sociable Weaver *Philetairus socius:* Summer observations. *Ibis* **118,** 402–410.

Beaver, D. L. (1976). Avian populations in herbicide treated brush fields. *Auk* **93,** 543–553.

Bendell, J. F., and Elliot, P. W. (1966). Habitat selection in the Blue Grouse. *Condor* **68,** 431–446.

Bengtson, S.-A. (1971). Habitat selection of duck broods in Lake Myvatn area, north-east Iceland. *Ornis Scand.* **2**(1), 17–26.

Bertin, R. I. (1977). Breeding habitats of the wood thrush and veery. *Condor* **79,** 303–311.

Birkhead, T. R. (1977). The effect of habitat and density on breeding success in the common guillemot (*Uria aalge*). *J. Anim. Ecol.* **46,** 751–764.

Bock, C. E., and Lynch, J. F. (1970). Breeding bird populations of burned and unburned conifer forest in the Sierra Nevada. *Condor* **72,** 182–189.

Brewer, R., and Harrison, K. G. (1975). The time of habitat selection by birds. *Ibis* **117,** 521–522.

Brownsmith, C. B. (1977). Foraging rates of starlings in two habitats. *Condor* **79,** 386–387.

Burger, J. (1974). Determinants of colony and nest site selection in the Silver Grebe, *Podiceps occipitalis,* and Rolland's Grebe, *Rollandia rolland. Condor* **76,** 301–306.

Burger, J., and Gochfeld, M. (1981). Nest site selection by kelp gulls in southern Africa. *Condor* **83,** 243–251.

Buskirk, W. H. (1975). Substrate choice of oxpeckers. *Auk* **92,** 604–606.

Caccamise, D. F. (1974). Competitive relationships of the Common and Lesser Nighthawks. *Condor* **76,** 1–20.

Calder, W. A. (1973). Microhabitat selection during nesting of hummingbirds in the Rocky Mountains. *Ecology* **54,** 127–134.

Catterall, C. P., Wyatt, W. S., and Henderson, L. J. (1982). Food resources, density, and reproductive success of an island Silvereye population *Zosterops lateralis. Ibis* **124,** 405–421.

Cody, M. L. (1970). Chilean bird distribution. *Ecology* **51,** 455–464.

Cody, M. L. (1974a). "Competition and the Structure of Bird Communities," Monographs in Population Biology, Vol. 7, pp. 1–318. Princeton Univ. Press, Princeton, New Jersey.

Cody, M. L. (1974b). Optimization in ecology. *Science* **18,** 1156–1164.

Cody, M. L. (1975). Towards a theory of continental species diversity: Bird distributions over Mediterranean habitat gradients. *In* "Ecology and Evolution of Communities" (M. L. Cody and J. M. Diamond, eds.), pp. 214–257. Harvard Univ. Press (Belknap), Cambridge, Massachusetts.

Cody, M. L. (1981). Habitat selection in birds: The roles of habitat structure, competitors, and productivity. *BioScience* **31,** 107–113.

Cody, M. L. (1983a). Continental diversity patterns and convergent evolution in bird communities. *In* "Mediterranean-Type Ecosystems" (F. J. Kruger, D. T. Mitchell, and J. U. M. Jarvis, eds.), Ecological Studies Vol. 43, pp. 357–402. Springer-Verlag, Berlin and New York.

Cody, M. L. (1983b). Bird distribution and density in the Afromontane woodlands of southern Africa. *Oecologia* **50** (in press).

Cody, M. L., and Cody, C. B. J. (1972). Territory size, food density and clutch size in island wren populations. *Condor* **74,** 473–477.

Collins, S. L. (1981). A comparison of nest-site and perch-site vegetation structure for seven species of warblers. *Wilson Bull.* **93,** 542–547.

Collins, S. L., James, F. C., and Risser, P. G. (1982). Habitat relationships of wood warblers (Parulidae) in northern central Minnesota. *Oikos* **39**(1), 50–58.

Confer, J. L., and Knapp, K. (1981). Golden-winged warblers and blue-winged warblers: The relative success of a habitat specialist and a generalist. *Auk* **98,** 108–114.

Constant, P., Eybert, M.-C., and Maheo, R. (1973). Recherches sur les oiseaux nicheurs dans les plantations de résineaux de la forêt de Paimpont (Bretagne). *Ardea* **41,** 371–384.

Cox, G. W., and Ricklefs, R. E. (1977). Species diversity and ecological release in Caribbean land bird faunas. *Oikos* **28**(1), 113–122.

Crowell, K. (1962). Reduced interspecific competition in the birds of Bermuda. *Ecology* **43,** 75–88.

Darwin, C. (1897). "The Voyage of the Beagle. Journal of Researches into the Natural History and Geology of the Countries Visited during the Voyage of H. M. S. 'Beagle' round the World." John Murray, London. (Originally published 1836.)

Davis, W. J. (1982). Territory size in *Megaceryle alcyon* along a stream habitat. *Auk* **99,** 353–362.

Dhondt, A. A. (1977). Interspecific competition between great and blue tit. *Nature (London)* **268,** 521–523.

Dhondt, A. A., and Eykerman, R. (1980a). Competition and the regulation of numbers in great and blue tit. *Ardea* **68,** 121–132.

Dhondt, A. A., and Eykerman, R. (1980b). Competition between the Great Tit and the Blue Tit outside the breeding season in field experiments. *Ecology* **61,** 1291–1296.

Diamond, A. W. (1973). Habitats and feeding stations of St. Lucian forest birds. *Ibis* **115,** 313–329.

Diamond, J. M. (1975). Distributional ecology and habits of some Bougainville birds (Solomon Islands). *Condor* **77,** 14–23.

Diamond, J. M., and Marshall, A. G. (1977). Niche shifts in New Hebridean birds. *Emu* **77,** 61–72.

Disney, H. J. de S., and Stokes, A. (1976). Birds in pine and native forests. *Emu* **76,** 133–138.

Dunning, J. B. Jr., and Brown, J. H. (1982). Summer rainfall and winter sparrow densities: A test of the food limitation hypothesis. *Auk* **99,** 123–129.

Dwernychuk, L. W., and Boag, D. A. (1972). Ducks nesting in association with gulls—an ecological trap? *Can. J. Zool.* **50,** 559–563.

Dyrcz, A. (1969). The ecology of the Song Thrush (*Turdus philomelos* Br.) and Blackbird (*Turdus merula* L.) during the breeding season in an area of their joint occurrence. *Ekol. Pol. Ser. A.* **17,** 735–793.

Dyrcz, A., Witkowski, J., and Okulewicz, J. (1981). Nesting of "timid" waders in the vicinity of "bold" ones as an anti-predator adaptation. *Ibis* **123,** 542–545.

Emlen, J. T. (1970). Habitat selection by birds following a forest fire. *Ecology* **51,** 343–345.

Emlen, J. T. (1972). Size and structure of a wintering avian community in southern Texas. *Ecology* **53,** 317–329.

Emlen, J. T. (1974). An urban bird community in Tucson, Arizona: Derivation, structure, regulation. *Condor* **76,** 184–197.

Enemar, A., and Sjöstrand, B. (1972). Effects of the introduction of pied flycatchers *Ficedula hypoleaca* on the composition of a passerine bird community. *Ornis Scand.* **3**(2), 79–89.

Enemar, A., Nyholm, E., and Persson, B. (1972). The influence of nest-boxes on the passerine bird community of Fågelsångsdalen, southern Sweden. *Vår. Fågelv.* **31,** 263–268.

Eriksson, M. O. G. (1981). Betydelsen av betade fuktängars aved och inbördes avstånd för häckfågelfaunan. *Vår. Fågelv.* **40,** 185–192.

Faaborg, J. (1976). Habitat selection and territorial behavior of the small grebes of North Dakota. *Wilson Bull.* **88,** 390–399.

Ficken, M., and Ficken, R. W. (1966). Notes on mate and habitat selection in the Yellow Warbler. *Wilson Bull.* **78,** 232–233.

Ficken, R. W., Ficken, M., and Morse, D. H. (1968). Competition and character displacement in sympatric pine-dwelling warblers. *Evolution* **22,** 307–314.

Fischer, D. H. (1980). Breeding biology of curve-billed thrashers and long-billed thrashers in southern Texas. *Condor* **82,** 392–397.

Fischer, D. H. (1981). Wintering ecology of thrashers in southern Texas. *Condor* **83,** 340–346.

Ford, H. A. (1979). Interspecific competition in Australian honeyeaters—depletion of common resources. *Aust. J. Ecol.* **4,** 145–164.

Ford, H. A. (1981). Territorial behaviour in an Australian nectar-feeding bird. *Aust. J. Ecol.* **6,** 131–134.

Ford, H. A., and Patton, D. C. (1976). Resource partitioning and competition in honeyeaters of the genus *Meliphaga. Aust. J. Ecol.* **1,** 281–287.

Foster, M. S. (1977). Ecological and nutritional effects of food scarcity on a tropical frugivorous bird and its fruit source. *Ecology* **58,** 73–85.

Fox, B. J. (1979). An objective method of measuring the vegetation structure of animal habitats. *Aust. Wildl. Res.* **6,** 297–303.

François, J. (1975). Contribution à la connaisance de l'avifaune d'Afrique du Nord. (Note sur l'extension éventuelle du milieu saharien por H. Heim de Balsac. *Alauda* **43,** 279–293.

Franzreb, K. E., and Ohmart, R. D. (1978). The effects of timber harvesting on breeding birds in a mixed-coniferous firest. *Condor* **80,** 431–441.

Fretwell, S. D. (1972). "Populations in a Seasonal Environment," Monographs in Population Biology, Vol. 5, p. 1–220. Princeton Univ. Press, Princeton, New Jersey.

Fretwell, S. D., and Lucas, H. L. (1969). On territorial behavior and other factors affecting habitat distribution in birds. I. Theoretical development. *Acta Biotheor.* **19,** 16–36.

Galli, A., Leck, C. E., and Forman, R. T. T. (1976). Avian distribution patterns in forest islands of different sizes in central New Jersey. *Auk* **93,** 356–364.

Gaston, A. J. (1974). Adaptation in the genus *Phylloscopus. Ibis* **116,** 432–450.

Gates, J. E., and Gysel, L. W. (1978). Avian nest dispersion and fledgling success in field-forest ecotones. *Ecology* **59,** 871–883.

Glas, P. (1960). Factors governing density in the chaffinch *Fringilla coelebs* in different types of wood. *Arch. Neerl. Zool.* **13,** 466–472.

Gochfeld, M, (1978). Ecological aspects of habitat selection by two sympatric mockingbirds, *Mimus* spp., in Patagonia. *Ibis* **120,** 61–65.

Göransson, G., Karlsson, J., Nilsson, S. G., and Ulfstrand, S. (1975). Predation on birds nests in relation to antipredator aggression and nest density: An experimental study. *Oikos* **26**(2), 117–120.

Gorman, M. L. (1975). Habitats of the land-birds of Viti Levu, Fiji Islands. *Ibis* **117,** 152–160.

Gorton, R. E., Jr. (1977). Territorial interactions in sympatric song sparrow and Bewick's wren populations. *Auk* **94,** 701–708.

Grant, P. R. and Nettleship, D. N. (1971). Nesting habitat selection by Puffins *Fratercula arctica* L. in Iceland. *Ornis Scand.* **2**(2), 81–87.

Greenberg, R. (1979). Body size, breeding habitat, and winter exploitation systems in *Dendroica. Auk* **96,** 756–766.

Grinnell, J. (1904). The origin and distribution of the chestnut-backed chickadee. *Auk* **21,** 364–382.

Grinnell, J. (1917). The niche-relationships of the California thrasher. *Auk* **34,** 427–433.

Gyllin, R., Kallander, H., and Sylven, M. (1977). The microclimate explanation of town centre roosts of Jackdaws *Corvus monedula. Ibis* **119,** 358–361.

Hanski, I. (1980). Exploitative competition in transient habitat patches. *In* "Quantitative Population Dynamics" (D. G. Chapman, V. Gallucci, and F. M. Williams, eds.), Statist Ecol. Vol. 13, pp. 25–38.

Hanski, I. (1981). Coexistence of competitors in patchy environments with and without predation. *Oikos* **37**(3), 306–312.

Harmeson, J. P. (1974). Breeding ecology of the dickcissel. *Auk* **91,** 348–359.

Harris, M. P. (1980). Breeding performance of Puffins *Fratercula arctica* in relation to nest density, laying date and year. *Ibis* **122,** 193–209.

Hebrard, J. J. (1978). Habitat selection in two species of *Spizella:* A concurrent laboratory and field study. *Auk* **95,** 404–410.

Helle, E., and Helle, P. (1982). Edge effect on forest bird densities on offshore islands in the northern Gulf of Bothnia. *Ann. Zool. Fenn.* **19**(3), 165–169.

Herlugson, C. J. (1981). Nest site selection in mountain bluebirds. *Condor* **83,** 252–255.

Herrera, C. (1978). Niche-shift in the genus *Parus* in southern Spain. *Ibis* **120,** 236–240.

Higuchi, H., and Hirano, T. (1983). Comparative ecology of White and Japanese Wagtails, *Motacilla alba* and *M. grandis,* in winter. *Tori* **32,** 1–11.

Higuchi, H., Tsukamoto, Y., Hanawa, S., and Takeda, M. (1982). Relationship between forest areas and the number of bird species. *Strix* **1,** 70–78.

Hildén, O. (1965). Habitat selection in birds: A review. *Ann. Zool. Fenn.* **2,** 53–75.

Holmes, R. T. (1975). Bird community dynamics and energetics in a northern hardwoods ecosystem. *J. Ecol.* **44,** 175–200.

Holmes, R. T., and Robinson, S. K. (1981). Tree species preferences of foraging insectivorous birds in a northern hardwoods forest. *Oecologia* **48,** 31–35.

Holmes, R. T., and Sturgis, F. W. (1975). Bird community dynamics in a northern hardwoods ecosystem. *J. Anim. Ecol.* **44,** 175–200.

Holmes, R. T. Bonney, R. E., Jr., and Pacala, S. W. (1979). Guild structure of the Hubbard Brook bird community: A multivariate approach. *Ecology* **60,** 512–520.

Horn, H. S., and MacArthur, R. H. (1972). Competition among fugitive species in a harlequin environment. *Ecology* **53,** 749–752.

Houston, D. C. (1975). Ecological isolation of African scavenging birds. *Ardea* **63,** 55–64.

Howe, R. W, (1979). Distribution and behavior of birds on small islands in northern Minnesota. *J. Biogeogr.* **6,** 379–390.

Hubbard, J. P. (1973). Avian evolution in the arid lands of North America. *Living Bird* **12,** 155–196.

Hurley, R. J., and Franks, E. C. (1976). Changes in the breeding ranges of two grassland birds. *Auk* **93,** 108–115.

Hutchins, H. E., and Lanner, R. M. (1982). The central role of Clark's Nutcracker in the dispersal and establishment of Whitebark Pine. *Oecologia* **55,** 192–201.

James, F. C. (1971). Ordinations of habitat relationships among breeding birds. *Wilson Bull.* **83,** 215–236.

James, F. C., and Shugart, H. H. (1970). A quantitative method of habitat description. *Audubon Field Notes* **24,** 727–736.

James, R. D. (1976). Foraging behavior and habitat selection of three species of vireos in southern Ontario. *Wilson Bull.* **88,** 62–75.

Järvinen, O. (1979). Geographical gradients of stability in European land bird communities. *Oecologia* **38,** 51–69.

Järvinen, O. and Ulfstrand, S. (1980). Species turnovers of a continental bird fauna: Northern Europe 1850–1970. *Oecologia* **46**(2), 186–195.

Järvinen, O., and Väisänen, R. A. (1978). Recent changes in forest bird populations in northern Finland. *Ann. Zool. Fenn.* **15,** 279–289.

Johnson, N. K., and Garrett, K. L. (1974). Interior bird species expand breeding ranges into southern California. *West. Birds* **5,** 46–56.

Karr, J. R. (1971). Structure of avian communities in selected Panama and Illinois habitats. *Ecol. Monogr.* **41,** 207–233.

Karr, J. R. (1976). On the relative abundance of migrants from the north temperate zone in tropical habitats. *Wilson Bull.* **88,** 433–458.

Keast, A. R. (1961). Bird speciation on the Australian continent. *Bull. Mus. Comp. Zool.* **123,** 305–495.

Keast, A. R. (1980). The evolution of habitat specializations in space and time. *Proc. Int. Ornithol. Congr., 17th, 1978* Vol. 2, pp. 1025–1030.

Kendeigh, S. C. (1945). Community selection by birds of the Heidelberg Plateau of New York. *Auk* **62,** 418–436.

Kilgore, B. M. (1971). Response of breeding bird populations to habitat changes in a giant sequoia forest. *Am. Midl. Nat.* **85,** 135–152.

Kolmodin, U., and Nilsson, S. G. (1982). Häckfåglorna på Dalarnas my rov tätheter och artrikedom i förhållande till myrstorlek och vöthet. *Vår. Fågelv.* **41,** 1–9.

Koskimies, J. (1957). Terns and gulls as features of habitat recognition for birds nesting in their colonies. *Ornis Fenn.* **34,** 1–6.

Kushlan, J. A. (1976). Site selection for nesting colonies by the American White Ibis *Eudocimus albus* in Florida. *Ibis* **118,** 590–593.

Lack, D. (1933). Habitat selection in birds. *J. Anim. Ecol.* **2,** 239–262.

Lack, D. (1940). Habitat selection and speciation in birds. *Br. Birds* **34,** 80–84.

Lack, D. (1944). Ecological aspects of species-formation in passerine birds. *Ibis* **86,** 260–286.

Lack, D. (1971). "Ecological Isolation in Birds" Blackwells Sci. Publ., pp. 1–404, Blackwell, Oxford and Edinburgh.

Lack, D., and Venables, L. S. V. (1939). The habitat distribution of British woodland birds. *J. Anim. Ecol.* **8,** 39–71.

Lack, D., Lack, E., Lack, P., and Lack, A. (1973). Birds on St. Vincent. *Ibis* **115,** 46–52.

Leck, C. F. (1979). Avian extinctions in an isolated tropical wet-forest preserve, Ecuador. *Auk* **96,** 343–352.

LeClercq, B. (1976). Etude experimentale des facteurs limitant la densité des mesanges en forêt. *Alauda* **44,** 301–318.

Lennerstedt, I. (1983). Födoområden hos lövsångare *Phylloscopus trochilus* och svartvit flugsnappare *Ficedula hypoleuca* i fjällsjörkskog. *Vår. Fågelv* **42,** 11–20.

Levin, S. A. (1974). Dispersion and population interactions. *Am. Nat.* **108,** 207–228.

Levin, S. A. (1977). Spatial patterning and the structure of ecological communities. *Lect. Math. Life Sci., V.,* 9 1977.

Levins, R. (1980). Coexistence in a variable environment. *Am. Nat.* **114,** 765–783.

Levins, R. and Culver, D. (1971). Regional coexistence of species and competition between rare species. *Proc. Natl. Acad. Sci. U.S.A.* **68,** 1246–1248.

Ligon, J. D. (1978). Reproductive interdependence of pinyon jays and pinyon pines. *Ecol. Monogr.* **48,** 111–126.

Lloyd, M., and White, J. (1980). On reconciling patchy mirospatial distributions with competition models. *Am. Nat.* **115,** 29–44.

Loman, J. (1979). Nest tree selection and vulnerability to predation among Hooded Crows *Corvus corone cornix. Ibis* **121,** 204–207.

Loman, J. (1980). Habitat distribution and feeding strategies of four south Swedish corvid species during winter. *Ekol. Pol. Ser. A* **28,** 95–109.

Loyn, R. H., Runnalls, R. G., Forward, G. Y., and Tyers, J. (1983). Territorial Bell Miners and other birds affecting populations of insect prey. *Science* **221,** 1411–1413.

Lyon, D. L., Crandall, J., and McKone, M. (1977). A test of the adaptiveness of interspecific territoriality in the Blue-throated hummingbird. *Auk* **94,** 448–454.

MacArthur, R. H. (1958). Population ecology of some warblers of northeastern coniferous forest. *Ecology* **39,** 599–619.

MacArthur, R. H. (1972). "Geographical Ecology" Prentice-Hall, Englewood Cliffs, New Jersey.

MacArthur, R. H., and MacArthur, J. (1961). On bird species diversity. *Ecology* **42,** 594–598.

MacArthur, R. H., and Wilson, E. O. (1967). "The Theory of Island Biogeography," Monographs in Population Biology, Vol. 1, Princeton Univ. Press, Princeton, New Jersey.

MacArthur, R. H., MacArthur, J., and Preer, J. (1962). On bird species diversity. II. Prediction of bird census from habitat measurements. *Am. Nat.* **96,** 167–174.

MacArthur, R. H., Recher, H., and Cody, M. L. (1966). On the relation between habitat selection and bird species diversity. *Am. Nat.* **100,** 319–332.

MacArthur, R. H., Diamond, J. M., and Karr, J. R. (1972). Density compensation in island faunas. *Ecology* **53,** 330–342.

McCrimmon, D. A., Jr. (1978). Nest site characteristics among five species of herons on the North Carolina coast. *Auk* **95,** 267–280.

MacKenzie, D. I., and Sealy, S. G. (1981). Nest site selection in eastern and western Kingbirds: A multivariate approach. *Condor* **83,** 310–321.

MacLaren, P. I. R. (1950). Bird–ant nesting association. *Ibis* **92,** 564–566.

Maher, W. J. (1973). Growth of ground-nesting passerines at Matador, Saskatchewan, Canada. *In* "Productivity, Population Dynamics and Systematics of Granivorous Birds" (S. C. Kendeigh and J. Pinkowski, eds.), pp. 85–101. Polish Sci. Publ., Warsaw.

Marion, W. R., and Ryder, R. A. (1975). Perch site preferences of four diurnal raptors in north-eastern Colorado. *Condor* **77,** 350–352.

Martin, S. G. (1971). Polygyny in the bobolink: Habitat quality and the adaptive complex. Ph.D. thesis, Oregon State Univ.

Martin, T. E. (1981). Limitation in small habitat islands: Chance or competition? *Auk* **98,** 715–734.

May, P. G. (1982). Secondary succession and breeding bird community structure: Patterns of resource utilization. *Oecologia* **55,** 208–216.

Mengel, R. (1964). Species formation in wood warblers. *Living Bird* **3,** 9–43.

Middleton, A. L. A. (1979). Influence of age and habitat on reproduction by the American Goldfinch. *Ecology* **60,** 418–432.

Mikkola, H. (1973). The Red-flanked Bluetail and its spread to the west. *Br. Birds* **66,** 3–12.

Miller, G. R., and Watson, A. (1978). Territories and the food plant of individual red grouse. *J. Anim. Ecol.* **47,** 293–305.

Minot, E. O. (1981). Effects of interspecific competition for food in breeding blue and great tits. *J. Anim. Ecol.* **50,** 375–385.

Moore, N. W., and Hooper, M. D. (1975). On the number of bird species in British woods. *Biol. Conserv.* **8,** 239–250.

Moreau, R. E. (1942). The nesting of African birds in association with other living things. *Ibis* **84,** 240–263.

Morse, D. H. (1968). A quantitative study of foraging male and female spruce-woods warblers. *Ecology* **49,** 779–784.

Morse, D. H. (1971). Effects of the arrival of a new species upon habitat utilization by two forest thrushes in Maine. *Wilson Bull.* **83,** 57–65.

Morse, D. H. (1976). Variables affecting the density and territory size of breeding spruce-woods warblers. *Ecology* **57,** 290–301.

Murray, B. (1971). The ecological consequences of interspecific territorial behavior in birds. *Ecology* **52,** 414–423.

Nilsson, L. (1972). Habitat selection, food choice and feeding habits of diving ducks in coastal waters of south Sweden during the non-breeding season. *Ornis Scand.* **3,**(1), 55–78.

Noon, B. R. (1981). The distribution of an avian guild along a temperate elevational gradient: The importance and expression of competition. *Ecol. Monogr.* **51,** 105–124.

Novy, F. O., and McGrew, A. D. (1974). Orange-breasted Bunting in Southern Texas. *Auk* **91,** 178–179.

Nudds, T. D. (1983). Niche dynamics and organization of waterfowl guilds in variable environments. *Ecology* **64,** 319–330.

Ogasawara, K. (1976). Breeding interspecific relations between Red-cheeked Myna and Grey Starling in Kawauchi residential area, Sendai. *J. Yamashina Inst. Ornithol.* **8,** 27–37.

Opdam, P. (1975). Inter- and intra-specific differentiation with respect to feeding ecology in two sympatric species of the genus *Accipiter*. *Ardea* **63,** 30–54.

Orians, G. H. (1969). On the evolution of mating systems in birds and mammals. *Am. Nat.* **103,** 589–603.

Osborne, P., and Osborne, L. (1980). The contribution of nest site characteristics to breeding-success among blackbirds *Turdus merula*. *Ibis* **122,** 512–517.

Pierotti, R. (1982). Habitat selection and its effect on reproductive output in the Herring Gull in Newfoundland. *Ecology* **63,** 854–868.

Pimm, S. L., and Pimm, J. W. (1982). Resource use, competition, and resource availability in Hawaiian honeycreepers. *Ecology* **63,** 1468–1480.

Pimm, S. L., and Rosenzweig, M. L. (1981). Competitors and habitat use. *Oikos* **37,** 1–6.

Pinkowski, B. C. (1979). Nest site selection in Eastern Bluebirds. *Condor* **81,** 435–436.

Price, T. (1981). The ecology of the Greenish Warbler *Phylloscopus trochilioides* in its winter quarters. *Ibis* **123,** 131–144.

Prŷs-Jones, R. P. (1980). Breeding periodicity, nesting success, and nest site selection among red-tailed tropic birds (*Phaeton rubricauda*) and white-tailed tropic birds (*P. lepturus*) on Aldabra Atoll. *Ibis* **122,** 76–81.

Rabenold, K. N. (1978). Foraging strategies, diversity and seasonality in bird communities of Appalachian spruce–fir forests. *Ecol. Monogr.* **48,** 397–424.

Rand, A. L. 1936. The distribution and habits of Madagascar birds. *Bull. Am. Mus. Nat. Hist.* **72,** 143–499.

Rappole, J. H., and Warner, D. W. (1976). Relationships between behaviour, physiology and weather in avian transients at a migration stopover site. *Oecologia* **26,** 193–212.

Rice, J. (1978). Ecological relationships of two interspecifically territorial vireos. *Ecology* **59,** 526–538.

Rice, J., Anderson, B. W., and Ohmart, R. D. (1980). Seasonal habitat selection by birds in the lower Colorado River valley. *Ecology* **61,** 1402–1411.

Roberts, R. C. (1979). Habitat and resource relationships in acorn woodpeckers. *Condor* **81,** 1–8.

Robins, J. D. (1971). Differential niche utilization in a grassland sparrow. *Ecology* **52,** 1065–1070.

Robinson, S. K. (1981). Social interactions and ecological relations of Philadelphia and Red-eyed Vireos in a New England forest. *Condor* **83,** 16–26.

Robinson, S. K., and Holmes, R. T. (1982). Foraging behavior of forest birds: The relationships among search tactics, diet, and habitat structure. *Ecology* **63,** 1918–1931.

Rosenzweig, M. L. (1979). Optimal habitat selection in two-species competitive systems. *Fortschr. Zool.* **25,** 283–293.

Rosenzweig, M. L. (1981). Theory of habitat selection. *Ecology* **62,** 327–335.

Rotenberry, J. T. (1978). Components of avian diversity along a multifactorial climatic gradient. *Ecology* **59,** 693–699.

Roth, R. R. (1976). Spatial heterogeneity and bird species diversity. *Ecology* **57,** 773–782.

Rothstein, S. J. (1971). High nest density and nonrandom nest placement in the Cedar Waxwing. *Condor* **73,** 483–485.

Rzedowski, J., and McVaugh, R. (1966). La vegetación de Nueva Galicia. *Contrib. Univ. Mich. Herb.* **9**(1), 1–123.

Sabo, S. R. (1980). Niche and habitat relations in subalpine bird communities of the White Mountains of New Hampshire. *Ecol. Monogr.* **50,** 241–259.

Salomonson, M. G., and Balda, R. P. (1977). Winter territoriality of Townsend's Solitaire (*Myadestes townsendi*) in a pinyon–juniper–Ponderosa pine ecotone. *Condor* **79,** 148–161.

Saunders, D. A. (1974). The occurrence of the white-tailed black cockatoo, *Calyptorhynchus boudinii,* in *Pinus* plantations in Western Australia. *Aust. Wild. Res.* **1,** 45–54.

Schaldach, W. J., Jr. (1961). The avifauna of Colima and adjacent Jalisco, Mexico. *Proc. West. Found. of Vert. Zool.* I **1,** 1–100.

Schemske, D. W., and Brokaw, N. (1981). Treefalls and the distribution of understory birds in a tropical forest. *Ecology* **62,** 938–945.

Shifflett, W. A. (1975). First photographic record of the Brown Jay in the United States. *Auk* **92,** 797.

Shugart, H. H., and James, D. (1973). Ecological succession of breeding bird populations in northwestern Arkansas. *Auk* **90,** 62–77.

Slagsvcld, T. (1980a). Habitat selection in birds: On the presense of other species with special regard to *Turdus Pilaris. J. Anim. Ecol.* **49,** 523–536.

Slagsvold, T. (1980b). Egg predation in woodlands in relation to the presence and density of breeding Fieldfares *Turdus pilaris. Ornis Scand.* **11,** 92–98.

Slatkin, M. (1974). Competition and regional coexistence. *Ecology* **55,** 128–134.

Smith, K. D. (1974). The utilization of gum trees by birds in Africa. *Ibis* **116,** 155–164.

Smith, K. G. (1977). Distribution of summer birds along a forest moisture of gradient in an Ozark watershed. *Ecology* **58,** 810–829.

Smith, K. G. (1982). Drought-induced changes in avian community structure along a moisture sere. *Ecology* **63,** 952–961.

Snow, B. K., and Snow, D. W. (1980). Relationships between hummingbirds and flowers in the Andes of Colombia. *Bull. Br. Mus. Nat. Hist. (Zool.)* **38,** 105–139.

Snow, D. W., and Teixeira, D. L. (1982). Hummingbirds and their flowers in the coastal mountains of southeastern Brazil. *J. Ornithol.* **123,** 446–450.

Southern, H. N. (1963). Mimicry in Cuckoos' eggs. *In* "Evolution as a Process" (J. Huxley, A. C. Hardy, and E. B. Ford, eds.) pp. 257–270. Collier Books, New York.

Stenger, J. (1958). Food habits and available food of ovenbirds in relation to territory size. *Auk* **75,** 335–346.

Stepney, P. H. R., and Power, D. M. (1973). Analysis of the eastward expansion of Brewer's-Blackbird plus general aspects of avian expansions. *Wilson Bull.* **85,** 452–464.

Stiles, E. W. (1978). Avian communities in temperate and tropical alder forest. *Condor* **80,** 276–284.

Stott, R. S., and Olson, D. P. (1973). Food–habitat relationships of sea ducks on the New Hampshire coastline, *Ecology* **54,** 996–1007.

Sturman, W. A. (1968). Description and analysis of breeding habitats of the chickadees *Parus atricapillus* and *P. refescens. Ecology* **49,** 418–431.

Svärdson, G. (1949). Competition and habitat selection in birds. *Oikos* **1,** 157–174.

Swingland, J. R. (1977). The social and spatial distribution of winter colonial rooting in Rooks (*Corvus frugeligus*). *J. Zool.* **182,** 509–529.

Tast, J., and Rassi, P. (1973). Roosts and roosting flights of wintering Jackdaws *Corvus monedula* at Tampere, Finland. *Ornis Fenn.* **50,** 29–45.

Terborgh, J. (1977). Bird species diversity on an Andean elevational gradient. *Ecology* **58,** 1007–1019.

Terborgh, J., and Faaborg, J. (1973). Turnover and ecological release in the avifauna of Mona Island, Puerto Rico. *Auk* **90,** 759–779.

Terborgh, J., and Weske, J. S. (1975). The role of competition in the distribution of Andean birds. *Ecology* **56,** 562–576.

Thiollay, J.-M. (1981). Ségrégation écologique et pression de prédation de deux buses sympatriques dans un désert Mexicain. *Le Gerfaut* **71,** 575–610.

Thompson, L. S. (1978). Species abundance and habitat relations of an insular montane avifauna. *Condor* **80,** 1–14.

Titus, K., and Mosher, J. A. (1981). Nest-site habitat selected by woodland hawks in the central Appalachians. *Auk* **98,** 270–281.

Tomiałłojć, L. (1978). The influence of predators on breeding Woodpigeons in London parks. *Bird Study* **25,** 2–10.

Tomoff, C. S. (1974). Avian species diversity in desert scrub. *Ecology* **55,** 396–403.

Trivelpiece, W., and Volkman, N. J. (1979). Nest-site competition between Adelie and Chinstrap Penguins: An ecological interpretation. *Auk* **96,** 675–681.

Tyler, J. D. (1976). Mexican crow range extension. *Am. Birds* **30,** 773.

Udvardy, M. D. F. (1951). The significance of interspecific competition in bird life. *Oikos* **3,** 98–123.

Ulfstrand, S., Alatalo, R. V., Carlson, A., and Lundberg, A. (1981). Habitat distribution and body size of the great tit *Parus major*. *Ibis* **123,** 494–499.

van Balen, J. H., Booy, C. J. H., van Franeker, J. A., and Osiek, E. R. (1982). Studies on hole-nesting birds in natural nest sites. 1. Availability and occupation of natural nest sites. *Ardea* **70,** 1–24.

vander Wall, S. B., and Balda, R. P. (1977). Coadaptations of the Clark's Nutcracker and the pinyon pine for efficient seed harvest and dispersal. *Ecol. Monogr.* **47,** 89–111.

van de Weghe, J.-P., and Monfort-Braham, N. (1975). Quelques aspects de la séparation écologique des Vanneaux du Parc National de l'Akagera. *Alauda* **43,** 143–166.

Wagner, J. L. (1981). Seasonal change in guild structure: Oak woodland insectivorous birds. *Ecology* **62,** 973–981.

Wakeley, J. S. (1978). Hunting methods and factors affecting their use by ferruginous hawks. *Condor* **80,** 316–326.

Walkinshaw, L. H. (1968). *Spizella pusilla pusilla:* Eastern Field Sparrow. *Bull. U.S. Nat. Mus.* **237,** 1217–1235.

Walsberg, G. E. (1981). Nest-site selection and the radiative environment of the warbling vireo. *Condor* **83,** 86–88.

Walsh, J. E., and Walsh, B. (1976). Nesting association between the Red-headed Weaver *Malimbus rubriceps* and raptorial birds. *Ibis* **118,** 106–108.

Welsh, D. A. (1975). Savannah sparrow breeding and territoriality on a Nova Scotia dune beach. *Auk* **92,** 235–251.

White, G, (1906). "The Natural History and Antiquities of Selbourne" Dent. London. (Originally published by Benjamin White, London, 1789.)

White, M. G., and Conroy, J. W. H. (1975). Aspects of competition between pygoscelid penguins at Sigmy Island, South Orkney Islands. *Ibis* **117,** 371–373.

Whitmore, R. C. (1975). Habitat ordination of passerine birds of the Virgin River valley, southwestern Utah. *Wilson Bull.* **87,** 65–74.

Wiklund, C. G. (1979). Increased breeding success for Merlins *Falco columbarius* nesting among colonies of Fieldfares *Turdus pilaris*. *Ibis* **121,** 109–111.

Wiklund, C. G., and Anderson, M. (1980). Nest predation selects for colonial breeding among Fieldfares *Turdus pilaris*. *Ibis* **122,** 363–366.

Williams, J. B., and Batzli, G. O. (1979). Competition among bark-foraging birds in central Illinois: Experimental evidence. *Condor* **81,** 122–132.

Williamson, P. (1971). Feeding ecology of the Red-eyed Vireo (*Vireo olivaceus*) and assorted foliage-gleaning birds. *Ecol. Monogr.* **41,** 129–152.

Willis, E. O. (1974). Populations and local extinctions of birds on Barro Colorado Island, Panama. *Ecol. Monogr.* **44,** 153–169.

Willson, M. F. (1974). Avian community organization and habitat structure. *Ecology* **55,** 1017–1029.

Winternitz, B. L. (1976). Temporal change and habitat preference of some montane breeding birds. *Condor* **78,** 383–393.

Wittenberger, J. F. (1980). Vegetation structure, food supply, and polygyny in Bobolinks (*Dolichonyx oryzivorus*). *Ecology* **61,** 140–150.

Woodall, P. F. (1975). The life history of the bronze mannikin. *Ostrich* **46,** 55–86.

Wright, S. J. (1980). Density compensation in island avifaunas. *Oecologia* **45,** 385–389.

Yeaton, R. I. (1974). An ecological analysis of chaparral and pine forest bird communities on Santa Cruz Island and mainland California. *Ecology* **55,** 959–973.

Yodzis, P. (1980). Competition for space and the structure of ecological communities. *Lecture Notes in Biomath.* **25,** 1–191.

Yom–Tov, Y. (1979). The disadvantage of low positions in colonial roosts: An experiment to test the effect of droppings on plumage quality. *Ibis* **121,** 331–333.

Yom–Tov, Y., Imber, A., and Otterman, J. (1977). The microclimate of winter roosts of the starling *Sturnus vulgaris*. *Ibis* **119,** 366–368.

Zang, H. (1982). Der einfluss der Höhenlage auf Alterzusamensetzung and Brutbiologie bei Kohl- und Blau-meise (*Parus major, P. caeruleus*) im Harz. *J. Ornithol.* **123,** 145–154.

Zimmerman, J. L. (1982). Nesting success of dickcissels (*Spiza americana*) in preferred and non-preferred habitats. *Auk* **99,** 292–298.

Part II

Habitat Selection in Specific Bird Taxa

Chapter 2

Habitat Selection in the Pied Flycatcher *Ficedula hypoleuca*

RAUNO V. ALATALO
ARNE LUNDBERG
STAFFAN ULFSTRAND

Department of Zoology
Uppsala University
Uppsala, Sweden

I. INTRODUCTION AND LIFE HISTORY

The Pied Flycatcher *Ficedula hypoleuca* is a small (11–13 gm) passerine bird ranging over most of northern and eastern Europe. It winters in Africa, the first

HABITAT SELECTION IN BIRDS

ISBN 0-12-178080-5

males returning to their breeding grounds in central Sweden at the end of April or the beginning of May. Males generally arrive about 1 week ahead of females, but all males do not arrive before all females. Pied Flycatchers nest in holes in trees and readily accept nest boxes. In fact, they prefer them over natural holes, and by erecting boxes one can control a whole breeding population, which makes the Pied Flycatcher especially suitable for population studies using experimental manipulation. The breeding biology of the species has been intensively studied in northern Europe during the past 30 years [e.g., von Haartman, 1949, 1951a, 1954, 1967a,b; Campbell, 1955 (and in Lack, 1966); Creutz, 1955; Curio, 1959a,b; Berndt and Winkel, 1967; Tompa, 1967; Källander, 1975; Askenmo, 1977; Järvinen, 1980; Alatalo *et al.*, 1981; Berndt *et al.*, 1981; Lundberg *et al.*, 1981].

Pied Flycatcher males are polygynous and try to attract females in succession to several more or less adjacent territories (mean separation, 200 m). After having attracted one female to a territory (the primary female) most males move into another territory and try to attract another female (the secondary female). Whether successful or not, males return to the primary female and help her feed the young while the secondary female has to raise her young almost without male aid. In secondary nests it often happens that some young die from undernourishment as a result of low male assistance, and secondary females raise significantly fewer offspring than do simultaneously laying primary and monogamous females. Polyterritorial successive polygyny in the Pied Flycatcher was first described in detail by von Haartman (1945, 1951b, 1956) but has also been dealt with by Trettau and Merkel (1943), Creutz (1955), Curio (1959b), Askenmo (1977), and Silverin (1980, 1983). We have proposed that polyterritoriality is likely to have evolved, because it enables males to hide the fact that they are already mated. As a result mated males can deceive females about their true mating status and thus improve their likelihood of becoming polygynous (Alatalo *et al.*, 1981, 1982a, 1984a,b; Alatalo and Lundberg, 1984a). Polyterritoriality necessarily means that females on frequent occasions are left unguarded during their fertile period, and as a consequence cuckoldry by neighboring males leading to multiple paternity within broods is possible and has been found to occur commonly (Björklund and Westman, 1983; Alatalo *et al.*, 1984c).

Pied Flycatchers are found in a variety of forest habitats from luxuriant deciduous woodland and subalpine birch forest to poor coniferous forest, although they clearly prefer deciduous over coniferous habitats. In Scandinavia, many males show breeding area fidelity, while females less frequently return to their prior nesting areas. The young only seldom return to their natal locality (Enemar, 1948; von Haartman, 1949; R. V. Alatalo, unpublished observations). Breeding area fidelity among females and fidelity to birthplace among young seem to be higher in west, central, and east European populations (von Haartman, 1960; Berndt and Sternberg, 1969; Winkel, 1982) than in Scandinavia.

In this chapter we present data on the reproductive success and morphometric characters of birds breeding in two habitat types, namely deciduous and coniferous forest. These data will allow us to discuss why Pied Flycatchers end up in different habitats and how the prospects of successful breeding for individual males and females are influenced by the habitat in which they have settled. We also discuss criteria for habitat selection in a more general way.

II. METHODS

Our data were collected in nest-box areas around Uppsala (59°50′N/17°40′E), central Sweden, during the five summers of 1979 to 1983. Nest boxes were put up in both deciduous and coniferous forest. The vegetation in deciduous areas is dominated by mature stands of oak (*Quercus robur* L.), elm (*Ulmus glabra* Hubs.), lime (*Tilia cordata* Mill.), ash (*Fraxinus excelsior* L.), and hazel (*Corylus avellana* L.), while in coniferous areas pine (*Pinus silvestris* L.) is prevalent with an admixture of spruce (*Picea abies* L.) and sometimes also birch (*Betula* spp.) and aspen (*Populus tremula* L.).

In the first year we designed an experiment to test Fretwell's model of habitat selection (Fretwell and Lucas, 1969; Fretwell, 1972), in which nest boxes were put up in excess. The results (based on 1 year's work) are presented in Lundberg *et al.* (1981). Subsequently, certain rearrangements of nest boxes were made, but we kept nest boxes in excess in both habitat types throughout the past 5-year period. In deciduous forest we put up, on average, 8 boxes/ha and in coniferous forest 4/ha. The reason for having twice as high nest-box density in deciduous forest is that we thus obtained approximately the same frequency of occupied nest boxes in both habitats. For experimental purposes we also maintained low-density nest-box areas in deciduous forest (Alatalo and Lundberg, 1984b) where most nest boxes invariably were occupied, but these data will only be discussed briefly in this presentation. In each year males and females were captured, measured, weighed, and individually ringed. All young were ringed, and starting in 1980 they were weighed at the age of 13 days, and their tarsi were measured. Most males were separated as yearlings or older by the amount of white on the tip of the outermost greater wing coverts, while females were aged on the basis of relative length of wing and first primary (see Alatalo *et al.*, 1984d). To measure the tarsus, in both adults and nestlings, we bent the foot at the intertarsal joint and at the toes and measured the distance between the extreme bending points with a dial caliper. This is not the true tarsus but affords a very precise and repeatable measurement closely correlated with real tarsus length. For wing length we used the maximum method (Svensson, 1975). Males vary in color on crown, mantle, and back from entirely black to brown (femalelike), and from

1980 we scored male color using Drost's (1936) scale from I to VII, on which I represent males fully black on back.

For each nest we recorded laying date of first egg, clutch size, hatching date, number of eggs hatched, and number of fledged young. More detailed methodological information is given under each section.

III. HABITAT DISTRIBUTION AND BREEDING SUCCESS IN AREAS WITH NEST BOXES IN EXCESS

A. Breeding Density

Pied Flycatchers breed in both deciduous and coniferous forest but reach higher densities in the former habitat. In our nest-box areas, nest boxes are provided in excess, and usually more than 50%, and never less than 30%, have remained unoccupied. Breeding density in deciduous forest was found to vary, between areas and years, from 0.4 to 3.1 pairs/ha, with a mean of 1.8 pairs/ha ($n = 14$). In coniferous forest density ranged from 0.2 to 1.3 pairs/ha, with an average of 0.6 pairs/ha ($n = 6$). These figures are in general agreement with breeding densities reported from elsewhere: deciduous forest in Germany, 5 pairs/ha (Creutz, 1955) and in southern Sweden, 1.0–2.3 pairs/ha (Källander, 1975); subalpine birch forest in northern Sweden, 1.4–3.3 pairs/ha (Enemar and Sjöstrand, 1972); coniferous forest in southern Sweden, 1.1–1.8 pairs/ha (Källander, 1975) and 0.44–0.75 pairs/ha (Askenmo, 1977); and in Finland, 0.31–2.08 pairs/ha (Virolainen, 1984).

For populations inhabiting forests without nest boxes, density data are fewer but do show that density is lower than in nest-box areas. In unmanaged, apparently optimal, deciduous habitat in southern Sweden population density was 0.25–0.40 pairs/ha (Nilsson, 1979; based on territory mapping). In a 40-ha deciduous woodland in central Sweden we found 0.5 pairs/ha (Alatalo and Lundberg, 1984a), while in an unmanaged part (8.8 ha) of the same woodland breeding density was 1.0 pairs/ha. (The estimate was based on nests found, and moreover, the males in the area were color ringed.) In coniferous forest in southern Sweden density was 0.15 pairs/ha (Nilsson, 1979; territory mapping). In Finland, Palmgren (1930) and Soveri (1940) found from 0.10 to 0.48 pairs/ha in deciduous and mixed forests compared to only 0.01–0.03 pairs/ha in coniferous forest. They based their estimates on the "homogeneous census plot method" where parallel line transects are censused. However, line-transect methods underestimate the size of breeding populations (Palmgren, 1930; Järvinen *et al.*, 1978), especially of the Pied Flycatcher (Helle and Pulliainen, 1983).

B. Territory Occupancy in Different Habitats

The timetable of territory occupancy of Pied Flycatcher males and the start of egg laying were followed in 1979 in both deciduous and coniferous forest. The arrival date was closely correlated with, and can be approximated by, the date for start of egg laying (Alatalo *et al.*, 1984b). We possess egg laying data for the years 1981–1983, while we did not check the nest boxes in coniferous forest in 1980. In 1979 all nest boxes were newly erected in areas previously lacking nest boxes. Of our two coniferous plots one was situated close to a deciduous nest-box area, while the other was far from any deciduous vegetation (see Lundberg *et al.*, 1981). We found that the coniferous area adjacent to deciduous forest became occupied at the same time as the deciduous forest, while in the remote coniferous area colonization was delayed by 1 week (Table I; Lundberg *et al.*, 1981). In later years when birds could return to previously inhabitated nest-box areas we found no difference in 1982 and only a tendency in 1981, with respect to timetable of habitat occupancy (as measured by time of egg laying; Table II). In 1983 we put up new nest boxes in coniferous forest 1 km away from an old nest-box area. The arrival timetable was significantly delayed in the new coniferous area compared to the old one (Table III). Our conclusion is that Pied Flycatchers, when not homing to familiar nest-box areas, first look for nest sites in deciduous habitat (and its surroundings) before they start searching in more or less remote coniferous forest. However, if some birds have acquired experience of nesting facilities in a coniferous forest, to which they can return from year to year, this habitat will be colonized as fast as deciduous habitat, although density will not become as high (see Section III,A). New males colonizing the coniferous habitat (possibly also deciduous forest) probably use the song of other males as a guide to finding empty nest holes (Alatalo *et al.*, 1982b; see also Section IV,C).

TABLE I

Timetable of the First Egg Laying in Deciduous (Dec) and Coniferous (Con) Forest in 1979[a]

		Date of egg laying			
Forest	First female	25%	50%	75%	n
Dec	22/5	26/5	27/5	30/5	33
Con-a	23/5	25/5	26/5	2/6	10
Con-f	28/5	31/5	2/6	7/6	10
Con-a + Con-f	23/5	26/5	31/5	4/6	20

[a] From Lundberg et al., 1981. Con-a denotes a coniferous area adjacent to a deciduous forest, while Con-f denotes an area far from such forest. Tests of significance, Mann–Whitney U-test. Dec against Con-a + Con-f, $U=437.5$, $p<0.05$; Con-a against Con-f, $U=21$, $p<0.05$.

TABLE II

Timetable of First Egg Laying in Deciduous Forest and in Coniferous Forest in 1981 and 1982[a]

Year and forest	First female	Date of egg laying 25%	50%	75%	n	$\bar{x}$	U-test
1981							
Deciduous	18/5	21/5	23/5	26/5	87	24.4	$z=1.66$
Coniferous	20/5	22/5	24/5	28/5	54	25.2	$p<0.10$
1982							
Deciduous	19/5	26/5	27/5	30/5	69	27.7	$z=0.15$
Coniferous	23/5	26/5	27/5	30/5	27	27.9	NS[b]

[a] In both these years birds could home to previously known nest-box areas.
[b] NS, not significant.

C. Morphometry and Age of Birds in Different Habitats

In our paper on morphometry and habitat distribution in the Pied Flycatcher (Lundberg *et al.*, 1981), which was based on a 1-year study, we showed that, in terms of bill length, wing length, and weight, males were larger and heavier in deciduous than in coniferous forest. In subsequent years we included tarsus length, tail length, and color type among the variables measured. Out of the differences found in the first year, wing length and weight, but not bill length, still remain significant between the two habitats. Among the added variables none was found to differ significantly between males in deciduous compared to coniferous forest (Table IV). Neither in the first-year sample (Lundberg *et al.*, 1981), nor subsequently, have we been able to find any morphometric difference between females in the two habitats respectively.

In the following we will examine in more detail wing length and weight in males, which are the characters in which males differ between habitats. Since, as is well known, wing length in passerine birds increases with age, the wing length

TABLE III

Median Arrival Date of Males and Females and Median Date of First Egg Laying in an Old and a New Nest-Box Area in Coniferous Forest in 1983

Sequence	Date of egg laying Old area	New area	U-test
Male arrival	8 May ($n=9$)	17 May ($n=7$)	$U=8$, $p<0.05$
Female arrival	13 May ($n=7$)	19 May ($n=5$)	$U=3$, $p<0.05$
First egg	21 May ($n=7$)	25 May ($n=5$)	$U=2$, $p<0.01$

TABLE IV

Morphometric Measurements of Pied Flycatcher Males and Females in Deciduous and Coniferous Forest[a]

	Measurements							
	Deciduous			Coniferous				
Characteristics	$\bar{x}$	S.D.	n	$\bar{x}$	S.D.	n	t-test	p
Males								
Wing (mm)	79.42	1.49	172	78.81	1.88	70	t=2.70	p<0.01
Tarsus (mm)	19.49	0.51	132	19.50	0.49	51	t=0.12	NS[b]
Bill (mm)	13.57	0.42	173	13.55	0.39	71	t=0.34	NS
Tail (mm)	53.02	1.46	131	52.80	1.40	51	t=0.92	NS
Weight (gm)	12.49	0.59	134	12.22	0.55	69	t=3.12	p<0.01
Color type	3.24	1.29	130	3.51	1.33	51	t=1.26	NS
Females								
Wing (mm)	77.58	1.55	158	77.48	1.46	93	t=0.50	NS
Tarsus (mm)	19.58	0.49	140	19.57	0.46	68	t=0.14	NS
Bill (mm)	13.61	0.43	184	13.53	0.41	94	t=1.49	NS
Tail (mm)	52.27	1.49	141	52.16	1.31	68	t=0.52	NS
Weight (gm)	12.95	0.96	156	12.90	0.87	83	t=0.40	NS

[a] Because weight varies with breeding cycle, we have included data from only the nestling period for males and from the last week of the nestling period for females. A lower value for color type indicates darker color.

[b] NS, not significant.

difference conceivably could be explicable simply in terms of age differences. This is, however, not the case, because the proportions of 1-year-old and older males, respectively, do not differ significantly between habitats (Table V). Also among the females the proportion of 1-year-old birds was similar in the 2 habitats (Table VI).

TABLE V

Age of Pied Flycatcher Males in Deciduous and Coniferous Forests[a]

Age	Deciduous	Coniferous	χ^2
One-year-old	34.6%	41.2%	0.97
Unknown age	11.5%	7.8%	
Older	53.8%	51.0%	NS[b]
Total number	130	51	

[a] Data from 1981 and 1982.

[b] NS, not significant.

TABLE VI

Age Index of Pied Flycatcher Females in Deciduous Forest and in Coniferous Forest[a]

Parameters	Age index		t-test
	Deciduous	Coniferous	
$\bar{x}$	40.92	40.77	t=0.51
S.D.	1.78	1.63	NS[b]
n	82	59	

[a] The age index is based on wing length minus first primary length. A smaller value indicates a greater proportion of one-year-old birds.

[b] NS, not significant.

We can think of two hypotheses to explain the morphometric variation observed in the Pied Flycatcher: (1) Birds inhabiting different habitats have different morphological adaptations for exploiting optimally those environments; either they select habitats according to their morphology, or birds in different habitats belong to separate subpopulations. (2) Social dominance affects habitat distribution of individual birds, with larger individuals being found in the preferred habitat and smaller birds in the less preferred habitat(s).

The factor most likely to differ between the habitats is foraging conditions, and, therefore, one would expect, above all, bill dimensions to differ between habitats (e.g., Lack, 1971). If there is a direct adaptive response between foraging conditions and bill dimensions, a relatively longer bill is to be expected in the coniferous habitat (Snow, 1954; Grant, 1979). This morphological adaptation should evolve in both males and females. However, we could only detect differences in wing length and weight, not in bill dimensions, between birds in the two habitats, and, moreover, these differences were confined to one sex, namely the males. These observations speak against the hypothesis of morphological adaptation to different foraging conditions. As an alternative, food abundance, which probably is higher in deciduous than in coniferous forest, may be thought to affect optimal body size and favor large individuals (Schoener, 1969; Case, 1978). Wing length and weight slightly and positively correlate with overall body size in the Pied Flycatcher (Lundberg *et al.*, 1981), but only among males did dimensions differ between habitats, and, thus, our data are not consistent with this explanation either. As yet another alternative, one might imagine that shorter wings could be an adaptation to better maneuverability needed in coniferous forest (for some unknown reason), but, again, this should apply to both sexes and not only to males. Furthermore, some polygynous males have their first female in the one habitat and the second in the other, or they (both males and females) may change habitats between years. There are thus many facts speaking against the hypothesis of body size differences being adaptive responses to different food-associated selection pressures across habitats or of birds actively choosing habitat on the basis of their morphology. Furthermore, individual birds do not exhibit strict habitat fidelity (see Section III,D).

If, by contrast, social dominance affects habitat distribution and large size confers an advantage in competitive interactions (Morse, 1974; Garnett, 1976), one would expect smaller males to be overrepresented in the suboptimal habitat(s). Furthermore, size differences should be found among males rather than among females since only males compete for territories (Lundberg *et al.*, 1981, Ulfstrand *et al.*, 1981). Weight and wing length measurements indicate that Pied Flycatcher males, but not females, are larger in deciduous than in coniferous forests. Since intraspecific competition for territories is much more prevalent in the more densely populated deciduous forest (Alatalo and Lundberg, 1984a), this observation is consistent with the hypothesis of social dominance. The size assortment of male Pied Flycatchers across habitats is likely to be a result of large individuals being superior to small ones in encounters. It is, however, difficult to understand why the males do not differ in more than two characters. Interestingly, these are the same ones as those in which Willow Warblers (*Phylloscopus trochilus*) were found to differ between suboptimal islands and the preferred mainland (Ebenman and Nilsson, 1981). Maybe weight and wing length are particularly important for fighting capacity and, hence, for the establishment of dominance relationships.

D. Habitat Fidelity

In Scandinavia, among surviving male and female Pied Flycatchers having bred once, 50–70% of the males and 16–28% of the females return to their previous breeding area (von Haartman, 1949, 1960; Nyholm and Myhrberg, 1983). Later in life almost all surviving birds return. The median distance between consecutive male nest sites was 100–400 m, while the corresponding figure for females was significantly longer (200–700 m; von Haartman, 1949; Nyholm and Myhrberg, 1983). Pied Flycatchers only seldom return to their natal area, and correcting for mortality, von Haartman (1949) estimated that only 7% of the males and 2% of the females return in their first year. In Germany, Winkel (1982) found higher returning rates, both for males and females returning after their first breeding and for young, than is the case in Scandinavia. Also in Germany, males and females returning for the first time settled, on average, 4–5 km from their birthplace, while the corresponding distance for older females was 2 km and for older males 140 m (Berndt and Sternberg, 1969).

Pied Flycatchers do not always return to their former breeding habitat. In our nest-box areas, 6 out of 57 returning males (10.5%) changed habitats between years. This is quite a high proportion compared to the proportion of males which changed between our study areas within deciduous woodland (6.5%, n = 46).

Moreover, 3 out of 25 polygynous males (12%) had their females in different habitats. In Germany, of Pied Flycatchers born in deciduous or coniferous (pine) forest and later recovered breeding more than 10 km away, 70–80% bred in deciduous and 20–30% in coniferous forest, irrespective of their natal habitat (Berndt and Winkel, 1975). All these observations show that Pied Flycatchers can change their nesting habitats between years, and, thus, these observations speak against the hypothesis of separate subpopulations being adapted to different habitats.

E. Reproductive Success

In the Pied Flycatcher, as in many other passerine birds, clutch size and fledging success decrease with the progress of the breeding season (Figs. 1 and 2). This is an important fact which must be taken into account in any com-

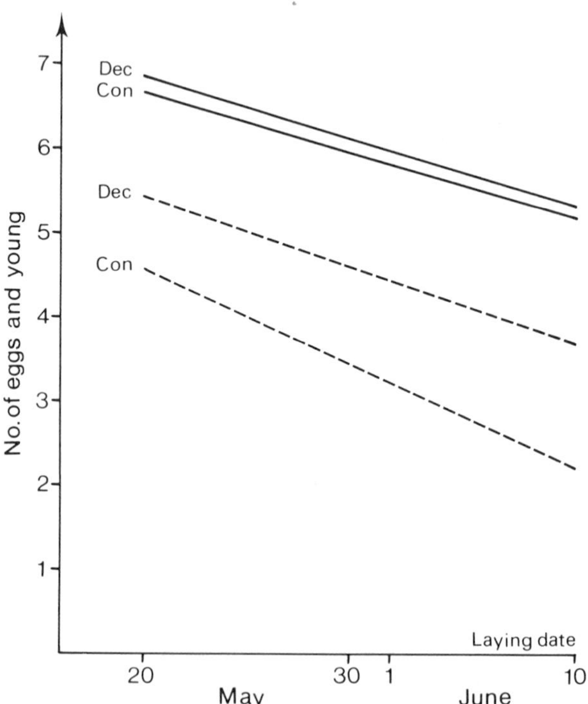

Fig. 1. Clutch size (solid line) and final brood size (dashed line) for all pairs in deciduous (Dec) and coniferous (Con) forest in relation to laying date. Regression equations for clutch size in deciduous, $y = -0.072x + 8.38$, $n = 199$; in coniferous, $y = -0.070x + 8.05$, $n = 106$; for brood size in deciduous, $y = -0.083x + 7.04$, $n = 196$; in coniferous $y = -0.112x + 6.78$, $n = 105$.

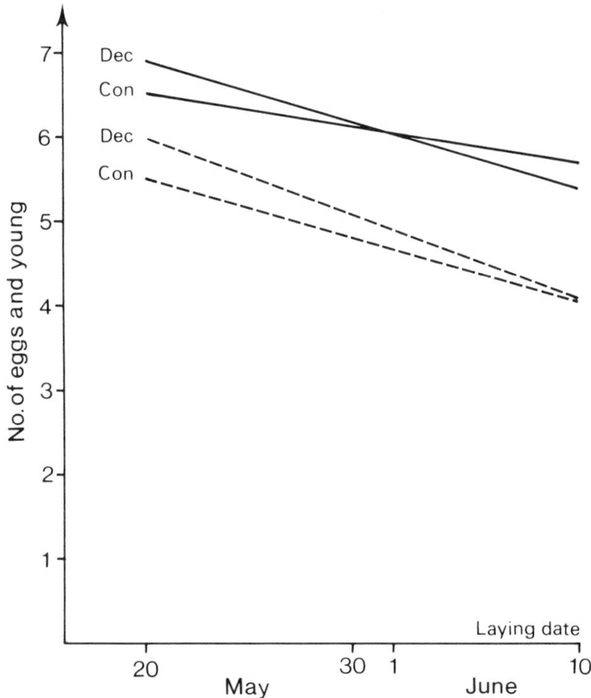

Fig. 2. Clutch size (solid line) and final brood size (dashed line) for monogamous pairs in deciduous (Dec) and coniferous (Con) forest in relation to laying date. Regression equations for clutch size in deciduous, $y = -0.073x + 8.40$, $n = 151$; in coniferous, $y = -0.041x + 7.35$, $n = 56$; for brood size in deciduous, $y = -0.091x + 7.80$, $n = 149$; in coniferous $y = -0.046x + 6.45$, $n = 56$.

parison. Therefore, when we have compared and tested the figures of clutch size and final brood size between the habitats, we have corrected for laying date.

Including all pairs we found both clutch size and fledgling number to be significantly higher in deciduous than in coniferous forest (Tables VII and VIII). However, clutch size and reproductive success are influenced by the mating status of the female. Both primary and secondary females were found to lay slightly smaller clutches than did simultaneously laying monogamous females (Alatalo and Lundberg, 1984a). Moreover, secondary females can only raise a reduced brood (64%) in comparison to concurrently laying monogamous females, because they often have to raise the clutch without any help from the male (Alatalo *et al.*, 1981, 1982a, 1984a,b). In our nest-box areas polygyny (i.e., the proportion of secondary females) was much more frequent in coniferous (28.7%, $n = 108$) than in deciduous forest (10.5%, $n = 366$; pooled average for both habitats 15.0%), but polygyny frequency also increased in cases

TABLE VII

Clutch Size and Fledgling Success in Deciduous and in Coniferous Forest[a]

Sampling	Deciduous			Coniferous			t- or U-Test	p
	$\bar{x}$	S.D.	n	$\bar{x}$	S.D.	n		
All pairs								
Clutch size	6.36	0.89	199	6.11	0.71	106	$t=2.43$	$p=0.05$
Fledglings	4.83	2.18	196	3.70	2.34	105	$z=4.31$	$p=0.001$
Monogamous pairs								
Clutch size	6.46	0.88	151	6.29	0.62	56	$t=1.33$	NS[b]
Fledglings	5.38	1.75	149	5.23	1.37	56	$z=1.32$	NS

[a] For clutch size we have used the t-test and for final brood size, the Mann–Whitney U-test.
[b] NS, not significant.

where nest boxes were provided in excess, in particular in the coniferous habitat (Table IX). After correcting for polygyny frequency and comparing only monogamous pairs, the differences between habitats with respect to clutch size and reproductive success disappear (Tables VII and VIII). These results differ slightly from what we found after the first year (given in Lundberg *et al.*, 1981), when we found a larger clutch size and a tendency ($p < 0.10$) for higher reproductive success of primary plus monogamous females in deciduous compared to coniferous forest. This discrepancy may have arisen, because in our early analysis we incorporated two low-density areas in the deciduous category, where reproductive success is higher (see Section IV,A) and because we did not correct for the 2 days' difference in laying date.

TABLE VIII

Average Fledgling Numbers in Deciduous and Coniferous Forests in Relation to Those in Deciduous Forest for Pairs Laying at the Same Time[a,b]

Sampling	Deciduous			Coniferous			t- or U-Test	p
	$\bar{x}$	S.D.	n	$\bar{x}$	S.D.	n		
All pairs								
Clutch size	1.000	0.126	199	0.971	0.098	106	$t=2.07$	$p<0.05$
Fledglings	1.000	0.444	196	0.771	0.486	105	$z=4.18$	$p<0.001$
Monogamous pairs								
Clutch size	1.000	0.122	151	0.972	0.093	56	$t=1.57$	NS[c]
Fledglings	1.000	0.316	149	0.971	0.260	56	$z=1.37$	NS

[a] As described by the regression equations in Figs. 1 and 2.
[b] For clutch size we have used the t-test and for final brood size the Mann–Whitney U-test.
[c] NS, not significant.

TABLE IX

Polygyny Frequency in Deciduous and Coniferous Forests at Different Breeding Densities[a]

Breeding site	Density/ha	% Polygyny		Test
Deciduous forest				
Boxes in deficit	<1	6.8	(n=44)	χ^2=0.53
Boxes in excess	1–3	11.3	(n=212)	NS[b]
Coniferous forest				
Boxes in deficit	<0.5	6.3	(n=16)	Fisher exact
Boxes in excess	0.5–1	39.4	(n=71)	p<0.01

[a] From Alatalo and Lundberg (1984).
[b] NS, not significant.

We have collected data about the quality of the young, as measured by their weight and tarsus length at 13 days' age, i.e., just prior to fledging. Comparisons including either all pairs or only monogamous pairs showed that offspring raised in deciduous and coniferous forest were not different, neither with respect to tarsus length nor to weight (Table X).

All in all, then, a monogamous Pied Flycatcher female will neither produce fewer offspring nor raise young of lower quality, if breeding in coniferous compared to deciduous forest. However, in nest-box areas, a female's probability of becoming a secondary female to an already mated male will be much higher in coniferous forest, and, as a result, females on average do worse in coniferous forest (Tables VII and VIII, Fig. 1). For the male the situation may be the other way around, for he is more likely to become mated with more than one

TABLE X

Tarsus Length and Weight of Offspring at 13 Days of Age at High and Low Breeding Density in Deciduous and Coniferous Forests

	All pairs			Monogamous pairs		
Parameter	$\bar{x}$	S.D.	n	$\bar{x}$	S.D.	n
Tarsus (mm)						
Deciduous	19.44	0.43	127	19.44	0.41	105
Coniferous	19.36	0.50	62	19.53	0.35	41
		t=1.14, NS[a]			t=1.24, NS	
Weight (gm)						
Deciduous	13.87	1.02	127	13.88	1.02	105
Coniferous	13.97	1.12	61	14.22	1.16	41
		t=0.61, NS			t=1.74, p<0.10	

[a] NS, not significant.

female in coniferous than in deciduous forest. However, we cannot give any precise data for males because some polygynous males have one female in each habitat, and, moreover, for several secondary nests we failed to identify the male. (Many bigamous males seldom visit the nest and therefore are difficult to catch.) Moreover, we do not know the exact proportion of unmated males in the two habitats, and, therefore, it is not possible to calculate an unbiased estimate of male reproductive success in different habitats. Whether the probability of becoming mated with several females and of becoming a secondary female are important factors in the habitat selection of Pied Flycatcher males and females, respectively, will be considered in Section IV,C.

IV. CRITERIA FOR HABITAT SELECTION

A. Food Factors

Food availability generally must be an important factor for the prospects of successful breeding. In deciduous forest Pied Flycatchers raised slightly more offspring when breeding density was low than when it was high, and offspring also grew better in the first situation (R. V. Alatalo and A. Lundberg, in preparation). This was especially pronounced in a year with unfavorable weather conditions. Since all our data derive from nest-box breeding pairs, nest-site quality should not differ, and, therefore, differences in reproductive success are safely explained in terms of density effects, i.e., presumably intraspecific competition for food (Alatalo and Lundberg, 1984b). A comparison between low population density in deciduous forest and coniferous forest with corresponding density shows fledging success to be higher in the former habitat (monogamous pairs: deciduous, 5.77, S.D. = 1.42, $n = 82$; coniferous, 5.23, S.D. 1.37, $n = 56$, $z = 3.04$, $p < 0.01$). In deciduous forest food probably is more abundant and average size of food items larger (Palmgren, 1932; Kuusisto, 1941; Lundberg *et al.*, 1981; Temrin, 1984). That food abundance is important for the reproductive success of the Pied Flycatcher was demonstrated, for example, by Järvinen (1982), who found that late-breeding pairs near the shore of a lake in Finnish Lapland achieved better reproductive success than pairs farther away because of the rich supply of swarming aquatic insects at that time.

In summary, the food factor is very likely one of the important criteria in the habitat selection of Pied Flycatchers. Our strongest evidence in favor of this view is that breeding density, when nest holes are not limiting, is almost always markedly lower in coniferous than in deciduous forest (Section III,A).

B. Nest-Hole Availability

Nest-hole availability limits the population size of the Pied Flycatcher, and by erecting nest boxes one can greatly increase population density (von Haartman,

1971). By providing nest boxes we obtained up to 3.1 pairs/ha in deciduous forest, while in a forest of similar structure and tree species composition, but without nest boxes, breeding density was lower (0.5 pairs/ha; in an unmanaged part, 1.0 pairs/ha). Without doubt there are many fewer natural nest sites in coniferous than in deciduous forest. In fact, in predominantly coniferous forest, many nest holes are actually situated in the few deciduous trees, for example, birch and aspen. In coniferous forest the population density of the Pied Flycatcher is much higher in unmanaged compared to managed plots (Haapanen, 1965; von Haartman, 1971), which probably reflects a difference in nest-hole abundance. At present, almost all coniferous forests in Scandinavia are strongly affected by forestry, while deciduous woodlands are much less affected. Therefore, the difference between the two habitat types with respect to nest-hole availability may be larger now than under a more natural situation.

Furthermore, it is quite likely that nest-hole quality is better in deciduous than in coniferous forest. In the former habitat many nest sites suitable for Pied Flycatchers have arisen because thin branches have fallen off tree trunks, or holes have been excavated by Lesser Spotted Woodpeckers (*Dendrocopos minor*) (R. V. Alatalo and A. Lundberg, personal observations). In coniferous forest, by contrast, larger holes probably are more frequent than smaller ones (Haapanen, 1965), because the smallest woodpecker in that habitat is the Great Spotted Woodpecker (*Dendrocopos major*). However, in coniferous forest Pied Flycatchers can take over nest sites of Willow Tits (*Parus montanus*) and Crested Tits (*P. cristatus*), which, however, often are situated in decaying trunks and are of poor quality. The poorer the nest-site quality, the higher will be the predation risks. These considerations lead to the conclusion that deciduous forests offer more and better nest sites than do coniferous forests. In fact, in the last-mentioned habitat Pied Flycatchers may have great difficulties finding suitable nest holes. This view is supported by our observation that new nest-box areas in coniferous forest are colonized much later in spring than old nest-box areas (Table III). Pied Flycatchers, unless homing to places of which they have experience from previous years, can be expected first to look for nest sites where nest holes are most abundant, which no doubt is in deciduous forest. This fits with the pattern of settlement we observed when putting up nest boxes in deciduous and coniferous areas previously lacking nest boxes (Table I; Lundberg *et al.*, 1981). Some of the cues birds may employ for finding nest sites in coniferous forest will be discussed in Section IV,C.

C. Social Environment

Both intra- and interspecific competition are likely to have a negative effect on the reproductive success of the Pied Flycatcher and may influence habitat selection. This will be discussed in more detail in Sections V,A and B (see also Section III,C).

However, the presence of other individuals may be a positive factor, too. In our nest-box areas we found that the distribution of nests was patchy, and this patchiness could not be explained solely by a patchy environment (Alatalo *et al.*, 1982b). Nest boxes very close to an occupied box were inhabited less often, and somewhat more distant boxes, more often than expected if nest box choice was random. By a playback experiment we found that the song of a male may attract rather than repel other males, and, therefore, Pied Flycatcher males probably make use of the song of other males as a cue when trying to locate suitable nest sites (Alatalo *et al.*, 1982b). This may be especially important in coniferous forest where nest holes are very scarce and may be difficult to find (see Section IV,B).

With respect to food availability and nest-site quality both males and females should (and probably do) use the same criteria when choosing a breeding territory (Alatalo *et al.*, 1984b). However, they may also have conflicting interests. For example, males may prefer to breed where their chances of becoming polygynous are highest (see Alatalo *et al.*, 1984c; Alatalo and Lundberg, 1984a), while females should avoid such places. Males therefore should prefer low-density areas with plenty of free nest sites well separated from each other, where the probability of achieving bigamy is comparatively high, as, for example, in our coniferous nest-box areas. Females should prefer deciduous areas with higher breeding density, where males have difficulties in finding free nest sites in which they can maintain secondary mates (Table IX). However, our nest-box areas, compared to the natural situation, may be somewhat artificial, because the increased breeding density which prevents males from taking up secondary territories in deciduous areas will suppress polygyny frequency below the natural level. On the other hand the superabundance of nest sites not claimed by any male in coniferous forest will give males easy access to secondary territories and thus facilitate the males' deception of females and higher polygyny (Alatalo and Lundberg, 1984a). In unmanaged deciduous forest we found polygyny frequency to be 15% (Alatalo and Lundberg, 1984a), and it is quite possible that it is of the same order in natural coniferous forests too. If so, females may follow some rule of thumb, which is or was appropriate in all natural habitats, with respect to polygyny frequency.

At present, we do not possess any evidence that the social environment influences the habitat choice of Pied Flycatcher males and females in different ways, although certain circumstances render such a difference conceivable.

V. INTRA- AND INTERSPECIFIC COMPETITION

A. Does Interspecific Competition Influence Habitat Selection?

In general, bird densities are much higher in deciduous forest (ca. 1000–1500 pairs/km^2) than in coniferous forest (ca. 400–500 pairs/km^2; e.g., Nilsson,

1979), which may influence the prospect of successful breeding for birds choosing to settle in the respective habitat. As pointed out in Section IV, an essential resource for prospecting Pied Flycatcher males and females is a suitable nest site. Furthermore, Pied Flycatchers prefer nest boxes, in which they probably score a higher breeding success than in natural tree holes (Alatalo and Lundberg, 1984a; Nilsson, 1984). Since tree holes in general represent a scarce and finite resource, one would expect it to be the subject of keen competition, intra- as well as interspecific. The primary importance of nest sites for this species is demonstrated by the invariable finding that as soon as nest boxes are put up, almost irrespective of habitat, a considerable population of Pied Flycatchers rapidly becomes established in them, and competition for nest sites is easily observed (Section V,B).

However, there are other hole nesters in northern European forests. The most abundant of the potential competitors of the Pied Flycatcher is the Great Tit (*Parus major*). The competition between these two species was discussed by Slagsvold (1975). Since Great Tits, especially the adult ones, usually spend the winter near the nest site and thus are able to occupy territories and nest boxes in early spring, the Pied Flycatcher, upon its arrival, finds some proportion of all nest sites occupied by Great Tits. The outcome of the ensuing contests for nest boxes between flycatchers and tits significantly depends, as Slagsvold argues, upon certain phenological relationships. As soon as the Great Tits have laid a full clutch and started incubating, they are relatively immune to the attempts by flycatchers to take over their nest sites, but during nest building and egg laying they are absent from the nest for such long periods that some Pied Flycatchers succeed in usurping their nest sites. Therefore, in spring, if the time interval between the start of egg laying of the Great Tit and the arrival of the Pied Flycatcher is short, the flycatchers stand a good chance of taking over nest boxes occupied by Great Tits, If, on the other hand, the tits are early and the flycatchers late, the result will be less favorable for the last-mentioned species. Like the Pied Flycatcher, the Great Tit prefers deciduous forest but occurs also in coniferous forest. Unfortunately, we do not know the relative importance of competition between these two species for nest sites in different habitats. Furthermore, in deciduous forest, Blue Tits (*P. caeruleus*) and Marsh Tits (*P. palustris*) may compete for nest holes, while the tit species (apart from the Great Tit) inhabiting coniferous forest excavate their own holes.

Practically all data sets relating to the Pied Flycatcher derive from nest-box areas where there is a surplus of breeding facilities. It is difficult to assess the significance of inter- and intraspecific competition for nest holes in natural forests, for nobody knows the abundance and dispersion of natural tree holes. It is clear that in modern forests nest sites for hole nesters are a limiting resource unless nest boxes are put up. On the other hand, all nest boxes are not occupied if they are put up in a density higher than a certain value which may differ between habitats.

Another essential resource for successful breeding is food, and while Slagsvold (1975) concluded that there is interspecific competition between Pied Flycatchers and Great Tits for nest holes, a view with which we concur, he found no unequivocal evidence for an effect of interspecific competition for food on Pied Flycatcher populations. Even if Pied Flycatchers and Great Tits have different foraging techniques, nestling diets broadly overlap (Dornbusch, 1981), which makes competition between the two species likely. However, the Pied Flycatcher probably diffusely competes with many other insectivorous passerines too. Under such conditions interspecific competition should become forceful and could result in density-dependent breeding success. It may be that the large number of insectivorous passerines, and hence total food consumption, in deciduous compared to coniferous forest suppresses actual food availability so that individual Pied Flycatchers have access to approximately equal amounts of food in both habitats.

We know of only a very few published studies providing quantitative evidence for the role of interspecific competition on the offspring production of a given bird species (Högstedt, 1980; Minot, 1981). In another study, though so far unpublished, L. Gustafsson (in preparation) deals with the Collared Flycatcher (*Ficedula albicollis*), a close relative of the Pied Flycatcher, and we wish to present his findings briefly here. Gustafsson's study of the Collared Flycatcher is conducted on the island of Gotland in the Baltic. By manipulating the density of breeding tits (both Great and Blue; the Marsh Tit is absent from this island) while holding that of the Collared Flycatcher constant, Gustafsson was able to demonstrate that Collared Flycatchers breeding among dense tit populations raised fewer and lighter young, which returned at a lower rate to the area for their first breeding and then achieved a lower reproductive success. These observations were made in comparison with observations about Collared Flycatchers breeding under much less severe competition from tits. Unfortunately a similar study is unfeasible for the Pied Flycatcher, whose young disperse widely from their birthplace.

Gustafsson's elegant experimental demonstration of the impact of interspecific competition on the offspring production in the Collared Flycatcher suggests that the presence of food competitors, negatively influencing reproductive success, may be incorporated among the set of criteria birds should utilize when selecting habitat for breeding purposes. However, it is not clear at present whether the flycatchers actually use that criterion in their choice of breeding habitat.

On the island of Gotland the Pied Flycatcher occurs alongside the Collared Flycatcher, the latter species making up about 90% of the two. These congeneric species, which in most respects closely resemble each other, are allopatric except on the islands of Gotland and Öland in the Baltic and in restricted parts of central Europe. Where sympatric they hybridize frequently (Alerstam *et al.*, 1978; Alatalo *et al.*, 1982c). What is interesting in the present context is that where they are sympatric they partially segregate in terms of habitat. In areas where the Collared Flycatcher is absent, Pied Flycatchers prefer deciduous forest (see Section IV,B). On Gotland Collared Flycatchers are almost 30 times as abundant as the Pied Flycatcher in deciduous forest, while they are only 3 times as abundant in coniferous forest (Alatalo *et al.*, 1982c). Apparently the Collared Flycatcher, being slightly larger, is dominant in controversies (Löhrl, 1955) and can prevent the Pied Flycatcher from taking up territories in the optimal habitat (deciduous forest). However an increase in Pied Flycatcher density as a consequence of the introduction of nest boxes does not seem to cause any noticeable changes in the population densities of other insectivorous bird species (Enemar and Sjöstrand, 1972).

B. Intraspecific Competition and Habitat Selection

Theoretical discussions on how individuals should distribute themselves across habitats are few (Huxley, 1934; Kluyver and Tinbergen, 1953; Lack, 1964), and these discussions have been synthesized in the models of "ideal free" and "despotic" distribution (Fretwell and Lucas, 1969; Fretwell, 1972). In the ideal-free model it is assumed that individuals are free to settle in the place of their choice and that they are ideal in their judgment of the habitat's profitability (suitability). It is further assumed that habitat profitability decreases with increasing breeding density and that, at some density level in the most preferred habitat, the profitability of the habitat next in order of preference will be equally high. As a result, payoff will be the same in all places (habitats) occupied by ideal-free birds. Thus, individuals end up in different habitats in relation to their respective profitabilities. According to the despotic model, by contrast, some individuals can monopolize resources through dominance and/or territorial behavior and prevent other individuals from exploiting them, thus relegating these less-dominant individuals to less favorable habitats where their payoff will be lower.

On the basis of morphometric comparisons and of information given in Sections III,C and D, we were able to exclude the possibility of Pied Flycatchers in deciduous and coniferous forest belonging to different locally adapted subpopulations. Thus, one assumption of the ideal-free model is fulfilled. To be able to discriminate between the two distribution models, however, one has to know the payoffs in different habitats and under different population densities. Moreover, the ideal-free distribution can only apply if birds are free to enter any habitat, which requires that no resource that cannot be shared, such as nest holes, be limiting. We therefore first consider the case when nest boxes have been available in excess to see if either of the two models will work.

In the ideal-free model, birds, by definition, should go to the habitat with highest potential success, and the payoff for a newly settling individual should be equal to the average of all individuals. A complicating factor, already pointed out by Brown (1975), is that the progress of the breeding season, as such, negatively influences breeding success, and one therefore has to compare individuals breeding at the same time (Figs. 1 and 2). Comparing the regression lines for reproductive success versus time (Fig. 1), female Pied Flycatchers in coniferous forest do worse throughout the season than do birds in deciduous forest, and, obviously, the prediction of equal reproductive success is violated. However, the difference in reproductive success between habitats is solely due to the higher polygyny frequency in coniferous forest, which in turn is due to low breeding density in combination with excess of nest sites, and, thus, differences may have been caused by an artificial situation. Since we almost certainly suppress polygyny frequency in deciduous and enhance it in coniferous forest (Section IV,C) and since breeding success is the same in both habitats for monogamous pairs, we cannot reject either model simply on the basis of differences in reproductive success. In the natural situation where breeding density is lower than in our nest-box areas it is very likely that pairs in deciduous forest have higher reproductive success than pairs in coniferous forest. We found, for example, higher reproductive success for monogamous pairs in low-density deciduous nest-box areas than in coniferous areas with similar low breeding density. Thus, our data on payoff in the two habitats in nest-box areas do not give unambiguous evidence for or against the ideal-free distribution. However, in a natural situation nest holes almost certainly are limiting, and therefore the ideal-free model is not applicable.

As mentioned above, we found that birds occupied coniferous forest, provided the presence of nest-box areas was known to some birds, as fast as deciduous woodland, while according to the assumption for ideal-free birds, if the ideal distribution changes between years and areas, individuals should first select habitats where their chances of success are highest (Fretwell and Lucas, 1969; Fretwell, 1972; see also Brown, 1969, 1975). This in our case means deciduous woodland. However, one could argue that if the ideal-free distribution prevails, the most preferred habitat will always be filled, and as a consequence its suitability will be lowered to the level of the next preferred habitat. If so, the first birds to arrive would be equally well-off in whichever of the two habitats they settled.

Another relevant piece of information is that, in comparisons between habitats, smaller males were found to be overrepresented in coniferous forest (Table IV). Pied Flycatchers do not defend exclusive feeding territories but rather a nest hole and its close vicinity (van Haartman, 1956). The first males to arrive at an area often advertise several nest holes but will later restrict their attention to a few of them. However, a male's territory is large enough to cover several nest sites at high population densities if boxes are provided at less than about 50-m intervals. Under such circumstances, unoccupied nest boxes close to an occupied one become inhabited significantly less often than expected from random choice, which can only be explained by territorial behavior and dominance (von Haartman, 1971; Alatalo *et al.*, 1982b). Thus, even if nest boxes are provided in excess, most space is defended by some male, and, consequently, new males are not free to settle in between established nest-box owners. However, such a pattern violates the ideal-free model only if some individuals can be shown to monopolize a more than equal share of the resources. If there were no difference in breeding success between habitats, why would larger males be overrepresented in deciduous forest? The reason probably is that large size usually confers an advantage in interindividual confrontations (Morse, 1974; Garnett, 1976). Since nest holes are the most essential resource for successful breeding and are more abundant in natural deciduous forest, this habitat is the most preferred one, and in the competition for a territory, larger than average males may be at an advantage and hence become overrepresented in this habitat. In the natural situation, moreover, the probability of high offspring production, for both males and females, is likely to be highest in deciduous forest (see earlier, this section and Sections IV,A and C).

According to the expectations of the despotic model, if residents in the most suitable habitat, by dominance and/or by territorial behavior, may be able to prevent newcomers from settling, then these newcomers, which are ideal in their judgment and try to maximize their payoff, would profit by settling in another habitat, although their expected payoff will be lower. If there exists an asymmetry in dominance and larger males can expel smaller ones, dominance hierarchies will become established, and the ideal-free distribution model does not hold. Thus, our finding of size assortment of males across habitats leads us to the conclusion that, in the Pied Flycatcher, habitat distribution appears to be ideal despotic, not ideal free.

VI. SUMMARY

At the present time our understanding of the Pied Flycatcher can be summarized as follows. Pied Flycatchers occupy coniferous forest, but not new nest-box areas, as fast as deciduous woodland but reach clearly higher densities in the latter habitat. In coniferous forest the flycatchers seem to have difficulties in finding nest sites. Males are larger in deciduous than in coniferous forest. This is probably a result of dominance rather than of adaptive responses to different habitat-associated selection pressures. Birds sometimes change habitats between years, supporting this view. Monogamous pairs produce as many offspring in coniferous as in deciduous forest, but owing to higher polygyny frequency in the former habitat, mean reproductive success per female is lower in coniferous forest. However, the observed difference in polygyny frequency between habitats is likely to be larger in our nest-box areas than in a natural situation. In deciduous woodlands with excess of nest boxes population density is often so high that many males lack an opportunity to take up secondary territories because free nest boxes are only to be found within other males' territories. By contrast, in coniferous forests free nest sites are available between territories, which will allow polyterritoriality and opportunities for males to deceive females and achieve polygyny. At low population density in deciduous forest breeding success is higher than in coniferous forest, after correcting for the effect of polygyny.

In a review of the breeding biology of the Pied Flycatcher, Lack (1966) suggested that its population density was determined by the density of insect food, a view with which von Haartman (1971) disagreed. Our conclusion from this study is that nest holes are the most essential resource and that in natural forests population density, in most cases, is determined by the availability of nest sites. However, if nest boxes are provided the population density increases up to a limit which is probably set by food. The habitat distribution of the Pied Flycatcher is despotic rather than ideal free. Further progress in the analysis of habitat selection requires more information about Pied Flycatcher populations breeding under natural conditions.

ACKNOWLEDGMENTS

We wish to thank L. Gustafsson for allowing us to quote his unpublished work. The study is being supported by grants from the Swedish Natural Science Research Council.

REFERENCES

Alatalo, R. V., and Lundberg, A. (1984a). Polyterritorial polygyny in the Pied Flycatcher *Ficedula hypoleuca*—evidence for the deception hypothesis. *Acta Zool. Fenn.* **21,** 217–228.

Alatalo, R. V., and Lundberg, A. (1984b). Density-dependence in breeding success of the pied flycatcher (*Ficedula hypoleuca*). *J. Anim. Ecol.* **53,** 969–978.

Alatalo, R. V., Carlson, A., Lundberg, A., and Ulfstrand, S. (1981). The conflict between male polygamy and female monogamy: The case of the pied flycatcher *Ficedula hypoleuca. Am. Nat.* **117,** 738–753.

Alatalo, R. V., Lundberg, A., and Ståhlbrandt, K. (1982a). Why do pied flycatcher females mate with already-mated males? *Anim. Behav.* **30,** 585–593.

Alatalo, R. V., Lundberg, A., and Björklund, M. (1982b). Can the song of male birds attract other males? An experiment with the pied flycatcher *Ficedula hypoleuca. Bird Behav.* **4,** 42–45.

Alatalo, R. V., Gustafsson, L., and Lundberg, A. (1982c). Hybridization and breeding success of collared and pied flycatchers on the island of Gotland. *Auk* **99,** 285–291.

Alatalo, R. V., Carlson, A., Lundberg, A., and Ulfstrand, S. (1984a). Male deception or female choice in the pied flycatcher *Ficedula hypoleuca:* A reply. *Am. Nat.* **123,** 282–285.

Alatalo, R. V., Lundberg, A., and Ståhlbrandt, K. (1984b). Female mate choice in the pied flycatcher *Ficedula hypoleuca*. *Behav. Ecol. Sociobiol.* **14,** 253–261.

Alatalo, R. V., Gustafsson, L., and Lundberg, A. (1984c). High frequency of cuckoldry in pied and collared flycatchers. *Oikos* **42,** 41–47.

Alatalo, R. V., Gustafsson, L., and Lundberg, A. (1984d). Why do young passerine birds have shorter wings than older birds? *Ibis* **126,** 410–415.

Alerstam, T., Ebenman, B., Sylven, M., Tamm, S., and Ulfstrand, S. (1978). Hybridization as an agent of competition between two bird allospecies: *Ficedula albicollis* and *F. hypoleuca* on the island of Gotland in the Baltic. *Oikos* **31,** 326–331.

Askenmo, C. (1977). Some aspects of the reproduction strategy of the pied flycatcher *Ficedula hypoleuca* (Pallas). Doctoral dissertation, Univ. of Gothenberg, Gothenberg.

Berndt, R., and Sternberg, H. (1969). Alters- und Geschlechtsunterschiede in der Dispersion des Trauerschnäppers (*Ficedula hypoleuca*). *J. Ornithol.* **110,** 22–26.

Berndt, R., and Winkel, W. (1967). Die Gelegegrösse des Trauerschnäppers (*Ficedula hypoleuca*) in Beziehung zu Ort, Zeit, Biotop und Alter. *Die Vogelwelt* **88,** 97–136.

Berndt, R., and Winkel, W. (1975). Gibt es beim Trauerschnäpper *Ficedula hypoleuca* eine Prägung auf den Biotop des Geburtsortes? *J. Ornithol.* **116,** 195–201.

Berndt, R., Winkel, W., and Zang, M. (1981). Über Legebeginn und Gelegestärke des Trauerschnäppers (*Ficedula hypoleuca*) in Beziehung zur geographishen Lage des Brutortes. *Die Vogelwarte* **31,** 101–110.

Björklund, M., and Westman, B. (1983). Extra-pair copulations in the Pied Flycatcher (*Ficedula hypoleuca*). A removal experiment. *Behav. Ecol. Sociobiol.* **13,** 271–275.

Brown, J. L. (1969). The buffer effect and productivity in tit populations. *Am. Nat.* **103,** 347–354.

Brown, J. L. (1975). "The Evolution of Behavior," W. W. Norton, New York.

Campbell, B. (1955). A population of pied flycatchers (*Muscicapa hypoleuca*). *Proc. Int. Ornithol Congr. 11th, 1954,* pp. 428–434.

Case, T. J. (1978). A general explanation for insular body size trends in terrestrial vertebrates. *Ecology* **58,** 1–18.

Creutz, G. (1955). Der Trauerschnäpper [*Muscicapa hypoleuca* (Pallas)]. Eine Populationsstudie. *J. Ornithol.* **96,** 241–326.

Curio, E. (1959a). Beiträge zur Populationsökologie des Trauerschnäppers (*Ficedula h. hypoleuca* Pallas). *Zool. Jahrb.* **87,** 185–230.

Curio, E. (1959b). Verhaltensstudien am Trauerschnäpper. *Z. Tierpsychol.* (suppl. 3).

Dornbusch, M. (1981). Die Ernährung einiger Kleinvogelarten in Kiefernjungbestockungen. *Beitr. Vogelkd., Jena* **27,** 73–99.

Drost, R. (1936). Über das Brutkleid männlicher Trauerfliegenfänger, *Muscicapa hypoleuca. Der Vogelzug* **6,** 179–186.

Ebenman, B., and Nilsson, S. G. (1981). Size patterns in Willow Warblers *Phylloscopus trochilus* on islands in a south Swedish lake and the nearby mainland. *Ibis* **123,** 528–534.

Enemar, A. (1948). Some experiences from 5 years' nesting box studies (in Swedish with English summary). *Vår Fågelv.* **7,** 105–117.

Enemar, A., and Sjöstrand, B. (1972). Effects of the introduction of Pied Flycatchers *Ficedula hypoleuca* on the composition of a passerine bird community. *Ornis Scand.* **3,** 79–89.

Fretwell, F. D. (1972). "Populations in a Seasonal Environment." Princeton Univ. Press, Princeton, New Jersey.

Fretwell, F. D., and Lucas, Jr., H. L. (1969). On territorial behavior and other factors influencing habitat distribution in birds. I. Theoretical development. *Acta Biotheor.* **19,** 16–36.

Garnett, M. C. (1976). Some aspects of body size in tits. Doctoral dissertation, Univ. of Oxford, Oxford.

Grant, P. R. (1979). Ecological and morphological variation of Canary Island Blue Tits *Parus caeruleus* (Aves: Paridae). *Biol. J. Linn. Soc.* **11,** 103–129.

Haapanen, A. (1965). Bird fauna of Finnish forests in relation to forest succession. I. *Ann. Zool. Fenn.* **2,** 153–196.

Helle, P., and Pulliainen, E. (1983). On the efficiency of the line transect method: A study based on nest searching. *Ornis Fenn.* **60,** 35–41.

Högstedt, G. (1980). Prediction and test of the effects of interspecific competition. *Nature* **283,** 64–66.

Huxley, J. S. (1934). A natural experiment on the territorial instinct. *Br. Birds* **27,** 270–277.

Järvinen, A. (1980). Population dynamics in the Pied Flycatcher *Ficedula hypoleuca* at subarctic Kilpisjärvi, Finnish Lapland. *Ornis Fenn.* **57,** 17–25.

Järvinen, A. (1982). Influence of the lake shore on the reproduction of the pied flycatcher *Ficedula hypoleuca,* and the redstart *Phoenicurus phoenicurus* in Finnish Lapland. *Ann. Zool. Fenn.* **19,** 171–174.

Järvinen, O., Väisänen, R. A., and Enemar, A. (1978). Efficiency of the line transect method in mountain birch forest. *Ornis Fenn.* **55,** 16–23.

Källander, H. (1975). Breeding data for the Pied Flycatcher *Ficedula hypoleuca* in southernmost Sweden. *Ornis Fenn.* **52,** 97–102.

Kluyver, H. N., and Tinbergen, L. (1953). Territory and regulation of density in Titmice. *Arch. Neerl. Zool.* **10,** 265–289.

Kuusisto, P. (1941). Studien über die Ökologie und Tagesrhythmik von *Phylloscopus trochilus acredula* (L.) mit besonderer Berücksichtigung der Brutbiologie. *Acta Zool. Fenn.* **3,** 1–120.

Lack, D. (1964). A long term study of the great tit (*Parus major*) *J. Anim. Ecol.* **33** (suppl.), 159–173.

Lack, D. (1966). "Population Studies of Birds." Oxford Univ. Press, (Clarendon), London and New York.

Lack, D. (1971). "Ecological Isolation in Birds." Blackwell, Oxford.

Löhrl, H. (1955). Beziehungen zwischen Halsband- und Trauerfliegenschnäpper *Muscicapa albicollis* und *M. hypoleuca* in demselben Brutgebiet. *Proc. Int. Ornithol. Congr., 11th, 1954,* pp. 333–336.

Lundberg, A., Alatalo, R. V., Carlson, and Ulfstrand, S. (1981). Biometry, habitat distribution and breeding success in the Pied Flycatcher *Ficedula hypoleuca. Ornis Scand.* **12,** 68–79.

Minot, E. O. (1981). Effects of interspecific competition for food in breeding blue and great tits. *J. Anim. Ecol.* **50,** 375–385.

Morse, D. H. (1974). Niche breadth as a function of social dominance. *Am. Nat.* **108,** 818–830.

Nilsson, S. G. (1979). Density and species richness of some forest bird communities in South Sweden. *Oikos* **33,** 392–401.

Nilsson, S. G. (1984). Clutch size and breeding success of the pied flycatcher *Ficedula hypoleuca* in natural tree-holes. *Ibis* **126,** 407–410.

Nyholm, N. E. I., and Myhrberg, M. E. (1983). Breeding area fidelity of the Pied Flycatcher *Ficedula hypoleuca* at Ammarnäs, Swedish Lapland. *Ornis Fenn.* **60,** 22–27.

Palmgren, P. (1930). Quantitative Untersuchungen über die Vogelfauna in den Wäldern Südfinnlands, mit besonderer Berücksichtigung Ålands. *Acta Zool. Fenn.* **7,** 1–218.

Palmgren, P. (1932). Zur Biologie von *Regulus r. regulus* (L.) und *Parus atricapillus borealis* Selys. Eine vergleichend-ökologische Untersuchung. *Acta Zool. Fenn.* **14,** 1–113.

Schoener, T. W. (1969). Models of optimal size for solitary predators. *Am. Nat.* **103,** 277–313.

Silverin, B. (1980). Effects of long-acting testosterone treatment on freeliving pied flycatchers, *Ficedula hypoleuca,* during the breeding period. *Anim. Behav.* **28,** 906–912.

Silverin, B. (1983). Population endocrinology and gonadal activities of the male pied flycatcher. *In*

"Avian Endocrinology: Environmental and Ecological Perspectives" (S. Mikami, ed.), pp. 289–305. Springer-Verlag, Berlin and Tokyo.

Slagsvold, T. (1975). Competition between the Great Tit *Parus major* and the Pied Flycatcher *Ficedula hypoleuca* in the breeding season. *Ornis Scand.* **6,** 179–190.

Snow, D. W. (1954). The habitats of Eurasian tits *Parus* spp. *Ibis* **96,** 565–585.

Soveri, J. (1940). Die Vogelfauna von Lammi, ihre regionale Verbreitung und Abhängigkeit von den ökologischen Faktoren. *Acta Zool. Fenn.* **27,** 1–176.

Svensson, L. (1975). "Identification Guide to European Passerines." Naturhistoriska Riksmuseet, Stockholm.

Temrin, H. (1984). Why are some Wood Warbler (*Phylloscopus sibilatrix*) males polyterritorial? *Acta Zool. Fenn.* **21,** 243–247.

Tompa, F. S. (1967). Reproductive success in relation to breeding density in pied flycatchers, *Ficedula hypoleuca* (Pallas). *Acta Zool. Fenn.* **118,** 1–28.

Trettau, W., and Merkel, F. (1943). Ergebnisse einer Planberingung des Trauerfliegenfängers (*Muscicapa hypoleuca* Pallas) in Schlesien. *Der Vogelzug* **14,** 77–90.

Ulfstrand, S., Alatalo, R. V., Carlson, A., and Lundberg, A. (1981). Habitat distribution and body size of the Great Tit *Parus major. Ibis* **123,** 494–499.

Virolainen, M. (1984). Breeding biology of the Pied Flycatcher *Ficedula hypoleuca* in relation to population density. *Ann. Zool. Fenn.* **21,** 187–197.

von Haartman, L. (1945). Några fall av polygami hos svart och vita flugsnapparen (Muscicapa h. *hypoleuca* Pall.). *Vår Fågelv.* **4,** 27–32.

von Haartman, L. (1949). Der Trauerfliegenschnäpper. I. Ortstreue und Rassenbildung. *Acta Zool. Fenn.* **56,** 1–104.

von Haartman, L. (1951a). Der Trauerfliegenschnäpper. II. Populationsprobleme. *Acta Zool. Fenn.* **67,** 1–60.

von Haartman, L. (1951b). Successive polygamy. *Behaviour* **3,** 256–274.

von Haartman, L. (1954). Der Trauerfliegenschnäpper. III. Die Nahrungsbiologie. *Acta Zool. Fenn.* **83,** 1–96.

von Haartman, L. (1956). Territory in the pied flycatcher *Muscicapa hypoleuca. Ibis* **98,** 460–475.

von Haartman, L. (1960). The *Ortstreue* of the Pied Flycatcher. *Proc. Int. Ornithol. Congr., 12th, 1958,* pp. 266–273.

von Haartman, L. (1967a). Clutch-size in the Pied Flycatcher. *Proc. Int. Ornithol. Congr., 14th, 1966,* pp. 155–164.

von Haartman, L. (1967b). Geographical variations in the clutch-size of the Pied Flycatcher. *Ornis Fenn.* **44,** 89–98.

von Haartman, L. (1971). Population dynamics. *In* "Avian Biology" (D. S. Farner and J. R. King, eds.), Vol. I, pp. 391–459. Academic Press, New York.

Winkel, W. (1982). Zum Ortstreue-Verhalten des Trauerschnäppers (*Ficedula hypoleuca*) im westlichen Randbereich seines mitteleuropäischen Verbreitungsgebietes. *J. Ornithol.* **123,** 155–173.

Chapter 3

Habitat Selection in the Sylviine Warblers of Western Europe and North Africa

MARTIN L. CODY

Department of Biology
University of California at Los Angeles
Los Angeles, California

HABITAT SELECTION IN BIRDS

ISBN 0-12-178080-5

I. INTRODUCTION

A. Species and Their Distributions

Current taxonomy combines the warblers Sylviinae with the thrushes Turdinae and the flycatchers Muscicapinae in the dominant superfamily of Old World land birds Muscicapoididae. Whereas the thrushes are generally larger and feed on or near the ground and the flycatchers usually sally after flying insects from elevated perches, the warblers are mostly gleaners of foliage and searchers for insects hidden in the vegetation. In this chapter, I briefly review information on breeding systems and territoriality in these warblers and then discuss habitat selection among species in different regions of western Europe and North Africa in the three large genera, *Phylloscopus, Sylvia,* and *Hippolais.* These genera comprise species of scrub and woodland; other warblers, especially *Acrocephalus* and *Locustella,* are common in grassland and marsh vegetation, and these are the subjects of morphological analyses relative to their habitats in Chapter 14. Other warblers, such as the woodland and forest kinglets *Regulus,* the grass- and marshland *Cisticola, Cettia,* and *Prinia* (all monospecific in Europe), and the North African desert-edge *Scotocerca,* will receive more casual mention. Excellent range maps of the species mentioned in this Chapter are given in Harrison (1982).

B. The Genus *Phylloscopus*

Four species of ''leaf warblers'' in the genus *Phylloscopus* have broad ranges in Europe: Willow Warblers (*P. trochilus*) extend from central Europe to the Arctic Ocean; Chiffchaffs (*P. collybita*) range from the Canary Islands and coastal areas of Algeria to central Scandinavia but are absent from southern Sweden; Wood Warblers (*P. sibilatrix*) occur throughout central Europe and much of Britain to southern Scandinavia; and Bonnelli's Warblers (*P. bonelli*) extend from central Europe south to North Africa, excluding Sardinia. Two additional species have spotty and variable distributions in the north: Greenish Warblers (*P. trochilioides*) breed in eastern Finland and occasionally in areas west of the Baltic, and Arctic Warblers (*P. borealis*) breed sporadically in extreme northern Sweden and Finland but have apparently retreated east from former breeding areas in the last 15 years (see Section VI).

C. The Genus *Sylvia*

The nominate genus *Sylvia,* with almost 20 species, is the largest genus of passerine birds in the western Palaearctic region, rivaled only by *Emberiza* buntings. In western central Europe (for example, central France) 6 species occur: Orphean Warblers (*S. hortensis*), Garden Warblers (*S. borin*), Blackcaps (*S. atricapilla*), Common Whitethroats (*S. communis*), Lesser Whitethroats (*S. curruca*), and Dartford Warblers (*S. undata*). All but the first-mentioned species breed in Britain, and those 5 species minus the Dartford Warbler, but together with the largely eastern European Barred Warbler (*S. nisoria*), breed in southern Sweden. To the north in Scandinavia, Barred Warblers, Blackcaps, Common Whitethroats, and Lesser Whitethroats drop out sequentially, and only Garden Warblers breed north to the arctic tree line.

In southern Europe and North Africa, the Subalpine Warbler (*S. cantillans*), the Spectacled Warbler (*S. conspicillata*), and the Sardinian Warbler (*S. melanocephala*) occur. Marmora's Warblers (*S. sarda*) have a more restricted distribution on Mediterranean islands and adjacent continental coasts. In North Africa the *S. undata–S. sarda*-related Tristram's Warbler (*S. deserticola*) and the *S. communis*-related Desert Warbler (*S. nana*) breed in scrub and desert-edge habitats. Three of these species, Blackcaps, Sardinian Warblers, and Spectacled Warblers, breed on the Canary Islands.

D. The Genus *Hippolais*

Two species of *Hippolais* are widely distributed in Europe, the Icterine Warbler (*H. icterina*), from central Europe north to southern Scandinavia, and the Melodious Warbler (*H. polyglotta*), from central Europe south to North Africa. In addition, the Olivaceous Warbler (*H. pallida*) breeds from central Spain south into North Africa; no *Hippolais* species occur in the Canary Islands.

E. Recent Range Changes

Bird distributions are not static but may change in response to several factors. One of these is the chance colonization of isolated areas, an example of which is the arrival and breeding of *Sylvia undata* on Minorca in 1975. Now nonmigratory residents there, Dartford Warblers have become abundant on the island (Bergman, 1978; Muntaner, 1980).

The ranges of several warbler species appear to be extending north in western Europe. While this may be a response to a general postglacial amelioration of climate, rapid range extensions in several species belie this view. Habitats change faster in response to human agency than to climatic shifts, and, also, genetic changes in expanding populations may facilitate a rapid range expansion; these are alternative explanations for such a phenomenon. Cetti's Warbler (*Cettia cetti*) took 50 years to reach Britain from the Mediterranean region, and it is now well established there (Bonham and Robertson, 1975; Bibby, 1982). The arrival of Fan-tailed Warblers (*Cisticola juncidis*) in Britain is apparently imminent (Ferguson-Lees and Sharrock, 1977), and Firecrests (*Regulus ignicapillus*) have colonized and spread in the last 20 years (Batten, 1973). Other species with largely southern European ranges, such as Bonelli's Warbler and Melodious Warbler, are also pushing north in continental Europe (Fourage, 1969; Peltzer, 1973; Paquet, 1978). Willow Warblers have become more abundant in northern Finland (Järvinen and Väisänen, 1978), and Blackcaps show an increasing tendency to overwinter in Belgium (Fouarge, 1980). Overall, the biogeographic picture of these warblers is dynamic; species composition in a given region is continuously changing over the longer term, with interesting implications for local species coexistence and community structure.

F. Sylviine Warblers in Winter

Some warblers winter quite far north in western Europe, including Chiffchaffs and Blackcaps, but year-round residency is common only around the Mediterranean and further south, with most northerly breeding species wintering in Africa, both north and south of the Sahara. Although different species may be constrained to use similar habitats (N. Boren, unpublished data) and eat similar foods (e.g., Laursen, 1978) during migration, there is in general a degree of habitat segregation on the wintering grounds similar to that in the breeding areas, and although quantitative information is scarce, there appear to be some similarities between breeding and wintering habitats. Whitethroats, for example, overwinter in habitats in the northern Cameroun comparable to those in which they breed in Europe (LeDant and Jacobs, 1981). Via leapfrog migration, species that breed together also winter in adjacent areas (Cody and Walter, 1976). It appears that many of these warbler species are territorial on their wintering grounds, as reported for Willow Warblers in Nigeria (Elgood *et al.*, 1966) and Greenish Warblers in southern India (Price, 1981) and as indicated by Moreau's (1966) many observations. I have seen Icterine Warblers wintering in karoo scrub north of Oudtshourn, South Africa, where they appear to be uniformly spaced and where they occasionally sing. These observations contrast to those of wintering parulid warblers in the New World, the majority of which are mobile and flocking in winter habitats (see Chapters 1 and 4, this volume).

II. BREEDING SYSTEMS AND TERRITORIALITY

A. Breeding Systems

Sylviine warblers are aggressively territorial during the breeding season, and most appear to be opportunistically polygynous. Orians (1969) predicted that polygyny should be common in species which inhabit high-productivity marsh habitats, and the Cetti's Warbler is a good example. In this species males are 30% heavier than females and have no nest duties; numbers of females per male are correlated with the size of the males and with the size of the territory (Bibby, 1982). Similarly, about one in six Marsh Warbler males that are already mated establish a second territory, and about half of these attract and breed with a second female (Dowsett-Lemaire, 1979). In Willow Warblers the number of females per male increases with the male's age, up to three in 3-year-old males (daPrato, 1982), and two females per territorial male are sometimes found in Whitethroats and Wood Warblers as well (Lawn, 1982; Eddington and Eddington, 1972). These data support the view of sylviines as species of highly productive but variable habitats, in which more resources are preempted by the stronger and more aggressive males and opportunistically converted to offspring via flexible breeding systems.

B. Interspecific Territoriality

A feature of sylviine warbler breeding systems is the frequency and ubiquity of interspecific territoriality. Various species defend territories not only against conspecifics but also against certain other, usually congeneric, species of warbler. This was noted and examined by Eliot Howard (1920), to whom the phenomenon made a good deal of sense: "Do these battles between different species contribute towards the attainment of the end for which the whole territorial system has been evolved?" Yes they do, he argued, since "many species require like conditions of existence and are bound to assemble wherever these conditions are suitable," and, so, "the territory would fail to serve its purpose if no restrictions were imposed upon the measure of such multi-species assemblies" (see Cody, 1978, for Howard's complete argument).

Since Howard's early but thorough observations, many authors have reported interspecific territoriality in sylviine warblers. In *Phylloscopus* species, the phenomenon occurs between Wood Warblers and Willow Warblers (Eddington and Eddington, 1972) and between Willow Warblers and Chiffchaffs in England

(Cody, 1978) and in Norway, where Saether (1983b) conducted removal experiments of singing territorial males of both species in Norwegian alder woods and found that the vacant territories were reoccupied by the two species indiscriminately, without regard to the identity of the previous occupant. In the genus *Acrocephalus,* Reed, Sedge, and Marsh Warblers (*A. scipaceus, A. schoenobaenus, A. palustris*) are interactive over territory (Catchpole, 1972, 1973; Dowsett-Lemaire, 1978; Svensson, 1978), and *Hippolais* species are interspecifically territorial where they overlap in France (Ferry and Deschaintre, 1966).

In *Sylvia* species there are many examples of interspecific territoriality, including the species pairs *S. borin–S. atricapilla* and *S. communis–S. curruca,* in England and Sweden; the trio *S. borin–S. communis–S. nisoria,* in Sweden; and the pair *S. atricapilla–S. cantillans* and the trio *S. undata–S. melanocephala–S. sarda,* in Sardinia, as well as other species combinations (Cody and Walter, 1976; Cody, 1978; and other references therein). These examples and others are discussed in the following sections.

Interspecific territoriality has not been recorded between sylviines of different genera, but the study of Crivelli and Blandin (1977) suggests this possibility between Wood Warblers and Blackcaps.

In the genus *Regulus,* territorial interactions between the two European species, Goldcrest (*R. regulus*) and Firecrest (*R. ignicapillus*), are variable. Territories overlap interspecifically in France (Fouarge, 1974) and in Austria (Thaler, 1976). However, Becker (1977) reported that the two species are interspecifically territorial in Spanish pine plantations. Where both occur in high densities in spruce–fir forests in central Europe, some Firecrests exclude Goldcrests from their territories (but not vice versa). Becker found that both avoid pines in central Europe and that there is no segregation of the two between deciduous and coniferous forest.

Interspecific interactions over territory are presumably effected through interspecific reactions to territorial songs. In Sweden, where the songs of the Barred and Garden Warblers differ chiefly in length rather than in structure, Barred Warblers respond aggressively to the playback of the Garden Warbler song in their territories, and vice versa (Cody, 1978). Garden Warbler and Blackcap territorial songs are very similar in England and different from the Blackcap song in Sardinia. In this latter area Blackcaps interact with Subalpine Warblers, and here again their territorial songs bear some resemblance to one another. Similarly, Dartford Warblers and Marmora's Warblers are strongly interspecifically territorial in Sardinia and sing very similar songs. Where *Regulus* species are interspecifically territorial in Spain, either species can sing species-specific or mixed songs, and most males react to the song of the other species (Becker, 1977).

Yet similarity in songs is not a prerequisite for interspecific territoriality;

Phylloscopus species are readily distinguished by their specific songs, and in at least some locations they respond aggressively only to conspecific and not to heterospecific song playbacks (Helb, 1973; Saether, 1983b). Although the Sardinian Warbler interacts with both the Dartford and the Marmora's Warbler over territory, its song is distinct from theirs. The interpretation of songs in these warblers is complicated by considerable local variation in song structure, including variation with habitat type (e.g., Jilka and Leisler, 1974, in *Acrocephalus* Warblers) and presumably also with the suite of coexisting congeneric species (Cody, 1978; Cody and Walter, 1976).

III. HABITAT SEGREGATION AND ADAPTIVE MORPHOLOGY

A. Habitat Segregation among Genera

Warblers in different genera occupy a habitat gradient from marshes and grassland to forest, to some extent, sequentially. Thus *Cisticola* is strictly a grassland bird (in Europe); *Acrocephalus* and *Locustella* are species of marshland, edge, and open habitats. *Sylvia* species range from low scrub to woodland and forest, *Hippolais* species from open woodland and edge habitats to forest, and *Phylloscopus* species from taller scrub to woodland to forest (Fig. 1). In some places this corresponds to a successional sequence, as in southern

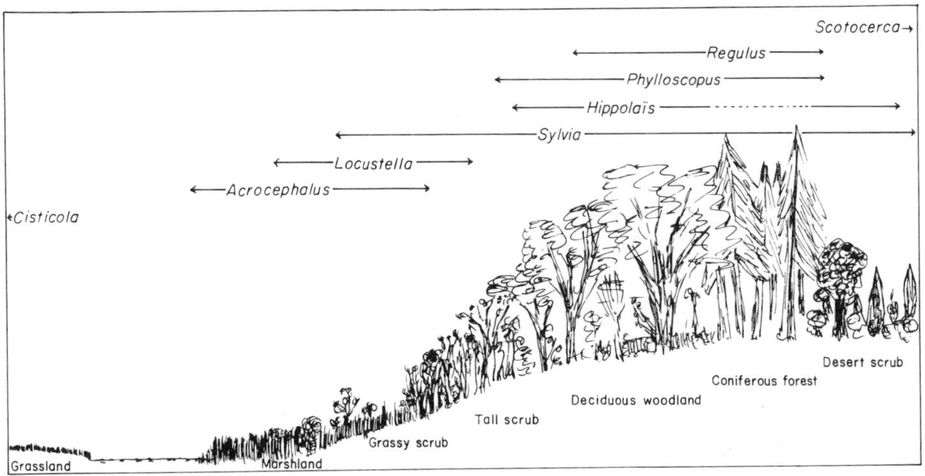

Fig. 1. Stylistic representation of the distribution over habitats of species in eight genera of western Palaearctic sylviine warblers. The habitat gradient from left to right is similar to that of riverine or postglacial succession in much of northwestern Europe.

Poland, where species of *Acrocephalus* first occupy old riverbeds and are followed by *Sylvia, Hippolais,* and last, *Phylloscopus* species (Głowaciński, 1975). In taller vegetation on such a habitat gradient, there are usually representatives of some three genera.

B. Habitat Segregation among Congeneric Species

A number of authors have shown that there are general differences in habitat selection among the congeneric warbler species of various regions. Bairlein *et al.* (1980) emphasized the differences in breeding habits of four central European *Sylvia* species, Mason (1976) summarized data on habitat segregation in the five English *Sylvia* species, and Affre (1975) presented a similar analysis for eight *Sylvia* species in southern France. Tiainen *et al.* (1983) described broad differences in breeding habitats in four *Phylloscopus* species plus *Regulus regulus* in Finland and the extent to which interspecific distributions are correlated.

In the following sections, such interspecific differences are summarized in a comparative and quantitative fashion. What becomes clear is that, while interspecific differences in mean habitat selection exist, most species find a rather broad range of habitats acceptable for breeding. Further, most species preferences overlap greatly. Berthold's (1978) study of Blackcaps in Germany, where the species occupies habitats from very small trees to tall forests, illustrated the wide habitat tolerance of many warblers. Thus species with broad habitat preferences meet other species with similarly broad tolerances over many habitat types. Interspecific territoriality may be seen as a mechanism to regulate habitat and space use where the number of qualified and potential occupants exceeds the number that can be supported as separate ecological entities by habitat and/or resource diversity. An illuminating early study is that of Wasilewski (1967), which documented that the co-occupancy of birds in forest habitats has a strong component of interspecific competition over and above factors of species-independent habitat selection and suitability.

Notwithstanding the broad range of habitats occupied by many breeding sylviine species (in some cases from low scrub to tall forest or from deciduous to coniferous vegetation), mean species differences exist, and morphological correlates of these mean habitat differences have been documented. At one level of resolution is the finding by Holyoak and Thibault (1977) that the body sizes of different *Acrocephalus* species on Polynesian islands are correlated with the average leaf size in the native vegetation. At a much finer level of resolution are the studies of Leisler (e.g., 1975 and Chapter 14, this volume) in which specific differences in locomotor and foot morphology in *Acrocephalus* and *Locustella* species are linked to specific structural attributes of the vegetation in which they spend their lives. A similar study (Leisler and Thaler, 1982) dealt with the two *Regulus* species and their separation on discriminant axes in three sets of morphological data: external characters, flight apparatus, and skeletal characters. Goldcrests, in relation to Firecrests, have a narrower bill width, shorter rictal bristles, longer and more rugose toes more suitable for clinging and hanging, and wings more appropriate for slower flight with shorter stops; these are adaptations for slower searching and for foraging for smaller, slower prey in coniferous rather than deciduous vegetation. These morphological differences correlate with the perch choices of Goldcrests in aviary situations, where the birds choose coniferous rather than broad-leafed vegetation, just as they do in the more northerly parts of their range. Where Goldcrests are sympatric with Firecrests, however, habitat choice shows no unequivocal difference.

At a somewhat intermediate level, Gaston (1974) reviewed various morphological attributes of 30 species of *Phylloscopus* and included comparisons with *Sylvia* species. A variety of correlations were apparent, in which the vegetation density, its deciduous versus coniferous character, and the migratory distances of the bird species could all be seen to influence various morphological traits, such as relative wing, tail, and tarsus length. Tiainen (1982) compared morphological attributes of the three *Phylloscopus* species in Finland and emphasized sexual dimorphism (which is greatest in the smallest species, *P. collybita*, and least in the largest species, *P. sibilatrix*) and the relative dissimilarity of *P. sibilatrix* from the other two species. He also found that *P. trochilus* males were significantly larger in deciduous-dominated versus spruce-dominated habitats, a result akin to that in the Pied Flycatcher (see Chapter 2, this volume), with most likely a similar explanation, the relegation of smaller and subordinate males to the poorer (coniferous) habitat.

Such morphological adaptation must act to place a limit on the range of habitats that might be occupied by each species, and this adaptation presumably renders ecological performance or efficiency as a function of vegetation structure and, as such, variable among species over habitats.

IV. BRITISH WARBLERS

Data on the habitat distributions of three *Phylloscopus* species and four *Sylvia* species were collected in Yorkshire (Cody, 1978) and analyzed together with those on Dartford Warblers from southern England in terms of discriminant functions. Using the vegetation density in each of 10 vertically stacked horizontal layers obtained from 171 territories of 8 species as variables in the analysis, species distributions are shown as contour ellipses on the ''habitat plane'' of the first two canonical variables generated by the analysis. The results are shown in Fig. 2a and b for *Phylloscopus* species and in Fig. 2c and d for *Sylvia* species.

The three *Phylloscopus* species segregate largely on the basis of the first axis, U_1, which ranks tall habitats with dense canopy to the left and low habitats with

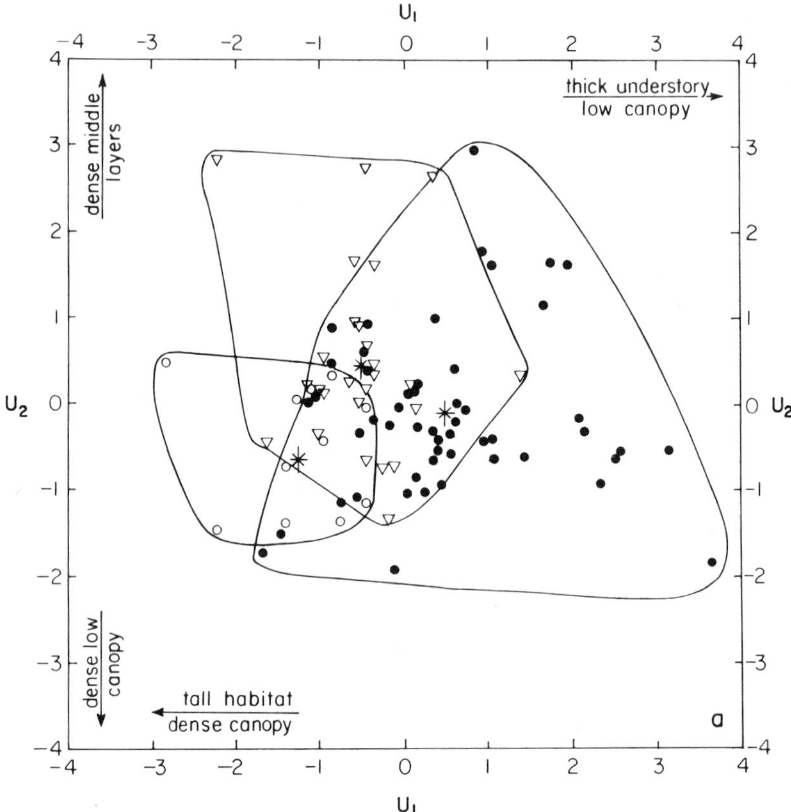

Fig. 2. (a) Ranges of three *Phylloscopus* species [*P. trochilus* (●), *P. collybita* (▽), *P. sibilatrix* (○)] in Yorkshire, England, on a discriminant function plane U_1–U_2. The first axis achieves 80% of the interspecific segregation, the second 10% more. The species' means are shown as stars. (b) The habitat plane U_1–U_2 is divided by heavy lines that separate different *Phylloscopus* species in Yorkshire, and frequency or probability ellipses are given for each species, such that, for example, 50% of *P. trochilus* territories are found in habitats represented by that within its 50% frequency ellipse. The dashed lines are 95% confidence ellipses around species' means. (c) Five species of *Sylvia* [*S. atricapilla* (●), *S. borin* (○), *S. communis* (■), *S. curruca* (▲), *S. undata* (+)] in England segregate into two groups along discriminant axis U_1, with two species in taller and three species in shorter vegetation. (d) The habitat plane is divided by heavy lines to show the most likely habitat occupant in different parts of the plane. Confidence ellipses are shown for each species as before. (*Continued on pp. 95–96.*)

thick understory to the right. This, however, leaves the mean values of habitats in Wood Warbler and Chiffchaff territories indistinct, a situation not entirely remedied by their further segregation along the second axis, U_2, on which vegetation with dense middle layers scores high and that with dense upper–

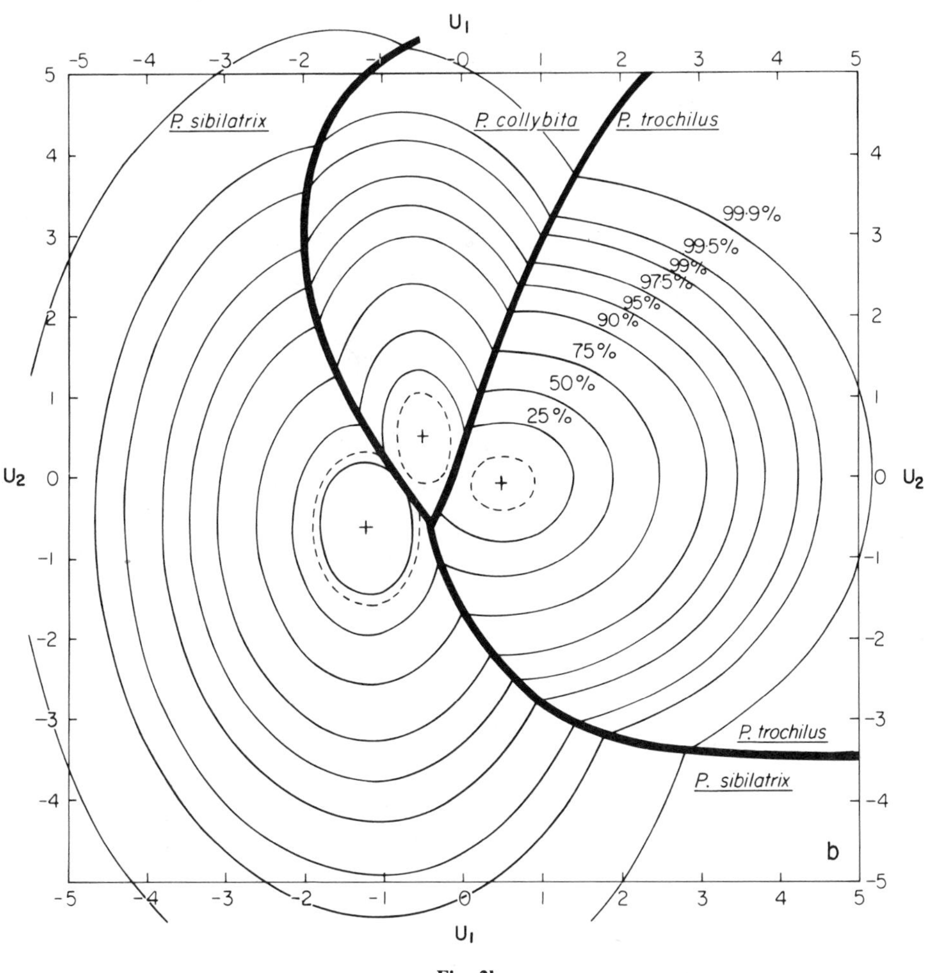

Fig. 2b

middle canopy scores low. The outcome shows a considerable habitat overlap among species, such that even mean habitat preferences, at least of Wood Warblers and Chiffchaffs, are not clearly different.

The *Sylvia* warblers (Fig. 2c,d) fall into two groups largely according to vegetation height (strongly correlated to the first axis U_1), with *S. borin* and *S. atricapilla* in taller habitats and *S. communis, S. curruca,* and *S. undata* in lower habitats. However, *S. undata* is restricted to the south coast heathlands, but the pairs *S. borin–atricapilla* and *S. communis–S. curruca* come into frequent contact, both geographically and in terms of preferred habitat. Both of these species

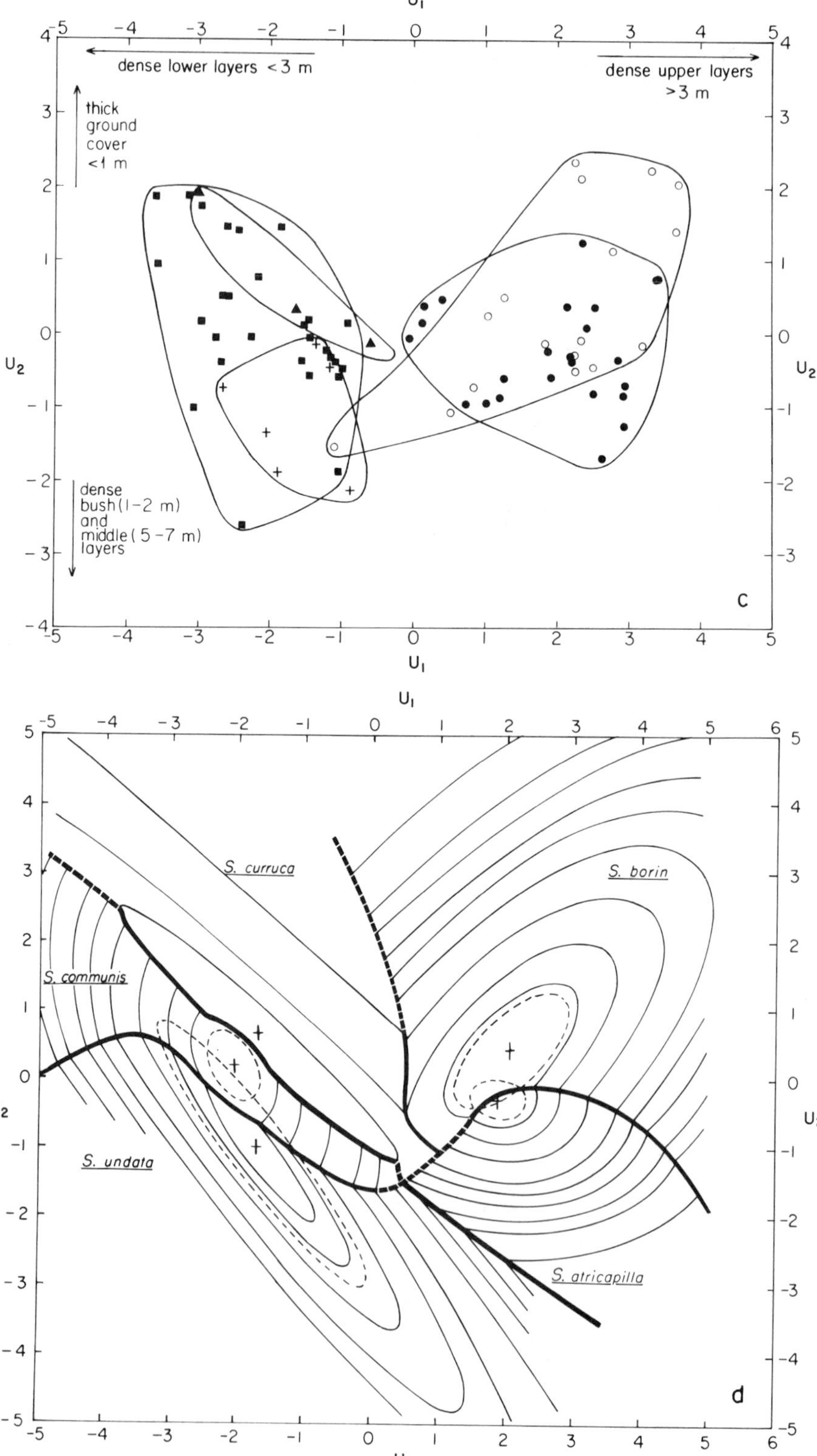

Fig. 2c and d (*Continued*)

pairs are interspecifically territorial; the other species combination that may occasionally occur is *S. borin–S. curruca,* a pair that shows an intermediate level of interspecific interactions (see Section V,D).

V. SOUTHERN SWEDEN

A. Species Composition and Discriminant Analysis

In southern Sweden, at the same latitude as Yorkshire, England, there are five breeding *Sylvia* species, the four Yorkshire species plus the Barred Warbler (*S. nisoria*). This last species has a more easterly distribution than its congenerics, but it has extended its range northwest and invaded southern Sweden in the last two decades. Its first recorded foothold in Scandinavia was on the Baltic island of Öland. Several breeding studies of the species have been published (e.g., Cambi, 1979, in northern Italy; Neuschulz, 1981, in northern Germany; see also Christie, 1975). The species has the reputation for nesting in association with the shrike *Lanius collurio* (Gotzmann, 1965).

Southern Sweden supports both *Phylloscopus trochilus* and *P. sibilatrix,* but the area lies within a somewhat anomalous range hiatus of *P. collybita,* which breeds on the continent to the south and in Sweden north of about 61° N. However, in this hiatus zone a third "upper-story" warbler is present, the Icterine Warbler (*Hippolais polyglotta*). The results of a survey of warbler territories in all habitats from low scrub to tall forest in southern Sweden were published earlier (Cody, 1978); this includes 92 *Sylvia* territories (5 species), 59 *Phylloscopus* territories (2 species), and 16 *Hippolais* territories. From vegetation data collected within these territories, a discriminant analysis was made as before, with the results shown in Fig. 3. The ranges of the 3 upper-story species are given in Fig. 3a, and means and confidence ellipses of their distributions, in Fig. 3b. The 3 species have interspecific overlaps comparable to the 3 species in Yorkshire. *Phylloscopus sibilatrix* occurs in similar habitat, but in the absence of *P. collybita* in the upper center of the habitat plane, *P. trochilus* occurs at lower U_1 and higher U_2 values, and *H. icterina,* at the lower U_2 values occupied by *P. trochilus* in Yorkshire. Thus, species numbers and the degree of interspecific overlap in habitat are similar between Yorkshire and Sweden, but the species composition differs, and the actual habitat niches on the U_1–U_2 plane are shifted in species common to both regions. Here and in Yorkshire, the predictability of species identity from vegetation measurements is about 65% (Cody, 1978).

Figures 3c and d give similar information for the five *Sylvia* species. Species segregate largely on the first discriminant axis, with *S. atricapilla* to the right in tall habitats with dense upper-canopy layers. Moving sequentially to the left are ranked *S. borin, S. curruca, S. nisoria,* and, last, *S. communis.* This ranking is

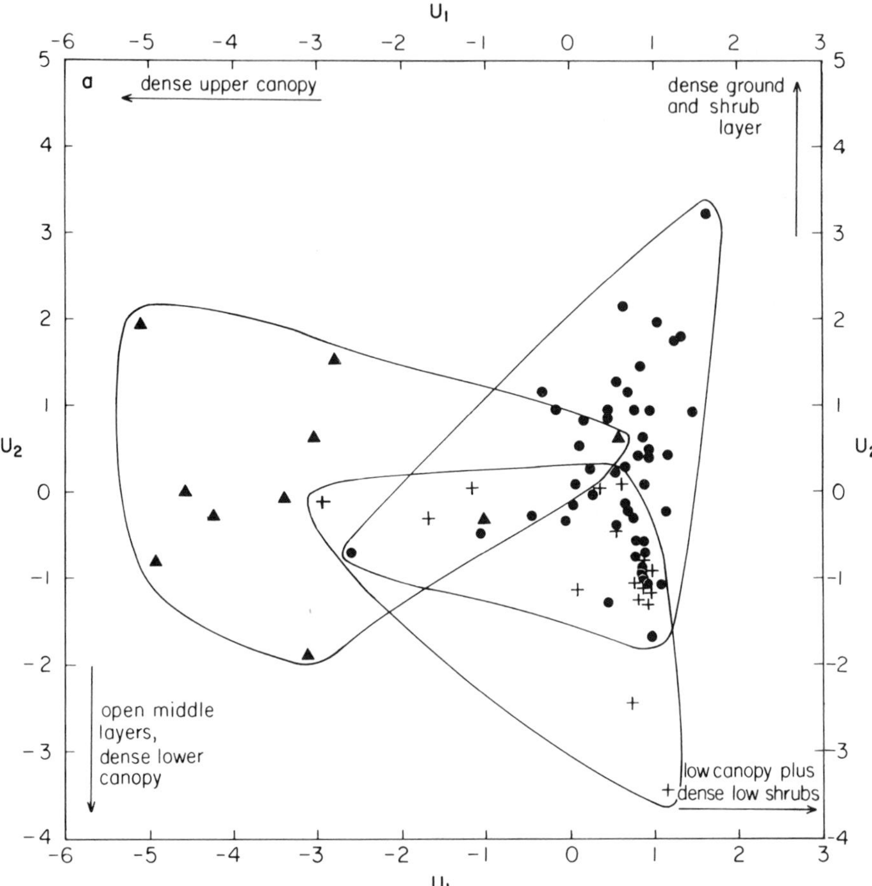

Fig. 3. (a) Two *Phylloscopus* species [*P. trochilus* (●), *P. sibilatrix* (▲)] plus *Hippolais icterina* (+) occupy taller habitats in southern Sweden, again represented here in a discriminant function plane U_1–U_2. (b) *Phylloscopus sibilatrix* (S) occurs in taller open woods to the left, and *P. trochilus* (T) and *H. icterina* (H) share lower habitats to the right. Confidence ellipses are shown as in Fig. 2. (c) Five species of *Sylvia* [*S. atricapilla* (●), *S. borin* (○), *S. curruca* (▲), *S. nisoria* (+), *S. communis* (■)] in southern Sweden represented on their habitat plane U_1–U_2. Note that all five species can occur in similar habitats, which represent 2–3 m scrub. (d) Species' separation curves and confidence ellipses for the five southern Swedish *Sylvia* species [*S. atricapilla* (A), *S. borin* (B), *S. communis* (C), *S. curruca* (L), *S. nisoria* (N)]. (*Continued on pp. 99–100.*)

very similar to that of these species in England, but habitat overlap in the species pairs *S. atricapilla*–*S. borin* and *S. curruca*–*S. communis* is greater in Yorkshire than in Sweden. The new species *S. nisoria* is situated somewhat between *S. communis,* to the left, and *S. curruca*–*S. borin,* to the right, in habitat representing lower ±2-m scrub that all four species may occupy. Interspecific overlaps

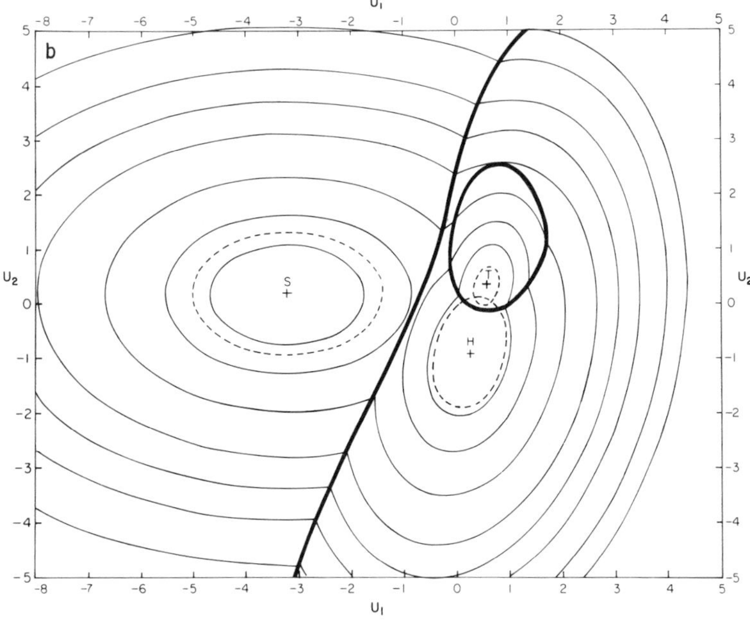

Fig. 3b

in habitat preference are large, and as in Yorkshire there is a low, 50% accuracy rate for predictions of *Sylvia* habitat occupants from vegetation measurements.

B. Interspecific Interactions

As with the English species, overlap in habitat requirements is matched by a variety of interspecific interactions that range from overt interspecific territoriality to a more subtle avoidance of one species' territories by another species, without overt and obvious behavioral interaction. Such interactions were studied at three sites in Sweden. The first two sites were in tall woodland at Höganäs, where *Sylvia atricapilla* and *S. borin* displayed interspecific territoriality, and in tall scrub plus boggy woodland near Hörby, where *S. borin* and *S. curruca* were more weakly interactive and in a nonsymmetrical fashion, with *S. borin* showing some dominance in habitat that both species preferred. The third site was at Bejershamn on Öland, which is treated in somewhat more detail next.

C. Bejershamn Warblers

At a site with vegetation ranging from low, open, grassy scrub to birch and oak woodland at Bejershamn, the territories of seven warbler species were

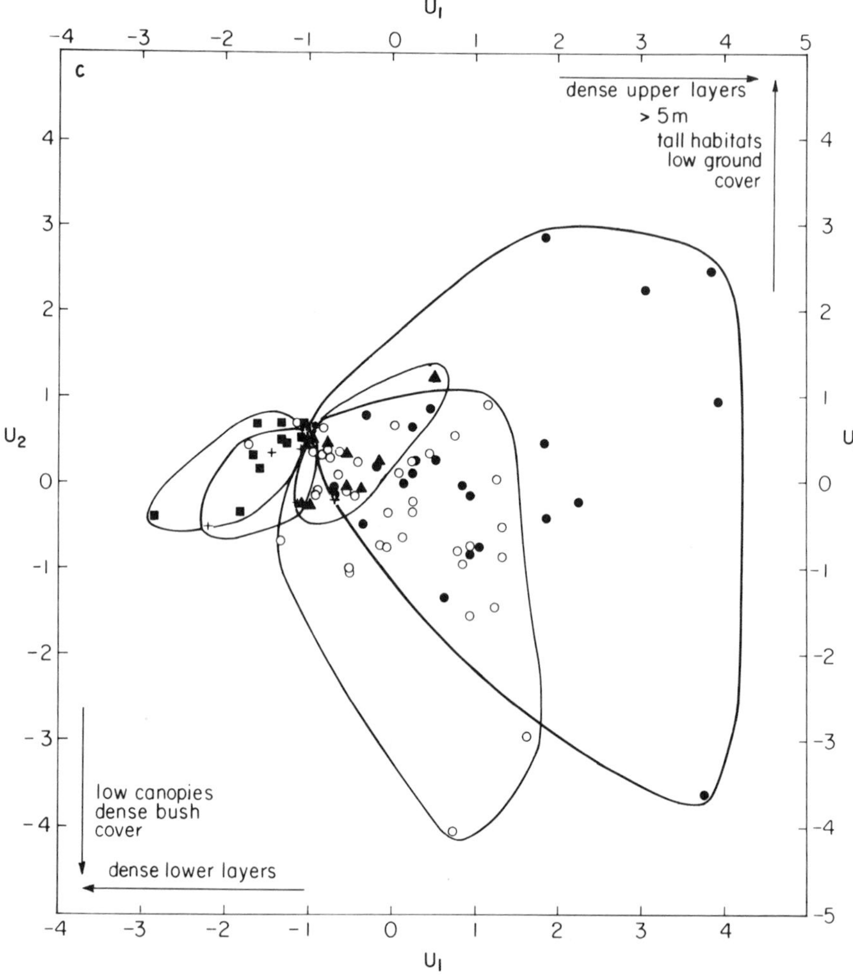

Fig. 3c (*Continued*)

mapped in 1975 (Cody, 1978). These included five pairs of *Hippolais icterina* and of *Phylloscopus trochilus,* two of *Sylvia atricapilla, S. nisoria,* and *S. communis,* five of *S. curruca,* and six of *S. borin.* Vegetation was mapped throughout the study area in contiguous 10 m × 10 m quadrats (erroneously given as 15 m × 15 m quadrats in Cody, 1978), and each quadrat was classified as a central, marginal, or remote habitat for each warbler species, based on whether it fell within the 50% probability ellipse, between the 50% and the 90% ellipses, or beyond the 90% ellipse. For each species pair, a quadrat was classi-

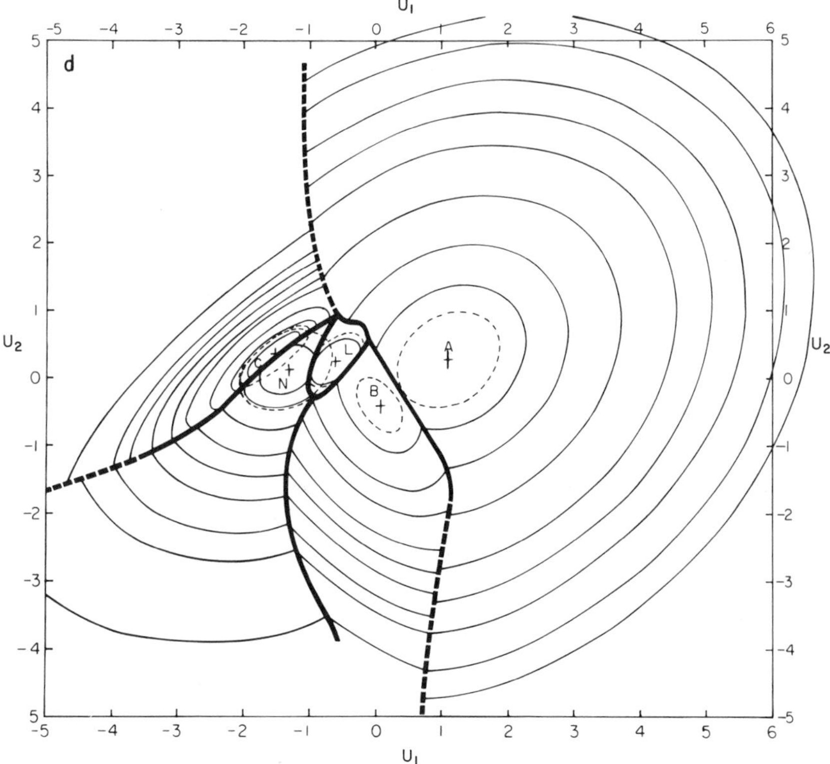

Fig. 3d *(Continued)*

fied as central, marginal, or remote for the two species independently and was occupied by both, one or the other, or neither. Thus, for each species pair the total number of quadrats, 1172, was broken down into 3 × 3 × 2 × 2 = 36 categories. In each category the number of observed quadrats was compared to the number expected by chance, if the two species set up territories independently of each other, and a level of significance assigned to any resultant differences.

The results of this analysis document strong interspecific interactions among all pairs of species in the foursome *Sylvia borin–S. nisoria–S. curruca–S. communis,* in habitats suitable to both member species of the pairwise comparison. The only exception is that *S. borin* and *S. curruca* display a weaker interaction, with some segregation of territories but also some amicable coexistence with territory overlap in certain habitats. Beyond these conclusions, there were further subtleties in the interspecific interactions, which varied in strength, symmetry, and habitat specificity among species pairs. As an illustration of the latter effect, I found that *S. borin* dominated *S. nisoria* in habitats that were both *S. borin* central and *S. nisoria* central and did not avoid *S. nisoria*-occupied habitats of this type. But in *S. borin*-marginal and *S. nisoria*-central habitats, the two interacted symmetrically, and both avoided habitats occupied by the other species.

D. Bejershamn Recensused

In the spring of 1983 I recensused the Bejershamn site, using 15 m × 15 m quadrats positioned independently of the previous census and grid system of 1975. The recensus covered a similarly sized area (11.43 ha versus 11.72 ha in 1975) positioned slightly differently (with about a 95% map overlap, to avoid recently disturbed sites around a parking lot and an observation tower). Further, the vegetation had changed somewhat in the 8-year interval, due to additional brush removal and reduced grazing. Some successional scrub and birch habitats had grown taller, and the distribution of habitats that are classified as central, marginal, and remote to different species was somewhat changed. Table I summarizes these changes and shows the proportions of habitats occupied by the *Sylvia* species.

According to my classification of the vegetation, preferred *Sylvia borin, S. atricapilla,* and *S. trochilus* habitats became more common between 1975 and 1983, and those for *S. curruca, S. communis,* and *S. nisoria,* less common. In fact *S. trochilus* and *S. borin* had increased in coverage at the site in 1983, from the proportions 0.189 and 0.142 to 0.258 and 0.189, respectively. *Sylvia*

TABLE I

Availability and Occupancy of Central, Marginal, and Remote Habitats to Five *Sylvia* Species at Bejershamn, Öland, Southern Sweden, in 1975 and 1983[a]

		Central		Marginal		Remote	
Species	Year	Ava	Occ	Ava	Occ	Ava	Occ
S. borin	1975	0.230	0.322	0.708	0.082	0.062	0.153
	1983	0.444	0.385	0.494	0.036	0.062	0.000
S. atricapilla	1975	0.002	0.500	0.020	0.125	0.978	0.039
	1983	0.248	0.119	0.679	0.000	0.073	0.000
S. curruca	1975	0.985	0.177	0.015	0.167	0.000	0.000
	1983	0.860	0.142	0.116	0.068	0.024	0.000
S. communis	1975	0.981	0.123	0.018	0.000	0.001	0.000
	1983	0.537	0.099	0.219	0.045	0.244	0.032
S. nisoria	1975	0.984	0.171	0.015	0.333	0.001	0.000
	1983	0.640	0.151	0.232	0.051	0.128	0.031

[a] Availability (Ava) is given as the proportion of the study site that falls within each habitat category, and its occupancy (Occ) by each species, as the proportion of available habitats occupied by territories.

atricapilla coverage had dropped slightly (from 0.49 to 0.34 ha), but this species is very rare at the site. The remaining species all decreased in coverage, from 0.177, 0.120, and 0.173, respectively, to 0.130, 0.071, and 0.112, respectively. These density shifts are in accordance with the changing vegetation and lend support to my classification of the vegetation as a measure of different species' preferences.

Despite the fact that species densities had changed at the site and that, in general, species occupied different areas within the site, the distribution of species in relation to the vegetation within their territories is similar to that measured previously. This is illustrated in Fig. 4, where each territory mapped in 1983 is represented by a mean and standard deviation in the U_1–U_2 plane, and the centers of territories mapped in 1975 are included for comparison. Given that a certain amount of vegetational change had taken place, species still exhibited the same vegetational preferences over the census interval.

A repeat of the analysis of interspecific interactions described previously (and in detail in Cody, 1978) yields very similar results to those of 1975. I will illustrate this similarity in interspecific interactions using just three species, *Sylvia borin, S. curruca,* and *S. nisoria*. The figures on which the 1983 analysis is based are given in the Appendix, and comparable figures from the 1975 censuses are shown in Table 15 of Cody (1978).

In 1975 at Hörby and at Bejershamn, *Sylvia borin* (*B*) and *S. curruca* (*C*) showed a weak tendency to segregate their territories in habitats suitable to both. In Bejershamn in 1983, 199 quadrats were of habitats classified as central (c) to both species. Since *S. curruca*'s occupancy of the C_c habitat was (62)/(62 + 375) = 0.142 and *S. borin*'s occupancy of the B_c habitat was (87)/(87 + 139) = 0.385, they were expected to occupy (0.142 × 199) = 28.3 and (0.385 × 199) = 76.6 quadrats in the B_cC_c, respectively. Actually they occupied just 12 and 32 quadrats, respectively. Thus, both showed a significant tendency ($\chi^2 = 35$, $p << 0.05$) to avoid the other's central habitat; however, the two species co-occupied 12 quadrats in the B_cC_c habitat and 15 quadrats in the $B_{cm}C_{cm}$ habitat, results that show no significant deviation from numbers expected by chance.

The species pair *Sylvia curruca–S. nisoria* (*N*) show no tendencies to avoid setting up territories in each other's central or marginal habitats ($p > 0.05$), but each avoids habitats actually occupied by the other. In the 312 quadrats that are C_cN_c, 5.6 are expected to be co-occupied and 72.8 occupied by just a single species; the observed figures are 0 and 84 quadrats, respectively ($\chi^2 = 7.48$, $p = 0.05$ that this mutual avoidance might happen by chance). In the 437 quadrats of $C_{cm}N_{cm}$ habitats co-occupancy is expected in 7.43 and single-species occupancy in 99.2 quadrats, respectively, numbers that differ significantly from the observed 0 and 114 quadrats ($p = 0.02$).

While interactions between *Sylvia borin–S. curruca* and *S. curruca–S. nisoria* remained the same between the 1975 and 1983 census dates, those between *S. borin–S. nisoria* did not. In 1975 *S. borin* showed no tendency to avoid B_c

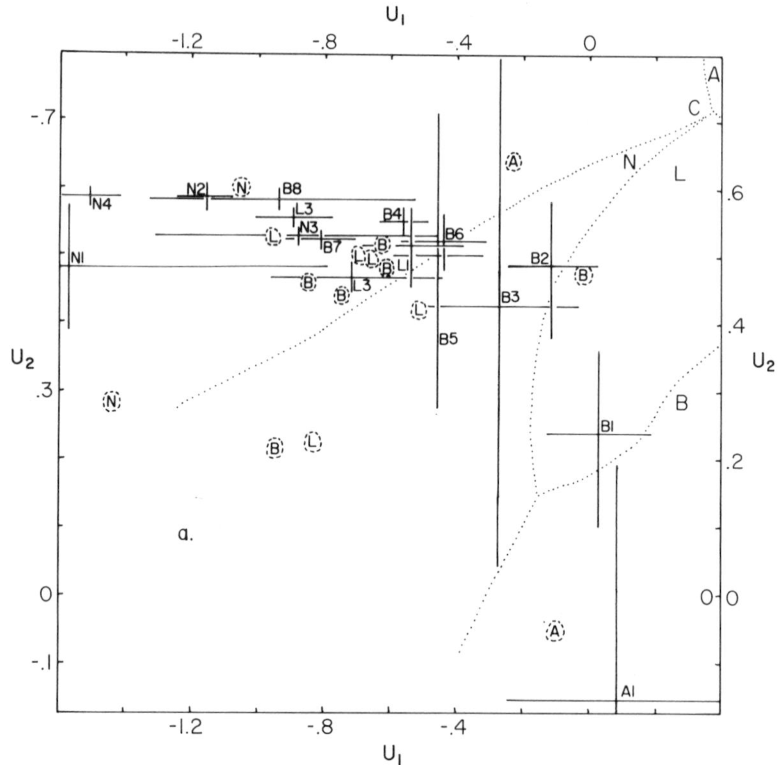

Fig. 4. (a) Vegetation characteristics in the territories of four *Sylvia* species [*S. nisoria* (N), *S. borin* (B), *S. curruca* (L), *S. atricapilla* (A)] at Bejershamn, Öland, southern Sweden. The site was censused in 1975 and in 1983. The mean U_1–U_2 values for 1975 territories are shown in dashed circles; the means and standard deviations of 1983 territories are shown as crossed lines. The vegetation and species abundances had changed somewhat between 1975 and 1983. The dotted lines and the larger letters A, C, N, L, and B refer to the species separation curves of Fig. 3d. (b) Territories of *Phylloscopus trochilus* (P), *Hippolais icterina* (H), and *Sylvia communis* (S) at Bejershamn, 1975 and 1983. Symbols have the same meaning as in Fig. 4a.

habitats whether occupied by *S. nisoria* or not, but both species avoided habitats actually occupied by the other in B_cN_m, and *S. nisoria* did avoid the B_c habitat. In 1983 these interactions had become more symmetrical, with both species avoiding habitats which were central both to themselves and to the other species, i.e., in B_cN_c. On the other hand, neither avoided equally suitable habitats in B_{cm} N_{cm} ($p > 0.05$), yet both avoided habitat co-occupancy in habitats of this type: in its 420 quadrats, co-occupancy is expected in 7.78, occupancy by one species or the other is expected in 100.4, and 311.82 quadrats are expected to be vacant. These figures are compared to the observations of 0, 116, and 304 quadrats, respectively ($\chi^2 = 10.5$, $df = 3$, $p = 0.02$). Thus, in the intervening 8 years,

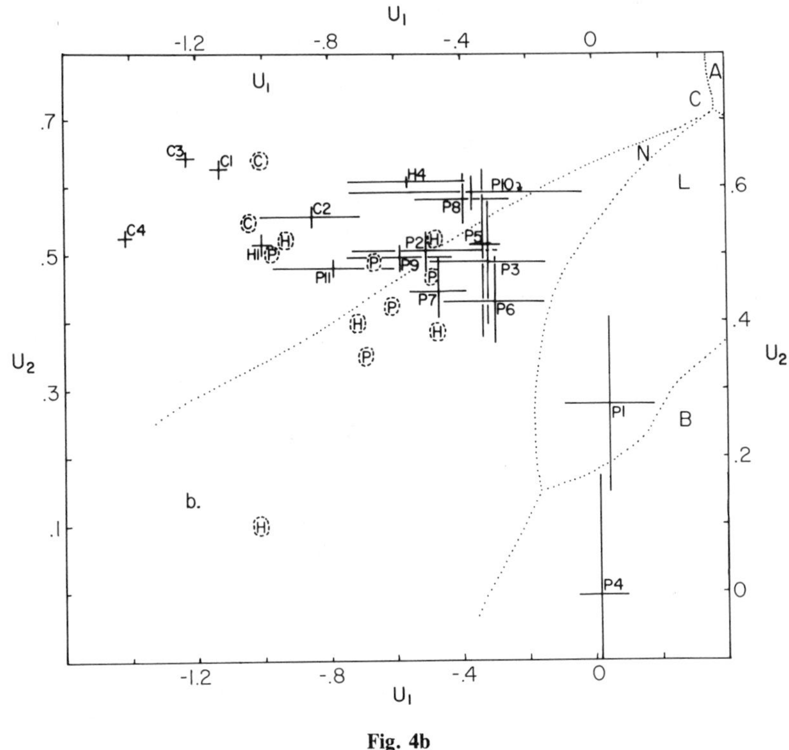

Fig. 4b

which is a large fraction of the time these 2 species have co-occurred on Öland, their interspecific relationships have shifted to a more symmetrical form and to a more mutual avoidance of the other's prime habitat. These species appear to be in the process of increased habitat subdivision and the evolution of narrower habitat preferences.

VI. NORTHERN SCANDINAVIA

A. Northeastern Finland

In the spring of 1983 I surveyed habitats in the vicinity of Kajaani, around 64° N in northeastern Finland, for warbler territories. This latitude is far enough north to be beyond the ranges of *Hippolais icterina, Sylvia atricapilla,* and *S. borin,* but the remaining three *Sylvia, S. borin, S. communis,* and *S. curruca,* as well as three *Phylloscopus* species, *P. trochilus, P. collybita,* and *P. sibilatrix,* are recorded from the area. The region is also on the western edge of the range of Greenish Warblers (*P. trochilioides*),

Habitats in the Kajaani area range from young to old conifer forest, with some broad-leafed birch and alder woods on steeper banks and streamsides. My survey showed that *Phylloscopus trochilus* was common over this range of habitats, *P. collybita* less common but predictable in certain sorts of habitat, and *Sylvia borin* scarce, found only in broad-leafed vegetation, with or without a coniferous overstory. The habitat distributions from 24 *P. trochilus,* 8 *P. collybita,* and 4 *S. borin* are shown in Fig. 5, in the habitat plane of southern Swedish upper-canopy warblers. From the distribution of these points, it appears that *Hippolais* warblers are absent because their preferred habitats, lower on the U_2 axis, are absent. The tall conifers lacking much lower- or middle-story vegetation are occupied by *P. collybita,* which is predominant in very similar habitats in England; habitats in the lower-left sector of the habitat plane favored by *P. sibilatrix* in England and southern Sweden were not recorded in my survey, nor was its usual occupant. Similarly, the lower and denser broad-leafed vegetation, in which *S. communis*

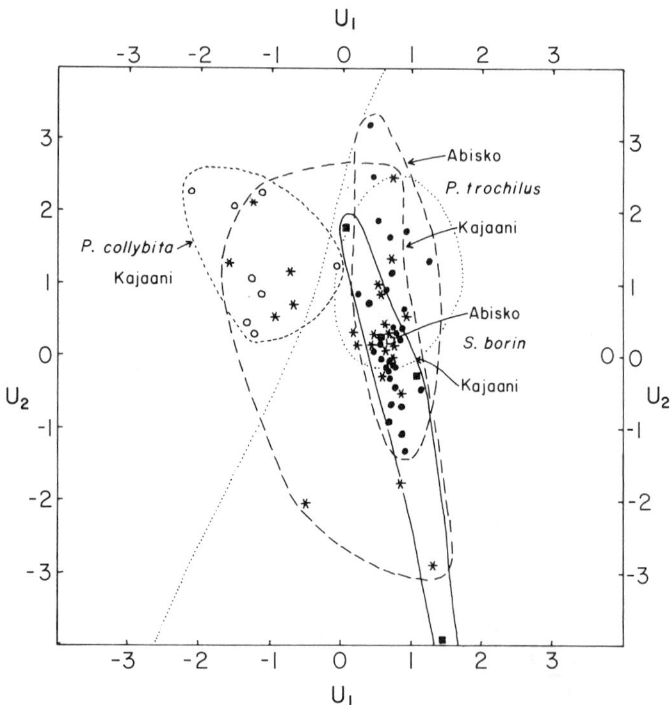

Fig. 5. The habitat distributions of warblers (*Phylloscopus collybita* [Kajaani (○)], *P. trochilus* [Abisko (●), Kajaani (∗)], *Sylvia borin* [Abisko (□), Kajaani (■)]) at two sites in northern Scandinavia, at Kajaani in northeastern Finland, and at Abisko in northern Sweden. See the text for further explanation.

and *S. curruca* are common to the south, is rare around Kajaani; these warblers must be rare also, since I encountered neither them nor *P. trochilioides,* whose occurrence in the region is apparently rather sporadic.

At a site in mixed pine–spruce–birch forest 12 km southeast of Kajaani, near Korholanmaki, I plotted the territories of five pairs of *Phylloscopus collybita* and six of *P. trochilus.* Territories of the two species were contiguous intra- and interspecifically, with no overlap between species; the interspecific territoriality that is typical of these species elsewhere in Scandinavia and in England also obtains here.

B. Northern Sweden

I conducted further work in the spring of 1983 around Lake Torneträsk at about 68° N in northern Sweden. The Arctic Warbler (*Phylloscopus borealis*), quite common in this area in the 1950s in meadow–birch woods (see, e.g., Swanberg, 1953), became increasingly scarce in the 1960s and has been seen only very sporadically since. My survey of habitats within 50 km of Abisko failed to uncover the species, which has apparently retreated north and east of this area, away from Gulf Stream influences and current warming trends.

The avifauna and vegetation of the Abisko region are well studied (see, e.g., Lundevall, 1952). The region is north of the coniferous forest and, besides tundra and lower heaths, supports meadow–birch woods, in which the ground-layer vegetation is a diverse array of herbaceous plants, and heath–birch, with a ground layer of low ericaceous shrubs. The two warblers regularly breeding here are *Phylloscopus trochilus,* which is ubiquitous, and *Sylvia borin,* which I found to be extremely localized in denser meadow–birch. The vegetation was measured in 33 *P. trochilus* territories, over the variety of habitats in which *P. trochilus* is found. The species is quite narrowly distributed here (Fig. 5) in relation to Kajaani, reflecting the narrower range of habitats around Abisko; the typical *P. collybita* habitat is absent, and the single *S. borin* territory falls in the center of the *S. borin* habitat at Kajaani.

The birch woods around Abisko support 8–10 passerine bird species; among these, *Phylloscopus trochilus* is certainly the most common species, comprising 25% (in meadow–birch) to 33% (in heath–birch) of the total bird density. A similar predominance of *P. trochilus* characterizes similar woodland to the south (Enemar and Sjöstrand, 1972). Additional species in both birch habitats are Brambling (*Fringilla montifringilla*), Redpolls (*Carduelis* spp.), Redwings (*Turdus iliacus*), Bluethroats (*Luscinia svecica*), and Fieldfares (*Turdus pilaris*), in order of relative abundance. Hooded Crows (*Corvus corone*) are widespread; Pied Flycatchers (*Ficedula hypoleuca*) and Hedge Sparrows (*Prunella modularis*) occur in meadow–birch and Tree Pipits (*Anthus trivialis*) in heath–birch.

Interspecific relations between *Phylloscopus trochilus* and Bramblings have been studied by Högstad (1975), who finds that year-to-year fluctuations in the densities of each are negatively correlated and that territories show less overlap than expected in uniform habitat. Bramblings become less common and *P. trochilus* more common on a habitat gradient from mixed pine–birch to tall birch to subalpine birch, and the foraging ecology of the two species is very similar in pure birch, especially after July, when the young are fledged (Angell-Jacobsen, 1980).

A second influence on territory selection in *Phylloscopus trochilus,* besides that of competition with Bramblings, comes from nest predation by the Hooded Crow and possible protection from such predation by location of warbler territories within those of the large, noisy, and aggressive Fieldfares. Slagsvold (1980a) has shown that egg predation increases with crow density and that egg survival is enhanced in nests within 40 m of Fieldfare nests, especially when fledgling Fieldfares are present. He has demonstrated a positive association between smaller passerines and Fieldfares, especially for Bramblings and Redwings (Slagsvold, 1980b).

To test for interspecific effects on territory site selection in *Phylloscopus trochilus,* including possible competition from Bramblings and protection from Fieldfares, I mapped territories at two sites, in meadow–brich on Nyulla and in heath–birch near Abisko Östra. Territories were mapped at a resolution of 15 m × 15 m quadrats. In meadow–birch, the territories of seven *Phylloscopus trochilus* pairs, four Brambling pairs, and parts of three Fieldfare pairs were mapped and measured in 3.8 ha, and in heath–birch those (or parts) of six *P. trochilus,* three Bramblings, and two Fieldfares, in 3.4 ha.

First, there is a possibility of within-habitat segregation of species based on variation in vegetational characteristics. Means, standard deviations, and sample sizes (of 15 m × 15 m quadrats occupied) are given in Table II for each species in the U_1–U_2 plane, from which it appears that differences in vegetation within

TABLE II

Vegetation Characteristics in the Territories of Three Passerine Species in Meadow–Birch (Nyulla) and Heath–Birch (Abisko Östra) in Northern Sweden

Location and plane	Willow Warbler	Brambling	Fieldfare
Abisko Östra			
U_1	0.676 + 0.071	0.663 + 0.032	0.669 + 0.037
U_2	−0.427 + 0.270	−0.468 + 0.170	−0.405 + 0.249
n	90	29	46
Nyulla			
U_1	0.535 + 0.212	0.457 + 0.249	0.527 + 0.225
U_2	0.819 + 0.606	0.997 + 0.748	0.886 + 0.781
n	93	60	79

TABLE III

Expected versus Observed Numbers of Quadrats Occupied by Various Species Combinations of Willow Warblers (WW), Bramblings (BR), and Fieldfares (FF) in Meadow–Birch and Heath–Birch Habitats

A. Abisko Östra

	Bird species			
	WW	BR	FF	
No. quadrats occupied	90	29	46	
Proportion of site	0.529	0.171	0.271	
	Quadrats			
	+BR	+FF	+BR,FF	−BR,FF
Quadrats +WW				
EXP	11.20	20.20	4.15	54.45
OBS	17	16	2	55
Quadrats −WW				
EXP	9.96	17.96	3.69	48.39
OBS	3	21	7	49
	$\chi^2 = 13.35$, $p = 0.07$			

B. Nyulla

	Bird species			
	WW	BR	FF	
No. quadrats occupied	93	60	79	
Proportion of site	0.447	0.289	0.380	
	Quadrats			
	+BR	+FF	+BR,FF	−BR,FF
Quadrats +WW				
EXP	16.64	25.13	10.19	41.04
OBS	6	38	7	42
Quadrats −WW				
EXP	20.58	31.08	12.60	50.74
OBS	30	17	17	51
	$\chi^2 = 26.64$, $p < 0.001$			

territories among species are minimal. In heath–birch *Phylloscopus trochilus* occupied 53% of the sample area, Brambling 17%, and Fieldfare 27%. Table III shows the expected numbers of quadrats occupied by all species combinations and compares these numbers to those actually observed. The data show only a weak deviation from random disposition of territories within the site, due largely to a coincidence of *P. trochilus* and Bramblings ($\chi^2 = 13.35$, $df = 7$; $p = 0.07$). Pairwise comparisons (rather than the full three-species comparison) likewise show no significant interspecific interactions over space.

In meadow–birch *Phylloscopus trochilus* occupies 45% of the site, Brambling 28%, and Fieldfare 38%, and analysis of expected versus observed species coincidences reveals a different result. Here *P. trochilus* avoids Bramblings, both in the full three-species analysis ($\chi^2 = 26.64$, $df = 7$; $p < 0.001$) and in pairwise analysis ($\chi^2 = 18.11$, $df = 3$, $p < 0.001$). There is a somewhat weaker but still significant tendency for *P. trochilus* territories to be located within Fieldfare territories ($\chi^2 = 7.74$, $df = 3$; $p = 0.05$). In contrast, there is no tendency for Brambling territories to coincide with Fieldfare territories, possibly because the latter are occupied preferentially by *P. trochilus,* which Bramblings avoid. This finding is in accordance with the observation by Angell-Jacobsen (1980) that Bramblings are more or less subordinate to *P. trochilus* in interspecific encounters.

With respect to why different results are obtained in meadow–birch and heath–birch, my only suggestion is that in the latter habitat, with lower bird densities and larger territories, productivity is likely lower, and the birds are therefore more constrained to locate territories similarly in the most productive areas. Such an effect might override the advantages of predator protection and the disadvantages of interspecific competition with species of similar foraging ecology and diet.

VII. SOUTHERN EUROPE: SARDINIA

A. Interspecific Distributions

In southern Europe the generally distributed *Phylloscopus* species is *P. bonelli; Sylvia curruca* and *S. borin* are absent, and the remaining *Sylvia* species, *S. atricapilla, S. undata,* and *S. communis,* are joined by *S. cantillans, S. sarda,* and *S. conspicillata.* In contrast to most of southern Europe (see Blondel, 1969, for representative censuses from southern France), *S. communis* and *S. hortensis* are absent from Sardinia, but *S. sarda* and *S. conspicillata* are common. Cody and Walter (1976) surveyed warbler distributions in the southwestern peninsula of Sardinia, south of the Orestano–Cagliari valley, where there are no breeding *Phylloscopus,* but six *Sylvia* species and Firecrests are widespread. The Firecrest was found in taller macchia, oak woodland, and pine forest (see Cody, 1983, for census results). Further comment here is restricted to the *Sylvia* species.

A discriminant analysis of *Sylvia* habitat preferences is shown in Fig. 6. *Sylvia melanocephala* has the broadest distribution over habitat, from low scrub to tall

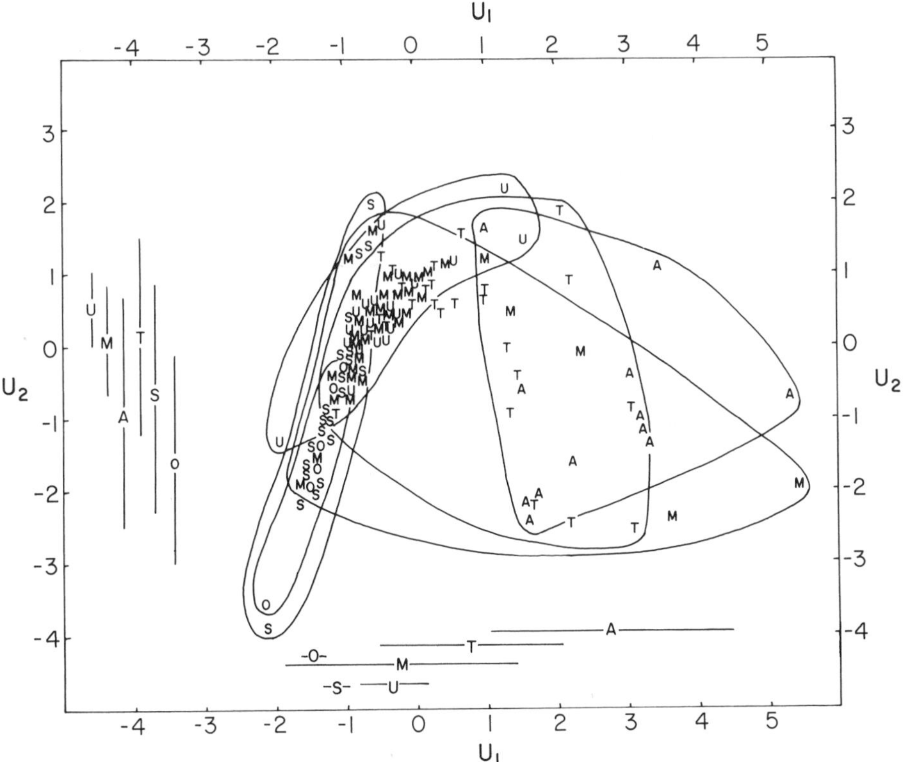

Fig. 6. Habitat distributions in the discriminant plane of six species of *Sylvia* [*S. atricapilla* (A), *S. melanocephala* (M), *S. conspicillata* (O), *S. sarda* (S), *S. cantillans* (T), *S. undata* (U)] in Sardinia. The first axis, U_1, segregates species mainly by vegetation height, the second, U_2, mainly by the density and contiguity of the vegetation. Species' means and standard deviations on each axis are included to the left and below.

woodland; *S. atricapilla* occupies rather taller habitats than does *S. cantillans,* but the two overlap broadly in tall macchia. In medium-to-tall scrub *S. melanocephala* and *S. undata* overlap greatly, as do the former and *S. sarda* in somewhat lower scrub; *S. conspicillata* overlaps little with the others, being restricted to coastal areas of garrigue—low, open, and degenerate coastal scrub.

A study of the habitat distributions of *Sylvia* warblers in Sicily by Massa (1981) shows several similarities to these results. There *S. sarda* was absent and *S. undata* too rare to document, but four of the five species studied (*S. melanocephala, S. cantillans, S. communis,* and *S. conspicillata*) reached maximum density in 1- to 2-m macchia, with the fifth, *S. atricapilla,* more restricted to deciduous oaks. The broadest habitat niches were measured in *S. melanocephala* and *S. cantillans,* and the largest habitat overlaps were between species pairs *S.*

melanocephala–S. communis (84%), *S. melanocephala–S. cantillans* (80%), and *S. communis–S. conspicillata* (76%).

B. Interspecific Interactions

Interspecific interactions among the *Sylvia* species in Sardinia were elaborated by Cody and Walter (1976) through detailed observations at three sites in mixed macchia (dense scrub) habitats: Narcao, with most vegetation around 1 m high. Bau Pressiu, with vegetation mostly 1–2 m high, and Terrubia, where vegetation was mostly 2–3 m high. All three sites were located in a single valley within 35 km of each other. At Narcao in low and rather uniform vegetation, three *Sylvia* species showed minimal differences in habitat use (four territories of *S. melanocephala,* three of *S. undata,* and three of *S. sarda*). Figure 7a shows habitat availability and species distributions over vegetation height. There was some segregation of species by foraging height, especially between the higher feeding *S. undata* and the lower *S. sarda; S. melanocephala* was intermediate (see Cody

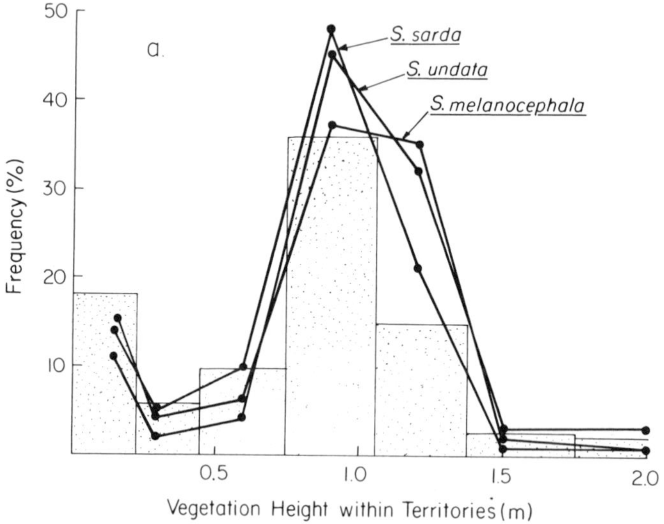

Fig. 7. (a) In low scrub at Narcao, Sardinia, three *Sylvia* species are present, with similar use of habitat. The availability of vegetation of different heights at the site is represented by the stippled histogram. (b) At Bau Pressiu, Sardinia, four *Sylvia* species held territories. Species-specific habitat preferences are shown, along with the vegetation availability (histogram). (c) At Terrubia, Sardinia, in taller shrub, four *Sylvia* species are present: A = *S. atricapilla* (●), T = *S. cantillans* (○); U = *S. undata* (■); and M = *S. melanocephala* (□). The first two species have very similar habitat preferences, matched by the site-specific vegetation (histogram).

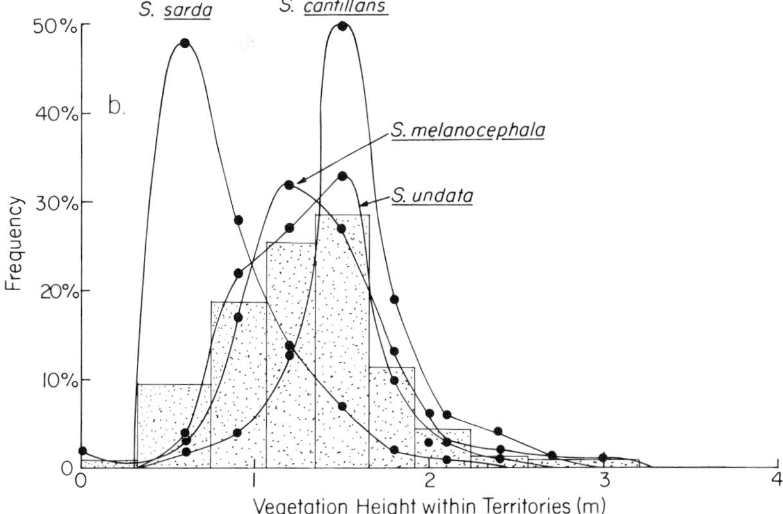
S. sarda
S. cantillans
50%
40%
30%
20%
10%
0
b.
S. melanocephala
S. undata
Frequency
0
1
2
3
4
Vegetation Height within Territories (m)

Fig. 7b

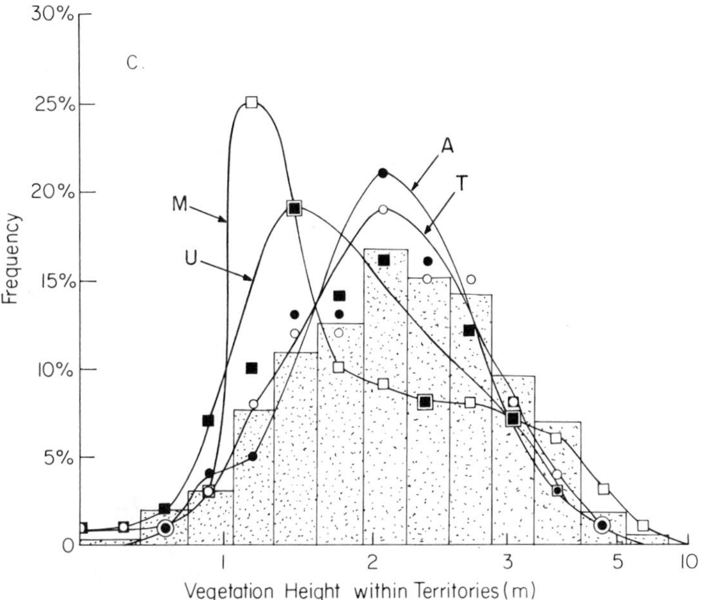
30%
25%
20%
15%
10%
5%
0
c.
A
T
M
U
Frequency
1
2
3
5
10
Vegetation Height within Territories (m)

Fig. 7c

and Walter, 1976, for these data). Here habitat occupancy was high in the aggressively dominant *S. melanocephala,* which occupied 94% of the habitat within its preference limits, but much lower in *S. undata* (64%) and *S. sarda* (48%), which were excluded from a good deal of their preferred habitat by interspecific aggression from *S. melanocephala*. In addition, *S. undata* managed to exclude *S. melanocephala* from habitats suitable to the latter but occupied by *S. undata* only; the interaction between *S. melanoceophala* and *S. sarda* was less symmetric, with no observed exclusion by the latter of the former in habitats preferred by both. A weaker interaction between *S. undata* and *S. sarda* was dominated by *S. undata*.

In taller vegetation at Bau Pressiu these three species were joined by *Sylvia cantillans,* where considerable habitat segregation characterized *S. sarda* and *S. cantillans,* but *S. undata* and *S. melanocephala* overlapped extensively (Fig. 7b). Here the interaction between *S. undata* and *S. melanocephala* is more symmetrical, with overall occupancy of suitable habitats at 75% and 78%, respectively; both species avoided otherwise suitable habitats where the other held territory. A similar and symmetrical interaction characterized *S. undata–S. sarda* and *S. undata–S. cantillans,* but in habitats equally suitable for each, *S. melanocephala* again exerted dominance over both *S. sarda* and *S. cantillans*. Foraging-height distributions paralleled habitat use as shown in Fig. 7b, i.e., *S. cantillans* foraged higher in the vegetation, *S. sarda* lower, with least overlap between *S. sarda–S. cantillans* and most between *S. melanocephala–S. undata*. However, *S. melanocephala* maintained a very broad foraging-height distribution in relation to the other species.

At the third site, Terrubia, *Sylvia sarda* was absent and *S. cantillans* present (Fig. 7c); overlap in habitat use was high among all species pairs, but especially between *S. undata–S. melanocephala* and *S. cantillans–S. atricapilla*. Again, interspecific interactions were common, with a tendency for *S. undata* to be excluded from occupiable territory by both *S. cantillans* and *S. atricapilla*. Especially strong interactions were observed between these last two, interactions dominated somewhat by *S. cantillans*. On the other hand, while *S. undata* and *S. melanocephala* territories overlapped extensively in their co-preferred habitat, we noticed no aggressive interactions between them here.

The overall pattern in these warblers is of species which, despite some general differences in habitat preferences, show broad habitat co-occupancy and flexibility in foraging-height distributions, especially in *S. melanocephala*. The warblers show a great variety of interactions among species, from complete to partial interspecific territoriality and from symmetrical to nonsymmetrical reciprocal exclusion from otherwise suitable habitats. Moreover, the interactions appear to be habitat specific, and species which interact in shorter or less productive habitats coexist peacefully in taller or more productive habitats. The stability of the system may be maintained by each species having a preferred habitat either outside the range preferred by others or at least in which it has the advantage of dominance.

VIII. NORTH AFRICA: THE HIGH ATLAS OF MOROCCO

A. Species Distributions over Habitats

Warbler distributions in Morocco were measured over a similar habitat range to those in Sardinia (including the pine forests of Corsica), in the spring of 1978 (Cody, 1980a,b, in part) and 1980 (Cody, 1984a,b). In contrast to Sardinia, *Phylloscopus bonelli* and *Hippolais polyglottus* join *Regulus ignicapillus* as upper-canopy warblers, and *Hippolais pallida* is present locally. In *Sylvia, S. sarda* is absent and *S. conspicillata* is coastal, but in addition to *S. atricapilla, S. cantillans, S. undata,* and *S. melanocephala,* four other species were encountered: *S. deserticola, S. communis, S. nana,* and *S. hortensis,* although the last mentioned two are rare.

Warblers were censused at 22 sites in the High Atlas and 6 in the Middle Atlas. These sites covered a habitat gradient from Mediterranean-climate succulent scrub and low maquis through taller maquis, oak woodland, and pine and cedar forests to desert scrub in the Anti-Atlas and the Saharan southern foothills of the mountains. *Sylvia* species were found throughout this habitat range: *Hippolais polyglotta* occurred in 7 of 28 sites, in mixed maquis, *Juniperus*-dominated maquis, *Quercus ilex*-dominated woodland to *Acacia*-dominated desert scrub; *Phylloscopus bonelli* occurred in 11 of 28 sites, including all but one (lower maquis) of the preceding, plus pines, *Quercus suber* woodland, and high-elevation cedar forest; *Regulus ignicapillus* was distributed over 11 of 28 sites very similarly to *P. bonelli;* and *H. pallida* was found only in the desert scrub.

A discriminant analysis of the eight *Sylvia* species is given in Fig. 8a. From taller to lower habitats on U_1 are ranked *S. atricapilla, S. cantillans,* and *S. melanocephala,* with increasing habitat overlap; *S. hortensis* was found at a single site only, an open juniper woodland on dry southern slopes, at which no other warblers occurred. The species of the series *S. nana–S. communis–S. deserticola–S. undata* all occur in low-scrub habitats, but the vegetation is most open and discontinuous in *S. nana* territory (one observation) and becomes increasingly dense and continuous up to *S. undata* habitat; this segregation occurs along the second axis U_2. A further perspective on these distributions is obtained by plotting the Moroccan sylviines on the discriminant plane of the Sardinian species. Figure 8b shows the result, a segregation of *S. atricapilla* from other species but almost complete overlaps among the remaining species; it appears that the wider range and variety of low, drier, and more open scrub

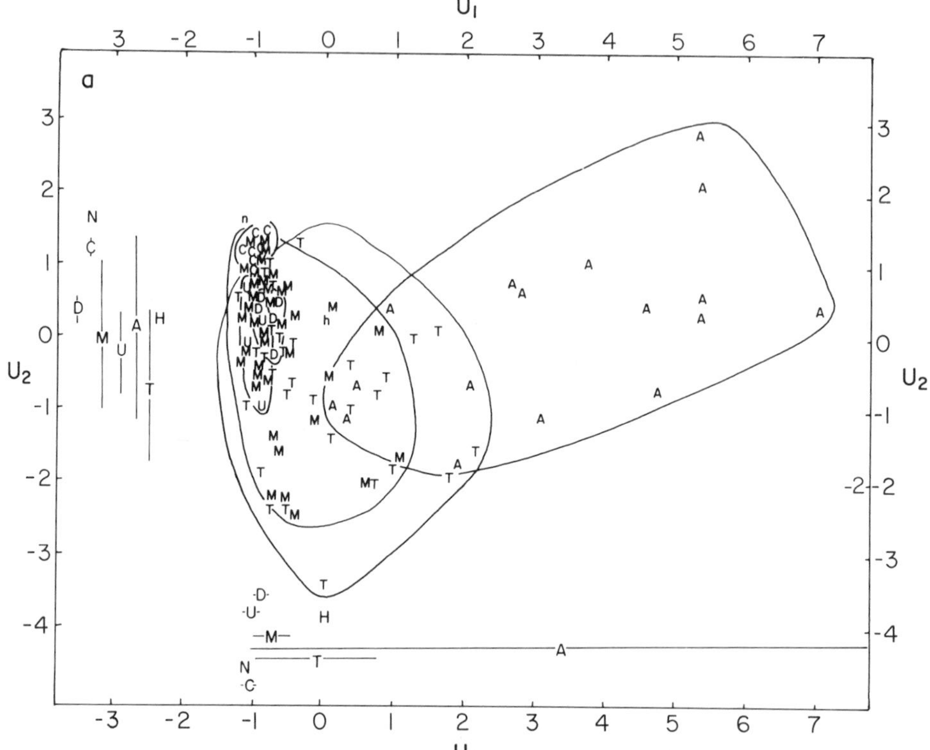

Fig. 8. (a) Habitat distributions of eight species of *Sylvia* warblers in Morocco [*S. atricapilla* (A), *S. communis* (C), *S. deserticoia* (D), *S. hortensis* (H), *S. melanocephala* (M), *S. nana* (N), *S. cantillans* (T), *S. undata* (U)]. Two species, *S. hortensis* and *S. nana*, are represented by measurements from a single territory each. The mean habitats U_1–U_2 of species *S. undata, S. deserticola*, and *S. communis* in the crowded area of the plane are in fact significantly different. The co-occupancy of habitats by warbler species is indicated by the bars to the left and below, for Fig. 8a and b. (b) When the Moroccan *Sylvia* species are plotted in the discriminant plane of the Sardinian species, there is considerably more interspecific overlap in preferred habitats. The habitat range is broader and species' habitat preferences are narrower in Morocco. The species are represented by the same letters as in Fig. 8a. (c) In taller habitats in Morocco, two species of *Hippolais, H. polyglotta* and *H. pallida*, plus *Phylloscopus bonelli* and *Regulus ignicapillus* are found, with interspecific overlaps as shown. The co-occupancy of these taller habitats by *Sylvia* species is indicated by the bars to the left and above. (*Continued on pp. 117–118.*)

habitats in Morocco promote an interspecific segregation by habitat that would not be possible in Sardinia.

B. Lack of Interspecific Interactions

Detailed observations on habitat use, foraging heights, and territory disposition were made at two sites, in mixed juniper-dominated maquis in the Ourika

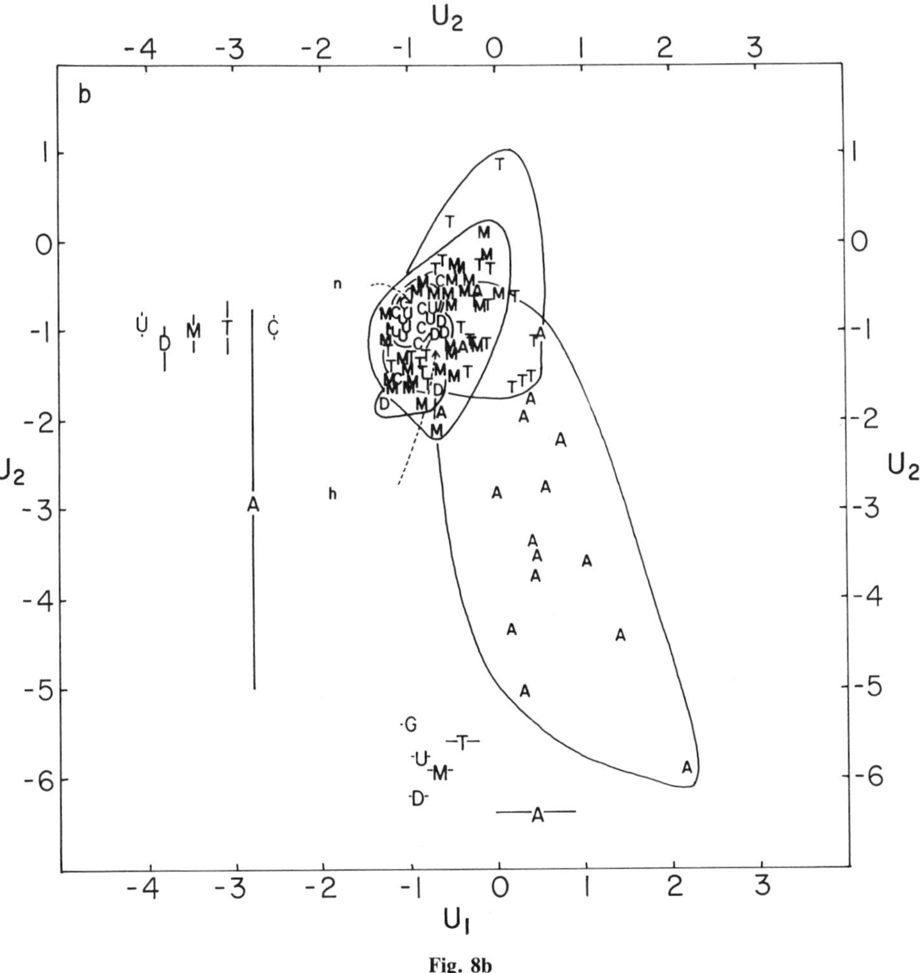

Fig. 8b

valley south of Marrakesh and in an oak–juniper maquis near Assaka northeast of Agadir. The first site supported *Phylloscopus bonelli, Hippolais polyglottus,* and the three *Sylvia* species *S. undata–S. melanocephala–S. cantillans;* the second site supported *S. melanocephala* and *S. deserticola* and the Bush Warbler *Scotocerca inquieta.* At both sites there was considerable subdivision of habitats such that different species and species combinations occurred in vegetation with different profiles (Cody, 1980a). At Assaka *S. melanocephala* and *S. deserticola* differed substantially in foraging height and overlapped in territories without any interspecific interaction; at Ourika various two- and three-species combinations occurred in different vegetation, but nowhere were direct interactions between

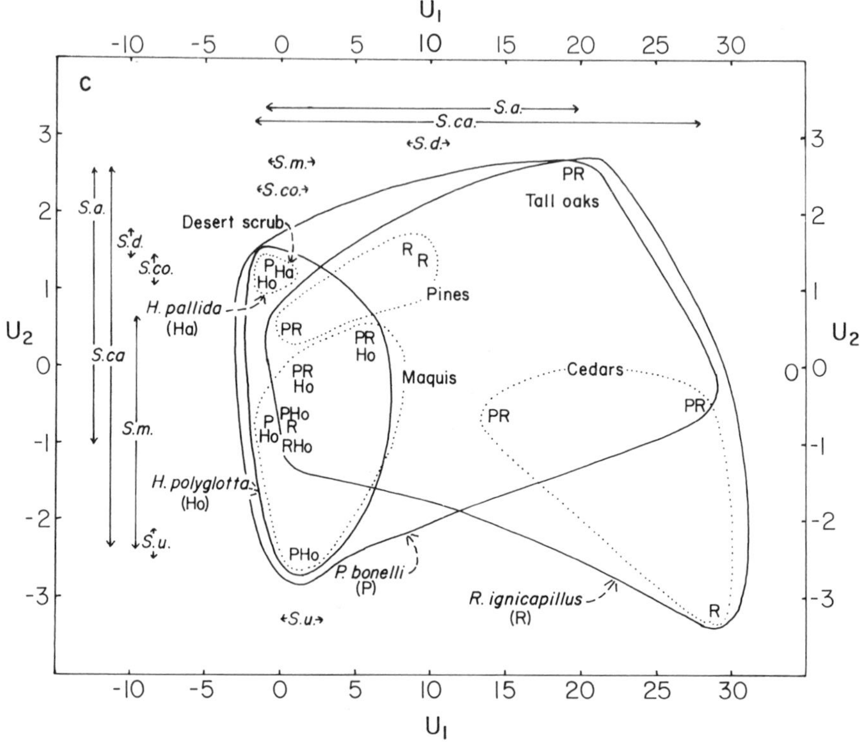

Fig. 8c (*Continued*)

species observed. Here interspecific subdivision of habitats, both among sites and within sites, and interspecific differences in foraging-height distribution apparently serve to segregate species ecologically without recourse to direct interaction; playbacks of different species' songs in the territories of others invariably elicited no response, and songs appear to the ear more distinct species-specifically than they are in Sardinia for example. In summary, by dint of more precise habitat segregation, both horizontally and vertically, *Sylvia* species seem to coexist amicably in Morocco, without direct interaction. This situation may owe a good deal to the longer time span in which species and habitats have remained relatively undisturbed throughout periods of glacial ebb and flow further north. The relation between wider habitat preferences and direct interspecific interactions in the north and narrower habitat use without direct interspecific interaction in the south corresponds to the longer time periods over which species have been in contact in North Africa (Cody, 1980a).

The only interspecific interaction over territory was noted in desert scrub, between the two *Hippolais* species, a segregation that appeared at the same time to polarize *Sylvia communis* with *Hippolais pallida* in sparser habitat and *S. cantillans* with *H. polyglotta* in denser habitat; *Phylloscopus bonelli* ranged throughout this site. The distribution of the upper-canopy warblers is given in Fig. 8c, which shows that up to six species of four genera are accommodated in the maquis vegetation in the Center of the habitat plane.

IX. CANARY ISLANDS

A. Bird Species, Habitats, and Distributions

Field work was conducted on four of the Canary Islands, Tenerife, Gran Canaria, La Palma, and Hierro, in the spring of 1980 (Cody, 1984a). Here there are no *Hippolais,* but *Phylloscopus collybita* is extremely common, and a *Regulus* species, patterned like *R. regulus* but reclassified most recently as *R. ignicapillus,* occurs. There are three *Sylvia* species present. *S. atricapilla, S. melanocephala,* and *S. conspicillata.*

On each of these four islands an effort was made to census structurally similar habitats of five types: *Euphorbia*-dominated scrub, *Cistus*-dominated scrub, macchia in which *Erica* is common, laurel forest with *Laurus, Ocotea, Apollonicus,* and *Prunus,* and *Pinus* woodland. Habitats were closely duplicated on all islands, except that no *Cistus* was censused on La Palma and the laurel forest on Gran Canaria is a very relict and sadly degraded couple of hectares. Besides these habitats, warblers were censused in two other habitats, in the *Spartocytissus* subalpine scrub on El Teide, the volcano on Tenerife, and the *Adenocarpus* scrub on Gran Canaria. These habits are shown in Fig. 9a in an elevation–habitat structure plane.

Like most oceanic islands, the Canary Islands are species poor, being occupied by a mix of endemic species, such as the chaffinch (*Fringilla teydea*), the chat (*Saxicola dacotiae*), the Canary (*Serinus canaria*), the pigeons (*Columba trocaz* and *C. junoniae*), and the pipit (*Anthus berthelotii*); North African-derived species, such as the Trumpeter Finch (*Rhodopechys githaginea*), the Lesser Short-toed Lark (*Calandrella rufescens*), the courser (*Cursorius cursor*), the partridge (*Alectoris barbara*), and the Houbara Bustard (*Chlamydotis undulata*); typical Mediterraean-region species, such as *Sylvia conspicillata* and *S. melanocephala,* the sparrows (*Passer hispaniolensis* and *Petronia petronia*), the hoopoe (*Upupa epops*), and the swift (*Apus pallidus*); and relict north-temperate forms, such as the warblers (*P. collybita* and *S. atricapilla*), the Woodcock (*Scolopax rusticola*), and the Common Buzzard (*Buteo buteo*). Further, some widely distributed forms have well-developed subspecies on the Canary Islands, such as the Blackbird (*Turdus merula*), the Robin (*Erithacus rubecula*), and the Blue Tit (*Parus caeruleus*).

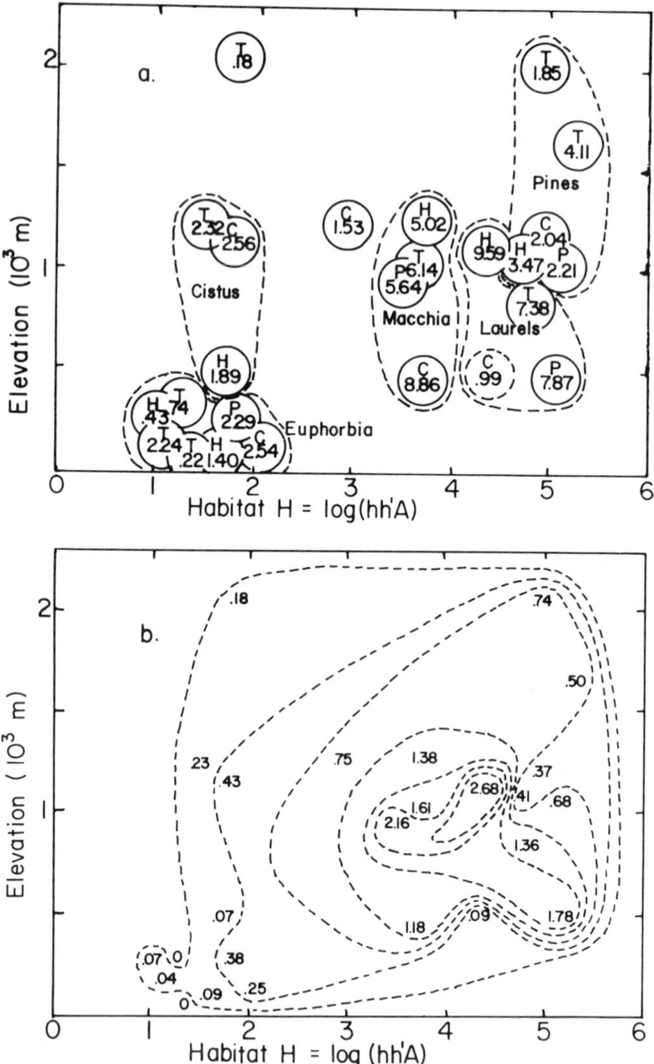

Fig. 9. (a) On the plane of elevation (ordinate) versus habitat H (abscissa), a measurement of vegetation structure that increases with vegetation height and density (see Fig. 1, Chapter 1, this volume). Different census sites on the Canary Islands are shown. Sites on different islands are: T = Tenerife; C = Gran Canaria; H = Hierro; P = La Palma. The five major habitat types, repeated on each island, are *Euphorbia* scrub, *Cistus* scrub, macchia, laurel forest, and pine forest. The numbers at each site represent the total bird density there, in pairs per acre, for Fig. 9a–f. (b) The Chiffchaff *Phylloscopus collybita* is one of the most common and widely distributed birds on the Canary Islands and occurs from sea level to over 2000 m in all but the most sparse habitats and highest densities in laurels and macchia. (c) The firecrest *Regulus "regulus"* is found in macchia, pine, and laurel forest. *(d–f)* Three *Sylvia* species occur on the islands, with *S. conspicillata* most common in the lowest and most open habitats (d), *S. melanocephala* with an intermediate distribution (e), and *S. atricapilla* most common in the taller and denser habitats (f). (*Continued on pp. 121–122.*)

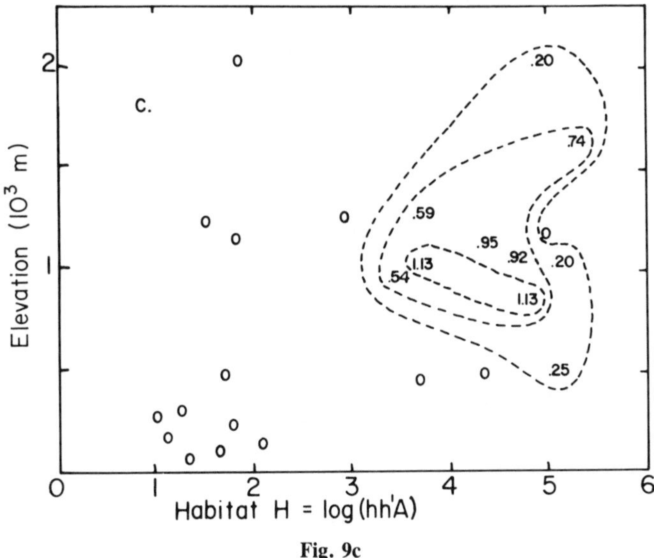
c.
Elevation (10³ m)
Habitat H = log(hh'A)
.20
.74
.59
.95
.92
.20
1.13
.54
1.13
.25
0
1
2
3
4
5
6

Fig. 9c

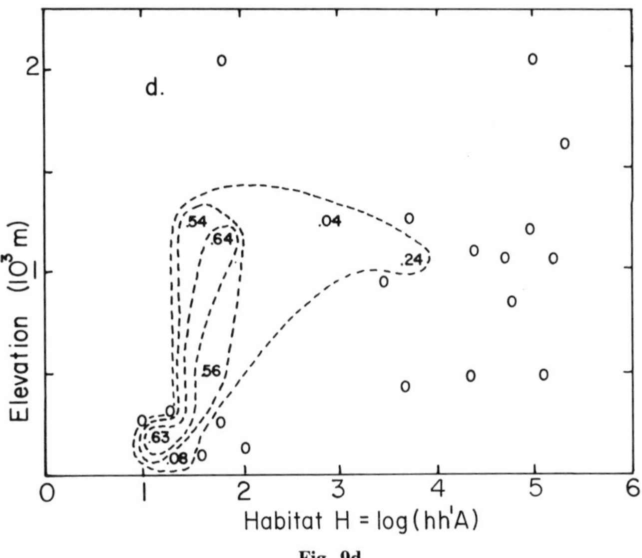
d.
Elevation (10³ m)
Habitat H = log(hh'A)
.54
.64
.04
.24
.56
.63
.08
0
1
2
3
4
5
6

Fig. 9d

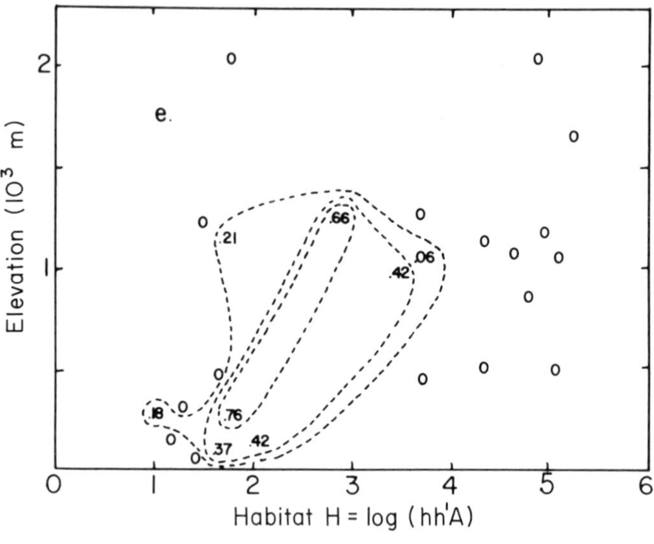

Fig. 9e *(Continued)*

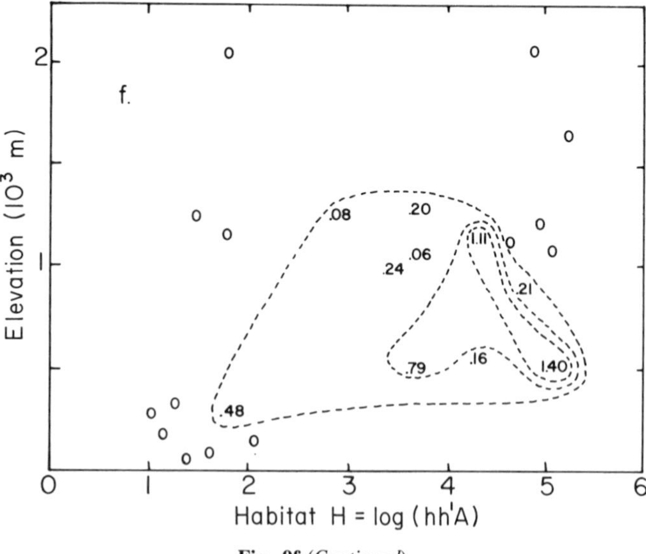

Fig. 9f *(Continued)*

Species distributions vary among islands in the Canaries. Examples are the absence of the Firecrest from Gran Canaria (although a specimen from there is so labeled in the British Museum), the occurrence of the Blue Chaffinch only on Tenerife and Gran Canaria, and the restricted distributions on the islands of the Chough (*Pyrrhocorax pyrrhocorax*), the woodpecker (*Dendrocopus major*), and the pigeons, among others. Further, some warblers differ among islands, and I found *Sylvia melanocephala* rarer, *S. conspicillata* more broadly distributed, and *S. atricapilla* less common on Tenerife than elsewhere.

B. Bird Densities and Density Compensation

Habitats on the Canary Islands have in general 50–75% of the species numbers of comparable Moroccan habitats (Cody, 1984a,b). In Table IV, three habitat types that are well matched between the Canary Islands and Morocco are compared in species numbers and total bird density. Although Morocco species numbers exceed those in the Canary Islands in 24 of 25 cases (expected by chance: 12.5 of 25, $p << 0.05$), total bird densities in Morocco exceed those in Canary Islands habitats in 15 of 25 cases (expected by chance: 12.5 of 25, $p >> 0.05$). Thus, there is an overall density compensation in the Canary Islands for the general reduction in bird species. Among the warblers, the Chiffchaff (*P. collybita*) shows the most spectacular competitive release, being common in all habitats from *Euphorbia* desert scrub to laurel and pine forest, from sea level to over 2000 m elevation (Fig. 9b). Firecrests are found from tall macchia to laurel and pine forests (Fig. 9c), and *Sylvia* species display complementary habitat ranges, with *S. conspicillata* occupying very low scrub to *S. melanocephala* in low scrub and macchia and *S. atricapilla* in macchia and laurel forest (Fig. 9d–f). Interspecific effects on habitat utilization is suggested from (1) the presence of *S. atricapilla* in *Euphorbia* scrub sites only on La Palma, (2) a wider habitat range and the highest-measured densities for *S. melanocephala* on La Palma, where the closest competitor *S. conspicillata* was found only on the southwest peninsula, and (3) the scarcity of *S. atricapilla* in macchia and laurels on Tenerife, where Firecrests are abundant.

However, interspecific influences among these warblers in general must be hardly significant, as illustrated by the following test. Chiffchaffs are numerically dominant in most habitats, and their density increases along the habitat abscissa H in Fig. 9a from low scrub to macchia and laurel forest and then falls off in the pines: Density $= 2.17H - 0.30H^2 - 2.40$ pairs per acre; $R^2 = 62\%$. The combined density of *Sylvia* species also follows this trend (but without statistical significance, $R^2 = 12\%$): Density $= 0.39H - 0.70H^2 - 0.07$; but the residuals from the first regression are positively, not negatively, correlated with

TABLE IV

Comparison of Species Numbers and Total Bird Densities[a] between the Canary Islands and Morocco in Three Habitat Types

Locality	Pine forest		Macchia		Euphorbia scrub	
	Species	Density	Species	Density	Species	Density
Canary Islands						
Tenerife	10	4.11	15	6.14	9	2.24
Tenerife	9	1.85				
Gran Canaria	13	2.04	13	8.86	12	2.54
La Palma	11	2.21	13	5.64	7	2.19
Hierro	8	3.47	10	5.02	5	1.30
Means	10.2	2.74	12.8	6.42	8.3	2.07
Number > Morocco mean	0/5	0/5	0/4	4/4	0/4	0/4
Morocco						
Immouzer	12	3.31				
Tifni	9	4.90				
Tounfliht	14	6.26				
Assaka			18	2.66		
Beni Mellal T			22	5.44		
Beni Mellal 1D			22	6.78		
Beni Mellal 1U			15	5.98		
Ourika			26	4.48		
O-Oukaimedan			22	1.59		
Cap Brise					11	1.61
Tabouga					13	3.92
Imi'n'ifri					17	3.22
Means	14.7	4.82	20.8	3.95	13.7	2.92
Number > Canary means	2/3	3/3	6/6	1/6	3/3	2/3
In total species numbers, Morocco>Canaries 24/25						
In total bird densities, Morocco>Canaries 15/25						

[a]Densities are breeding pairs per acre.

the residuals of the second regression ($r = 0.783$, $n = 21$; $p < 0.05$) and also with Firecrest abundance ($r = 0.25$, $p > 0.05$). Thus, Chiffchaffs, Firecrests, and *Sylvia* warblers are all particularly common or particularly rare in certain habitats, presumably following local variations in productivity and showing no tendency among islands or habitats for density compensation. The main effects of low species numbers in the islands are in the broad habitat range of the Chiffchaff and in the broader but still complementary ranges of the three *Sylvia* species. Given such habitat expansion, are warbler density and habitat use directly affected by other species on the islands? Although some local but minor shifts in warbler density and distribution may be attributable to density variations in other species, such direct interspecific effects must play but a minor role on the islands, unlike their prime importance in the species-rich continental situations previously discussed.

X. CONCLUSION

In conclusion, it appears that there are two circumstances in which sylviine habitat preferences are distinct and in which there are no direct interspecific interactions: (1) on species-poor islands, such as the Canaries, where there are half as many warbler species over a similar habitat range as on Sardinia and on the mainland, and (2) in Morocco, where the relevant circumstance appears to be the age of the bird communities. Habitats and their occupants were presumably far less affected by Pleistocene glaciations south of the Mediterranean Sea than to the north of it, and, thus, the warbler communities in Morocco may represent older and more extensively coevolved species groups. Warbler associations must be progressively younger to the north and lack the extra thousands of years' selection for coadaptation and evolution of distinct habitat preferences that characterize the Moroccan birds (Cody, 1980a). If the latitudinal gradient parallels community age, then it appears that precise habitat selection is a feature of warbler species in long-extant and stable associations. Natural selection appears to favor a plasticity in habitat use in conjunction with direct interspecific interactions at higher latitudes where ranges are dynamic, new species associations are continually being generated, and glacial influences are recent or, indeed, still felt.

APPENDIX

Availability and occupancy of habitats at Bejershamn, Öland, southern Sweden, in 1983 for the three *Sylvia* species *S. borin* (*B*), *S. curruca* (*C*), and *S. nisoria* (*N*) is shown in the tabulation following. Subscripts "O" as in C_O refer to habitat actually occupied by species *C*, and subscripts "X" as in C_X refer to habitat not occupied by species *C*. Habitats are independently classified for each species as central (*c*), marginal (*m*), or remote (*r*), depending on whether the quadrat lies within the 50% confidence ellipse, between the 50 and 90% confidence ellipses, or beyond the 90% confidence ellipse, respectively for each species.

	S. curruca						*S. nisoria*					
	$C_O = 66$			$C_X = 442$			$N_O = 57$			$N_X = 451$		
	c	m	r	c	m	r	c	m	r	c	m	r
Species	62	4	0	375	55	12	49	6	2	276	112	63
S. borin												
B_O=96;												
c=87,	12	0	0	61	14	0	0	0	0	17	43	27
m=9,	2	1	0	4	2	0	0	0	0	6	2	1
r=0	0	0	0	0	0	0	0	0	0	0	0	0
B_X=412;												
c=139,	32	2	0	94	11	0	5	1	0	56	52	25
m=242,	16	1	0	214	11	0	41	1	0	192	4	4
r=31	0	0	0	2	17	12	3	4	2	5	11	6
S. curruca												
C_O=66;												
c=62,							0	0	0	38	21	3
m=4,							0	0	0	0	0	4
r=0							0	0	0	0	0	0
C_X=442;												
c=375,							46	2	0	228	78	20
m=55,							3	4	0	9	8	31
r=12							0	0	2	0	5	5

REFERENCES

Affre, G. (1975). Denombrement et distribution géographique des fauvettes du genre *Sylvia* dans une région du Midi de la France. II. Resultats. *Alauda* **43,** 229–262.

Angell-Jacobsen, B. (1980). Overlap in feeding pattern between Willow Warbler *Phylloscopus trochilus* and Brambling *Fringilla montifringilla* in two forest habitats in western Norway. *Ornis Scand.* **11,** 146–154.

Bairlein, F., Berthold, P., Querner, U., and Schlenker, R. (1980). Die Brutbiologie der grasmüchen, *Sylvia atricapilla, borin, communis* und *curruca* im Mittel- und N-Europa. *J. Ornithol.* **121,** 325–369.

Batten, L. A. (1973). The colonisation of England by the Firecrest. *Br. Birds* **66,** 159–166.

Becker, P. H. (1977). Verhalten auf Lautäusserungen der Zwillingsart interspecifische Territorialität und Habitat ansprücke von Winter- und Sommer-Goldhähnchen (*Regulus regulus, R. ignicapillus*). *J. Ornithol.* **118,** 233–260.

Bergmann, H.-H. (1978). Étude d'une population de Fauvettes pitchous *Sylvia undata* sur l'île de Minorque (Baléares). *Alauda* **46,** 285–291.

Berthold, P. (1978). Brutbiolische Studien am Grassmüchen: Über die nestplatswahl der Mönchsgrasmucke *Sylvia atricapilla* im Pichten-*Picea abies*-Wald. *J. Ornithol.* **119,** 287–329.

Bibby, C. J. (1982). Polygyny and breeding ecology of the Cetti's Warbler *Cettia cetti. Ibis* **124,** 288–301.

Blondel, J. (1969). "Synécologie des Passereaux Résidents et Migrateurs dans le Midi Méditerranéen Français." Centre Regional de Documentation Pedagogique, Marseille.

Bonham, P. E., and Robertson, J. C. M. (1975). The spread of Cetti's Warbler in north-west Europe. *Br. Birds* **68,** 393–408.

Cambi, D. (1979). Contributo allo studio sulla biologia riproduttiva e sulla distribuzione di *Sylvia nisoria* (Bigia padovana) in Italia. *Riv. Ital. Ornithol.* **49,** 208–229.

Catchpole, C. K. (1972). A comparative study of territory in the Reed Warbler (*Acrocephalus scirpaceus*) and Sedge Warbler (*A. schoenobaenus*). *J. Zool.* **166,** 213–231.

Catchpole, C. K. (1973). Conditions of coexistence in sympatric breeding populations. *J. Anim. Ecol.* **42,** 623–635.

Christie, D. A. (1975). Studies of less familiar species. 176. Barred Warbler. *Br. Brids* **68,** 108–114.

Cody, M. L. (1978). Habitat selection and interspecific territoriality among the sylviid warblers of England and Sweden. *Ecol. Monogr.* **1978,** 351–396.

Cody, M. L. (1979). Resource allocation patterns in palaearctic warblers (Sylviidae). *Fortschr. Zool.* **25,** 223–234.

Cody, M. L. (1980a). Evolution of habitat use: Geographic perspectives. *Proc. Int. Ornithol Congr., 17th, 1978,* pp. 1013–1018.

Cody, M. L. (1980b). Species packing, bird density, and productivity in insectivorous bird communities. *Proc. Int. Congr. Ornithol. 17th, 1978,* pp. 882–889.

Cody, M. L. (1983). Continental diversity patterns and convergent evolution in bird communities. *In* Mediterranean-Type Ecosystems" (F. J. Kruger, D. T. Mitchell, and J. U. M. Jarvis, eds.), Ecological Sutdies, Vol. 43, pp. 357–402. Springer-Verlag, Berlin and New York.

Cody, M. L. (1984a). Bird distribution and density on four Canary Islands. (In preparation).

Cody, M. L. (1984b). Bird distributions in the High Atlas of Morocco. (In preparation).

Cody, M. L., and Walter, H. (1976). Habitat selection and interspecific interactions among Mediterranean sylviid warblers. *Oikos* **27,** 210–238.

Crivelli, A., and Blandin, P. (1977). L'organisation spatiale d'un peuplement de passereaux forestiers. *Alauda* **45,** 219–230.

daPrato, S. R. D. (1982). Polygamy by Willow Warblers. *Br. Birds* **75,** 406–411.

Dowsett-Lemaire, F. (1978). Étude d'une population nicheuse de Rouserolles Verderolles (*Acrocephalus palustris*) dans la vallée mosane (Liége). *Aves* **15,** 41–50.

Dowsett-Lemaire, F. (1979). The sexual bond in the Marsh Warbler *Acrocephalus palustris. Le Gerfaut* **69,** 3–12.

Eddington, J. M., Eddington, M. A. (1972). Spatial patterns and habitat partition in the breeding birds of an upland wood. *J. Anim. Ecol.* **41,** 331–357.

Elgood, J. H., Sharland, R. E., and Eward, P. (1966). Palaearctic migrants in Nigeria. *Ibis* **108,** 84–116.

Enemar, A., and Sjöstrand, B. (1972). Effects of the introduction of pied flycatchers *Ficedula hypoleuca* on the composition of a passerine bird community. *Ornis Scand.* **3,** 79–89.

Ferguson-Lees, I. J., and Sharrock, J. T. R. (1977). When will the Fantailed Warbler colonise Britain? *Br. Birds* **70,** 152–159.

Ferry, C., and Deschaintre, A. (1966). *Hippolais icterina* et *polyglotta* dans leur zone de sympatrie. *Abstr. Int. Congr. Ornithol. 14th,* pp. 57–58.

Fouarge, J. (1969). Le pouillot de Bonelli (*Phylloscopus bonelli*): Étend-il son aire de nidification vers le Nord? *Aves* **6,** 34–139.

Fouarge, J. (1974). Étude de la densité de Roitelets huppés (*Regulus regulus*) et triple-bandeau (*R. ignicapillus*) dans une pessiere âgée; premier resultats. *Aves* **11,** 151–156.

Fouarge, J. (1980). Le point sur les cas d'hiverage de la Fauvette à tête noir (*Sylvia atricapilla*) en Belgique. *Aves* **17,** 17–27.

Gaston, A. J. (1974). Adaptation in the genus *Phylloscopus*. *Ibis* **116,** 432–450.

Głowaciński, Z. (1975). Succession of bird communities in the Niepotomice forest (southern Poland). *Ekol. Pol.* **23,** 231–263.

Gotzmann, J. (1965). Die transspezifischen räumlichen Beziehungen zwischen dem Neuntöter (*Lanius collurio* L.) und der Sperbergrasmücke (*Sylvia nisoria* Bechst.). *Ekol. Pol.* **13,** 1–22.

Harrison, C. (1982). "An Atlas of the Birds of the Western Palaearctic." Princeton Univ. Press, Princeton, New Jersey.

Helb, H.-W. (1973). Analyse der artisolierenden Parameter im Gesang des Fitis (*Phylloscopus t. trochilus*) mit untersuchungen zur Objektivierung der analytischen Methode. *J. Ornithol* **114,** 145–206.

Högstad, O. (1975). Interspecific relations between Willow Warbler (*Phylloscopus trochilus*) and Brambling (*Fringilla montifringilla*) in subalpine forests. *Norw. J. Zool.* **23,** 223–234.

Holyoak, D. T., and Thibault, J. C. (1977). Habitats, morphologie, et inter-actions des oiseaux insectivores de polynesie orientale. *L'Oiseaux et la Rev. Fran. de l'Orn.* **47,** 115–147.

Howard, E. (1920). "*Territory in Bird Life.*" Collins, London.

Järvinen, O., and Väisänen, R. A. (1978). Recent changes in forest bird populations in northern Finland. *Ann. Zool. Fenn.* **15,** 279–289.

Jilka, A., and Leisler, B. (1974). Die Einpassung drier Rohrsängerarten (*Acrocephalus schoenobaenus, A. scirpaceus, A. arundinaceus*) in ihre Lebensräume in bezug auf das Frequenzspektrumihrer Reviergesänge. *J. Ornithol* **115,** 192–212.

Laursen, K. (1978). Interspecific relationships between some insectivorous passerine species, illustrated by their diet during Spring migration. *Ornis Scand.* **9,** 178–192.

Lawn, M. R. (1982). Pairing systems and site tenacity of the Willow Warbler *Phylloscopus trochilus* in southern England. *Ornis Scand.* **13,** 193–199.

LeDant, J.-P., and Jacobs, P. (1981). Observations sur l'écologie de la Bergeronette printanière, de la Fauvette Grisette, et du Franquet Motteau hivernant au Nord-Cameroun. *Le Gerfaut* **71,** 433–442.

Leisler, B. (1975). Die Bedentung der Fussmorphologie für die ökologische Sonderung mittelseuropäischer Rohrsänger (*Acrocephalus*) und Schwirle (*Locustella*). *J. Ornithol* **116,** 117–153.

Leisler, B., and Thaler, E. (1982). Differences in morphology and foraging behaviour in the goldcrest *Regulus regulus* and firecrest *R. ignicapillus*. *Ann. Zool. Fenn.* **19,** 277–284.

Lundevall, C.-F. (1952). "The Bird Fauna in Abisko National Park and Its Surroundings." Almqvist and Wiksell, Stockholm.

Mason, C. F, (1976). Breeding biology of the *Sylvia* warblers. *Bird Study* **23,** 213–232.

Massa, B. (1981). Primi studi sulla nicchia ecologica di cinque silvidi (Genere *Sylvia*) in Sicilia. *Riv. Ital. Ornithol.* **51,** 167–178.

Moreau, R. E. (1966). "The Bird Faunas of Africa and Its Islands." Academic Press, New York.

Muntaner, J. (1980). Sur la colonisation récente de l'île de Minorque (Baléares) pur la Fauvette pitchou *Sylvia undata*. *Alauda* **48,** 185–192.

Neuschulz, F. (1981). Brutbiologie einer population der Sperbergrasmücke (*Sylvia nisoria*) in Norddeutschland. *J. Ornithol.* **122,** 231–257.

Orians, G. (1969). On the evolution of mating systems in birds and mammals. *Am. Nat.* **103,** 589–603.

Paquet, A. (1978). Un couple d'hypolais polyglotte (*Hippolais polyglotta*) cantonné dans l'Entre-Sambre-et-Meuse. *Aves* **15,** 81–83.

Peltzer, R. (1973). *Hippolais polyglotta* nicheuse dans les Ardennes. *L'Oiseaux et la Rev. Fran. de l'Orn.* **43,** 80–81.

Price, T. (1981). The ecology of the Greenish Warbler *Phylloscopus trochilioides* in its winter quarters. *Ibis* **123,** 131–144.

Saether, B.-E. (1983a). Habitat selection, foraging niches and horizontal spacing of willow warbler (*Phylloscopus trochilus*) and chiffchaff (*P. collybita*) in an area of sympatry. *Ibis* **125,** 24–32.

Saether, B.-E. (1983b). Mechanism of interspecific spacing out in a territorial system of the chiffchaff *Phylloscopus collybita* and the willow warbler *P. trochilus. Ornis Scand.* **14,** 154–160.

Slagsvold, T. (1980a). Egg predation in woodland in relation to the presence and density of breeding Fieldfares *Turdus pilaris. Ornis Scand.* **11,** 92–98.

Slagsvold, T. (1980b). Habitat selection in birds: On the presence of other species with special regard to *Turdus pilaris. J. Anim. Ecol.* **49,** 523–536.

Svensson, S. (1978). Territorial exclusion of *Acrocephalus schoenobaenus* by *A. scirpaceus* in reed beds. *Oikos* **30,** 467–474.

Swanberg, P. O. (1953). Om nordsångaren (*Phylloscopus borealis* Blas.). *Vår Fågelv.* **12,** 49–78.

Thaler, E. (1976). Nest und nestbau von Winter- und Sommer-goldhähnchen (*Regulus regulus* und *R. ignicapillus*). *J. Ornithol.* **117,** 121–144.

Tiainen, J. (1982). Ecological significance of morphometric variation in three sympatric *Phylloscopus* warblers. *Ann. Zool. Fenn.* **19,** 285–295.

Tiainen, J., Wickholm, M., Pakkala, T., Piiroinen, J., and Virolainen, E. (1983). The habitat and spatial relation of breeding *Phylloscopus* warblers and the goldcrest *Regulus regulus* in southern Finland. *Ann. Zool. Fenn.* **20,** 1–12.

Wasilewski, A. (1967). The effect of interspecific competition on the number and distribution of birds in forest biotopes. *Ekol. Pol.* **15,** 641–695.

Chapter 4

Habitat Selection in North American Parulid Warblers

DOUGLASS H. MORSE

Division of Biology and Medicine
Brown University
Providence, Rhode Island

I. INTRODUCTION

The New World Warblers (Parulidae) are a family of about 120 species of small insectivorous birds ranging over most of North and South America (Van

HABITAT SELECTION IN BIRDS

ISBN 0-12-178080-5

Tyne and Berger, 1976). These birds are 10–18 cm long, and most are arboreal, frequenting either deciduous or coniferous forests. However, others are primarily brush or ground dwellers, and one species crawls about on tree trunks in the manner of a nuthatch. Their regional diversity, at least in breeding species, is highest in eastern North America. Most warblers retreat in winter from high latitudes to tropical or subtropical areas, where insects are likely to be constantly available.

These birds provide several excellent opportunities for studies of habitat selection, or choice of macrohabitats. Warblers are notable among temperate-zone species in that as many as 5–6 congeners may occur together in a single habitat. Densities of individuals are also high, and sometimes warblers make up a majority of the breeding avifauna in forests of eastern North America. Other habitats and geographical areas support far lower numbers and provide excellent comparisons. Species composition differs between habitats, and individual species differ in the numbers of habitats and the sizes of geographical areas that they exploit.

I will concentrate in this chapter on the effects of different habitats and of interactions among species on warbler habitat selection, because most of the available information deals with these subjects. Almost all of the data are descriptive or comparative, and much fieldwork remains to be done. Thus, I will discuss also some ideas on how this work should progress. Other factors, such as predation pressures, may be important in the habitat selection of warblers, but there is very little information in the literature specific to warblers on this subject. For the most part I will discuss between-species and between-population differences, although I must emphasize that between-individual differences are important for understanding how species-specific or population-specific differences might evolve in the first place or how they might change over time.

II. ROLE OF HABITAT STRUCTURE IN HABITAT SELECTION

A wide range of studies provides information on the role of habitat structure in habitat selection. These include correlative studies performed in different habitats, multivariate analyses, natural experiments, and manipulative studies. Habitat structure may play an important role in habitat selection, and the morphology of the birds themselves may limit in part the habitat range that they are able to exploit.

A. Comparative Studies

One can observe easily that different warbler species occupy markedly different habitats. Some are confined to tall-growth forests, either coniferous or deciduous, others to brushy habitats. In the Grand Tetons, Common Yellowthroats (*Geothlypis trichas*) concentrate in habitats with dense vegetation between 0–2 ft; Yellow Warblers *(Dendroica petechia),* 3–9 ft; Wilson's Warblers *(Wilsonia pusilla),* 9–12 ft; and MacGillivray's Warblers *(Oporornis tolmiei),* 12–15 ft or higher (Cody, 1974). Some species segregate by height or alternative parameters more precisely than others, suggesting that their requirements differ not only in kind, but quantitatively. Of course, widely distributed species may be those with wide ranges of habitat tolerance or may be merely those with quite specific but widely distributed requirements.

1. Qualitative Differences in Habitat Distribution

Studies comparing the diversity and abundance of birds in disturbed and undisturbed forests have been made in both eastern North American (Titterington *et al.,* 1979; Moore, 1980; Apfelbaum and Haney, 1981; Freedman *et al.,* 1981) and western North America (Hagar, 1960; Kilgore, 1971; Franzeb, 1978; Szaro and Balda, 1979; Morrison, 1981). The disturbances range from clear-cutting to light selective thinning and effects of fire, and the plots have been examined for varying numbers of years afterward. Except for the papers by Moore and by Morrison on warblers, these are broad-based studies that consider all of the birds present; I will concentrate on the warblers in this chapter.

Some warbler species are much more sensitive to differences in vegetation than others, and their sensitivity may vary with the vegetational parameter examined. In a successional sequence in spruce–fir forests, the Magnolia Warbler *(Dendroica magnolia)* was the only species to occupy all habitats ranging from newly cut to mature forests, although its abundance differed greatly (Titterington *et al.,* 1979). This pattern differs from the one found by Beals (1960), who compared the presence of species across six different types of mature forests (pine, aspen, birch–oak, etc.). He found that Magnolia Warblers had a narrow range of tolerance across habitats and occupied just two of them. Clearly this species is viewing the two parameters, successional sequence and between-habitat comparisons, as distinct. Ovenbirds (*Seiurus aurocapillus*), on the other hand, occupied all of Beals's forest types but only the two most mature seral habitats in the study of Titterington *et al.* Most other species occupied intermediate numbers of sites along these two habitat axes. Interestingly, the species occupying the largest number of sites in both studies, the Black-throated Green Warbler (*Dendroica virens*), is extremely sensitive to yet another variable, habitat size, and does not occupy small habitat islands which are routinely occupied by two other co-occurring warbler species, the Parula Warbler (*Parula americana*) and Yellow-rumped Warbler (*Dendroica coronata*) (Morse, 1971a, 1977).

In breeding-bird censuses of deciduous habitats subjected to different forestry practices (strip-cut, clear-cut, thinned, control), Freedman *et al.* (1981) examined the response of some of these warblers to another vegetational parameter.

Forestry practices in their study did not greatly affect the niche dimensions of the species present. Each warbler species occupied either two or three of the four differently managed forest habitats, although the ones they used differed among species. But there was variation in the abundance of different species over the differently manipulated habitats, as in the studies by Titterington *et al.* and by Beals.

Thus, warblers are differingly sensitive to a variety of factors associated with vegetational characteristics. They seem to be considerably more sensitive to forest type and successional state than to various silvicultural manipulations, which is not surprising, since seral stage and especially the type of mature forest occupied have been the habitat variables more frequently recognized than the various thinning schemes. Unfortunately, most single studies do not provide the opportunity to compare a full range of variables, and it must be noted that Beals' study was conducted in Wisconsin, Titterington *et al.*'s in northern Maine, Morse's in coastal Maine, and Freedman *et al.*'s in Nova Scotia. Therefore, such a combination of studies cannot possibly control for potential regional differences in birds or habitats.

One consequence of the different responses by warblers to these variables of forest structure is that a mosaic of species sets can coexist within a geographic area. However, if interspecific interactions also have a major effect on habitat selection (see Section III), these interactions may limit adaptations to ones that are viable over a variety of conditions.

Patterns of warbler distribution in western forests generally parallel those in the east, although they are less complex, since the number of species per habitat is much smaller. In the pine forests subjected to varying degrees of thinning (Szaro and Balda's study), one species, the Red-faced Warbler (*Cardellina rubrifrons*), did not occupy any of the thinned areas, although it occupied all of the undisturbed pine forests at low densities. Grace's Warblers (*Dendroica graciae*) were sensitive to the complete removal of large conifers (Franzeb, 1978) but resilient to a thinning regime that merely lowered the density of the large pines (Szaro and Balda, 1979).

2. Quantitative Differences in Habitat Distribution

The population density of warblers may not always correlate closely with vegetation density (e.g., Freedman *et al.*, 1981), indicating that birds do not always respond directly to simple quantitative parameters, e.g., canopy volume for arboreal species and shrub volume for brush-inhabiting species. Some probably actually prefer intermediate vegetation densities. For instance, the Yellow-rumped Warbler becomes more abundant if the tree density of mature conifers is reduced (Franzeb, 1978; Szaro and Balda, 1979; Apfelbaum and Haney, 1981), a change that fosters its predeliction for hawking insects in openings. Numbers were predictably low in Szaro and Balda's plots containing only 10% of the original forest cover, but this species attained its highest density on plots with tree densities only 40% of those of the undisturbed plots.

Morrison's (1981) work on MacGillivray's, Orange-crowned (*Vermivora celata*), and Wilson's Warblers in a young Douglas-fir (*Pseudotsuga menziesii*) plantation provides further insight into why numbers of all species do not show close fits to simple differences in vegetation density. Deciduous trees and shrubs were abundant in the younger plantations, and these were used heavily by the warblers. The Orange-crowned exploited both shrubs and deciduous trees, the MacGillivray's concentrated on shrubs, and the Wilson's exploited deciduous trees. If the decidous cover was reduced, the MacGillivray's was unaffected, and the Orange-crowned shifted its activity more heavily into the shrubs, maintaining its previous population density. The Wilson's, however, did not change its foraging patterns, and its density declined to about half that of the control plots. Thus, both the plasticity of these species and the nature of the vegetation change affected their abundance in the modified habitat.

In other instances, however, a reasonably close correlation has been shown between the quantity of vegetation and abundance of a species. Apfelbaum and Haney (1981) found that numbers of Bay-breasted Warblers (*Dendroica castanea*) and Blackburnian Warblers (*D. fusca*) declined to 25 and 50% of their former density after a forest fire reduced the canopy to one-half its previous density. Similarly, Moore (1980) noted that the Prairie Warbler (*Dendroica discolor*), a species exploiting Pitch Pine (*Pinus rigida*) in Massachusetts, was five times as abundant in an undisturbed forest as in a forest from which all canopy-level pines had been killed by a forest fire.

Summarizing these data, quantitative relationships between warbler density and simple vegetation variables are not easily predicted and often are obscure. This suggests that other factors frequently play a role, and I will discuss these in subsequent sections.

B. Experimental Manipulations

I have discussed studies in Section II,A as if they were standard experimental designs; however, almost none included censuses both before and after manipulation as well as simultaneous censuses in an undisturbed control plot. Apart from schemes which add artificial cavities for hole-nesting birds, I am aware of few field studies of habitat selection in small birds which incorporate all factors of basic experimental design. I will discuss next two studies that include more than an occasional parulid warbler.

One of Szaro and Balda's four manipulated plots meets the requirements outlined above. Their lightly thinned plot, from which about 20% of the trees were removed, was not cut until after the first year of their study and was censused for 2 subsequent years. I have incorporated these results within my discussion of Szaro and Balda's study in Section II,A. Bird densities were much lower in the initial year of their study than in the following 2 years, which makes a detailed interpretation of their figures difficult; however, the Red-faced Warbler, a species that occupied only uncut forests, was present in the plot to be lightly thinned the year before the manipulation but did not appear afterward.

Kilgore (1971) compared bird populations of California sequoia groves in which dense young second growth had been allowed to develop and sites from which this growth was cleared. Yellow-rumped, Hermit (*Dendroica occidentalis*), and Nashville (*Vermivora ruficapilla*) Warblers occupied these forests; however, only the Nashville was characteristic of the undergrowth and subcanopy trees, and only it disappeared after second-growth removal.

Conclusions from these studies do not differ from those of Section II,A, but one may interpret their results with more confidence. No comparable studies have been made in eastern North America, where warblers make a much greater contribution to the avifauna than they do in the west.

C. Ordination and Multivariate Analyses

1. Origin and Background

Other workers have taken a rather different approach, borrowing and building on the ordination techniques of plant ecologists to find correlations between birds and the structural features of the sites that they occupy. This methodology, sometimes referred to as the "habitat–niche" approach (James, 1971), has been numerically dominant throughout the past 15 years of North American studies of habitat selection in small terrestrial birds. Most of these studies consider the bird community as a whole, but a few treat warblers exclusively. Although it is beyond the scope of this chapter to discuss the methodology of the various techniques used, the references provided permit an access to key papers.

Initially, Bond (1957) adapted the continuum concept of plant communities to identify bird species that exploited vegetation of similar physical structure and to order the species along a vegetational gradient. Beals (1960) modified this technique to perform two-dimensional ordinations of several forests, based on the similarities of their bird populations. Tree species distribution and physical structure of the vegetation were both of major importance in explaining the variance among bird populations in his study.

Subsequently, a variety of multivariate statistical methods has come to dominate ordination analyses of bird populations. The most frequently used methods have been principal component analysis, which reduces a multidimensional data set to new, orthogonal variables ordered to explain the greatest possible amount of variance, and discriminant function analysis, which constructs axes from an original data set along which the differences among populations are maximized. James (1971) discussed in detail the rationale for using these techniques.

Both the methodology associated with the multivariate techniques and its aims differ markedly from the ordination schemes used by Bond and Beals. Rather than exhaustively measuring the vegetation of large areas (8–16 ha), many multivariate studies use measurements of many vegetation and physical variables taken within very small areas (0.04–0.08 ha), usually centering around the singing perch of a male. Thus, large numbers of small sites are substituted for small numbers of large sites, to eliminate the influence of vegetational heterogeneity in the analysis. Studies of this type are now too numerous to permit exhaustive review, and I will confine my comments to several that illustrate the utility of such analyses and their basic conclusions.

2. *Selected Studies*

Beals (1960) censused several forests in northern Wisconsin and, based on the habitats exploited by the different bird species, selected two indicator species, the Black-throated Green Warbler and White-throated Sparrow (*Zonotrichia albicollis*). Both indicator species occupied certain habitats in high densities, but their distributions were significantly negatively correlated with each other. Black-throated Green Warblers were especially common in old growth with some conifers present, White-throated Sparrows in young deciduous forests. These are termed "polar" species, and the technique is now called "Wisconsin Polar" ordination (see, e.g., Whittaker, 1975). Beals then classified the other bird species into four groups: those commonly associated with one or the other, both, or neither of the indicator species. This distinction, although subjective, clearly segregated 11 of the 15 warblers along a vegetational gradient: 8 were strongly associated with the mature forests typical of Black-throated Green Warbler habitat, and only 3, with pioneer growth. Of the others, Black-and-white Warblers and Ovenbirds were ubiquitous along the entire vegetation gradient, from pioneer shrubs to climax forest. Nashville Warblers and Pine Warblers (*Dendroica pinus*) were not strongly associated with either indicator, appearing as habitat specialists with a strong affinity for both pine and cedar forests. These results suggest that different warbler species specialize along different niche variables; for example, some specialize on habitat type and some on foraging sites, over a wide range of habitats (see Cody, 1974). The two species ubiquitous over habitats are strong foraging specialists, the Black-and-white Warbler (*Mniotilta varia*) on trunks and the Ovenbird on the ground. Few other warblers exploit such foraging sites, although other birds do. The Nashville and Pine Warblers differ from the others in exhibiting extreme habitat specialization. This analysis is not sensitive to other species, such as the Magnolia Warbler, that are uncommon in the censuses. Nevertheless, it succeeds in separating many species on the basis of their preferred habitats and identifies others as desirable subjects for further study.

James (1971) used mutivariate statistical techniques to characterize the species-specific habitat relationships of breeding songbirds in structurally complex habitats in Arkansas. Most subsequent studies have followed her methodology to varying degrees. James measured 15 vegetation variables (percentage canopy cover, canopy height, number of tree species, etc.) in over 400 small (0.04 ha) plots centered on the singing perches of territorial males of 46 bird species; 11 of these species were warblers. The first principal component described a gradient between open country with heavy ground cover and shaded forests with little ground cover and accounted for 65% of the variation in vegetation characteristics corresponding to variation in bird singing sites. This first principal component and the next three (secondary ordinations corresponding to other vegetation axes) accounted for 90% of the total site variance. A plot of the first three components (James, 1971, Fig. 7) suggests that the warblers are randomly scattered among the other species. The wide separation observed in this three-dimensional figure is exactly what one would predict if species were segregating in space; however, as Collins *et al.* (1982) cautioned, such correlational data can be attributed equally well to the direct effect of resources such as food, nest sites, or certain elements of vegetation structure. If the habitats sampled were expanded to include grasslands, as in Whitmore's (1975, 1977) study, warblers would exhibit habitat clumping.

Using linear discriminant function analysis, James found that warblers and other species separated widely along the first axis, which ranged from xeric to mesic, upland to bottomland, low to high biomass, and open country to closed forests. The similarity between this and Bond's (1957) univariate ordination, which also ran across a xeric–mesic continuum, is strong, although Bond dealt with only three warbler species. A second axis, which ranged from large isolated trees to trees with considerable understory, further separated the different warbler species. James (1971, Fig. 9) plotted these two axes against each other. This analysis is of particular interest, in that it includes plots that use only 3 variables and then 10 variables, as well as the entire 15. It shows the striking effect on the separation of species of adding habitat variables. Considerable differences occurred among the plots for 3 and 10 variables, illustrating the importance of incorporating several vegetational measures in studies of structurally complex habitats and pointing to possible weaknesses in other schemes of vegetational analysis (see Section II,D).

In studies of forest birds in Tennessee, Anderson and Shugart (1974) found that warblers exploited a narrower range of habitat variables than the permanently resident birds that were their nearest ecological equivalents, chickadees and titmice (*Parus* spp.), a pattern confirmed by Smith (1977) in Arkansas. Anderson and Shugart's warblers nevertheless varied considerably among themselves in terms of habitat discrimination. There were both species that exploited none of the habitat variables more frequently than predicted [Cerulean Warblers

(*Dendroica cerulea*) and Hooded Warblers (*Wilsonia citrina*)] and species that exhibited a strong habitat selectivity [Pine Warblers, Yellow-breasted Chats (*Icteria virens*), and Ovenbirds]. Most species with similar ecological requirements were separated by habitats. In the few instances in which such separation did not occur, species were vertically separated within habitats. Thus, although Cerulean and Hooded Warblers responded to similar habitat criteria, the Cerulean was a canopy species, the Hooded an undergrowth specialist.

Whitmore (1977) found that the seven warblers occupying riparian woodland in Utah were evenly distributed along a gradient from forested to dense shrub areas, with the exception of the Common Yellowthroat and the Yellow-breasted Chat, two species of extremely different sizes and feeding behaviors, which were ranked together. By using the same methods as James (1971), Whitmore (1975) was also able to compare common species in the two geographically distant areas, Utah and Arkansas. Unfortunately, the two studies had only two warblers in common, the yellowthroat and chat, yet both occupied very similar positions in discriminant function analyses in the two studies, indicating that these birds make similar vegetational choices over large geographical areas. Such repeated standardized studies provide an opportunity to identify sources of variation that would not be apparent within the scope of a single study.

Sabo's (1980) work on bird distribution along an altitudinal gradient in the White Mountains of New Hampshire is of special interest, because he measured both habitat and niche relationships and, in that way, provided some possible interpretations for the habitat patterns exhibited. Of particular note is his principal components analysis of correlations between habitat and niche variables. The first component was significantly correlated with both the first habitat and niche axes derived in principal components analyses for habitat and niche variables themselves (altitudinal gradient, terrestrial to arboreal foraging segregation), and the second component was significantly correlated with both the second habitat and niche axes [coniferous to deciduous gradient, from fine (leaves) to coarse (trunk) substrate gradient]. Although Sabo's analysis was based on the entire bird community, 8 of the 20 species were warblers. The different warblers exhibited considerable overlap over most of the altitudinal gradient but, nevertheless, differed clearly in the altitudes and habitats in which they concentrated their activities.

Using discriminant function analysis, Hendricks (1981) separated the habitat variables of five warblers occupying spruce-clad islands along the Maine coast, although both MacArthur (1958) and Morse (1968, 1971a) had already shown that these species separate by foraging differences. The only two species not separating substantially on habitat characters, the Yellow-rumped and Black-throated Green Warblers, characteristically differ markedly in foraging patterns. Of particular interest, several of the outlier points on her analyses involved males that did not attract females or pairs that did not successfully rear young.

3. *Caveat*

One of the virtues espoused for multivariate techniques is that they provide a potentially objective way of evaluating multivariate data sets (James, 1971). Nevertheless, these techniques must be used with care. James noted that the choice and number of habitat variables may affect the results obtained, which argues that, in order to characterize the habitat adequately, one should gather a broad range of variables. Sensitivity analyses on currently available data sets would provide insight into the type and amount of information needed. However, only James (1971) has performed such an analysis on this kind of data.

The intent of the habitat–niche approach is to determine whether distributions can be explained by species-specific habitat affinities, without resorting to inferences of competitive displacement (Collins *et al.*, 1982). Although the habitat–niche approach is a healthy alternative to inferring only a primarily competitive basis for distributions, it must be emphasized that by itself this technique cannot preclude the possibility of competition determining species-specific affinities. This problem can only be completely resolved by experimental manipulation.

Most workers have gathered vegetation data from areas much smaller than a breeding territory. A small area minimizes introduction of irrelevant or even avoided habitat characters (Anderson and Shugart, 1974; Sabo, 1980). By this method, large numbers of plots can be measured, a statistical advantage for these techniques. However, a potential problem arises in establishing which criteria to use for selecting sampling sites. Most workers have used male singing sites upon which to center their vegetation sampling. James (1971) warned of this technique's limitations if birds use a structurally heterogeneous territory, for example, ground and brush-dwelling species that require a singing perch. Collins (1981) found that vegetational data centered on singing perches and nest sites of more open-country warblers often differed; thus, data based on different criteria should not be mixed. Further, Collins's discovery highlights the importance of determining what role these different criteria play in habitat selection.

Although multivariate techniques are useful tools for sorting out correlations between birds and vegetation, it is the investigator's task, and responsibility, to interpret the resulting axes, i.e., to determine how each of the original habitat variables contributes to the new ones. In some studies individual habitat factors sometime significantly separate species (e.g., James, 1971; Whitmore, 1977); in other studies, they do not (e.g., Smith, 1977). Smith consequently suggested that species must respond to more than one habitat factor collectively; alternatively, James and Whitmore may have succeeded in classifying vegetation factors, or picked more appropriate ones, in a way that differs from Smith's.

Proximate factors are ones that serve to identify important features quickly and accurately, although not of direct importance themselves (Morse, 1980a). Several authors assert that birds use specific search images to select certain parts of the environment and that their results argue for the importance of proximate factors in habitat selection (James, 1971; Anderson and Shugart, 1974; Smith, 1977). This interpretation accords with James's (1971) "niche-gestalt," a combination of factors or aspects of the ecological niche that elicits from the birds a settling response. However, care should be exercised in use of the concept of a niche-gestalt, since it could be simply an artifact of the observer. Scientists may recognize distinct habitats or positions along environmental gradients, but birds may not make the same distinctions (Whitmore, 1977).

D. Other Methods

In the 1960s Robert MacArthur and his colleagues attempted to characterize the habitats of bird species by the proportions of foliage present in different horizontal layers, i.e., the foliage-height diversity of habitats. Initially, the amount of foliage was measured in three layers, 0–2 ft, 2–25 ft. (or 2–15 ft), and over 25 ft (or 15 ft), roughly corresponding to herb, shrub, and tree layers (MacArthur and MacArthur, 1961; MacArthur *et al.*, 1962). MacArthur *et al.* (1962) presented the vegetational profiles occupied by several species in three-coordinate triangular plots that depicted the proportions of vegetation in the three layers as a single point. They plotted foliage-height distributions of many singing males from several habitats, including several understory or brushland warbler species. Each was confined to a small part of the triangular plot. Then they predicted, with some success, the distribution of these warblers and other species in newly censused habitats of known foliage-height distribution but noted that certain species may have special requirements in addition to those of characteristic foliage height (e.g., a nest hole). Further, birds in structurally complex habitats (e.g., sites with patches of deciduous and coniferous forest or of sparse and dense vegetation) may use more than simple vegetational profiles (MacArthur, 1964).

The latter conclusions are consistent with those from the multivariate analyses (see Section II,C). Although foliage-height distributions are often variables in the multivariate analyses, seldom do they account for a majority of the variance in the bird distributions. Therefore, it should not come as a surprise that the predictive ability of simple foliage-height diversity measures may be somewhat limited (e.g., Willson, 1974).

E. Morphological Correlates of Habitat Selection

What, if any, role do morphological characteristics play in habitat selection, or vice versa? For the most part, warblers are a morphologically homogenous group; however, certain species do not follow the foliage-gleaning habit that characterizes most of them. For instance, the Ovenbird and Waterthrushes

(*Seiurus* spp.) are ground foragers that have converged on a thrushlike existence and appearance. These species are likely to occupy only forests with a relatively open floor, upon which they can forage easily. Black-and-White Warblers characteristically forage on trunks and large limbs, in a manner akin to nuthatches, and exhibit morphological characteristics convergent with them (Osterhaus, 1962). Therefore, they would not be at all common in the absence of tall, woody vegetation. Others, such as the American Redstart (*Setophaga ruticilla*), exhibit morphological characteristics associated with flycatching (e.g., flattened beak and large rictal bristles; Sherry, 1979), which should restrict their distribution to vegetation with some openings.

A variety of other morphological differences may play an important role in determining the foraging niche but have little effect in determining whether a given habitat itself will be occupied. They include differences in size, which dictate whether species will be successful branch-tip foragers (e.g., *Parula, Vermivora*) or will concentrate their activities more proximally to the foliage and branch tips (e.g., *Dendroica;* Morse, 1967a). Osterhaus (1962) demonstrated that most of the traits associated with the foraging niche of a species are correlated with differences in leg structure, and a more extensive study would doubtless reveal additional morphological correlates of habitats and foraging sites.

III. ROLE OF BIOTIC INTERACTIONS IN HABITAT SELECTION

Although multivariate analyses of vegetation can separate quantitatively most bird species in the studies discussed previously, some investigators view habitat selection as a factor potentially or actually influenced by interactions with conspecifics and other species. Their work focuses on potential competitors of the same species (e.g., Fretwell and Lucas, 1969) or of other species (e.g., Cody, 1974; Rosenzweig, 1981) and, to a much lesser extent, on predation (see Hildén, 1965).

Although there is much general support for the competition interpretation in the literature, studies that include warblers are observational, comparative, or, in a few instances, are natural experiments. In other studies, patterns that seem competition mediated may often have alternative explanations, as Colwell and Futuyma (1971) and Dayton (1973) cautioned.

Fretwell and Lucas (1969) and Rosenzweig (1981) have constructed models of habitat selection which predict the choices of habitats that individuals will make when subjected to various constraints, especially including density-dependent and interspecific factors. Fretwell and Lucas's model deals solely with intraspecific factors, Rosenzweig's with both intraspecific and interspecific factors. Fretwell and Lucas predicted that individuals would cease to select a normally preferred habitat after the density of conspecifics in that habitat had reached a critical level, which would be defined by the quality of that and other available habitats. At this critical level, it would be as profitable to occupy a secondary habitat as the normally preferred one, and, subsequently, one would predict that both habitats would be exploited. Rosenzweig's model incorporates the work of Fretwell and Lucas and others and accommodates for population densities, at both the intraspecific and interspecific levels, and behavior and other costs of habitat selection. Unfortunately, no studies on warblers (or other organisms, to the best of my knowledge) currently provide the data sets necessary to test Rosenzweig's comprehensive model, which requires detailed information on foraging costs, measurement of carrying capacity of a species when by itself, and measurement of its equilibrium density when other species are present.

A. Intraspecific Competition

Some of the best information in support of intraspecific competition among warblers is seen in relation to year-to-year changes of population densities. The relationship between a species' habitat–niche breadth and its population density has been well known for some time; it was discussed by Svärdson (1949), and it formed the basis for predictions of population densities in adjacent habitats of different quality by Brown (1969) and Fretwell and Lucas (1969). Ideally, removal experiments are needed to test this hypothesis, but a number of natural experiments exist, including ones on warblers.

In 1974 the breeding density of Black-throated Green Warblers along the coast of Maine, previously very stable, decreased markedly following two extremely poor breeding seasons (Morse, 1976a). However, this decline in density was confined to white spruce forests, and populations in red spruce forests showed no similar drop. During periods of peak population density, numbers of individuals in White Spruce (*Picea glauca*) forests never were as high as those in Red Spruce (*P. rubens*) forests, and it is tempting to suggest that birds shifted from white into adjacent red spruce forests, in response to the lowered pressure in preferred red spruce forests. This suggestion is strengthened by the behavior of several Black-throated Green Warblers following initial territorial establishment in white spruce forests during the spring of 1974. After establishing territories in the usual sites in white spruce forests, most individuals deserted those areas, and those remaining expanded their territory sizes. Although I have no direct evidence that the birds abandoning the white spruce forest switched to red spruce forests, population numbers in the red spruce forests reached their peak more slowly than usual that year. In as far as the data go, they are completely in accordance with the Fretwell–Lucas hypothesis. Shifts in distribution during the 1974 breeding season may have been a direct response to lowered competitive pressure in a preferred habitat and an adjustment of densities across habitats.

American Redstarts furnish another possible example of the role of intraspecific competition in habitat selection. Redstarts are unusual in that first-year males have a distinctive plumage, providing a convenient age–class marker. Habitats occupied by first-year males and older males differ strikingly in some areas. First-year males arrive later than the adult-plumage birds; they do not compete successfully with older birds for favored sites (Ficken and Ficken, 1967; Morse, 1973; Proctor-Gray and Holmes, 1981) and are frequently aggressively displaced by the latter. Adult-plumage birds typically nest in deciduous forests, but first-year birds often nest in coniferous forests (Ficken and Ficken, 1967; Morse, 1973). First-year birds also exhibit a strong tendency to nest on small, isolated islands, which like conifers may be also suboptimal habitats. Birds on these islands did not attract mates as frequently as did the first-year males occupying habitats similar to those frequented by adult males (Morse, 1973, 1976a). It would not be surprising if age-related differences in habitat selection occur regularly in populations of other species of warblers, but the difficulties of aging them preclude any easy comparisons.

B. Interspecific Competition

The distribution of breeding warblers within a habitat, at least in species-rich habitats, often looks as if it were mediated by interspecific interactions, and some plausible examples of segregation by habitat exist as well. High levels of aggressive behavior at certain times (Morse, 1976b) provide the strongest argument in favor of an interpretation of competition-mediated segregation.

The breeding distribution of Parula Warblers in the spruce forests along the coast of Maine provides a likely example of competition-mediated habitat selection. This species forages heavily in the tips of spruce foliage, in a way closely resembling the Golden-crowned Kinglet (*Regulus satrapa*), another common resident of spruce forests. Parula Warblers seldom occupy unbroken spruce forests, although they are common along their fringes or in openings. If Parula Warblers and kinglets come into contact, they may interact aggressively with each other (Morse, 1967a). On small islands not occupied by kinglets, Parula Warblers forage regularly through the interior of the spruce forests (Morse, 1976a, 1977). These data suggest that the Parula Warbler's habitat range is usually restricted by the kinglets. Several species of *Dendroica* warblers may in turn make it impossible for Parula Warblers to undergo niche shifts, such as frequenting the more interior foliage along branches, which might facilitate its coexistence with kinglets.

Yellow Warblers exhibit strong shifts in habitat selection on small islands, where they may establish territories with extensive spruce or mature deciduous growth, in vegetation that they seldom exploit on the adjacent mainland (Hebard, 1961; Morse, 1973). This shift is correlated with an absence of certain other warbler species, including Magnolia Warblers, Chestnut-sided Warblers (*Dendroica pensylvanica*), and American Redstarts, species with which they frequently interact aggressively on the mainland (Morse, 1966, 1977). In particular, these shifts are confined to islands on which redstarts are absent; almost all smaller islands of under 1 ha support either Yellow Warblers or redstarts, but not both species.

Cody (1968, 1974, 1978) has used a variety of multivariate techniques to test for habitat separation between various sorts of bird species and concluded that habitat segregation in ecologically related species is often slight. Although the mean habitat preferences of species are generally distinct, as stressed by most workers exploring the habitat–niche (see Section II,C), the ranges of habitats occupied can overlap greatly (Cody, 1978). Cody therefore concluded that other factors, including interspecific competitive interactions, are often responsible for much of the observed habitat partitioning between species, including grassland birds, parulid warblers, and, especially, sylviid warblers.

Attractive as it may be to relate these habitat-segregation distributions to interspecific competition, or to its consequences, the basic problem remains that such studies usually provide only correlations. Although Diamond (1978) argued that such patterns are so pervasive that they make the case for competition beyond reasonable doubt, Dayton (1973) presented arguments and examples that illustrate how easily results of this sort may be misinterpreted, and Wiens (1977) further emphasized the dangers of drawing more than tentative conclusions from this sort of study. I feel that the wealth of consistent data sets is too pervasive to ignore, especially if accompanying aggressive interactions are documented. However, it now seems essential to move beyond this type of inference and to test representative relationships more rigorously.

C. Predation

Potential competitors are not the only selective factor that might affect the suitability of an area for occupation. Predators can affect the selection of living sites by other animals (Hildén, 1965; Morse, 1980a), and the presence of bird-hunting hawks (*Accipiter* spp.), for example, could seriously affect the suitability of a site for warblers. Unfortunately, I know of no studies of parulid warblers that permit analysis of the role of predation on habitat selection. Neither do other studies on small forest birds provide much insight. Although European Sparrowhawks (*Accipiter nisus*) annually captured up to one-fourth of the tits (*Parus* spp.) in an English forest, neither Geer (1978) nor Perrins (1979) found any strong evidence that tits avoided particularly dangerous areas.

IV. CUES ASSOCIATED WITH HABITAT SELECTION

Sometimes one can demonstrate strong correlations between certain environmental factors and the presence or absence of a warbler species, and, consequently, it may be possible to identify specific habitat attributes to which these warblers are responding. Here I discuss several such correlations, but emphasize, however, that these patterns cannot establish *sine dubia* that the warblers respond to the same variables as do the humans observing them. Nevertheless, since experiments of such variables may be difficult to perform in the field, the only really relevant place in which to conduct such studies (see Partridge, 1978), these examples may suggest important opportunities for testing habitat selection in warblers.

A. Structural Characteristics of the Habitat

During the breeding season Blackburnian Warblers almost always occupy tall conifers (Bent, 1953) and have been regarded as extremely stereotyped in their activities in these sites (MacArthur, 1958; Morse, 1967b, 1968, 1971a). Most of their activities are carried out in the upper parts of the trees (MacArthur, 1958; Morse, 1968). These observations support the hypothesis that Blackburnian Warblers require tall coniferous vegetation, although the vegetation itself may not be the precise cue to which they respond or may be only part of it. The choice itself is unlikely to be as subtle as exact tree shape or needle texture; individuals of this species are attracted both to spruce–fir vegetation and, in the more southerly parts of their range, hemlock, although they are less likely to associate with pines. It is unclear whether the apparent demand for high vegetation is an absolute one or one mediated by competition with Black-throated Green Warblers, with which they have been invariably associated wherever their niche exploitation patterns have been studied. Black-throated Green Warblers are generally socially dominant to Blackburnian Warblers in encounters (Morse, 1974, 1976b). The absence of Blackburnian Warblers in a medium-height spruce forest, an area with very high densities of other spruce-forest warblers but with a completely closed canopy and little emergent vegetation (Morse, 1976a), could thus be a consequence of either inadequate habitat features or competitive exclusion. Holmes and Robinson (1981) noted that Blackburnian Warblers entered the northern hardwood forests at Hubbard Brook in extremely low density and entered only where occasional large spruce trees grew, in which they nested and did part of their foraging. However, these birds conducted some of their foraging in deciduous trees, which suggests that their requirements for nesting may be more exact than for foraging.

Singing perches may be limiting requirements for species living in open areas. Kendeigh (1941) suggested that the presence of a tree was a primary requirement for successful nesting by Yellow Warblers in prairie regions, which otherwise have few singing sites. In Maine, Yellow Warblers breed in marshes being invaded by Speckled Alders (*Alnus rugosa*), as long as considerable numbers of these bushes, up to 3 m in height, are present (Morse, 1966). The alders also provide nest sites and foraging areas. Although from different geographical regions, these two observations suggest that Yellow Warblers require some minimum vegetation height and that, in lieu of such vegetation, an elevated song perch is required. This hypothesis could be readily tested by adding artificial song perches to sites similar to Kendeigh's.

In addition to tall growth satisfactory for singing perches, Yellow Warblers seem dependent on the presence of understory vegetation. In New York, Ficken and Ficken (1966) found that none of five males setting up territories in tall riparian vegetation were successful in obtaining mates, but in otherwise similar forest with denser understory, 8 of 10 males obtained mates. These observations suggest that females are more selective than males in choosing sites, a pattern reported for several other warblers. Alternatively, males and females may select habitats with somewhat different features, and only when both coincide is a site successfully colonized.

Species with special nesting requirements seem among the most likely to be site limited by a single factor. The Prothonotary Warbler (*Protonotaria citrea*) and Lucy's Warbler (*Vermivora luciae*) are unique among warblers in being hole nesters (Bent, 1953), and Prothonotary Warblers readily occupy nest boxes (Bent, 1953). However, Prothonotary Warblers seem to require more than a satisfactory nest hole, for they characteristically inhabit low-lying swampy forests. Their nests are usually over water or less than 30 m from the water's edge, although a few may be 100 m or more distant (Walkinshaw, in Bent, 1953). Given their dependence on specific nest sites, one could readily design nest-box experiments that provided insight into other habitat requirements. By erecting boxes along vegetational gradients from desirable bottomland habitats, it should be possible to assess the quality of these areas independently of the presence or absence of nest holes, by measuring their nesting success in the occupied habitats.

B. Food Supply

Bay-breasted Warblers appear to respond directly to their food supply, as is characteristic of species that specialize on prey with well-marked cycles of abundance (see Pitelka *et al.*, 1955). These birds aggregate quickly at outbreak areas in spruce forests of Common Budworms (*Choristoneura fumiferana*) and Black-headed Budworms (*Acleris variana*), which is consistent with the tendency of individuals to wander many kilometers from known nesting areas during the breeding season (D. H. Morse, unpublished observation). Sealy (1979) noted an isolated nesting of Bay-breasted Warblers in southern Manitoba, well south of their normal breeding range but within their migratory pathway. Rather than responding to budworms on coniferous vegetation, these birds were found exploiting an outbreak of the Forest Tent Caterpillar (*Malacosoma disstria*) on deciduous vegetation. Such observations suggest that Bay-breasted Warblers respond directly to caterpillars, rather than to vegetational characteristics, although Sealy notes that defoliation was extensive in this area, which in its own right could provide a habitat cue to these warblers.

More often, however, birds appear to use indirect cues in selecting habitats with adequate food supplies (Hildén, 1965). Little specific information on this subject exists for wood warblers. However, Black-throated Green Warblers, which colonize red and white spruce forests at different densities (see Section III,A), provide an instance in which vegetation cues appear to determine the densities to which sites are occupied. When Black-throated Green Warblers arrive on their breeding grounds, the biomass of insects in white spruce foliage is higher than in red spruce foliage, a difference that disappears later in the season, when these warblers' food demands are at their maximum. Red spruce is the favored substrate, probably because the birds can forage in it more easily (Morse, 1976a). If Black-throated Green Warblers respond directly to the conditions existing when they arrived, their foraging opportunities would be more limited just when food demand is highest, at the time of fledging.

C. Social Stimulation

Individuals of some species may settle in response to the presence of others, a phenomenon well established in certain colony-nesting aquatic birds (Klopfer and Hailman, 1965). Evidence for such a phenomenon among warblers is sparse and circumstantial. However, intensive studies on local populations of Kirtland's Warblers (*Dendroica kirtlandii*) (see Ryel, 1979) suggest strongly that they concentrate in local areas and that these groups often are separated from other such groups by large stretches of seemingly identical but unoccupied vegetation. Further, studies on banded Kirtland's Warblers have shown regular movement among these "colonies" by 1-year-old returning birds. Similarly, Sealy (1979) noted that members of his disjunct population of Bay-breasted Warblers in southern Manitoba nested much closer to each other than would be predicted by chance. Such behavior is to be predicted in species likely to find themselves at low densities, if only as a way of increasing the probability of obtaining a mate. The social factor requires more attention before it will be possible to comment on its prevalence in warblers. Clumped patterns most likely would occur at the edges of species' ranges, where densities fall well below the carrying capacity of the habitat.

D. Conclusions

Although the data presented in this section are largely anecdotal, they suggest that as a group warblers employ a wide variety of cues in habitat selection. These include apparent examples of responses to direct factors (food) as well as a variety of indirect factors. Comparing these observations with Hildén's (1965) itemization of factors involved in habitat selection, it is clear that the traits noted within this morphologically and phylogenetically homogeneous group span a large range of the cues noted for birds in general.

V. OTHER SEASONS

The studies discussed so far have considered habitat selection in the breeding season only, but at other seasons warblers may be required to select habitat characteristics that differ from those of their breeding grounds (see Chapter 15, this volume). The breeding season is regarded traditionally as a critical period, and if so, selection for accurate choices of breeding habitats should be strong. Further, for species such as warblers with all-purpose territories, the site chosen must fill a variety of requirements: at a minimum, adequate foraging opportunities, nest site, and, possibly, display site. Even so, strong arguments have been advanced in support of the winter period being the more critical one, especially for species spending that season in the temperate zone (Lack, 1968; Fretwell, 1972). And although most warblers migrate to climatically "benign" climates, they may nevertheless encounter a host of problems in those areas also, including possible competition from residents and from other migrants and major differences between wintering and breeding areas (Morse, 1971b, 1980b; Keast, 1980; Rabenold, 1980).

Observations of the most northerly wintering warbler, the Yellow-rumped Warbler, suggest that habitat selection in the winter differs markedly from that in the breeding season. Although characteristically a breeding bird of the spruce–fir forests in eastern North America (MacArthur, 1958; Morse, 1968; Sabo, 1980), northern wintering populations of this species in the east are primarily confined to areas with substantial berry sources [Bayberry (*Myrica pennsylvanica*), Wax Myrtle (*M. cerifera*), etc.; Morse, 1970; Wilz and Giampa, 1978]. Another hardy species, the Pine Warbler, although not wintering as far north as the Yellow-rumped Warblers, retains a very close association with the same pine habitat throughout the season (Morse, 1967c, 1970). Information from these two species alone thus suggests that other parts of the yearly cycle could constrain habitat selection during the breeding season in some instances but not in others.

The tendency of breeding warblers to differ from permanent residents in their patterns of habitat exploitation (Anderson and Shugart, 1974; Smith, 1977) could be a consequence of demands during other parts of the year. However, since warblers often attain high species diversity on both their breeding and wintering grounds, their habitat niches could be influenced by niche partitioning in either or both areas.

Additionally, migratory species encounter a plethora of habitats on their migratory pathways, and they often may be far from habitats closely resembling either their breeding or wintering areas (Morse, 1971b, 1980b; Hutto, 1980; and Chapter 16, this volume). At other times the number of migrants may exceed the resources available in accustomed habitats (Rappole and Warner, 1976). Both factors must select strongly for plasiticity in habitat selection and resource exploitation patterns. To what degree they constrain the performances of these individuals at other seasons remains an open question.

VI. DIRECTIONS FOR FUTURE WORK

Comparative and natural experimental techniques have led to many profitable predictions. Similarly, multivariate studies, such as those carried out by James (1971) and many subsequent workers, have produced testable predictions and at the same time challenged the necessity of incorporating competition into schemes attempting to explain bird habitat distribution within an ecological time frame. However, correlative studies may already have taken us as far as we may realistically expect them to do. As Wiens and Rotenberry (1981) suggested, manipulations of habitat features, resources, or populations may be required to carry us past this level of understanding. Techniques such as those of Rosenzweig (1973), who carefully manipulated the habitat of desert grassland rodents and observed their subsequent habitat use, species distributions, etc., provide useful possibilities. Yet there are fundamental difficulties in using such techniques with birds: they are far more vagile than small mammals or most other terrestrial animals. Given the necessity of obtaining adequate sample sizes of individuals, the appropriate manipulations appear to be more massive than time, resources, and, most importantly, land use permit.

Need this mean that experimental procedures are impossible? I believe not; however, ingenuity will be required in order to perform them. Several feasible systems come to mind that are appropriate for warblers; other possibilities may suggest themselves for other avian groups.

A. Suggestions for Study

1. Manipulations of Old-Field Succession

Old-field habitats are widespread in eastern North America, a consequence of agriculture moving to more profitable regions. These fields exist in all stages from recent disuse to second-growth forest. Although the early successional stages do not support a high diversity of warblers, Blue-winged Warblers (*Vermivora pinus*), Prairie Warblers, Yellow-breasted Chats, and Common Yellowthroats often frequent them (Nolan, 1978).

Some of these areas are owned by conservation or scientific interests, which have an interest in maintaining them in varying stages of succession for educational, scientific, or aesthetic reasons. Management schemes for old-field successional areas require regular treatment; by incorporating research projects into the land-management plans of the governing organizations well before the actual manipulations are performed, it may be possible to design programs that follow conventional experimental procedures. Study areas should be large enough to support several pairs of the common species. Censuses should be made both before and after the manipulation in both the experimental area and in a paired control area. If possible, these censuses should be made more than 1 yr both before and after the manipulation, in order to minimize the possibility that the results from a single year coincide with an unusual population density (e.g., Morse, 1976a; Szaro and Balda, 1979; Cody, 1981). Replicate areas would also be highly desirable, but here we may be dealing with a scale exceeding that of most such manipulations.

The Kirtland Warbler Recovery Project (Ryel, 1979) uses land-management techniques analogous to the one suggested. Although that project is too strongly committed to a specific conservation goal to justify further experimental manipulation, similar schemes could provide valuable tests of the mechanisms responsible for habitat selection.

2. *Forest Manipulations*

A regular pattern of timber harvest takes place in managed forests, from clear-cutting to selectively removing over-aged trees. Collaboration with foresters provides an excellent opportunity to take advantage of the massive habitat modifications that they create, modifications which would be impossible for an individual researcher to produce for studies of habitat selection. Even if the harvesting regimes do not follow the exact lines that the ecologist would naturally choose, they may permit the testing of questions that would otherwise be inaccessible to experimental methods. Some modifications may be so severe (e.g., complete clear-cutting) that the result, in terms of warbler populations, is apt to be trivial. Others provide better opportunities; for instance, in a mixed coniferous–deciduous forest, the scheme of the timber interests may be to remove the coniferous trees only. Since many warblers exhibit a strong preference for one forest type or the other, the results of this type of manipulation would be highly instructive. Experience to date suggests that the greatest successes may be obtained in collaboration with government-sponsored forestry studies (see Franklin, 1983), although other opportunities should be explored as well.

3. *Modifying the Species Composition*

To test for the role of other species in habitat selection, it may be possible to manipulate the species composition. This technique would not be feasible in many instances, especially if large numbers of nonbreeding "floaters" were present. There may be additional difficulties in acquiring the permits necessary to remove substantial numbers of birds from a study area, not to mention the technical difficulties inherent in doing so. Still, under certain circumstances this technique provides a very powerful method for assessing the role of potential competitors in habitat selection and, simultaneously, permits the testing of niche-partitioning models.

Populations of warblers on small islands may provide the greatest opportunity for removal experiments. I have studied such populations along the coast of Maine (Morse, 1971a, 1977), and other workers have subsequently studied them on small islands there or elsewhere, either coastal islands (Hendricks, 1981) or islands in large lakes (Howe, 1979; Rusterholz and Howe, 1979). Habitat use by warblers can be readily observed on these islands, and selected individuals can be removed with relative ease. If it is not permissible to collect these birds, they can be captured by playing their songs on a recorder located adjacent to mist nets. These birds might be introduced onto other islands, permitting simultaneous removal and addition experiments. Success in adding birds will probably be lower than for removing them; however, the isolation of the islands may enhance the probability of inducing colonization. Temporary removal and subsequent reintroduction of the same individuals deserves attention, also. Another major advantage of these small islands is that immigration rates to them are not high.

4. Important Concurrent Work

If labor permits, other studies should be carried out concurrently. Observations of the resource exploitation patterns of the different species should be made in each habitat. This information might help to explain changes in habitat selection associated with differences in vegetational composition or species composition. It could also be extremely profitable to incorporate the multivariate techniques discussed in Section II,B into the experimental design. In this context one could advance causal explanations for differences, which are not strictly permissible using the multivariate analyses performed to date. I have not proposed any laboratory studies on warblers here, because warblers are harder to maintain than many other birds, such as granivorous species.

B. Opportunities for Collaboration

Scientists might profitably collaborate with enthusiastic amateurs on habitat-manipulation studies. Amateurs already conduct large numbers of censuses on breeding-bird populations, such as the Breeding-bird Censuses published in *American Birds* and the North American Breeding-Bird Survey (Robbins and

Van Velzen, 1967). Amateurs also participate at local or state levels in assessments of sanctuary bird populations, etc. Manipulative studies with implications for conservation and resource management are, if properly introduced and interpreted, likely to appeal to many amateurs. Such studies may have more practical significance than many projects currently being carried out by amateur groups.

VII. CONCLUSIONS

To date habitat selection has been a largely descriptive subject. Although efforts have been made to model habitat selection (Fretwell and Lucas, 1969; Rosenzweig, 1981), little attention has been paid to them by most field workers. Thus, we are still at a point at which very little serious evaluation of theory has taken place, much less quantitative testing (Rosenzweig, 1981), and I am unaware of any data sets for warblers that would permit the testing of models that require calculations of benefits and costs of foraging in two different habitats. Further, some warblers appear to respond strongly to factors not directly associated with energy gathering. In particular, nest-site selection appears to be of major importance. Levels of discrimination also often differ between males and females. Females may be primarily involved in choosing areas with satisfactory nest sites and males in choosing satisfactory display sites. These observations raise questions about models that emphasize the primacy of food exploitation.

In addition to development of theory, we need rigorous, quantitative, and, preferably, manipulative field studies to test such basic questions as which sets of variables are critical for habitat selection in a species. Given our current lack of knowledge at this level, such studies may provide clear signposts to the directions that the general theory should take.

The reader may protest that manipulations are neither profitable nor even possible with parulid warblers, or any other species of bird for that matter. If this charge is well founded, then these birds may offer little opportunity to advance this area of ecological theory, at least at its current stage. Yet with ingenuity on the part of avian field ecologists, and a willingness to work resourcefully and cooperatively, the problem may be resolved. In fact, it may be possible to work at a grand scale on organisms that, if manipulated, may be able to tell us a great deal about habitat selection and other ecological subjects. Parulid warblers may be among the most attractive birds for this type of study, because of their rich species diversity and abundance, even within a single habitat. I have suggested several situations in which manipulations of the habitat of warbler species are feasible and believe that these regimes deserve careful attention; we may now have reached the point at which further descriptive and correlative studies can tell us little more that is new.

ACKNOWLEDGMENTS

I thank J. K. Waage for comments on a draft of this paper.

REFERENCES

Anderson, S. H., and Shugart, H. H., Jr. (1974). Habitat selection of breeding birds in an east Tennessee deciduous forest. *Ecology* **55,** 828–837.

Apfelbaum, S., and Haney, A. (1981). Bird populations before and after wildfire in a Great Lakes pine forest. *Condor* **83,** 347–354.

Beals, E. W. (1960). Forest bird communities in the Apostle Islands of Wisconsin. *Wilson Bull.* **72,** 156–181.

Bent, A. C. (1953). Life histories of North American wood warblers. *Bull. U.S. Nat. Mus.* **203,** 1–734.

Bond, R. R. (1957). Ecological distribution of breeding birds in the upland forests of southern Wisconsin. *Ecol. Monogr.* **27,** 351–384.

Brown, J. L. (1969). The buffer effect and productivity in tit populations. *Am. Nat.* **103,** 347–354.

Cody, M. L. (1968). On the methods of resource division in grassland bird communities. *Am. Nat.* **102,** 107–137.

Cody, M. L. (1974). "Competition and the Structure of Bird communities." *Monographs in Popular Biology,* Vol. 7, pp. 1–318. Princeton Univ. Press, Princeton, New Jersey.

Cody, M. L. (1978). Habitat selection and interspecific territoriality among the sylviid warblers of England and Sweden. *Ecol. Monogr.* **48,** 351–396.

Cody, M. L. (1981). Habitat selection in birds: The roles of vegetation structure, competitors, and productivity. *BioScience* **31,** 107–113.

Collins, S. L. (1981). A comparison of nest-site and perch-site vegetation structure for seven species of warblers. *Wilson Bull.* **93,** 542–547.

Collins, S. L., James, F. C., and Risser, P. G. (1982). Habitat relationships of wood warblers (Parulidae) in northern central Minnesota. *Oikos* **39,** 50–58.

Colwell, R. K., and Futuyma, D. J. (1971). On the measurement of niche breadth and overlap. *Ecology* **52,** 567–576.

Dayton, P. K. (1973). Two cases of resource partitioning in an intertidal community: Making the right prediction for the wrong reason. *Am. Nat.* **107,** 662–670.

Diamond, J. M. (1978). Niche shifts and the rediscovery of interspecific competition. *Am. Sci.* **66,** 322–331.

Ficken, M. S., and Ficken, R. W. (1966). Notes on mate and habitat selection in the Yellow Warbler. *Wilson Bull.* **78,** 232–233.

Ficken, M. S., and Ficken, R. W. (1967). Age-specific differences in the breeding behavior of the American redstart. *Wilson Bull.* **79,** 188–199.

Franklin, J. (1983). Invitation to participate in experimental manipulations. *Bull. Ecol. Soc. Am.* **64,** 25.

Franzreb, K. E. (1978). Tree species used by birds in logged and unlogged mixed-coniferous forests. *Wilson Bull.* **90,** 221–238.

Freedman, B., Beauchamp, C., McLaren, I. A., and Tingley, S. I. (1981). Forestry management practices and populations of breeding birds in a hardwood forest in Nova Scotia. *Can. Field-Nat.* **95,** 307–311.

Fretwell, S. D. (1972). "Populations in a seasonal environment." *Monographs in Population Biology,* Vol. 5, pp. 1–217.

Fretwell, S. D., and Lucas, H. L., Jr. (1969). On territorial behavior and other factors influencing habitat distribution in birds. I. Theoretical development. *Acta Biotheor.* **19,** 16–36.

Geer, T. A. (1978). Effects of nesting sparrowhawks on nesting tits. *Condor* **80,** 419–422.

Hagar, D. C. (1960). The interrelationships of logging, birds, and timber regeneration in the Douglas-fir region of northwestern California. *Ecology* **41,** 116–125.

Hebard, F. V. (1961). Yellow Warblers in conifers. *Wilson Bull.* **73,** 394–395.

Hendricks, E. S. (1981). "Niche Relationships of Spruce-woods Warblers: A Multivariate Statistical Analysis." Ph.D. thesis. Indiana Univ, Bloomington.

Hildén, O. (1965). Habitat selection in birds. *Ann. Zool. Fenn.* **2,** 53–75.

Holmes, R. T., and Robinson, S. K. (1981). Tree species preferences of foraging insectivorous birds in a northern hardwoods forest. *Oecologia* **48,** 31–35.

Howe, R. W. (1979). Distribution and behavior of birds on small islands in northern Minnesota. *J. Biogeogr.* **6,** 379–390.

Hutto, R. L. (1980). Winter habitat distribution of migratory land birds in Western Mexico with special reference to small, foliage-gleaning insectivores. *In* "Migrant Birds in the Neotropics: Ecology, Behavior, Distribution, and Conservation" (A. Keast and E. S. Morton, eds.), pp. 181–203. Random House (Smithsonian Inst. Press), Washington, D.C.

James, F. C. (1971). Ordinations of habitat relationships among breeding birds. *Wilson Bull.* **83,** 215–236.

Keast, A. (1980). Spatial relationships between migratory parulid warblers and their ecological counterparts in the neotropics. *In* "Migrant Birds in the Neotropics: Ecology, Behavior, Distribution, and Conservation" (A. Keast and E. S. Morton, eds.), pp. 109–132. Random House (Smithsonian Inst. Press), Washington, D.C.

Kendeigh, S. C. (1941). Birds of a prairie community. *Condor* **43,** 165–174.

Kilgore, B. M. (1971). Response of breeding bird populations to habitat changes in a giant sequoia forest. *Am. Midl. Nat.* **85,** 135–152.

Klopfer, P. H., and Hailman, J. P. (1965). Habitat selection in birds. *Adv. Study Behav.* **1,** 279–303.

Lack, D. (1968), "Ecological Adaptations for Breeding in Birds." Methuen, London.

MacArthur, R. H. (1958). Population ecology of some warblers of northeastern coniferous forests. *Ecology* **39,** 599–619.

MacArthur, R. H. (1964). Environmental factors affecting bird species diversity. *Am. Nat.* **98,** 387–397.

MacArthur, R. H., and MacArthur, J. W. (1961). On bird species diversity. *Ecology* **42,** 594–598.

MacArthur, R. H., MacArthur, J. W., and Preer, J. (1962). On bird species diversity. II. Prediction of bird census from habitat measurements. *Am. Nat.* **96,** 167–174.

Moore, M. C. (1980). Habitat structure in relation to population density and timing of breeding in Prairie Warblers. *Wilson Bull.* **92,** 177–187.

Morrison, M. L. (1981). The structure of western warbler assemblages: Analysis of foraging behavior and habitat selection in Oregon. *Auk* **98,** 578–588.

Morse, D. H. (1966). The contexts of songs in the Yellow Warbler. *Wilson Bull.* **78,** 444–455.

Morse, D. H. (1967a). Competitive relationships between Parula Warblers and other species during the breeding season. *Auk* **84,** 490–502.

Morse, D. H. (1967b). The contexts of songs in the Black-throated Green and Blackburnian warblers. *Wilson Bull.* **79,** 62–72.

Morse, D. H. (1967c). Foraging relationships of Brown-headed Nuthatches and Pine Warblers. *Ecology* **48,** 94–103.

Morse, D. H. (1968). A quantitative study of foraging of male and female spruce woods warblers. *Ecology* **49,** 779–784.

Morse, D. H. (1970). Ecological aspects of some mixed-species foraging flocks of birds. *Ecol. Monogr.* **40,** 119–168.

Morse, D. H. (1971a). The foraging of warblers isolated on small islands. *Ecology* **52,** 216–228.
Morse, D. H. (1971b). The insectivorous bird as an adaptive strategy. *Annu. Rev. Ecol. Syst.* **2,** 177–200.
Morse, D. H. (1973). The foraging of small populations of Yellow Warblers and American Redstarts. *Ecology* **54,** 346–355.
Morse, D. H. (1974). Niche breadth as a function of social dominance. *Am. Nat.* **108,** 818–830.
Morse, D. H. (1976a). Variables determining the density and territory size of breeding spruce-woods warblers. *Ecology* **57,** 290–301.
Morse, D. H. (1976b). Hostile encounters among spruce-woods warblers (*Dendroica,* Parulidae). *Anim. Behav.* **24,** 764–771.
Morse, D. H. (1977). The occupation of small islands by passerine birds. *Condor* **79,** 399–412.
Morse, D. H. (1980a), ''Behavioral Mechanisms in Ecology.'' Harvard Univ. Press, Cambridge, Massachusetts.
Morse, D. H. (1980b). Population limitation: Breeding or wintering grounds? *In* ''Migrant Birds in the Neotropics: Ecology, Behavior, Distribution, and Conservation'' (A. Keast and E. S. Morton, eds.), pp. 505–516. Random House (Smithsonian Inst. Press), Washington, D.C.
Nolan, V., Jr. (1978). The ecology and behavior of the Prairie Warbler, *Dendroica discolor. Ornithol. Monogr.* **26,** 1–595.
Osterhaus, M. B. (1962). Adaptive modification in the leg structure of some North American warblers. *Am. Midl. Nat.* **68,** 474–486.
Partridge, L. (1978). Habitat selection. *In* ''Behavioural Ecology. An Evolutionary Approach'' (J. R. Krebs and N. B. Davies, eds.), pp. 351–376. Sinauer Assoc., Sunderland, Massachusetts.
Perrins, C. M. (1979), ''British Tits.'' Collins, London.
Pitelka, F. A., Tomich, P. Q., and Treichel, G. W. (1955). Ecological relations of jaegers and owls as lemming predators near Barrow, Alaska. *Ecol. Monogr.* **25,** 85–117.
Proctor-Gray, E., and Holmes, R. T. (1981). Adaptive significance of delayed attainment of plumage in male American Redstarts: Tests of two hypotheses. *Evolution* **35,** 742–751.
Rabenold, K. N. (1980). The Black-throated Green Warbler in Panama: Geographic and seasonal comparison of foraging. *In* ''Migrant Birds in the Neotropics: Ecology, Behavior, Distribution, and Conservation'' (A. Keast and E. S. Morton, eds.), pp. 297–307. Random House (Smithsonian Inst. Press), Washington, D.C.
Rappole, J. H., and Warner, D. W. (1976). Relationships between behavior, physiology, and weather in avian transients at migration stopover sites. *Oecologia* **26,** 193–212.
Robbins, C. S., and Van Velzen, W. T. (1967). The breeding bird survey, 1966. *U.S. Fish Wildl. Serv. Spec. Sci. Rep.: Wildl.* **102.**
Rosenzweig, M. L. (1973). Exploitation in three trophic levels. *Am. Nat.* **107,** 275–294.
Rosenzweig, M. L. (1981). A theory of habitat selection. *Ecology* **62,** 327–335.
Rusterholz, K. A., and Howe, R. W. (1979). Species–area relations of birds on small islands in a Minnesota lake. *Evolution* **33,** 468–477.
Ryel, L. A. (1979). On the population dynamics of Kirtland's Warbler. *Jack-Pine Warbler* **57,** 76–83.
Sabo, S. R. (1980). Niche and habitat relations in subalpine bird communities of the White Mountains of New Hampshire. *Ecol. Monogr.* **50,** 241–259.
Sealy, S. G. (1979). Extralimital nesting of Bay-breasted Warblers: Response to forest tent caterpillars? *Auk* **96,** 600–603.
Sherry, T. W. (1979). Competitive interactions and adaptive strategies of American Redstarts and Least Flycatchers in a northern hardwoods forest. *Auk* **96,** 265–283.

Smith, K. G. (1977). Distribution of summer birds along a forest moisture gradient in an Ozark watershed. *Ecology* **58,** 810–819.

Svärdson, G. (1949). Competition and habitat selection in birds. *Oikos* **1,** 157–174.

Szaro, R. C., and Balda, R. P. (1979). Bird community dynamics in a ponderosa pine forest. *Stud. Avian Biol.* **3,** 1–66.

Titterington, R. W., Crawford, H. S., and Burgason, B. N. (1970). Songbird responses to commercial clear-cutting in Maine spruce–fir forests. *J. Wildl. Manage.* **43,** 602–609.

Van Tyne, J., and Berger, A. J. (1976), "Fundamentals of Ornithology," 2nd ed. Wiley, New York.

Whitmore, R. C. (1975). Habitat ordination of passerine birds of the Virgin River Valley, southwestern Utah. *Wilson Bull.* **87,** 65–74.

Whitmore, R. C. (1977). Habitat partitioning in a community of passerine birds. *Wilson Bull.* **89,** 253–265.

Whittaker, R. H. (1975). "Communities and Ecosystems," 2nd ed. Macmillan, New York.

Wiens, J. A. (1977). On competition and variable environments. *Am. Sci.* **65,** 590–597.

Wiens, J. A., and Rotenberry, J. T. (1981). Habitat associations and community structure of birds in shrubsteppe environments. *Ecol. Monogr.* **51,** 21–41.

Willson, M. F. (1974). Avian community organization and habitat structure. *Ecology* **55,** 1017–1029.

Wilz, K. J., and Giampa, V. (1978). Habitat use by Yellow-rumped Warblers at the northern extremities of their winter range. *Wilson Bull.* **90,** 566–574.

Chapter 5

Habitat Selection in Raptorial Birds

STEWART W. JANES
Department of Biology
University of California at Los Angeles
Los Angeles, California

I. INTRODUCTION

Habitat selection among raptors is apparent even to the casual observer. Comments on species that inhabit "rocky, wooded ravines above the small streams" (Burns, 1911) or inhabit "the densest parts of fir forests" (Belding, 1890) or are "prairie birds" (Bendire, 1892) have long been common. However, it has been

HABITAT SELECTION IN BIRDS

ISBN 0-12-178080-5

only within the last 20 years or so that biologists have begun to examine raptor habitat selection in a more detailed manner. Quantitative descriptions of habitats and studies designed to decipher habitat features that influence a species' presence or its reproductive success are now common.

Many studies have attempted to understand raptor habitat selection through studies of the habitat preferences of their prey. This consideration receives greater emphasis among raptors than among habitat studies of many other taxanomic groups. One reason for this approach is perhaps the tendency to view raptor populations more as predator–prey systems than is customary in warblers and waterfowl, for example. Further, the prey of a raptor species in an area often consists of predominantly a single species for which habitat preferences may be obvious. The habitat of the Black Eagle (*Aquila verrauxi*), for example, is often described in terms of hyrax habitats, namely rocky outcrops (Gargett, 1975). An even more evident example involved a study on the Ferruginous Hawk (*Buteo regalis*) in which Lardy (1980) attempted to relate habitat selection to soil types, which he found, in turn, corresponded to the distribution of the hawk's principal prey species, the Townsend's Ground Squirrel (*Spermophilus townsendii*).

This is not to say that raptor habitat selection independent of the habitat requirements of the prey does not occur. There are many aspects of the environment that can influence raptors more directly. These include factors that influence foraging behavior, the detection and capture of prey, the location of nesting sites, and the presence or absence of predators and competitors, as well as factors that relate to physiology via the thermal environment. Each has been found important with respect to one or more raptors. Patterns of habitat selection have been examined not only at the species level but also in the context of interspecies interactions, as well as within species, for example, among sexes (Koplin, 1973; Schipper *et al.*, 1975) and even among color morphs (Preston, 1980). This chapter describes a study of habitat selection among raptors and discusses some of the biological issues involved.

II. THE STUDY

A. *Buteo* Species in Western North America

Three species of the genus *Buteo*, Ferruginous, Red-tailed (*B. jamaicensis*), and Swainson's Hawks (*B. swainsoni*), breed in the grassland and shrub–steppe regions of western North America. Each species is more or less opportunistic, with respect to prey. Wherever coexisting species have been studied, a high degree of dietary similarity has usually been noted (Smith and Murphy, 1973; Schmutz *et al.*, 1980; Cottrell, 1981), though Thiollay (1981) and Thurow *et al.* (1980) have noted substantial dietary differences. Although there is considerable evidence for competition among these species in the form of common resource (food) limitations during the breeding season, agonistic behavior, and interspecific territoriality (Smith and Murphy, 1973; Schmutz *et al.*, 1980; Thiollay, 1981; Cottrell, 1981; Janes, 1984, 1985b) the partitioning of prey does not appear to play a large role in permitting coexistence.

Differences in habitat selection play a larger role in this regard. In Alberta, Schmutz *et al.* (1980) have suggested that tree density is important. The Red-tailed Hawk was found more frequently in wooded parkland, while the Ferruginous Hawk occupied the more open prairie. The Swainson's Hawk resided most often along the ecotone between the two. Smith and Murphy (1973) observed a similar pattern in Utah but suggested that topographic differences and specific plant associations were also important. Cottrell (1981), on the other hand, made no mention of differences with respect to tree or perch density but instead found the species to differ with respect to plant associations and site aspects, as well as choice of nesting sites.

It is clear that habitat differences occur among these species. However, the results to date suggest that habitat relations among these species are complex. Many questions remain, including which particular features of the environment are being chosen, how these factors relate to the biology of these organisms, and how interspecies interactions influence habitat occupancy. I examined habitat selection and interspecific habitat relations among these species at six sites in the Columbia River basin and the Great Basin.

B. Study Areas

My study areas were located in northeastern Siskiyou County, California (Tulelake); southeastern Wasco County, Oregon (Antelope); northern Morrow County, Oregon (Boardman); central Morrow County, Oregon (Heppner); southern Cassia County, Idaho (Strevell); and southeastern Tooele County, Utah (Vernon). These areas are characterized by a mosaic of shrubland dominated by various species of sage (*Artemisia* spp.) and grassland dominated by bunchgrasses (*Agropyron spicatum, Festuca idahoensis*). A long history of grazing has reduced the abundance of the perennial grasses in many areas, and they have been replaced to a large degree by annual grasses.

An important component of each of the areas, except the Heppner site, are junipers (*Juniperus* spp.), which constitute the principal nesting sites and perches for each species. Utility poles and, in the Heppner area, trees associated with the riparian areas, comprise in large part the remainder of the potential perches.

Topographically, the sites vary from the relatively flat floodplains of the Columbia River, to the rolling foothills of the Blue Mountains, to the basin and range landform typical of the Great Basin. More detailed descriptions of the study areas and their locations may be found in Janes (1985b).

C. Methods

Each study area was divided into 40-acre (16.2-ha) blocks based on township and section lines. Within each of these blocks, a series of measures were taken pertaining to features potentially important to the raptors, either directly or indirectly through their prey. The variables were chosen for their potential relations to foraging behavior and the concealment of prey and to habitat features associated specifically with the presence and abundance of prey.

Perch availability can influence foraging. Potential perches were defined as utility poles and trees 2 m tall or taller and were assigned to one of five density classes, based on the number of perches within the block (0, 1–4, 5–9, 10–49, ≥ 50). Except perhaps at the highest density class (> 3.1 perches/ha), it is doubtful that perches influence prey distribution or interfere with the detection and capture of prey to any great degree. No relation was found among perch or juniper density and the distribution of Belding's Ground Squirrels (*Spermophilus beldingi*) at the Antelope site (S. W. Janes, unpublished observation).

Topography, cliffs, and outcrops can also affect foraging activities. Topographic relief can influence the aerial environment as it relates to flight. Hills deflect winds upward, enabling slope soaring (Pennycuick, 1972a). Hills are also "excellent thermal producers" (Cone, 1962). In both instances topographic relief is positively associated with favorable flight conditions. Hills and especially cliffs and outcrops can also enhance perch foraging. Cliffs were defined as outcrops with a vertical face of 5 m or more. Topographic relief was measured using USGS topographic maps by counting the number of 20-ft (6.1-m) contour lines or their equivalent intersected by a circle inscribed within each block and tangent to its boundaries.

Vegetation structure can influence the ability to detect prey. All habitats, except nonvegetated areas, cropland, and habitations, were assigned to one of four categories with respect to vegetation structure: wooded (>10% tree cover), shrubland (>20% shrub cover), sparse shrubland (5–20% shrub cover), and grassland (<5% shrub cover).

Diets were assessed in two ways: (1) observed captures and unconsumed prey found at the nest and (2) the analysis of castings.

The distribution of prey species is often influenced by specific plant associations and other features. The plant associations considered here include habitats dominated by sage (*Artemisia tridentata* and *A. arbuscula*), those by native grassland (*Agropyron spicatum* and *Festuca idahoensis*), riparian habitats, and cropland. In addition, cliffs and outcrops can influence the distribution of prey. For example, at the Antelope site, both Belding's (*Spermophilus beldingi*) and Townsend's Ground Squirrels were rare on slopes with inclines greater than 20%. Outcrops are often indicators of shallow soils. Shallow soils exert a negative influence on the distribution of burrowing prey species, such as ground squirrels and pocket gophers (*Thomomys* spp.). The home range of a pair is taken as the set of 16.2-ha blocks that conform most closely to the observed home range, for statistical purposes (see Janes, 1984, 1985b, for a more detailed description of the methods employed).

D. Results

Two *Buteo* species were common at each of the sites, except at the Heppner site, where all three species were common (Table I). Home-range size varies both interspecifically and intraspecifically between sites. Population means ranged from 3.14–8.09 km^2 for Ferruginous Hawks, 2.18–3.08 km^2 for Red-tailed Hawks, and 1.09–3.81 km^2 for Swainson's Hawks. The home ranges of Swainson's Hawks were one-third to two-thirds smaller than the home ranges of the other buteos at a site.

A considerable diversity was apparent in the composition of home ranges within each species, both within and between sites. Despite this wide variation, there are constants among populations as well as important differences, and these are discussed next.

1. Habitat Selection: Perches and Topography

The habitat in the 16.2-ha blocks occupied by buteos at a site was compared with the habitat for the entire site to assess species-specific habitat selection. Though the term "habitat selection" is used, it is recognized that the habitat each species occupies results from choice, as well as from other factors such as competition.

Each hawk occupies territories with particular and specific perch densities and avoids other blocks. The home ranges of Ferruginous Hawks show a charac-

TABLE I

Density of Breeding Pairs of Ferruginous, Red-tailed, and Swainson's Hawks at Six Sites in the Intermountain West[a]

Site	Area (km^2)	Number of pairs			
		Ferruginous	Red-tail	Swainson's	Total
Antelope	137	0.1	32.5	22.3	54.9
Tulelake	205	0.0	31.5	21.5	53.0
Boardman	185	7.5	0.5	18.0	26.0
Strevell	273	13	3	13	29
Vernon	274	17	7	3	27
Heppner	449	13	23	24	60
Total		50.6	97.5	101.8	249.9

[a] The mean number of pairs is presented for sites with data from more than one season.

teristic perch density relative to overall perch availability, and this feature varies little among sites (Fig. 1c–f). The territories include relatively large areas without perches and tend to avoid blocks with 50 or more perches/block.

Red-tailed Hawks, on the other hand, occupy areas with relatively high perch densities (Fig. 1a,b,e,f). However, there is much greater variation among populations of Red-tailed Hawks than among those of Ferruginous Hawks. At most sites 40% or more of the average Red-tailed Hawk home range contained at least 10 perches/block, while a mean of 80% of the blocks comprising a home range contained at least one perch. However, at the Heppner site the incidence of blocks with 10 or more perches/block averaged only 9.8%, and only 34.7% contained one or more perches. The distribution of blocks with respect to perch density at this site coincided closely with that of a typical Ferruginous Hawk home range.

The distribution of blocks with different perch densities also varied markedly among Swainson's Hawk pairs (Figure 1a–d,f). Some pairs were observed in the most open areas and others in areas with some of the highest perch densities. Several territories had mean perch densities of between 100 and 250 perches/block. However, there is a trend in Swainson's Hawks to occupy savanna-like habitats (low density and widespread perches, 1–4 perches/block).

Ferruginous Hawks commonly inhabited relatively level and unbroken habitats and selected ranges with low topographic relief at the Strevell site (Fig. 2d). Ferruginous Hawks were also common at the relatively flat Boardman and Vernon sites. However, the pattern is not universal; at the Heppner site with the greatest mean elevation change per block of any of the study sites ($\bar{x}$ = 57.6 m/block), Ferruginous Hawks preferentially occupied areas of high topographic relief ($\bar{x}$ = 77.7 m/block). In part, this reflects the preemption of much of the relatively level areas for wheat farming and the avoidance of such areas by Ferruginous Hawks, but it also illustrates that areas of high topographic relief are apparently acceptable to this species.

Red-tailed Hawks tend to avoid areas of low topographic relief, though many pairs occupy home ranges with very low relief, both at the Tulelake and the Vernon sites. In general, Swainson's Hawks avoid areas of high topographic relief.

Red-tailed Hawks preferentially include both outcrops and cliffs in their home ranges, while Swainson's Hawks avoid these habitat features (Table II). Ferruginous Hawks show no consistent preference, for or against, the inclusion of outcrops, but they avoid cliffs.

2. *Habitat Selection: Ground Cover and Prey Distribution*

Apart from species preferences for perch density, vegetation structure did not exert a strong influence on the habitat occupied by the hawks. Where habitat use is nonrandom with respect to vegetation structure, it is usually related to the

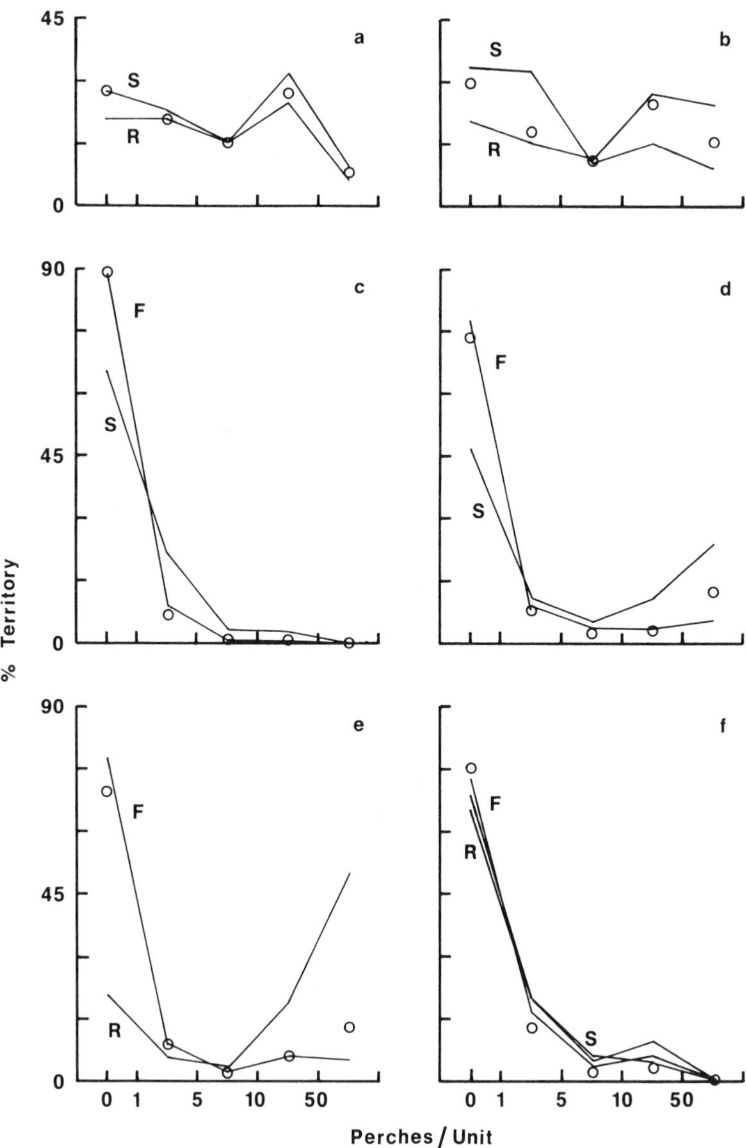

Fig. 1. Composition of Ferruginous, Red-tailed, and Swainson's Hawk home ranges, with respect to perch densities on the 16.2-ha blocks comprising the average home range at six sites in the intermountain west: (a) Antelope, (b) Tulelake, (c) Boardman, (d) Strevell, (e) Vernon, (f) Heppner. Circles indicate perch availability.

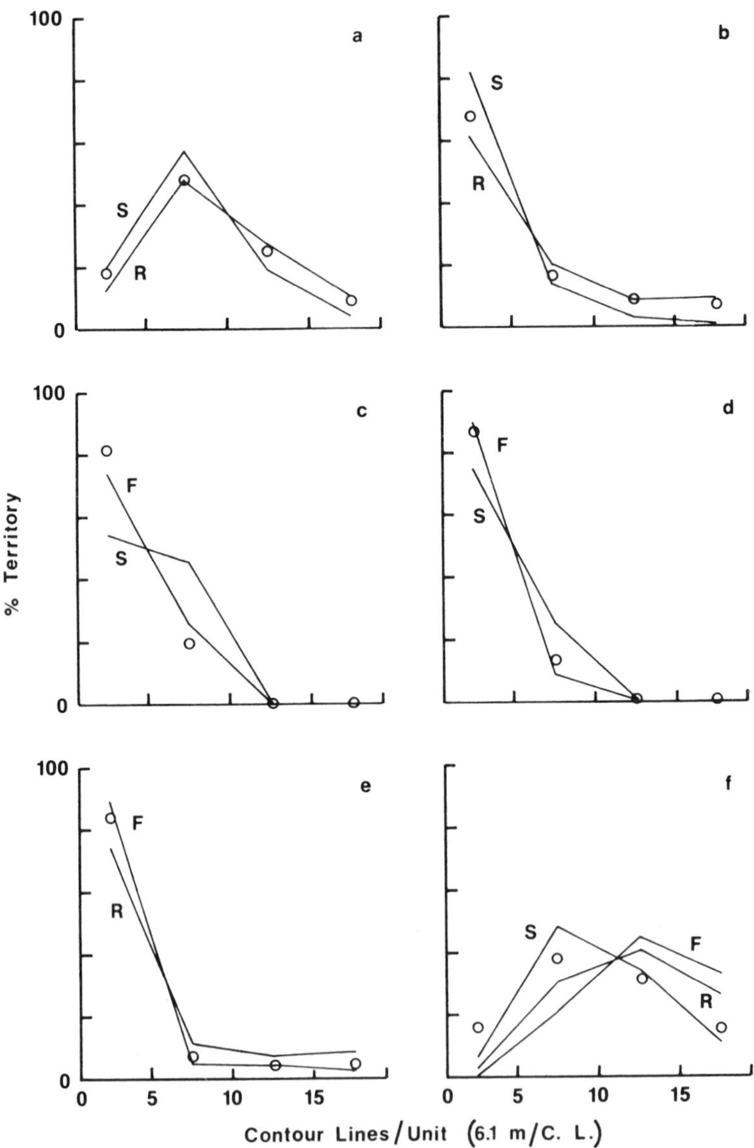

Fig. 2. Composition of Ferruginous, Red-tailed, and Swainson's Hawk home ranges, with respect to topographic relief on the 16.2-ha blocks comprising the average home range at six sites in the intermountain west. Topographic relief was measured as the number of 6.1-m contour lines intersected by a circle inscribed within each block and tangent to its boundaries. Sites are as in Fig. 1. Circles indicate availability.

TABLE II

Means of Habitat Variables for Ferruginous, Red-tailed, and Swainson's Hawk Home Ranges at Six Sites in the Intermountain West and Habitat Composition of the Six Sites Combined[a]

Variable	All sites	Ferruginous Hawk (50)[b]	Red-tailed Hawk (102)[b]	Swainson's Hawk (101)[b]
0 perches	67.0	77.3	30.9	48.1
1–4 perches	12.3	11.3	18.5	24.7
5–9 perches	4.5	3.0	11.1	8.5
10–49 perches	8.5	5.1	24.3	11.9
≥50 perches	7.6	3.2	15.3	6.7
Perch density (Number/block)	29.5	9.0	33.0	19.7
0–4 contour lines	55.8	62.5	32.7	48.0
5–9 contour lines	23.4	14.6	33.0	36.3
10–14 contour lines	13.7	13.4	21.7	12.0
≥15 contour lines	7.1	9.5	12.7	3.5
Topographic relief (6.1 m contour lines/block)	5.5	5.5	7.8	5.5
Incidence of outcrops	9.6	8.3	25.8	4.1
Incidence of cliffs	2.8	0.4	7.2	0.4
Woodland area	1.3	0.4	1.8	1.1
Shrubland area	23.9	31.5	29.2	20.6
Sparse shrub area	12.3	9.0	5.4	5.3
Grassland area	42.5	51.1	45.9	50.5
Sagebrush area	20.8	31.5	21.8	18.1
Bunchgrass area	18.8	17.9	25.9	25.7
Cropland area	19.5	6.5	12.4	19.3
Riparian area	1.5	1.9	2.4	2.0
Incidence of permanent water sources	8.2	8.5	19.6	10.9
Willow density (Number/block)	0.5	0.4	1.0	0.7

[a] Values represent the percentage of a home range containing such features except where noted.
[b] Numbers in parentheses indicate the number of pairs.

association of specific ground cover with potential nesting sites. However, this does not fully explain all of the hawk–ground cover associations: Ferruginous Hawks on the Strevell and Vernon sites display a preference for shrubland, while at the Heppner site they prefer native grasslands. This intersite difference is more readily explained by the habitat relations of the prey than by vegetation influences upon prey detection and capture.

The principal prey of hawks at the Vernon and Strevell sites is the Black-tailed Jackrabbit (*Lepus californicus*), and its distribution is closely associated with sage, which makes up most of the shrubland at both these sites (Stoddart and Anderson, 1972; Janes, 1985b). At the Heppner site hawk diets include Belding's and Townsend's Ground Squirrels, White-tailed Jackrabbits (*Lepus town-*

sendii), and Northern Pocket Gophers (*Thomomys talpoides*); the center of abundance for the latter three was in the native grasslands.

Prey-related habitat selection is also observed among Red-tailed Hawks, which incorporate floodplains into their home ranges at the Antelope site, in excess of the overall frequency. The Belding's Ground Squirrel, the principal prey species at this site, is strongly associated with low-lying areas adjacent to riparian habitat (Janes, 1984). Swainson's Hawks at this site display a similar tendency to prefer floodplains.

A similar association is observed at the Tulelake site, where Swainson's Hawks select grasslands over other habitats, with greater shrub cover. Belding's Ground Squirrels are most abundant in the grassland habitat. Here the situation regarding Red-tailed Hawks, which prefer sparse shrubland, is less clear, though competition with Swainson's Hawks is likely involved (see later this section).

Perhaps the strongest pattern of habitat selection with respect to prey distribution involves cropland, primarily dryland wheat. Dryland wheat farming involves annual plowing and the fallowing of fields on alternate years; this practice effectively eliminates the major prey species. Each hawk species avoids this habitat to some degree.

In general, each hawk species responds differently to a wide variety of habitat features and each occupies habitats in a nonrandom fashion. Only the avoidance of cropland is expressed by all species alike. The preferential use of particular perch densities, outcrops, and topographic relief produced the strongest and most consistent patterns of habitat selection among sites and species.

3. *Habitat Selection: Multivariate Analysis*

To assess the relative importance of habitat variables to interspecific differences in home ranges and to detect the importance of habitat features whose effects may be relatively small and thus easily masked, a stepwise discriminant function analysis was employed, within each site and with all sites pooled. These analyses combine and weigh variables in a linear fashion so as to best display the differences in habitat use among species.

Significant differences are found between all three species, although a fair amount of interspecific overlap still remains (Fig. 3). The variables that are important in the discriminant function analysis are generally the same as those discussed earlier and considered important in the habitat selection of each species alone. Perches, topographic relief, outcrops, and cliffs figure prominently (Table III); vegetation structure and habitat associated with prey are relatively unimportant.

The first discriminant axis, DF1, is approximately twice as important as the second axis with respect to the variance in hawk ranges it explains (Table III). The incidence of cliffs and perches at high densities contributes most to DF1, and, thus, this axis appears to correspond largely to a gradient of perch availability.

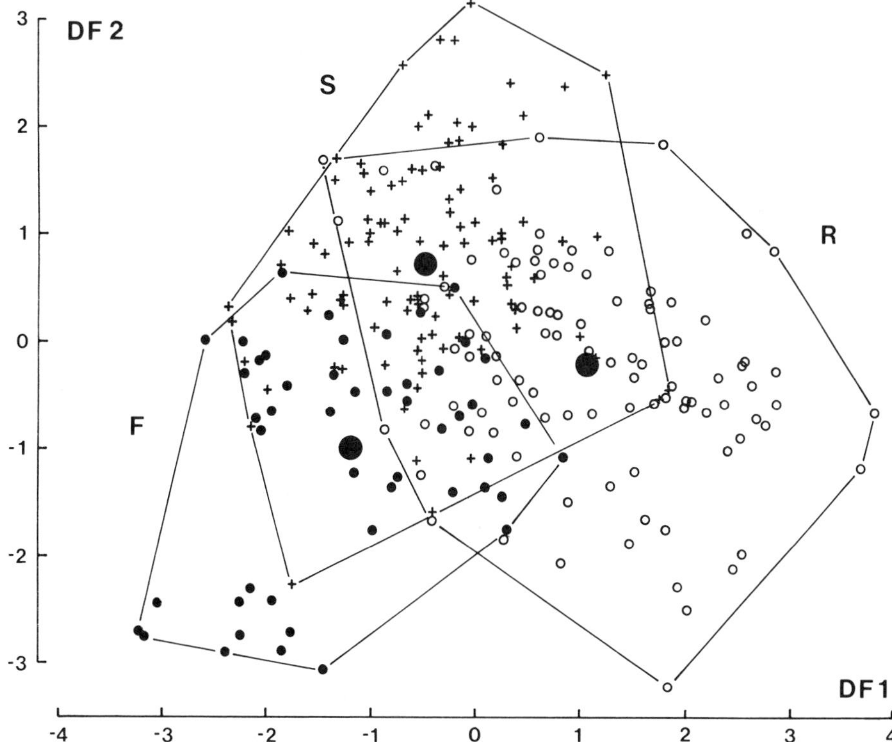

Fig. 3. Plot of discriminant scores, indicating breeding habitat distributions of Ferruginous (F), Red-tailed (R), and Swainson's Hawks (S) at six sites in the intermountain west.

The minor contributions to DF1 of the incidence of outcrops and of perches at moderate densities support this interpretation. Correlations of perch-related variables with this axis also suggest a perch-availability gradient.

Perch foraging involves the search for prey from a fixed and usually elevated point and may be coupled with frequent relocations to new perches (e.g., Janes, 1984). Aerial foraging typically involves slow gliding flight between 5 and 100 m above the ground but is energetically more expensive than perch foraging. However, aerial foraging has the advantage of not being restricted by the availability of perches. The initial height is usually achieved with the aid of rising air, in the form of declivity winds or thermals. Upon occasion, hovering is interspersed with gliding flight in each species.

Perch foraging is associated especially with the Red-tailed Hawk, while habitats least suitable for perch foraging are associated with the Ferruginous Hawk. Both native grassland and cropland are important minor components of this axis,

TABLE III

Summary of Stepwise Discriminant Function Analysis of Ferruginous, Red-tailed, and Swainson's Hawk Breeding Habitats over All Sites (n=253 pairs)[a]

Variable	Discriminant function DF1	Discriminant function DF2
Eigenvalue	0.837	0.444
Percentage of eigenvalue associated with the function	65.4	34.6
Canonical correlation	0.675	0.544
χ^2 statistic	238.45	89.76
Degrees of freedom	24	11
Significance	$p < 0.00005$	$p < 0.00005$
Standardized discriminant function coefficients		
0 perches/block	−0.248	−0.726
5–9 perches/block	0.248	−0.107
10–49 perches/block	0.217	−0.173
≥50 perches/block	0.413	−0.387
0–4 contour lines/block	0.273	0.743
5–9 contour lines/block	0.068	0.588
10–14 contour lines/block	0.292	0.148
Incidence of outcrops	0.302	−0.504
Incidence of cliffs	0.400	−0.052
Shrubland area	0.270	−0.022
Bunchgrass area	0.375	0.450
Cropland area	0.367	0.286

[a] Measures represent the proportion of a home range in 16.2-ha blocks possessing a given feature. Variables pertaining to vegetative associations represent proportional occurrence within a home range, independent of the 16.2-ha blocks.

and both are negatively correlated with Ferruginous Hawk habitats, suggesting an association with less disturbed and less productive habitats.

The distribution of the three species along DF1 is not regular. The Swainson's Hawk is situated between the other species but overlaps the Ferruginous Hawk to a much greater extent [$\alpha = 0.867$ using MacArthur's, (1972, p. 42) overlap measure]. It overlaps less with the Red-tailed Hawk ($\alpha = 0.548$), and the Ferruginous and Red-tailed Hawk's overlap even less ($\alpha = 0.330$).

These results correspond with the foraging behavior of the three species. Red-tailed Hawks are largely perch foragers (Fitch *et al.*, 1946, Thiollay, 1981, Janes, 1984), though they can and do forage while in flight (Orde and Harrell, 1977; Thiollay, 1981; Janes, 1984). Aerial foraging appears to be an unpopular substitute for perch foraging in these hawks, as only a limited increase in aerial foraging is observed in territories with few or poorly dispersed perches (Janes, 1984). In contrast, both the Ferruginous and Swainson's Hawks typically rely heavily on aerial foraging (Bowles and Decker, 1934; Bent, 1937; Weston, 1969; Smith and Murphy, 1973; Wakeley, 1978; Fitzner, 1978; Thiollay, 1981; Bechard, 1982).

While the first discriminant axis reflects an axis related to foraging behavior, the second axis, DF2, serves in large part to define the Swainson's Hawk's habitat, particularly as it differs from that of the Ferruginous Hawk. Topographic relief is strongly correlated to this axis, which reflects the avoidance of high topographic relief by the Swainson's Hawk (although not necessarily the preference by the other species for high topographic relief; see Fig. 2). It is not immediately clear why the Swainson's Hawk is restricted to areas of relatively low topographic relief. In Mexico, Thiollay (1981) also found Swainson's Hawks in association with low topographic relief, but topographic relief was also strongly and positively correlated with perch abundance.

The second major contribution to DF2 is open space. At both the Boardman and Strevell sites, the extent of open area was the most important habitat difference between Ferruginous and Swainson's Hawks. This reflects, in large part, differences in home-range size. Both species nest in relatively close proximity in these areas, due at least in part to the poor dispersion of potential nesting sites. The Ferruginous Hawk's home ranges are two to three times the size of those of the Swainson's Hawk, and the Ferruginous Hawk extends its activities further into the open areas away from the nesting sites than does the Swainson's Hawk. Otherwise, their habitats are similar. Thus, this habitat feature reflects the scarcity of potential nesting sites for the Swainson's Hawk in these vast open areas. The habitat differences between the birds are due to this feature rather than to some basic difference related to foraging.

Another factor in DF2 is the incidence of outcrops and native grassland. The combination of low topographic relief, presence of native grassland, and absence of outcrops is associated with high positive values on DF2. This particular set of habitat features suggests a relation to the distribution of burrowing prey, particularly ground squirrels and pocket gophers. The pattern is even more obvious at the Heppner site, where the first discriminant axis is composed largely of the incidence of outcrops and sparse *Artemisia rigida*-shrub habitat (Janes, 1985a). *Artemisia rigida* is associated with very shallow, rocky soils (Franklin and Dyrness, 1973). Thus, the primary habitat axis at this site is best described as a soil-depth gradient, and the three *Buteo* species occupy the same relative positions along this axis as they do along the second discriminant axis in the pooled analysis.

4. Diet and Prey Preferences

Of the six sites in this study, burrowing prey of the genera *Spermophilus* or *Thomomys* are common at five. Only at the Vernon site are such prey rare, and this is also the only site at which the Swainson's Hawk is rare. Most references to the Swainson's Hawk's diet mention these prey species as important during the breeding season (e.g., Fisher, 1893; Sharp, 1902; Gabrielson, 1922; Bowles and Decker, 1934; Bent, 1937; Beebe, 1974; Sprunt, 1955; Dunkle, 1977; Fitzner, 1978; Schmutz *et al.*, 1980; Cottrell, 1981). However, see Thurow *et al.* (1980) and Thiollay (1981) for exceptions.

The association of Swainson's Hawks with burrowing prey probably does not reflect a preference for these rodents per se but rather for some characteristic they possess. Both ground squirrels and gophers undergo minor annual population fluctuations (e.g., Slade and Balph, 1974; Boag and Murie, 1981), in relation to many of the alternative prey, such as *Microtus, Lepus,* and *Sylvilagus* (Krebs *et al.,* 1973; Gross *et al.,* 1974).

The importance of relatively stable and predictable prey populations may tie in to the migratory patterns of the Swainson's Hawk, which winters in southern South America. This migration route contains little habitats characteristic of wintering and breeding Swainson's Hawks; they may feed little on route and arrive on the breeding area in relatively poor physiological condition. Upon their arrival, the pair must establish or reestablish a territory in a habitat which is often occupied and vigorously defended by other buteos (Janes, 1984, 1985b). The female must also produce a clutch of eggs relatively soon after her return. Thus, the energetic demands on Swainson's Hawks at the time of their arrival on the breeding area may be severe in relation to other buteos, which migrate much shorter distances and do not have to evict other *Buteo* species from their eventual territory. Interspecific encounters involving the Swainson's Hawk involve higher levels of aggression and greater intensity than intraspecific encounters among any of the three species, and Swainson's Hawks are dominant over both Red-tailed and Ferruginous Hawks in these encounters (Janes, 1984, 1985b).

It appears that Swainson's Hawks are necessarily tied to reliable prey populations. At the Tulelake site each of 24 Swainson's Hawk territories contained Belding's Ground Squirrels, while 5 of 30 Red-tailed Hawk territories lacked them; these proportions are significantly different [$p \sim 0.001$; Sokal and Rolhf (1969, p. 607)]. The dominance of Swainson's Hawks over other species permits them to gain access to areas with some of the highest prey densities, and this in turn correlates to this species' relatively small territories. Prey densities averaged higher on Swainson's Hawk territories than on the territories of other buteos at each site in which prey densities were assessed (Janes, 1985b).

5. Morphological Correlates of Hawk Foraging Behavior

The principal habitat differences in the three buteo species correspond to basic differences in foraging behavior, especially to the dichotomy between perch and aerial foraging. There is little evidence that these competing species take different prey to any great extent, and differences in foraging behavior and, consequently, in habitat selection are likely to be a major factor in the ecology and evolution of these species. Therefore, we might also reasonably expect to see morphological adaptations that correspond to the dominant foraging behaviors in each species.

Flapping flight is uncommon in all of these species. Typically, the energy to remain aloft is gained from declivity winds (the winds deflected off hills) and thermals. In both cases the energy is utilized while holding the wings more or less stationary. Thermals are utilized in two major ways: a bird may remain in a thermal and circle (intrathermal soaring), or it may glide directly through them, passing from one to another (transthermal soaring; Pennycuick, 1972b). Both of these flight behaviors are practiced by each buteo species.

Gliding performance (and transthermal soaring) is influenced primarily by two values related to morphology: wing loading (w), the ratio of body mass to wing area (N/m^2), and aspect ratio (A, span^2/wing area). The ratio of forward velocity to sinking velocity in a glide, the glide ratio, is important in aerial foraging. A high glide ratio implies that a bird may cover large horizontal distances, while losing relatively little height, and the maximum glide ratio is proportional to $A^{1/2}$. Further, the forward velocity at the maximum glide ratio is also critical to foraging, a slower gliding velocity undoubtedly enhancing the ability to detect prey. The forward velocity at the maximum glide ratio is proportional to both $w^{1/2}$ and $A^{-1/4}$ (Janes, 1985b).

Intrathermal soaring is enhanced by a different combination of morphological features. The ability to utilize relatively small thermals and the rate of ascent within thermals is proportional to w^{-1}. Aspect ratio influences intrathermal soaring performance to a smaller degree (Cone, 1962).

The Swainson's Hawk possesses the best flight morphology for aerial foraging of the three species in terms of both aspect ratio and wing loading (Fig. 4). The Ferruginous Hawk, with its higher aspect ratio, also has a favorable maximum glide ratio in relation to the Red-tailed Hawk. However, the higher wing loading in the Ferruginous Hawk largely offsets any benefit in reduced gliding velocity derived from a higher aspect ratio. The higher wing loading of Ferruginous Hawks in relation to Red-tailed Hawks also implies poorer intrathermal soaring performance by the former. It should be noted, however, that the data for Red-tailed and Ferruginous Hawks were obtained largely from museum skins and, thus, may be subject to some error (though the relative positions of species and sexes on Fig. 4 should not change).

The males of each species tend to possess lower wing loadings and greater aspect ratios than the females. Thus, males possess superior capabilities, both in gliding and soaring flight. The males of each species perform the majority of hunting and territorial defense duties during the breeding season (Janes, 1984, 1985a,b).

Different hawk flight behaviors are associated to varying degrees with foraging. Slope soaring is a common foraging technique (Janes, 1984), whereas

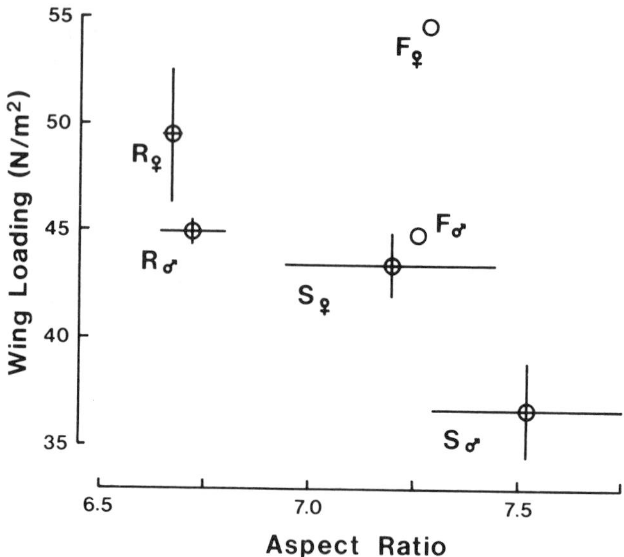

Fig. 4. Mean aspect ratio and wing loading (±SE) for the sexes of Ferruginous (F), Red-tailed (R), and Swainson's Hawks (S). The values for Swainson's Hawks were taken from breeding adults on the Tulelake site (2 each sex). The values for Ferruginous and Red-tailed Hawks were taken from skins at the Los Angeles County Museum (Ferruginous Hawk, 1 each sex; Red-tailed Hawk, 2 each sex). Additional values for Red-tailed Hawk wing loading from Poole (1938).

intrathermal soaring is not commonly associated with foraging (Ballam, 1984; Janes, 1984; Wakeley, 1978). Intrathermal soaring serves more to gain altitude. Thus, aspect ratio should be more relevant to aerial foraging than wing loading. Therefore, Swainson's and Ferruginous Hawks appear to be better adapted to aerial foraging than are Red-tailed Hawks.

III. INTERRELATIONS OF HABITAT AND RAPTOR PHENOTYPE

A. Morphology–Foraging Behavior–Habitat Selection Syndrome

1. Western North American Buteos

A relationship exists among morphology, foraging behavior, and habitat selection among *Buteo* species inhabiting the semiarid regions of western North America. At one extreme is the Red-tailed Hawk. It is a bird possessing a relatively low aspect ratio and thus is less well adapted to searching for prey while in flight, in relation to the other sympatric buteos in this region. Its primary hunting method is to survey the ground from an elevated perch. Consequently, the Red-tailed hawk occupies habitats that provide a relative abundance of potential perching sites. It occupies habitats relatively poor in perches only when conditions for aerial foraging are most favorable, i.e., areas with high topographic relief as seen in the Heppner site (Figs. 1f and 2f). An inverse correlation exists between the proportion of a Red-tailed Hawk territory posessing perches and the mean topographic relief ($r = -0.383$, $p < 0.01$).

At the other extreme is the Swainson's Hawk. The Swainson's Hawk possesses both a relatively high aspect ratio and low wing loading, which enables relatively slow gliding flight with little loss of altitude, enhancing the detection of prey. Though the Swainson's Hawk does search for prey from elevated perches, it relies much more on aerial foraging than the Red-tailed hawk. Consequently, it is not tied to habitats containing an abundance of perches and often occupies habitats containing few or no potential perches >2 m, except the nest tree.

The Ferruginous Hawk is similar to the Swainson's Hawk in aspect ratio, but its higher wing loading renders it less able to utilize thermals as efficiently and causes it to glide at a faster velocity at the maximum glide ratio. Like the Swainson's Hawk, it too forages from perches but relies on the aerial detection of prey to a far greater extent than the Red-tailed Hawk. Likewise, the Ferruginous Hawk also occupies open habitat, generally lacking perches.

2. *Other Raptors*

The relation among morphology, foraging behavior, and habitat selection can also be observed among other raptors. Wintering Rough-legged Hawks (*Buteo lagopus*) select shorter and relatively isolated perches, compared to sympatric Red-tailed Hawks (Schnell, 1968). The Rough-legged Hawk forages on the wing more than the Red-tailed Hawk, and the Rough-legged Hawk may perch more often, simply to rest, where height is less important. Red-tailed Hawks, relying more on perch foraging, seek out taller perches that provide a better vantage point, in areas with a relative abundance of perches. In accordance with these differences in habitat selection, the Rough-legged Hawk possesses a lower wing loading (Poole, 1938) and apparently a higher aspect ratio as well.

An analagous situation was observed during the breeding season on the Antelope site among Red-tailed and Swainson's Hawks, which have similar morphological and foraging differences. The Red-tailed Hawk perched above 2 m 79.6% of the time ($n = 269$), while the Swainson's Hawk did so only 40.3% of the time ($n = 283$; S. W. Janes, unpublished observations).

The same pattern has also been noted among Great Horned (*Bubo virginianus*) and Common Barn Owls (*Tyto alba*) in northern California. The perch-foraging Great Horned Owl preyed on tethered prey only in habitats containing utility poles, while the Common Barn Owl, which forages to a greater extent while in flight, also captured tethered prey in open habitats. Further, the Common Barn Owl possesses a lower wing loading and apparently a higher aspect ratio as well, in comparison to the Great Horned Owl (Rudolph, 1978). Selection for specific perches has been observed in many raptors (Fitch *et al.*, 1946; Fitch, 1974; Marion and Ryder, 1975; Sylvèn, 1978; Bildstein, 1978; Stalmaster and Newman, 1979).

In the southwestern United States, both Red-tailed and Harris' Hawks (*Parabuteo unicinctus*) occupy desert habitats. The Harris' Hawk is common in saguaro–paloverde habitats, while the Red-tailed Hawk is common in the more open adjacent habitats (Mader, 1978). Mader suggested that the more agile Harris' Hawk, owing to its shorter wings and longer tail, may be more successful in capturing prey in the relatively highly structured habitats.

In general, raptors of grasslands and other open habitats forage from the air and have both relatively high aspect ratios and low wing loadings, for example, harriers (Brown and Amadon, 1968), Short-eared Owls (*Asio flammeus;* Clark, 1975), and Long-eared Owls (*Asio otus;* Marti, 1976), in addition to the examples cited previously. Forest-dwelling raptors or raptors inhabiting more structured habitats, in contrast, tend to possess relatively low aspect ratios. Wingbeat frequency is positively correlated with maneuverability (Rüppell, 1975), which enhances movement in highly structured environments. Wingbeat frequency is also inversely correlated with wing length (Pennycuick, 1978). Thus, maneuverability is inversely correlated with aspect ratio.

Morphological and habitat differences have also been described for African vultures. The larger Ruppell's Griffon (*Gyps rüppellii*) is confined to hill country where slope soaring is an option when conditions may not permit intrathermal soaring. The White-backed Vulture (*Gyps africanus*) occurs on the flat savanna and, due to its smaller size, is able to maintain sustained flapping flight when conditions for thermal soaring are poor (Houston, 1975).

The flight environment is also important to forest dwelling raptors. "Flight space" below the canopy, as influenced by stem and branch density, has been mentioned as a potentially important factor in Spotted Owl (*Strix occidentalis*) habitat selection (Gould, 1977). A similar effect may be important to other forest raptors, especially accipiters.

B. Prey Detection and Capture

1. A Simple Model

The interaction between morphology, foraging behavior, and habitat selection aids in understanding raptor habitat selection. However, there are additional factors that can influence the general pattern described above. Ground-level vegetation can affect the ability to detect prey and hence may influence the success of particular foraging behaviors and, consequently, habitat selection. A simple model is shown in Fig. 5 to illustrate some of the obvious effects of vegetation structure on prey concealment.

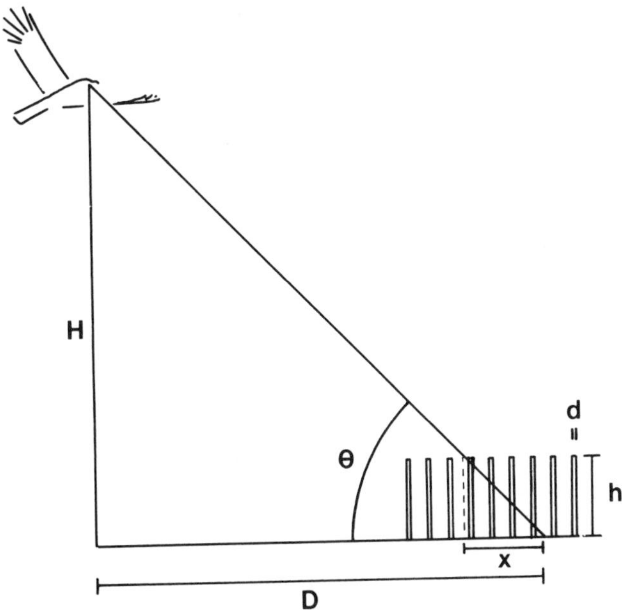

Fig. 5. The diagram depicts a foraging raptor and measures pertaining to a model concerning prey detection (see text for explanation). The measures include the height of observation H, height of vegetation h, diameter of vegetation d, horizontal distance to point at which prey is totally obscured D, corresponding distance through vegetation x, and the angle associated with this distance θ.

Consider a raptor perched or flying at height H above the ground (Fig. 5). With increasing horizontal distance D from the raptor, the raptor scans the ground at an increasingly acute angle through increasing amounts of vegetation until a value of D is reached at which prey is no longer detectable. There is some horizontal distance x, associated with angle θ and height of vegetation h, at which the ground becomes totally obscured. The expected distance between the nearest plants is related to their density ρ by the expression $1/(2\rho^{1/2})$ (Clark and Evans, 1954). The distance x is proportional to this value and at the same time inversely proportional to the diameter d of the plants,

$$x = k/2d\rho^{1/2}$$

where k is a proportionality constant. By similar triangles $h/x = H/D$; substituting for x, we obtain the expression

$$H/D = 2dh\rho^{1/2}/k$$

But prey also have dimensions and, consequently, influence the relative concealment value of the vegetation. We can express prey height as a proportion m of vegetation height, or *mh;* the effective height of the vegetation becomes $h - hm$ or $h(1 - m)$. Substituting this value into the above equation and solving for D, we obtain the expression

$$D = kH/2dh(1 - m)\rho^{1/2}$$

The area A in which a raptor can detect prey is $\pi\ D^2$ or

$$A = 0.785(kH/dh(1 - m))^2/\rho$$

Thus, the search area decreases slowly with increasing vegetation density, but relatively small changes in foraging height and in vegetation height lead to larger changes in the area that can be scanned for prey. One therefore predicts changes in foraging behavior and in raptor distribution in accordance with differences in vegetation structure. For example, a raptor may compensate for the increased prey concealment in denser and taller vegetation by increasing the height at which it forages. Schipper *et al.* (1975) and Schipper (1977) observed harriers (*Circus* spp.) to increase the height of their foraging flights with increasing vegetation height and density.

2. *Raptor Foraging and Ground Cover Vegetation*

Selection of habitats in relation to the structure of ground-level vegetation has been widely noted among raptors. In each case raptors utilized habitat consisting of shorter and/or sparser vegetation than expected based on the distribution of prey (Pearson and Pearson, 1947; Craighead and Craighead, 1956; Getz, 1961; Cavè, 1968; Southern and Lowe, 1968; Wakeley, 1978; Baker and Brooks, 1981; Jaksìc *et al.*, 1981; Bechard, 1982) or experienced greater reproductive success in such environments (Southern and Lowe, 1968, Howell *et al.* 1978).

It can be seen from the model that, if prey are sufficiently small or if vegetation is sufficiently tall or dense, perch foraging may not be feasible. Differences in foraging behavior with respect to vegetation have been noted among Snowy (*Nyctea scandiaca*) and Short-eared Owls (Lien and Webber, 1979; Lien and Boxall, 1979). Snowy Owls are perch hunters, and they select relatively short, sparsely vegetated habitats, though these support fewer prey. Short-eared Owls are aerial foragers and fly some distance above the substrate; this species favors the taller and more densely vegetated habitats, in which prey populations are denser. The more elevated viewpoint of Short-eared Owls presumably enhances visibility and makes foraging in the taller vegetation profitable.

Wakeley (1979), however, was unable to detect any differences in the foraging behavior of individual Ferruginous Hawks in different vegetation types, nor did I find strong or consistent relations between vegetation structure and the distribution of hawks with different foraging behaviors. This may be a result of the overall short stature and low density of vegetation at my study sites and relatively small difference in shrub and grass heights. In addition, lagomorphs and spermophilines are relatively tall prey compared to microtines and other mice, and this reduces the effective vegetation height even more.

The structure of vegetation may also influence prey capture, as well as prey detection. Hector (1981) found that vegetation density influences habitat quality and thence influences reproductive success in Aplomado Falcons (*Falco femoralis*). He suggested that a dense vegetation structure may assist fleeing avian prey, which might avoid capture by using trees and shrubs as obstacles.

Conversely, habitats may influence the capture of prey through concealment afforded the predator. Harriers are surprise hunters, and Schipper *et al.* (1975) described interactions among morphology, behavior, and habitat selection in these birds. Harriers forage on the wing, coursing low over the top of the vegetation, and pounce on prey as they flush in their path. Female harriers are less maneuverable than the males, as a result of differences in morphology. They are also brownish in color and contrast sharply against the sky more than the light-gray males. These authors concluded that females have a greater need to utilize habitats that offer a higher degree of concealment when foraging, and, indeed, females occupied taller and denser vegetation than males.

Preston (1980) described an apparently analagous situation in different color morphs of Red-tailed Hawks. Each color morph selected perch sites that corresponded to the concealment value afforded. Light-phase birds tended to perch in exposed positions where they were more likely to have the relatively light sky behind them from the viewpoint of the prey. Dark-phase birds perched more often against a relatively dark background.

IV. INFLUENCES OF COMPETITION AND PREDATION

A. Direct Interspecific Competiton

Raptors may not occupy the full range of available habitats that they are otherwise capable of occupying, because of interactions with other species. Both competitors and predators may restrict the habitat range of a species. Agonistic behavior and interspecific territoriality among raptors are common (e.g., Dixon, 1928; Stewart, 1949; Smith and Murphy, 1973; Fitch, 1974; Ogden, 1975; Schipper *et al.*, 1975; Sylvèn, 1978; Bildstein, 1978; Schmutz *et al.*, 1980; Thiollay, 1981; Cottrell, 1981; Janes, 1984, 1985b,c). Further evidence for interspecies effects on the use of space is observed in the regular spacing of nests (Baumgartner, 1939; Schmutz *et al.*, 1980; Cottrell, 1981; Smith and Murphy, 1982) and in the changes in reproductive success (Schmutz *et al.*, 1980). Red-tailed Hawks occupy areas early in the season, from which they are later excluded by the Swainson's Hawk, upon their arrival (Thiollay, 1981; Janes, 1984, 1985a,b). The areas abandoned by the Red-tailed Hawks are those relatively poor in perches.

Competition also may be detected on a geographic scale, through niche shifts coinciding with changes in the competitive environment. Luttich *et al.* (1970) observed that Red-tailed Hawks in Alberta occupy riparian habitats characteristic of the Red-shouldered Hawk (*Buteo lineatus*); the Alberta site is north of the range of the Red-shouldered Hawk.

B. Indirect Inter- and Intraspecific Competition

Differential habitat use among ecologically similar sympatric species has been widely observed and may also indicate the effects of competition. Interspecific differences have been noted in the selection of nesting sites (Smith and Murphy, 1973; Schipper, 1978; Titus and Mosher, 1981; Cottrell, 1981; Green and Morrison, 1983) and nesting habitats (Stewart, 1949; Thiollay, 1967; Smith and Murphy, 1973; Henny *et al.*, 1973; Fitch, 1974; Reynolds *et al.*, 1982; Schmutz *et al.*, 1980; Cottrell, 1981). Interspecific differences in wintering habitats have also been widely observed (Weller, 1964; Thiollay, 1967; Schnell, 1968; Opdam, 1975; Sylvèn, 1978; Bildstein, 1978; Lien and Boxall, 1979).

Intraspecific habitat differences among raptors have also been recorded, especially sex-specific habitat selection. Male American Kestrels (*Falco sparverius*) tended to winter in more structured habitats than females (Koplin, 1973; Mills, 1975). Mills suggested that females may exclude the males from the more open habitats. Opdam (1975) observed habitat differences between sexes of wintering European Sparrowhawks (*Accipiter nisus*), though he attributed these differences to the distribution of the preferred prey of each sex, rather than to intersexual competition.

C. Effects of Predation

Raptors are predators, sometimes of each other, and differential use of habitats may result from actual or potential interspecific predation. Baumgartner (1939) attributed the restriction of Barred (*Strix varia*) and Common Barn Owls to "less favorable" home ranges in the presence of the Great Horned Owl to this cause; Great Horned Owls have been known to prey on both other owls, as well as act as competitors to them. Rudolph (1978) described a similar situation between Great Horned and Common Barn Owls in northern California, though in this case the Common Barn Owl tended to forage in habitats lacking tail perches (and thus removed from the perch-foraging Great Horned Owl) and not necessarily in poorer habitat. Four instances of probable Great Horned Owl predation upon Common Barn Owls were noted out of a population of about 30 individuals.

Golden Eagles (*Aquila chrysaetos*) occasionally (but consistently) prey upon smaller raptors, including buteos [Carnie, 1954; Smith and Murphy, 1973; Olendorff, 1976; U. S. Department of the Interior (USDI), 1979] and thus may be responsible for the high levels of agonistic behavior directed toward eagles by other raptors (Fitch *et al.*, 1946; Janes, 1984d) and the resultant pattern of regularly spaced nests (Smith and Murphy, 1983) and interspecific territoriality (Janes, 1985c).

V. OTHER FACTORS IN RAPTOR HABITAT SELECTION

A. Prey Habitats

So far I have considered raptor habitat selection in relation to direct influences of the habitat and as mediated by competition. But, raptor habitat selection must surely be influenced strongly by the influences of habitats upon their prey. This indirect effect on raptor habitat selection is particularly apparent in areas of low prey diversity or where prey abundance is unevenly distributed with respect to habitat. A well-known example, besides Black Eagles and hyraxes (see Section I; Gargett, 1975), is the Snail Kite (*Rostrhamus sociabilis*), whose distribution in North America coincides closely with the distribution of the Apple Snail (*Pomacea paludosa*; Snyder and Snyder, 1969).

Prey-related habitat selection is also seen in raptors with less specialized diets. Red-shouldered Hawk habitat is characterized in part by the inclusion of marshland and wet meadows within floodplain forests (Bednarz and Dinsmore, 1981; Titus and Mosher, 1981). These more open and moist habitats are those in which the preferred prey is most abundant: mice, crayfish, reptiles, and amphibians.

In Sweden, wintering Common Buzzards (*Buteo buteo*) and Rough-legged Hawks select the habitats with the highest vole densities (Sylvèn, 1978). Habitat selection of raptors dependent on the habitat relations of prey has been widely implicated: Red-tailed and Rough-legged Hawks (Schnell, 1968), Golden Eagles (Marion and Ryder, 1975), Black-shouldered Kites (*Elanus caeruleus;* Warner and Rudd, 1975), European Sparrowhawks (Opdam, 1975), Spotted Owls (Gould, 1977), Harris' Hawks (Mader, 1978), Ferruginous Hawks (Howard and Wolfe, 1976; Lardy, 1980), Snowy Owls (Boxall and Lein, 1982), and, generally, in raptor communities (Craighead and Craighead, 1956; Jaksìc *et al.*, 1981).

Differential reproductive success among habitats corresponding to variations in prey density has also been observed in Tawny Owls (*Strix aluco*; Southern and Lowe, 1968), Common Buzzards (Picozzi and Wier, 1974), European Sparrowhawks, (Newton and Marquiss, 1976), and Red-tailed Hawks (Howell *et al.*,

1978; Janes, 1984). Such habitat-related reproductive success must be an important selective force in the evolution of habitat selection.

Raptor habitat selection for suitable prey habitats can also be detected by the inclusion of a relatively constant amount of prey-producing habitat within home ranges. The size of Eurasian Kestrel (*Falco tinnunculus*) home ranges varies much more than the extent of prey-rich grassland they encompass (Pettifor, 1983); the vole *Microtus arvalis* is chiefly associated with the grassland. Gargett (1975) noted a similar pattern within Black Eagle territories; although absolute territory size varied markedly in the eagles, the extent of rocky outcrops varied less.

In areas of wide prey diversity, or where prey are evenly distributed across habitat types, habitat selection by raptors may be less specifically tied to either prey or structural characters. In several studies in Alberta, both Red-tailed Hawks and Great Horned Owls occupied areas randomly, with respect to the surrounding cover types, and took prey from the various cover types in proportion to their abundance (Luttich *et al.*, 1970; Rusch *et al.*, 1972; McInvaille and Keith, 1974).

B. Nest-Site Selection

The selection of nest sites may involve a host of rather specific concerns, ranging from local factors such as the thermal environment in the nest to broader factors including but not limited to the proximity of foraging areas.

Some raptors utilize a variety of substrates for nesting; the Ferruginous Hawk nests in trees, in small cliffs, and on the ground (e.g., Smith and Murphy, 1973), though trees are apparently preferred (Woffinden and Murphy, 1983). Others, such as the Prairie Falcon (*Falco mexicanus*) and Swainson's Hawk, nest almost exclusively on a single substrate, cliffs and trees, respectively (Bent, 1937). Exceptions are rare enough to merit publication (e.g., Woffinden and Mosher, 1979).

Preferences are sometimes expressed at an even finer level. For example, Red-tailed hawks preferred elms over other tree species in Wisconsin (Orians and Kuhlman, 1956), and Broad-winged Hawks (*Buteo platypterus*) selected Yellow Birch in New England (Matray, 1974). In Utah, Ferruginous Hawks sought out the taller trees in an area for nesting (Woffinden and Murphy, 1983).

The placement of the nest with respect to a particular substrate is often quite specific. Sharp-shinned Hawks (*Accipiter striatus*) typically situate their nests in the denser portions of the ower canopy, next to the trunk (Reynolds *et al.*, 1982). Matray (1974) and Keran (1978) found that Broad-winged Hawks place their nests in the lower portion of the canopy, most often in the first available crotch up from the base of the canopy. Titus and Mosher (1981) found that four sympatric raptor species (Red-tailed, Red-shouldered, Broad-winged, and Cooper's Hawks (*Accipiter cooperii*)) differed significantly in the placement of their nests. These differences were quantified by trunk diameter (*dbh*), height of the nest, and relative height of the nest within the nest tree.

Nest placement on rocky substrates may also be nonrandom. The cliff nests of Golden Eagles in Alaska tend to face in a southeasterly direction, while nests in Utah tend to face in a northwesterly direction, presumably in response to differences in the thermal environment at the nest (Mosher and White, 1976). Since Golden Eagle young are subject to heat-related mortality (Beecham and Kochert, 1975), such orientation may be critical.

Apart from the nest site itself, the habitat in the immediate vicinity of the nest may be important for a variety of reasons. Ferruginous Hawk nest sites are usually associated with an elevated observation site, such as a tree or hill, within 100 m of the nest (Lokemoen and Duebbert, 1976). Lokemoen and Duebbert also found the majority of ground nests on west-facing slopes, while a majority of Ferruginous Hawk nests in Utah faced east (Weston, 1969). In each area these slopes face the prevailing wind during the breeding season and may offer a favorable environment for taking flight (Lokemoen and Duebbert, 1976). However, Lardy (1980) found little support for this idea among Oregon birds.

Goshawks (*Accipiter gentilis*) and Cooper's Hawks select trees on north-facing slopes for nesting, possibly for thermal considerations (Reynolds *et al.*, 1982). Red-shouldered Hawks, on the other hand, choose nest sites randomly with respect to topographic features (Morris and Lemon, 1983). The association of a particular vegetation type or structure with the location of nests has been routinely observed (Tubbs, 1967; Platt, 1976; Schipper, 1977, 1978; Reynolds *et al.*, 1982; Bednarz and Dinsmore, 1981; Titus and Mosher, 1981; Morris *et al.*, 1982; Gilmer and Stewart, 1983).

Though a nest site in raptors is an important concern, it is doubtful if this aspect of habitat selection plays a major role in the distribution of raptors or in their evolution. Given an area encompassing an adequate prey population, it is reasonable to expect that a suitable nest site can be found in most cases, specific nest site requirements, such as those of cavity-nesting raptors (Cavè, 1968; Hamerstrom *et al.*, 1973; Lundberg, 1979; Village, 1983) or of grassland raptors (Olendorff and Stoddart, 1974; Howard and Hilliard, 1980), notwithstanding.

C. Roosting Requirements

Raptors also display habitat preferences when roosting for a variety of concerns, including the threat of mobbing, predation, and thermal stresses. Short-eared Owls select dense, relatively light-colored grassland habitats that match their plumage (Craighead and Craighead, 1956). In coastal Washington, Bald Eagles (*Haliaetus leucocephalus*) roost in dense stands of conifers (Stalmaster and Newman, 1979), sites avoided by the eagles in the daytime. The choice of the relatively sheltered roost sites may provide a favorable thermal environment or concealment from enemies and pests. However, Bald Eagles wintering in the Klamath Basin of California and Oregon select larger and more openly structured trees for roosting (Keister and Anthony, 1983). Thermal considerations appear to be important in the selection of summer roosting sites among Spotted Owls, which are sensitive to heat stress. They tend to select roosts on north slopes, beneath dense canopy, and near water (Gould, 1977; Barrows and Barrows, 1978; Barrows, 1981). Such sites are between 4° and 6°C cooler than adjacent open habitats. Winter roosts are much more variable (Barrows, 1981).

ACKNOWLEDGMENTS

This study was funded in part by the Oregon Field Office of The Nature Conservancy, with the support of the Rose Tucker and Tectronix foundations.

REFERENCES

Baker, J. A., and Brooks, R. J. (1981). Distribution patterns of raptors in relation to density of voles. *Condor* **83,** 42–47.

Ballam, J. M. (1984). The use of soaring by the Red-tailed Hawk (*Buteo jamaicensis*). *Auk* **101,** 519–524.

Barrows, C. W. (1981). Roost selection by Spotted Owls: An adaptation to heat stress. *Condor* **83,** 302–309.

Barrows, C. W., and Barrows, K. (1978). Roost characteristics and behavioral thermoregulation in the Spotted Owl. *West. Birds* **9,** 1–8.

Baumgartner, F. M. (1939). Territory and population in Great Horned Owls. *Auk* **56,** 274–282.

Bechard, M. J. (1982). Effect of vegetative cover on foraging site selection by Swainson's Hawk. *Condor* **84,** 153–159.

Bednarz, J. C., and Dinsmore, J. J. (1981). Status, habitat use, and management of Red-shouldered Hawks in Iowa. *J. Wildl. Manage.* **45,** 236–241.

Beebe, F. L. (1974). Field studies of the Falconiformes of British Columbia. B. C. Prov. Mus. Occas. Pap., no. 17. Br. Columbia Prov. Mus., Victoria.

Beecham, J. J., and Kochert, M. N. (1975). Breeding biology of the Golden Eagle in southwestern Idaho. *Wilson Bull.* **87,** 506–513.

Belding, L. (1890). Land birds of the Pacific district. Occas. Pap. Calif. Acad. Sci. **2,** 1–274.

Bendire, C. E. (1892). Life histories of North American birds. U.S. Nat. Mus. Spec. Bull., Vol. 1.

Bent, A. C. (1937). Life histories of North American birds of prey. Part I. *Bull. U.S. Nat. Mus.* **170.**

Bent, A. C. (1938). Life histories of North American birds of prey. Part II. *Bull. U.S. Nat. Mus.* **170.**

Bildstein, K. L. (1978). Behavioral ecology of Red-tailed Hawks (*Buteo jamaicensis*), Rough-legged Hawks (*Buteo lagopus*), Northern Harriers (*Circus cyaneus*), American Kestrels (*Falco sparverius*), and other raptorial birds wintering in south-central Ohio. Ph.D. dissertation. Ohio State Univ., Columbus.

Boag, D. A., and Murie, J. O. (1981). Population ecology of Columbian Ground Squirrels in southwestern Alberta. *Can. J. Zool.* **59,** 2230–2240.

Bowles, J. H., and Decker, F. R. (1934). Swainson's Hawk in Washington State. *Auk* **51,** 446–450.

Boxall, P. C., and Lein, M. R. (1982). Territoriality and habitat selection of female Snowy Owls (*Nyctea scandiaca*) in winter. *Can. J. Zool.* **60,** 2344–2350.

Brown, L., and Amadon, D. (1968). "Eagles, Hawks, and Falcons of the World." McGraw-Hill, New York.

Burns, F. L. (1911). A monograph on the Broad-winged Hawk (*Buteo playtypterus*). *Wilson Bull.* **23,** 139–320.

Carnie, S. K. (1954). Food habits of Golden Eagles in the coast ranges of California. *Condor* **56,** 3–12.

Cavè, A. J. (1968). The breeding of the Kestrel, *Falco tinnunculus* L., in the reclaimed area, Oostelijk Flevoland. *Neth. J. Zool.* **18,** 343–407.

Clark, P. J., and Evans, F. C. (1954). Distance to nearest neighbor as a measure of spatial relationships in populations. *Ecology* **35,** 445–453.

Clark, R. J. (1975). A field study of the Short-eared owl, *Asio flammeus* (Pontoppidan), in North America. *Wildl. Monogr.* **47,** 1–67.

Cone, C. D., Jr. (1962). Thermal soaring in birds. Am. Sci. **50,** 180–209.

Cottrell, M. J. (1981). Resource partitioning and reproductive success of hawks (*Buteo* spp.) in an Oregon prairie. M.S. thesis. Oregon State Univ., Corvallis.

Craighead, J. J., and Craighead, F. C., Jr. (1956). "Hawks, Owls, and Wildlife." Stackpole, Harrisburg, Pennsylvania.

Dixon, J. B. (1928). Life histories of the Red-bellied Hawk. *Condor* **30,** 228–236.

Dunkle, S. W. (1977). Swainson's Hawk on the Laramie Plains, Wyoming. *Auk* **94,** 65–71.

Fisher, A. K. (1893). The hawks and owls of the United States in their relation to agriculture. U.S. Dept. Agric. Div. Ornithol. and Mamm. Bull, Vol 3.

Fitch, H. S. (1974). Observations on the food and nesting of the Broad-winged Hawk (*Buteo platypterus*) in northern Kansas. *Condor* **76,** 331–337.

Fitch, H. S., Swenson, F., and Tillotson, D. F. (1946). Behavior and food habits of the Red-tailed Hawk. *Condor* **48,** 205–237.

Fitzner, R. E. (1978). Behavioral ecology of the Swainson's Hawk (*Buteo swainsoni*) in southeastern Washington. Ph.D. dissertation. Washington State Univ., Pullman.

Franklin, J. F. and Dyrness, C. T. (1973). Natural vegetation of Oregon and Washington. *U.S.D.A. For. Serv. Gen. Tech. Rep. PNW,* **8.**

Gabrielson, I. N. (1922). From field and study. Some hawks of Harney County, Oregon. *Condor* **24,** 33–34.

Gargett, V. (1975). The spacing of Black Eagles in the Matopos, Rhodesia. *Ostrich* **46,** 1–44.

Getz, L. L. (1961). Hunting areas of the Long-eared Owl. *Wilson Bull.* **73,** 79–82.

Gilmer, D. S., and Stewart, R. E. (1983). Ferruginous Hawk populations and habitat use in North Dakota. *J. Wildl. Manage* **47,** 146–157.

Gould, G. I., Jr. (1977). Distribution of the Spotted Owl in California. *West. Birds* **8,** 131–146.

Green, G. A., and Morrison, M. L. (1983). Nest-site characteristics of sympatric Ferruginous and Swainson's Hawks. *Murrelet* **64,** 20–22.

Gross, J. E., Stoddart, L. C., and Wagner, F. H. (1974). Demographic analysis of a northern jackrabbit population. *Wildl. Monogr* **40,** 1–68.

Hamerstrom, F., Hamerstrom, F. N., Jr., and Hart, J. (1973). Nest boxes: An effective management tool for Kestrels. *J. Wildl. Manage.* **37,** 400–403.

Hector, D. P. (1981). The habitat, diet, and foraging behavior of the Aplomado Falcon, *Falco femoralis* (Temminck). M.S. thesis. Oklahoma State Univ., Stillwater.

Henny, C. J., Schmid, F. C., Martin, E. L., and Hood, L. L. (1973). Territorial behavior, pesticides, and the population ecology of Red-shouldered Hawks in central Maryland 1943–1971. *Ecology* **54,** 545–554.

Houston, D. C. (1975). Ecological isolation of African scavenging birds. *Ardea* **63,** 55–64.
Howard, R. P., and Hilliard, M. N. (1980). Artificial nest structures and grassland raptors. *Rap. Res.* **14,** 41–45.
Howard, R. P., and Wolfe, M. L. (1976). Range improvement practices and Ferruginous Hawks. *J. Range Manage.* **29,** 33–37.
Howell, J., Smith, B., Holt, J. B., Jr., and Osborne, D. R. (1978). Habitat structure and productivity in Red-tailed Hawks. *Bird Banding* **49,** 162–170.
Jaksìc, F. M., Greene, H. W., and Yàñez, J. L. (1981). The guild structure of a community of predatory vertebrates in central Chile. *Oecologia* **49,** 21–28.
Janes, S. W. (1984). Influences of territory composition and interspecific competition on Red-tailed Hawk reproductive success. *Ecology* **65,** 862–870.
Janes, S. W. (1985a). Habitat relations among three hawk species in western North America. Unpublished manuscript.
Janes, S. W. (1985b). Territorial relations among three species of *Buteo* hawks: Some implications of hawk morphology and prey population dynamics and density. Unpublished manuscript.
Janes, S. W. (1985c). Territorial relations among the Golden Eagle and three hawk species in western North America. Unpublished manuscript.
Keister, G. P., Jr., and Anthony, R. G. (1983). Characteristics of Bald Eagle communal roosts in the Klamath Basin, Oregon and California. *J. Wildl. Manage.* **47,** 1072–1079.
Keran, D. (1978). Nest site selection by the Broad-winged Hawk in north-central Minnesota and Wisconsin. *Rap. Res.* **12,** 15–20.
Koplin, J. R. (1973). Differential habitat use by sexes of American Kestrels wintering in northern California. *Rap. Res.* **7,** 39–42.
Krebs, C. J., Gaines, M. S., Keller, B. L., Myers, J. H., and Tamarin, R. H. (1973). Population cycles in small rodents. *Science* **179,** 35–41.
Lardy, M. E. (1980). Raptor inventory and Ferruginous Hawk breeding biology in southeastern Oregon. M.S. thesis., Univ. of Idaho, Moscow.
Lien, M. R., and Boxall, P. C. (1979). Interactions between Snowy and Short-eared Owls. *Can. Field-Nat.* **93,** 411–414.
Lien, M. R., and Webber, G. A. (1979). Habitat selection by wintering Snowy Owls (*Nyctea scandiaca*). *Can. Field-Nat.* **93,** 176–178.
Lokemoen, J. T., and Duebbert, H. F. (1976). Ferruginous Hawk nesting ecology and raptor populations in northern South Dakota. *Condor* **78,** 464–470.
Lundberg, A. (1979). Residency, migration and a compromise: Adaptations to nest-site scarcity and food specialization in three Fennoscandian owl species. *Oecologia* **41,** 273–281.
Luttich, S., Rusch, D. H., Meslow, E. C., and Keith, L. B. (1970). Ecology of Red-tailed Hawk predation in Alberta. *Ecology* **51,** 190–203.
MacArthur, R. H. (1972). "Geographical Ecology." Harper, New York.
McInvaille, W. B., Jr., and Keith, L. B. (1974). Predator–prey relations and breeding biology of the Great Horned Owl and Red-tailed Hawk in central Alberta. *Can. Field-Nat.* **88,** 1–20.
Mader, W. J. (1978). A comparative nesting study of Red-tailed Hawks and Harris' Hawks in southern Arizona. *Auk* **95,** 327–337.
Marion, W. R., and Ryder, R. A. (1975). Perch-site preferences of four diurnal raptors in north-eastern Colorado. *Condor* **77,** 350–352.
Matray, P. F. (1974). Broad-winged Hawk nesting and ecology. *Auk* **91,** 307–324.
Mills, G. S. (1975). American Kestrel sex ratios and habitat selection. *Auk* **93,** 740–748.
Morris, M. M. J., and Lemon, R. E. (1983). Characteristics of vegetation and topography near Red-shouldered Hawk nests in southwestern Quebec. *J. Wildl. Manage.* **47,** 138–145.
Morris, M. M. J., Penak, B. L., Lemon, R. E., and Bird, D. M. (1982). Characteristics of Red-shouldered Hawk, *Buteo lineatus,* nest sites in southwestern Quebec. *Can. Field-Nat.* **96,** 139–142.

Mosher, J. A., and White, C. M. (1976). Directional exposure of Golden Eagle nests. *Can. Field-Nat.* **90,** 356–359.

Newton, I. and Marquiss, M. (1976). Occupancy and success of nesting territories in the European Sparrowhawk. *Rap. Res.* **10,** 65–71.

Ogden, J. C. (1975). Effects of Bald Eagle territoriality on nesting Ospreys. *Wilson Bull.* **87,** 496–505.

Olendorff, R. R. (1976). The food habits of North American Golden Eagles. *Am. Midl. Nat.* **76,** 231–236.

Olendorff, R. R., and Stoddart, J. W., Jr. (1974). Potential for management of raptors in western grasslands. *In* "Management of raptors." (F. N. Hamerstrom, Jr., B. E. Harrell, and R. R. Olendorff, eds.), Rap. Res. Rep., No. 2, pp. 47–88. Raptor Research Foundation, Inc. Vermillion, South Dakota.

Opdam, P. (1975). Inter- and intra-specific differentiation with respect to feeding ecology in two sympatric species of the genus *Accipiter. Ardea* **63,** 30–54.

Orde, C. J., and Harrell, B. E. (1977). Hunting techniques and predatory efficiency of nesting Red-tailed hawks. *Rap. Res.* **11,** 82–85.

Orians, G., and Kuhlman, F. (1956). Red-tailed Hawk and Horned Owl populations in Wisconsin. *Condor* **58,** 371–385.

Pearson, O. P., and Pearson, A. K. (1947). Owl predation in Pennsylvannia, with notes on the small mammals of Delaware County. *J. Mammal.* **28,** 137–147.

Pennycuick, C. J. (1972a). "Animal Flight." Arnold, London.

Pennycuick, C. J. (1972b). Soaring behaviour and performance of some east African birds, observed from a motor glider. *Ibis* **114,** 178–218.

Pennycuick, C. J. (1978). Mechanics of flight. *In* "Avian Biology." D. S. Farner, and J. R. King, eds., Vol. V, pp. 1–76.

Pettifor, R. A. (1983). Territorial behaviour of Kestrels in arable fenland. *Br. Birds* **76,** 206–214.

Picozzi, N., and Wier, D. (1974). Breeding biology of the Buzzard in Speyside. *Br. Birds* **67,** 199–210.

Platt, J. B. (1976). Sharp-shinned Hawk nesting and nest site selection in Utah. *Condor* **78,** 102–103.

Poole, E. L. (1938). Weights and wing areas in North American birds. *Auk* **55,** 511–517.

Preston, C. R. (1980). Differential perch-site selection by color morphs of the Red-tailed Hawk (*Buteo jamaicensis*). *Auk* **97,** 782–789.

Reynolds, R. T., Meslow, E. C., and Wight, H. M. (1982). Nesting habitat of coexisting accipiters in Oregon. *J. Wildl. Manage.* **46,** 124–138.

Rudolph, S. G. (1978). Predation ecology of coexisting Great Horned and Barn Owls. *Wilson Bull.* **90,** 134–137.

Rüppell, G. (1975). "Bird Flight." Van Nostrand-Reinhold, Princeton, New Jersey.

Rusch, D. H., Meslow, E. C., Doerr, P. D., and Keith, L. B. (1972). Response of Great Horned Owl populations to changing prey densities. *J. Wildl. Manage.* **36,** 282–296.

Schipper, W. J. A. (1977). Hunting in three European harriers (*Circus*) during the breeding season. *Ardea* **65,** 53–72.

Schipper, W. J. A. (1978). A comparison of breeding ecology in three European harriers (*Circus*). *Ardea* **66,** 77–102.

Schipper, W. J. A., Buurma, L. S., and Bossenbroek, P. (1975). Comparative study of hunting behaviour of wintering Hen Harriers *Circus cyaneus* and Marsh Harriers *Circus aeruginosus. Ardea* **63,** 1–29.

Schmutz, J. K., Schmutz, S. M., and Boag, D. A. (1980). Coexistence of three species of hawks (*Buteo* spp.) in the prairie-parkland ecotone. *Can. J. Zool.* **58,** 1075–1089.

Schnell, G. D. (1968). Differential habitat utilization by wintering Rough-legged and Red-tailed Hawks. *Condor* **70,** 373–377.

Sharp, C. S. (1902). Nesting of Swainson's Hawk. *Condor* **4,** 116–118.

Slade, N. A., and Balph, D. F. (1974). Population ecology of Uinta Ground Squirrels. *Ecology* **55,** 989–1003.

Smith, D. G. (1971). Population Dynamics, Habitat Selection, and Partitioning of Breeding Raptors in the Eastern Great Basin of Utah. Ph.D. dissertation, Brigham Young Univ., Provo, Utah.

Smith, D. G., and Murphy, J. R. (1973). Breeding ecology of raptors in the eastern Breat Basin of Utah. Sci. Bull. Biol. Ser., Vol. 18, pp. 1–76. Brigham Young Univ., Provo, Utah.

Smith, D. G., and Murphy, J. R. (1982). Spatial relationships of nesting Golden Eagles in central Utah. *Rap. Res.* **16,** 127–132.

Snyder, N. F. R., and Snyder, H. A. (1969). A comparative study of mollusc predation by Limpkins, Everglade Kites, and Boat-tailed Grackles. *Living Bird* **8,** 177–223.

Sokal, R. R., and Rolhf, F. J. (1969). "Biometry." W. H. Freeman, San Francisco, California.

Southern, H. N., and Lowe, V. P. W. (1968). The pattern of distribution of prey and predation in Tawny Owl territories. *J. Anim. Ecol.* **37,** 75–97.

Sprunt, A., Jr. (1955). "North American Birds of Prey." Harper, New York.

Stalmaster, M. V., and Newman, J. R. (1979). Perch-site preferences of wintering Bald Eagles in northwest Washington. *J. Wildl. Manage.* **43,** 221–224.

Stewart, R. E. (1949). Ecology of a nesting Red-shouldered Hawk population. *Wilson Bull.* **61,** 26–35.

Stoddart, L. C., and Anderson, R. D. (1972). Biomass density of lagomorphs. U.S. Int. Biol. Program Curlew Valley Validation Site Report, no. RM72-1.

Sylvèn, M. (1978). Interspecific relations between sympatrically wintering Common Buzzards, *Buteo buteo,* and Rough-legged Buzzards, *Buteo lagopus. Ornis Scand.* **9,** 197–206.

Thiollay, J. M. (1967). Ecologie d'une population de rapaces diurnes en Lorraine. *Terre et Vie* **21,** 116–183.

Thiollay, J. M., (1981). Comparisons entre les peuplements de Falconiformes des plaines cotières du Méxique et de côte-d'ivoire. *Le Gerfaut* **68,** 139–162.

Thurow, T. L., White, C. M., Howard, R. P., and Sullivan, J. F. (1980). Raptor ecology of Raft River Valley, Idaho. EEG-2054. E. G. and G. Idaho, Idaho Falls, Idaho.

Titus, K., and Mosher, J. A. (1981). Nest site habitat selected by raptors in the central Appalachians. *Auk* **98,** 270–281.

Tubbs, C. R. (1967). Population study of Buzzards in the New Forest during 1962–1966. *Br. Birds* **60,** 381–395.

U.S. Department of the Interior (USDI) (1979). Snake River Bird of Prey Special Report. U.S. Dept. of the Int. Bur. of Land Manage., Washington, D.C.

Village, A. (1983). The role of nest-site availability and territorial behaviour in limiting the breeding density of Kestrels. *J. Anim. Ecol.* **52,** 635–645.

Wakeley, J. S. (1978). Factors affecting use and hunting methods of Ferruginous hawks. *Condor* **80,** 316–334.

Wakeley, J. S. (1979). Use of hunting methods by Ferruginous Hawks in relation to vegetation density. *Rap. Res.* **13,** 116–119.

Warner, J. S., and Rudd, R. L. (1975). Hunting by the White-tailed Kite *Elanus leucurus. Condor* **77,** 226–230.

Weller, M. W. (1964). Habitat utilization of two species of Buteos wintering in central Iowa. *Iowa Bird Life* **34,** 58–62.

Weston, J. B. (1969). Nesting ecology of the Ferruginous Hawk (*Buteo regalis*). Sci. Bull. Biol. Ser., Vol. 10, 25–36. Brigham Young Univ., Provo, Utah.

Woffinden, N. D., and Mosher, J. A. (1979). Ground-nesting and aggressive behavior in the Swainson's Hawk. *Great Basin Nat.* **39,** 253–254.

Woffinden, N. D., and Murphy, J. R. (1983). Ferruginous Hawk nest site selection. *J. Wildl. Manage.* **47,** 216–219.

Part III

Habitat Selection in Specific Habitat Types

Chapter 6

Habitat Selection in Grassland and Open-Country Birds

MARTIN L. CODY

Department of Biology
University of California at Los Angeles
Los Angeles, California

I. INTRODUCTION

Grasslands occur in most temperate and subtropical regions of the world where annual temperatures average 6–16°C and rainfall averages 40–80 cm (see, e.g., Risser *et al.*, 1981, for an introduction to grassland climate and distribution). Grassland merges into shrub steppe and then desert, with decreasing precipitation, and into brush, woodland, and then forest, with increasing precipitation,

HABITAT SELECTION IN BIRDS

ISBN 0-12-178080-5

but temperature, fire, substrate, and historical factors also affect the dominance of grasses. The birds that inhabit grasslands are usually quite characteristic of and restricted to this vegetation; in the Old World such passerine bird families as larks (Alaudidae), pipits (Motacillidae), and buntings (Emberizidae) are dominant, with the addition of weaverfinches (Ploceidae and Estrildidae), starlings (Sturnidae), and some sylviine warblers (*Acrocephalus* and *Cisticola*) at lower latitudes. In the New World there are few larks and pipits, but many buntings and Icteridae (meadowlarks and blackbirds), and, in South America, some grassland Tyrannidae and Furnariidae.

Besides passerine birds, grasslands are the home of certain "waders," plovers, and sandpipers (Charadriidae and Scolopacidae), although these birds are more diverse in wetter tundra habitats at higher latitudes. This chapter reviews some of the characteristic patterns of habitat selection in grassland and open-country birds in the breeding and nonbreeding seasons. The emphasis is on the breeding-bird communities of North American grasslands (Fig. 1), but other temperate and grassland systems are discussed, as well as desert-plains birds in North and South Africa. The aim is to present whatever generalities appear to be viable and to evaluate the extent to which fluctuating climates, bird populations, and their food resources combine to generate the observed patterns of habitat selection.

A. Bird Distribution, Diversity, and Density

Visibility is normally good for the ornithologist in grasslands, and since most species have a conspicuous territorial defense, by song and often aerial flights, accurate data can be collected without enormous efforts. Both diversity and density are low in grassland birds compared to most other habitats. Census sites in homogeneous vegetation usually produce 2–6 passerine species, with perhaps as many again of nonpasserines, and around 0.5–2 pairs/ha. But as I showed earlier (Fig. 4 in Cody, 1974a), species diversity is really a function of the habitat diversity included in the census area; censuses over structurally diverse areas might include a dozen passerine species. Some of these birds are specific to certain sorts of grasslands with certain structural or other features, and others are less selective. Henslow's Sparrows (*Ammodramus henslowii*) are found only in broomsedge fields with song perches in the northeastern United States (Robins, 1971), whereas Savannah Sparrows (*Passerculus sandwichensis*) occur over northern North America in tallgrass prairies, weedy field edges, swamps, tundras, and coastal marshes. Some species have narrow geographic ranges, and others are widespread. Rufous-winged Sparrows (*Aimophila carpalis*) occur only in scrubby-desert grasslands in southern Arizona and extreme northwestern Mexico, but Grasshopper Sparrows (*Ammodramus savannarum*), while most typical of mixed-grass prairies in the central United States, breed from east to west coast

Fig. 1. The birds of North America grasslands and related habitats.

and from Canada to Mexico and Central America; the congeneric Baird's Sparrows (*A. bairdii*) are restricted to northern mixed prairies in southern Canada and the adjacent Dakotas.

The only native lark in the New World is the Horned Lark (*Eremophila alpestris*), which breeds from the Arctic Ocean to Colombia, South America, in tundra, prairie, and desert habitats, with vegetation predictably very low and very open. In the Old World, where lark species are diverse, Horned Larks occur in arctic and alpine tundra but not in grasslands, where their presence is preempted by such species as the Skylark (*Alauda arvensis*). This latter species was introduced into North America on Vancouver Island 80 years ago; it has since undergone a modest increase (from 150 to 1000 birds by 1962) and expanded its range to the adjacent San Juan Islands (Weisbrod and Stevens, 1974).

The ranges and densities of grassland birds show considerable fluctuations, in response to both short- and long-term climatic fluctuations, to local factors such as fires and food supply variations (see Sections II and VII), and also to man's modification of the environment (see Chapter 1, this volume; also see Stepney and Power, 1973; Hurley and Franks, 1976, for northeasterly expansions of prairie birds following deforestation in the United States). A recent assessment of changes in prairie birds over half a century was published by Blankespoor and Krause (1982), with the general conclusion that ranges and densities were up in species associated with water (Short-billed Marsh Wrens and Swamp Sparrows), but down in a variety of true grassland species.

Breeding-bird densities may vary at a site between years and among sites both locally with habitat differences and regionally with changes in a wider variety of factors. For example, Horned Lark territory size may vary locally by a factor of 5 (Beason and Franks, 1974) or a factor of 12 (Pickwell, 1931) with habitat quality, the largest territories being on more recently plowed fields. In Grasshopper and Savannah Sparrows, variability in territory size may be much less, around 25% as measured by Wiens (1973), with smaller territories being centrally located and established earlier in the season and perhaps of higher quality. Breeding density is variable in Dickcissels (*Spiza americana*), and as density increases, less preferred habitat is included in territories and reproductive success declines (Zimmerman, 1971). Raitt and Pimm (1976) discussed the general decrease in breeding-bird density from north to south in the Great Plains grasslands, the decrease parallel to the gradient of decreasing annual rainfall, and the general increase in wintering-bird densities in these grasslands, from north to south. The southern wintering populations exploit seeds produced by late summer rains (see Section VII).

B. Habitats and Foraging Ecology

Grasslands vary in structure with climate, especially rainfall, from tallgrass prairies, with up to 80-cm annual precipitation, to mixed-grass prairies, with about half this rainfall, to shortgrass plains and desert grasslands, with annual rainfall again 50% lower. These grassland types in North America generally follow a northeast to southwest distribution, but further modifications follow grazing, fire, and other local edaphic variables. Within a particular grassland, vegetation may appear superficially uniform, but only rarely is it so. Thus, grassland birds may exhibit habitat selection on a regional scale, since most are summer residents only in their breeding habitats, and also on a local scale, at which the differences from one part of a field to another may be significant. Species may be widespread throughout one grassland type but local and habitat restricted in another. Thus, Horned Larks are common throughout the shortgrass plains in eastern Colorado and Wyoming, but occur only in upland mixed-grass prairies in eastern Kansas (Cody, 1968), and only after fires in mixed-grass prairies in south-central North Dakota (Johnson, 1972–1982). Conversely, Grasshopper Sparrows are more widely distributed in mixed-grass prairies in Kansas and North Dakota, but occur in the shortgrass plains only in wetter years, when some low-lying areas are temporarily waterlogged and their pastures unusually dense (M. L. Cody, personal observation, in Weld County, northeastern Colorado).

In general, habitat selection appears to be more precise in the bird species of taller grasslands, which differentiate among sites on the basis of vegetation height and density, and less characteristic of short-grassland species, which appear to be less responsive to structural cues and more opportunistic in just where their territories are located (Cody, 1968); mixed-grass birds are intermediate in this respect.

Most grassland bird species appear to be generalized omnivores, in terms of their diets, and feed on both insects and seeds. There is a higher proportion of insects taken in the breeding season, and a preponderance of insects are fed to nestlings; figures of 60–85% insects, versus seeds, have been published (Baldwin, 1973; Maher, 1979; Wiens and Rotenberry, 1979). These same studies have shown that a good deal of diet overlap is typical among different species at the same site, that common food items may vary from year to year in diets as well as at large, and that interspecific diet overlaps likewise vary between years. A study of three species of sparrows in a brushy Michigan field [Vesper Sparrows (*Pooecetes gramineus*), Field Sparrows (*Spizella pusilla*), and Chipping Sparrows (*S. passerina*)] likewise found no dietary differences among species (Evans, 1964). In a similar habitat in eastern Texas, the two *Spizella* species coexist with Bachman's Sparrows (*Aimophila aestivalis*), where all three species eat primarily grass seeds (Allaire and Fisher, 1975); however, Bachman's Sparrows are rather different from the other two, in specializing on *Panicum* seeds taken at ground level. Wiens and Rotenberry (1980) found only weak correlations between bill morphology and diet and emphasized the opportunistic nature of foraging grassland birds.

Although diets may be similar or unpredictable in grassland birds, foraging behavior can be very different among species, and similar food items may be captured using quite different techniques. Here the structure of the grassland vegetation appears to be extremely important, for in tall and dense grassland little more than slow searching is possible, whereas in shorter and more open grassland a wider variety of behaviors is practiced. Thus, in tallgrass prairies, bird body size is smaller, and bill morphologies (mostly short and deep) and foraging behaviors (slow searching at or near ground level) are more uniform among species. Such birds include Savannah Sparrows, LeConte's Sparrows (*Ammospiza lecontei*), Henslow's Sparrows, Dickcissels (*Spiza americana*), and Bobolinks (*Dolichonyx oryzivorus*); territory overlap among these species is usually low, because of species-specific habitat preferences, as well as direct interspecific interactions (see Section II,A).

In contrast, in shortgrass vegetation, body sizes are more variable and include larger species, bill morphologies are less uniform, and feeding behaviors range widely from slow searching and probing to rapid darting and short aerial flights after disturbed insects (see Fig. 2). Typical species are the Horned Lark, the longspurs *Calcarius,* the Lark Bunting (*Calamospiza melanocorys*), the meadowlarks *Sturnella* and pipits *Anthus,* as well as a variety of plovers [Mountain Plovers (*Eupoda montana*), Killdeers (*Charadrius vociferus*)] and sandpipers [Upland Sandpiper (*Bartramia longicauda*), Marbled Godwits (*Limosa fedoa*), and Willets (*Catoptrophorus semipalmatus*)] that do not venture into taller grasslands.

The extent to which foraging behavior is constrained by vegetation density is seen in Starlings (*Sturnus vulgaris*), which forage twice as fast, stop half as often, and spend half as much time stationary in short versus tall grassland (Brownsmith, 1977). Similarly, Horned Larks, which normally nest in short

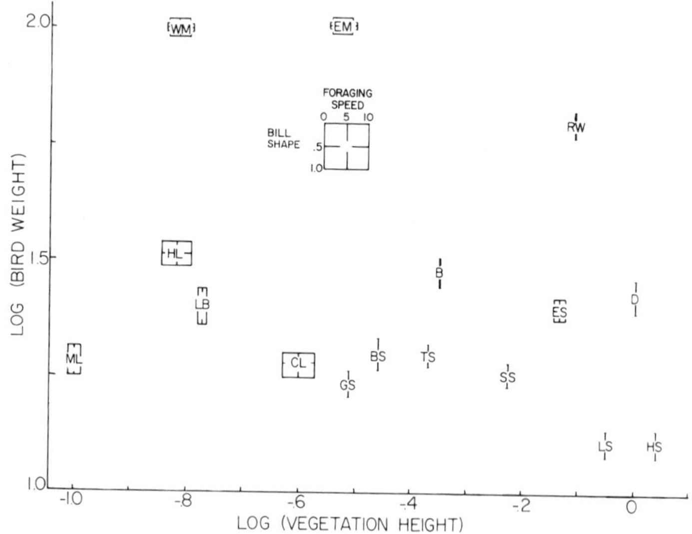

Fig. 2. Body weight (gm), bill shape (depth/length), and foraging speed (m/min) of grassland birds as a function of vegetation height (m). Birds in shorter grassland to the left occur in a wider range of morphologies and display a wider range of foraging behavior. Species abbreviations: WM, Western Meadowlark; EM, Eastern Meadowlark; HL, Horned Lark; ML, McCown's Longspur; LB, Lark Bunting; CL, Chestnut-collared Longspur; GS, Grasshopper Sparrow; BS, Baird's Sparrow; TS, Sharp-tailed Sparrow; B. Bobolink; SS, Savannah Saprrow; ES, Seaside Sparrow; RW, Red-winged Blackbird; D, Dickcissel; LS, LeConte's Sparrow; HS, Henslow's Sparrow. Data on bill shape, foraging speed, and vegetation height from Cody (1968); bird weights from the literature.

grassland or in grazed or burned mixed-grass prairie in the Great Plains, were found to forage inefficiently in ungrazed mixed-grass prairie in Saskatchewan, where their chicks starved in the nest; chicks grew normally in nests in adjacent grazed grassland.

C. Population and Resource Fluctuations

Climates are certainly severe in the Central Plains grasslands, with bitterly cold winters and frequent blizzard conditions, and hot, often dry, summers broken by occasional violent storms, during which small birds are endangered by golfball-sized hailstones. Of course, there are few birds present in the winter time, and the extent to which these climates are unpredictable rather than merely seasonal must be qualified. It does appear that the Central Plains weather is indeed unpredictable; of seven western North American sites for which climate data were analyzed earlier (see Appendix A in Cody, 1974a), the shortgrass prairie weather was least predictable in all three variables, rainfall variance, temperature variance, and bad-weather probability. In the absence of a great deal of direct measurement, it seems reasonable to assume that resources for grassland birds will also be variable and somewhat unpredictable. This in turn may mean that species and densities are not accurately predicted from measures of vegetation structure and that recruitment rates of species are variable from year to year.

Further, grassland birds may overwinter in areas of unpredictable climate and therefore unpredictable resources. Many plains birds utilize the seeds of annual plants in southern desert areas in the winter time; this seed production varies with the summer rainfall. Both Dunning and Brown (1982) and Pulliam and Parker (1979) found that the numbers of wintering sparrows in Arizona were proportional to the production of grass seeds from summer rains; in years of poor production the sparrows were forced to migrate facultatively further south for overwinter resources. There is thus a variety of variable environmental factors that aggravate a finely tuned, resource-coupled community structure in grassland-bird communities.

In view of climatic unpredictability and concommitant unreliability of resources in grassland habitats, how will natural selection mold the habitat selection behavior of the bird species? One option is that birds might not be site or locality specific on their return to breeding areas from wintering grounds but, rather, might breed in any location that satisfies certain requirements of vegetation structure and food supply. Thus, we might expect not philopatry, but an opportunistic use of more productive areas in any given year, a "tracking" of resource hot spots. Second, this tracking might be somewhat independent of habitat structure, since in some years certain habitats might be unproductive. Thus, a flexibility in the habitat selection process might be favored, such that criteria of structural suitability alone might be insufficient, and futher criteria of resource productivity might have to be satisfied.

The major questions in grassland-bird community structure and habitat selection become: (1) To what extent can these bird populations track a variable and shifting resource base, given that this base may be below thresholds for successful breeding at some times in some places and, at other times and places, provide essentially a superabundance of food? (2) To what extent do grassland-bird communities become ecologically saturated, such that selection can reinforce interspecific differences in foraging behavior and morphology where resources are contested sufficiently that these differences affect rates of food harvest and, consequently, reproductive success? The evidence that grassland-bird communities do become at least locally saturated comes from observations of regularities in resource partitioning as previously reported (Cody, 1968), but evidence to the contrary coming largely from unpredictable species composition, broad dietary overlaps, and poor correlations between diet and morphology certainly exists (Wiens, 1974; Wiens and Rotenberry, 1980; Rotenberry, 1980). Evidence that resource or habitat tracking is at least imprecise has been emphasized by Wiens and Rotenberry (1981) and Rotenberry and Wiens (1980a,b), but the species changes in grasslands correlated with changing climate and changing habitat structure support the view that some degree of resource tracking does take place (see Section II,A,B).

Perhaps some degree of resolution of the two extreme views would follow (1) the collection of more detailed information on food availability and the extent to which varying food availability affects reproductive success, (2) an evaluation of the extent to which food supplies are climate dependent and independent of habitat structure, and (3) the realization that selection for interspecific phenotypic differences in traits associated with food gathering need not operate at all times in all places.

D. Territoriality and Breeding Systems

Grassland habitats and the bird territories established within them are likely to be quite variable in quality, in terms of food supplies and expected reproductive success. This basic fact was used by Orians (1969) to develop a model relating the fitness of females mated monogamously versus polygynously to males on territories of variable quality. Although initially written for marshland habitats, the model is widely applicable and predicts that polygyny will be common wherever territory quality is widely variable, and females will be better off as second mates to mated males on premium territory than as sole mates of males on poorer territories. Emlen and Oring (1977) added an additional factor of the temporal availability of mates as a governing force in polygyny.

In grassland birds polygyny is relatively common, adding support to the notion of variable resource distribution in these habitats. There is sequential polygyny in

Bobolinks, in which structural habitat features are related to reproductive success. The second female feeds her significantly smaller brood alone, feeds closer to the nest, and feeds from a wider variety of food items (Martin, 1974). Wittenberger (1980) found that in this species successfully polygynous males select territories with higher caterpillar and grasshopper densities in areas with a higher percentage cover of forbs than the territories of unmated males. Similarly, food availability was higher in the territories of polygynously mated Dickcissels than in bachelor territories (Harmeson, 1974); levels of polygyny are higher in old fields than in prairies, and males, but not necessarily females, selectively chose the old-field habitats (Zimmerman, 1982).

In Nova Scotia some male Savannah Saprrows mate polygynously, and these tend to be the males with the larger territories (Welsh, 1975). However, polygyny is less advantageous to females than to males, even though females were able to double their feeding rates to nestlings when no assistance from the male was forthcoming. Near Hudson Bay, Manitoba, this species is monogamous, perhaps because the breeding season is too short for sequential polygyny, mates are of limited availability, and the variance of territory quality is equalized by the presence of Tree Sparrows (*Spizella arborea*) in the tall-willow, but not the short-willow, habitats (Weatherhead, 1979). Many grass- and marshland sylviine warblers are likewise polygynous (see Chapter 3, this volume), as are grassland ploceids such as *Euplectes* (Lack 1935; Crook, 1962). A grassland warbler of tropical Africa, *Cisticola anonyma,* has been shown to defend group territories, and the young of previous broods help to feed the young in later nests (Gowthorpe, 1977).

From shorter to taller grassland and thence to more productive marshlands there is a trend from monogamous to polygynous territoriality and thence to coloniality. All shortgrass species are apparently monogamous, and polygyny appears only in the taller grasslands occupied by such species as Bobolinks and Savannah Sparrows. Seaside Sparrows (*Ammospiza maritima*) defend only small "activity spaces" and feed colonially (Post and Greenslaw, 1975), and Sharp-tailed Sparrows (*Ammospiza caudacuta*) of marshes and wet grasslands exhibit a loose coloniality (Cody, 1968; Murray, 1969); the colonial breeding systems of icterids (Orians, 1978) and the convergently similar ploceids of African marshes complete the trend. It might appear that not only variable territory quality and a ready availability of mates are prerequisite for polygyny but so is a higher level of productivity than that usually found in shorter grassland.

E. Interspecific Interactions

Interactions between species may be direct and involve open aggression and interspecific territoriality or indirect via more subtle effects on each other's use of space or food resources. The former is easily recognized and documented and is often referred to as "interference competition"; the latter is not nearly so easily documented, and such "exploitation competition" must often be inferred from patterns of resource use, niche shifts in variable competitive environments, and other correlative techniques (Cody, 1974a). For example, the Song Sparrow (*Melospiza melodia*) is not normally a grassland bird, but it occupies grassland on Mandarte Island in the Georgia Strait, where the usual grassland birds of the adjacent mainland, Savannah Sparrows, Skylarks, Western Meadowlarks (*Sturnella neglecta*), and Brewer's Blackbirds (*Euphagus cyanocephalus*) are absent (Cody, 1974a). Similarly, the effects of subtle interactions between Field Sparrows and Chipping Sparrows, which are often labeled as strong competitors on the basis of similar diets and foraging sites (e.g., Allaire and Fisher, 1975), were shown by Hebrard (1978) to directly affect each other's use of space. Laboratory and field studies showed differences in habitat use by these two species, with Field Sparrows in more open, less arboreal habitats; yet when Chipping Sparrows are not present, Field Sparrows shift to a heavier use of arboreal habitats.

Often inferences on actual or potential resource competition, from observation of similar foraging ecology or of habitat segregation between species, are difficult to substantiate, to which resulting uncertainty a large literature now attests. Experimental data are of course far less equivocal; the removal of Golden-crowned Sparrows (*Zonotrichia atricapilla*) from willow thickets by Davis (1974) resulted in their increased use by Oregon Juncos (*Junco hyemalis*). In the presence of the sparrows, the juncos occur only in bordering areas, and their exclusion from the willows is related to the former's preference for and dominance in habitats close to water supplies.

One approach to demonstrating the existence of indirect interactions among species over habitats is a comparison of interspecific overlap in space with interspecific overlap in habitat as measured by its structural characteristics. This was done for four sparrows breeding in a Wyoming sagebrush–grassland area: Brewer's Sparrows (*Spizella breweri*), Vesper Sparrows (*Pooecetes gramineus*), White-crowned Sparrows (*Zonotrichia leucophrys*), and Savannah Sparrows (see Table 16 in Cody, 1974a). Overlap in space and overlap in habitat features were very similar in all species pairs except those involving Savannah Sparrows, which showed a much lower territorial overlap with the other three species than might be expected based on its habitat preferences. Although no overt aggression among species was noted, Savannah Sparrows are presumably excluded from habitats they would otherwise occupy by subtle interactions with the other three species.

Direct interactions between species are relatively common among the birds of taller grasslands, species which are relatively similar in foraging ecology and segregate largely by differences in habitats. Such interactions are not found in short-grassland birds; these are species which differ largely in foraging behavior but are similar in habitat use. Direct aggression or interspecific territoriality has been observed between Baird's and Grasshopper Sparrows (Cody, 1974a), Grasshopper and Henslow's Sparrows (Robins, 1971), LeConte's (*Ammospiza lecontei*) and Sharp-tailed Sparrows (Murray, 1969), Sharp-tailed and Seaside Sparrows (Post and Greenslaw, 1975), Savannah and Song Sparrows (Welsh 1975), and Field and Chipping Sparrows (Evans, 1978). Interspecific territoriality is also common in marshland birds, which led Orians and Willson (1964) to write the first substantial paper on the subject.

II. NORTH AMERICAN GRASSLANDS

A. Tallgrass Prairies

Tallgrass prairie is characterized by Big Bluestem (*Andropogon gerardi*) and Switch (*Panicum virgatum*) Grasses and a host of forbs, and its main distribution is from northeastern Oklahoma due north to the Red River valley on the Minnesota–North Dakota border. In the south, typical and common bird species are the Dickcissel, the Eastern Meadowlark, and the Grasshopper Sparrow; in the north, the Bobolink, the Savannah Sparrow, and LeConte's Sparrow. The Red-winged Blackbird and the Common Yellowthroat are widely distributed in the wetter sites (see, e.g., Zimmerman and Tatschl, 1975, for typical census data). Henslow's Sparrows and Field Sparrows occur in brushier prairies in the north and east of this range.

Bird diversity varies with census area and its homogeneity, but 3–4 common passerine species are usual, with total densities averaging 0.8 ± 0.4 pairs/acre (1.8 pairs/ha); some two dozen censuses in tallgrass prairies reported in the last decade in *American Birds* produced these figures. Species composition varies not only with geography, but with vegetation structure in the census area, following natural variations, burning, and grazing. For example, Knodel (1980) censused four sites in eastern Kansas, two of which were annually burned. Henslow's Sparrows occurred only in unburned prairies and Kildeers only in burned prairies. Further, Grasshopper Sparrows were more common in the former and Upland Sandpipers and Mourning Doves more common in the latter, more open habitats. Total diversity and density were not, however, much affected by the fires. At a site in northeastern Oklahoma, both Eastern Meadowlarks and Dickcissels were much more common in grazed grassland, and ungrazed grassland was lacking Grasshopper Sparrows (Risser *et al.*, 1981).

A series of 9 years of census data have been reported by Cink (1974), Lowther and Seibel (1982) and their associates from floodplain tallgrass prairie in Douglas County, eastern Kansas. These censuses show a remarkable repeatability between years, especially impressive since only one of the breeding birds overwinters there (e.g., Cink, 1983). The four common breeding birds are Dickcissels, Red-winged Blackbirds, Common Yellowthroats (*Geothlypis trichas*), and Eastern Meadowlarks, in that order of abundance in every year except one, when Common Yellowthroats outnumbered the blackbirds. Only two other species have been recorded in the 25-acre site in the 9 years 1974–1982, Bobwhite Quails (*Colinus virginianus,* twice) and Short-billed Marsh Wrens (*Cistothorus platensis,* once). Bird densities are shown in Table I.

Total bird density varied twofold during this 9-year period, mostly attributable to variable Dickcissel abundance, the species with the most extensive migration, to Central America, and a species which figures prominently in Fretwell's (1972) theory of the limitation of sparrow breeding density by overwintering resources. In conformance with that view, there are no correlations between Dickcissel breeding densities and year-to-year weather variations.

This is not true, however, for the other three common species at the site. The density of Red-winged Blackbirds (DENRWB) is predicted by the amount of year-to-date rainfall (YTDRFL; see Table I), which accounts for 44% of the observed variation in density (DENRWB = 2.852 + 0.164(YTDRFL); $F = 5.44$, $p = 0.05$). Both Common Yellowthroat and Eastern Meadowlark (DENEML) densities are positively correlated with spring rainfall (AMRFLL) and negatively correlated with rainfall over the previous 12 months (TOTRFL); both species evidently benefit from wet springs following dry years. These two factors account for 27% of the warbler's variability (not significant) and 57% of that of the meadowlark: DENEML = 3.084 + 0.401(AMRFLL) −0.108(TOTRFL); $F = 3.96$, $p = 0.06$. Thus, it appears that together with a high species predictability at the site, there are year-to-year density variations in three species that reflect recent weather and, presumably, current conditions of resource availability there.

The Dickcissel, however, appears to be influenced by conditions outside the immediate vicinity of this grassland. In this context, it is this species that is largely granivorous in the breeding season, in contrast to the greater amount of insect prey of the other three species (see, e.g., Risser *et al.*, 1981). Conceivably, seed production at the site is less strongly correlated to recent weather than is insect production.

B. Mixed-Grass Prairies

Where precipitation decreases, tallgrass prairies give way to mixed-grass prairies, and the ≥1-m-tall *Andropogon* bluestems are replaced by shorter grasses, such as wheatgrasses (*Agropyron*), needlegrasses (*Stipa*), June grass (*Koeleria*), and brome grasses (*Bromus*).These grasslands usually support an emberizid or two such as the Grasshopper, the Baird's or the Savannah Sparrow, a meadowlark, and one of the more actively foraging species such as the Chestnut-collared Longspur (*Calcarius ornatus*) or the Horned Lark. In Pottawatomie County, eastern Kanses, I found Grasshopper Sparrows, Eastern Meadowlarks,

TABLE I

Bird Census and Weather Data for 9 Years from an Eastern Kansas Floodplain Tallgrass Prairie[a]

Year	Dickcissel	Red-winged Blackbird	Common Yellowthroat	Eastern Meadowlark	Other	Rainfall				Excess spring temperatures[f]
						12 months[b]	YTD[c]	MAMJ[d]	AM[e]	
1974	10	5	2	2	—	52.79	17.56	14.45	9.00	+2.1
1975	14	7	5	4	—	36.28	17.86	14.20	8.60	−1.1
1976	15	5	5	4	—	25.33	13.28	12.86	8.70	−9.2
1977	18	6	5	3	—	34.46	26.16	24.50	8.26	+6.2
1978	24	5	7	1	1[g]	44.99	16.60	14.66	8.54	+1.2
1979	15	6	5	1	—	35.92	19.99	15.82	5.77	−3.4
1980	15	5	3	1	—	28.69	14.06	10.63	3.52	+5.0
1981	14	8	6	2	+[g]	42.27	22.97	22.06	9.72	+6.2
1982	14	7	6	2	3[h]	46.38	24.37	20.21	12.14	−5.3

[a] Bird data from Cink, Lowther *et al.* (1982 and preceding years); weather data from Lawrence, Kansas.
[b] Preceding 12 months precipitation July–June.
[c] Year-to-date rainfall.
[d] March, April, May, and June rainfall.
[e] April and May rainfall, all in inches.
[f] Deviation, in °F, of April, May, and June monthly temperatures from long-term average.
[g] Bobwhite Quail.
[h] Short-billed Marsh Wren.

and Horned Larks in mixed-grass prairie (Cody, 1968), and these are just the three species most typical of fields and early successional stages throughout the central United States (e.g., Shugart and James, 1973, in northwestern Arkansas). Mixed-grass prairies in the northern plains in Saskatchewan are similarly occupied (Cody, 1968, 1974a; Maher, 1979), except that the Western Meadowlark replaces the eastern species, the Baird's Sparrow replaces the Grasshopper Sparrow, and a longspur would be present. Breeding-bird densities in these prairies are generally lower than in tallgrass prairies; 14 censuses from *American Birds* give a mean of 0.59 ± 0.29 pairs acre (1.46 pairs/ha).

One of the most complete series of grassland-bird censuses has been conducted by Johnson and associates (Johnson and Springer, 1972 and successive years) in Stutsman County, east-central North Dakota, in mixed-grass prairies of various types. Seven of these sites were censused through 1983 (census data available up to 1982), including a control site on which no grazing or burning has been allowed, as well as sites on which various experimental burning and grazing treatments have been conducted. The 10-ha control site has some brushy vegetation, consisting mainly of Chokecherry (*Prunus virginianus*) and Wolfberry (*Symphoricarpos occidentalis*), the woody species *Rosa woodsii,* and *Eleaegnus commutata* and approximately 1 ha of seasonal wetlands.

Figure 3a shows the bird community divided into two components, one associated more with the wetlands (including Red-winged Blackbirds, ducks, and coots) and a second of more typically grassland and edge species. The latter includes, in order of abundance, Clay-colored Sparrows, Common Yellowthroats, Western Meadowlarks, Savannah Sparrows, Bobolinks, Grasshopper Sparrows, and a number of minor species; note that the first two species in the list are brushy-edge rather than true grassland species. Rainfall and temperature data are also included in Fig. 3a; the "wetland" birds are significantly correlated with year-to-date rainfall ($r = 0.63$, $p < 0.05$), while higher densities in the remaining species are associated with drier years ($r = 0.20$) followed by wet springs ($r = 0.24$, $p > 0.05$).

Figure 3b,c shows how the densities of some of the more common species vary between years. To assess the degree to which these birds respond to recent weather conditions, I correlated the density of the more common species with five weather variables: precipitation (1) in the past year July–June, (2) year-to-date, (3) March–June, (4) April and May, and (5) excess spring temperature over the long-term average. In 8 of the 11 most common species, the highest correlations are obtained with spring (April and May) rainfall. Densities are also well correlated with spring temperatures (which in turn are strongly negatively correlated with rainfall), and one of the remaining four species, the Common Yellowthroat, is most highly correlated with this variable. Of the remaining three species, Grasshopper Sparrow density is better correlated with yearly totals ($r = -0.56$) than spring ($r = -0.42$) rainfall, Eastern Kingbird (*Tyrannus tyrannus*)

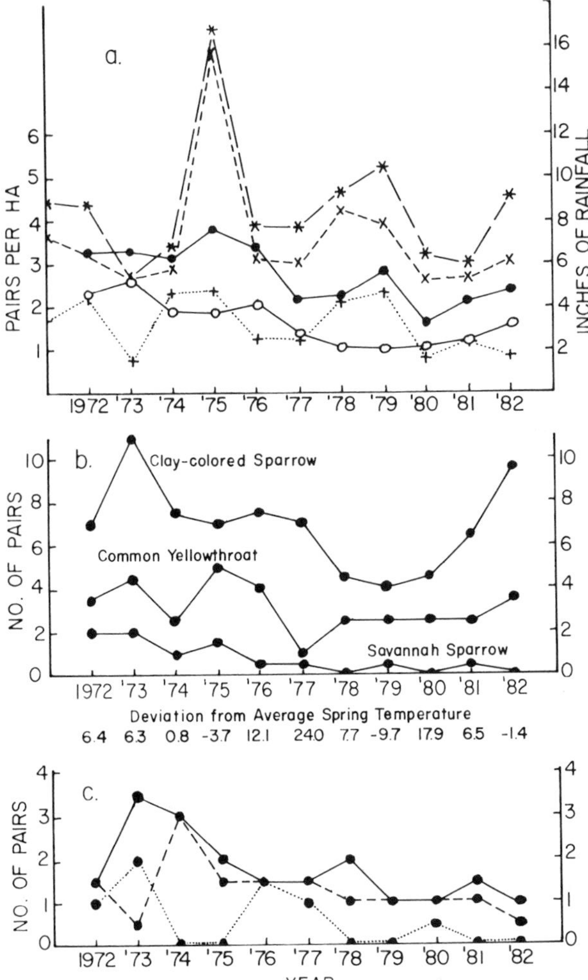

Fig. 3. (a) Census and weather data from mixed-grass prairie, North Dakota. Bird census data from Johnson (e.g., 1972) and associates. Birds are grouped into wetland (●——●) and grassland (○——○) species, rainfall is given as year-to-date (∗——∗), March–June (x- -x), and April–May (+···+) totals. (b) Variations in numbers of breeding Clay-colored Sparrows, Common Yellowthroats, and Savannah Sparrows from 1972–1982. Deviations from mean spring temperatures are given below (°F), April–June. (c) Variations in numbers of breeding Western Meadowlarks (●——●), Grasshopper Sparrows (●···●), and Bobolinks (●- -●) over 11 years in mixed-grass prairie.

density is most highly correlated ($r = -0.38$) with March–June rainfall, and Western Meadowlark density is poorly correlated with all variables, but least poorly with April–May rainfall ($r = +0.10$) and year-to-date rain ($r = -0.19$). Last, the sum of Blue-winged Teal (*Anas discors*) and American Coot (*Fulica americana*) densities is most strongly correlated with spring rainfall ($r = +0.80$) and with spring temperatures ($r = -0.67$).

These relations are shown in Fig. 4a,b, which represents the way in which various species' densities respond to differences in spring rainfall and temperatures. While there are some obvious effects, such as the higher densities of ducks, coots, and Red-winged Blackbirds following wet and cool springs, there are others not quite so intuitive but nevertheless empirically reasonable. The two grassland birds that occupy tallgrass as well as mixed-grass prairie are Bobolinks and Savannah Sparrows, and both are more common following wet springs. The two species whose distributions lie chiefly west of eastern North Dakota, Clay-colored Sparrows and Baird's Sparrows, are more common after unusually dry springs, weather more in accord with that further west of this site. Finally, Grasshopper Sparrows are most common after unusually warm and dry weather. Figure 4a,b shows that different species respond to different aspects of variable weather and that, in general, these responses conform to species' overall ecology and general distributional characteristics.

I used the five weather factors as independent variables in multiple-repression equations, and the results are shown in Table II. Table II gives the regression equation for the highest F-value (ratio of explained to error sums of squares), the coefficient values of the 1–3 independent variables in the regression with the highest F-value and the value of the constant, multiple R and R^2 in that regression equation, and the value of multiple R^2 for the best 3-factor regression. Table II shows: (1) Springtime rainfall is overwhelmingly the most important weather factor for these birds. (2) 9–64% of the variation in species densities is explained using a single weather variable and 17–79% using three weather variables; in two-thirds of the species more than one-third of the variation is explained. (3) Red-winged Blackbird density is correlated with spring rainfall ($r = 0.537$) and year-to-date rainfall ($r = 0.330$) as in eastern Kansas (see Section II.A), but 29% rather than 44% of its density is accounted for by this single variable. (4) Western Meadowlarks as was the case with Eastern Meadowlarks in eastern Kansas, prefer wet springs following dry years. (5) Common Yellowthroat density correlates most highly with wet spring weather as in eastern Kansas, but here in North Dakota, where long-term average precipitation is just 50% of what it is at the Kansas site, this species is positively rather than negatively correlated with annual rainfall.

The explained variance in many species' densities can be significantly increased using nonlinear models, but in the absence of a biological rationale for such models, such exercises seem academic.

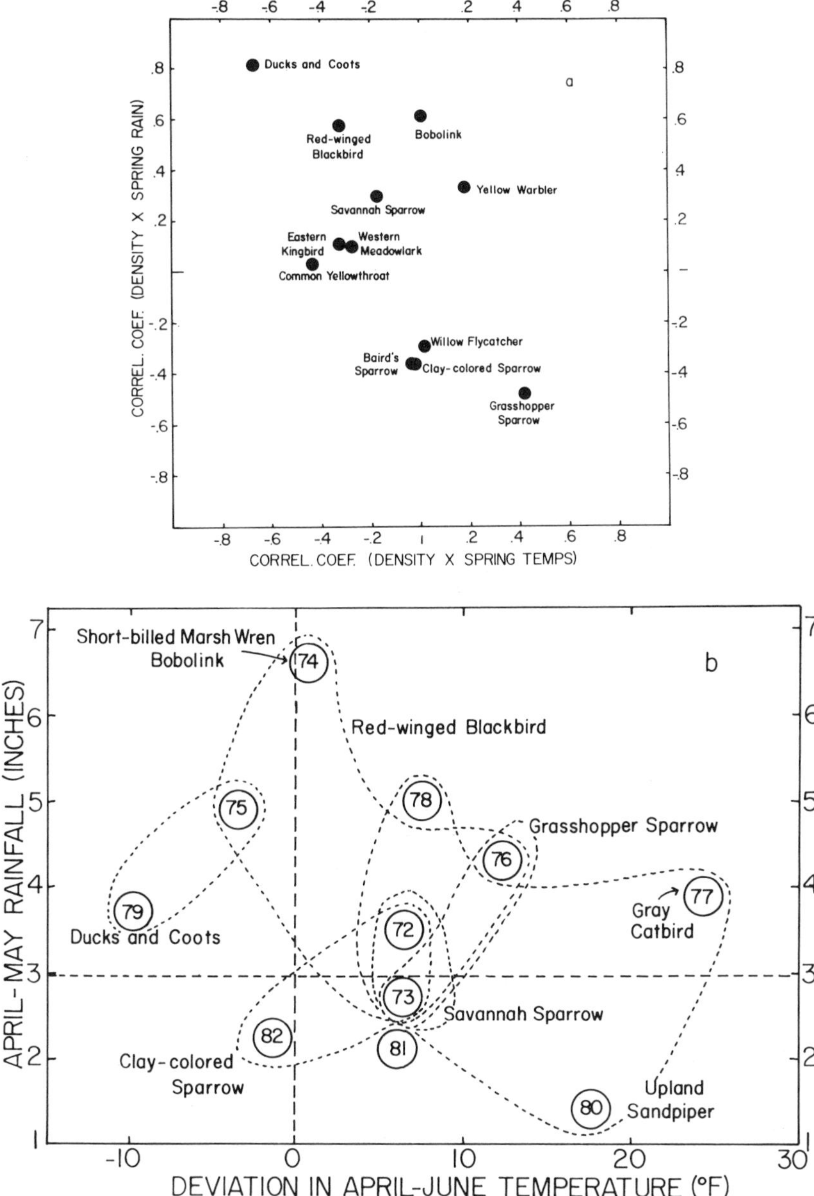

Fig. 4. (a) Correlations between bird numbers, spring rainfall, and spring temperature on mixed-grass prairie, North Dakota. Values of coefficients greater than 0.576 are significant at the 5% level. (b) Different bird species respond differently to variations in spring rainfall and temperature. Short-billed Marsh Wren and Bobolink reach peak densities after wet springs, ducks and coots after cool, wet springs, Grasshopper Sparrow after warm, wet springs, Upland Sandpiper after warm, dry weather, and Gray Catbird only after the warmest spring weather.

TABLE II

Summary of Linear Multiple Regression Statistics for the Influence of Five (Independent) Weather Variables on the Densities of Birds Breeding in Mixed-Grass Prairies, Stutsman County, East-Central North Dakota[a]

Species	Weather coefficients[b]						Highest		R^2	Highest
	YEAR	YTD	MAMJ	AM	TEMP	CONST	F	Multiple R	(%)	R^2
Clay-colored Sparrow				−.059		.890	1.45	0.372	14	26
Common Yellowthroat			.017			.195	2.05	0.430	18	36
Savannah Sparrow				.018		.027	0.85	0.293	9	17
Western Meadowlark		−.066	.065			.273	3.11	0.661	44	56
Grasshopper Sparrow	−.008					.198	4.12	0.56	31	33
Bobolink	−.010			.053		.136	9.30	0.836	70	73
Red-winged Blackbird				.073		.135	4.39	0.537	29	39
Eastern Kingbird			−.008	.014		.109	1.80	0.557	31	34
Willow Flycatcher				−.010		.143	0.86	0.295	9	18
Yellow Warbler				.008	.001	.055	1.47	0.518	27	30
Baird's Sparrow		−.023	.019		−.001	.082	1.82	0.662	44	44
Duck and Coot				.173		−.270	16.21	0.801	64	79

[a] Bird census data from control site I of Johnson (e.g., 1972) and associates.

[b] Weather variables as in Table I.

These relations are quite fascinating and reveal the great value in long-term census data; they show us, first, the extent to which censuses are repeatable year to year and, second, the degree to which nonrepeatability is due to year-to-year weather and presumably food-resource variation. Although a thorough analysis of the data is pending (D. H. Johnson, in preparation), a few more points can be noted here.

The regression analyses used weather data from a station ~55 km from the census sites (Jamestown), since a closer station (Medina, approximately 27 km distant) does not record temperature. I noted that correlations with rainfall data are some 10% higher using Medina figures and, presumably, would be higher still if weather data were actually collected at the census site.

At other sites censused by Johnson and colleagues, controlled burns were conducted, and in a grassland dominated by Kentucky Bluegrass (*Poa pratensis*), these burns occurred in the fall of 1972, June 1977, and June 1982. Some species were not dramatically affected by the fires; Western Meadowlarks, for example, were represented by a single pair in each year, except the year following the spring burns. In other species, the effects of the burns clearly outweigh those of recent weather variations. Some, such as Horned Larks and Chestnut-collared Longspurs, were present only in years immediately following fires when the vegetation was very sparse; these are species more broadly distributed in western North Dakota in shorter vegetation. Bobolinks reached peak densities some 21 months following the fall burn and declined in density as the vegetation recovered and its structure increased (Fig. 5). Savannah Sparrows reach high densities only some years after a fire, whereas Upland Sandpiper and Marbled Godwit densities decline rapidly with postfire regeneration of the vegetation (Fig. 5). These changes show the extent to which grassland birds are sensitive to vegetation structure and to the foraging opportunities such structure controls, and the degree to which these opportunities may be tracked regionally and between years by grassland birds. The further elucidation of such resource tracking will depend on more detailed analyses (D. H. Johnson, in preparation) and measures of vegetation structure and its affects on bird food and foraging ecology.

C. Shortgrass Prairies and Plains

Shorgrass prairies are found from eastern New Mexico and northwestern Texas north through eastern Colorado to southeastern Wyoming. The characteristic bird species are Western Meadowlarks, Lark Buntings, Mountain Plovers, Horned Larks, and longspurs, in particular McCown's Longspurs. Bird densities in these grasslands are about half those of tallgrass prairies (average 0.41 ± 0.13 pairs/acre; 1.00 pair/ha, from sources as in Section II,A and B). The lower productivity and presumed lower predictability of shortgrass prairies might be associated with reduced census repeatability from year to year, but there are few

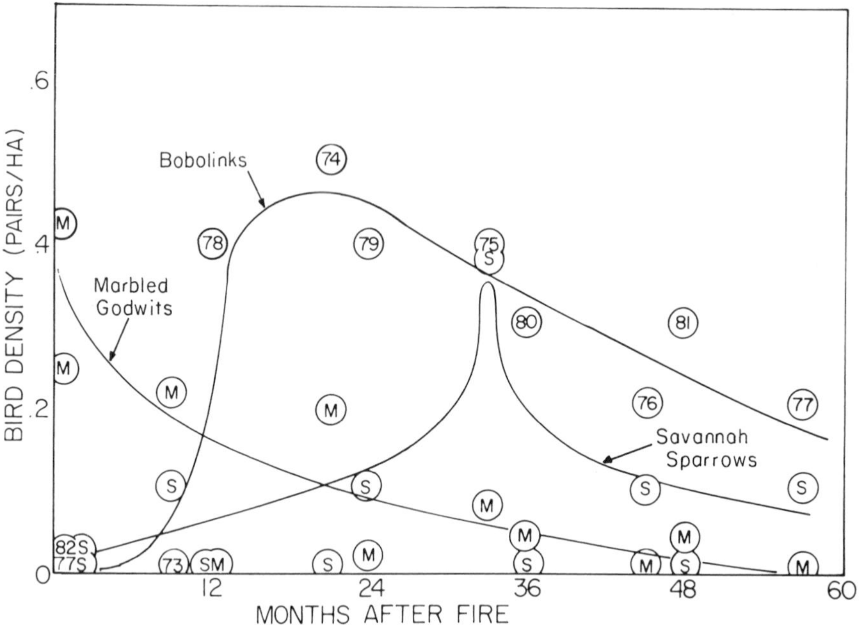

Fig. 5. Bird densities in Kentucky Bluegrass-mixed prairie, North Dakota, as a function of time since fires (which were in the fall of 1972, June 1977, and June 1982). Upland Sandpipers and Marbled Godwits reach peak densities immediately following the fires, Boblink in the spring 21 months after fall burns, and Savannah Sparrows some 3 years after fires, when the vegetation has extensively regenerated. Symbols: Circled M represents Marbled Godwits and Upland Sandpipers, circled numbers Bobolink density and census year, and circled S, Savanah Sparrow density.

long-term data sets on which this might be assessed. Data from Weld County, northeastern Colorado (see also Cody, 1968), collected by Ryder and associates (see, e.g., Ryder, 1980), in fact show a census repeatability comparable to that in other grasslands. In four censuses over 6 years, Horned Larks varied from 24 to 27 pairs, McCown's Longspurs from 17.5 to 19.5 pairs, and Mountain Plovers from 2 to 3 pairs. Three rarer species, the Western Meadowlark, the Lark Bunting, and the Mourning Dove, showed more variability. A comparably high census repeatability over 3 years was reported by Finzel (1964) for Horned Larks, McCown's Longspurs, and Mountain Plovers from sites near Cheyenne and Laramie, Wyoming, at a higher latitude and elevation.

I found no differences in habitat use by four passerine species at a shortgrass site in Weld County, Colorado, although point census results showed an average of three rather than four species, indicating that the habitat may not have been filled to capacity by the birds (Cody, 1968). These species did show considerable differences in foraging ecology, with Horned Larks moving more rapidly through the vegetation (~7 m/min), Lark Buntings and McCown's Longspurs searching more slowly (~2 m/min), with the latter taking food items higher off the ground, and Western Meadowlarks moving at an intermediate speed (~4 m/min), taking food at heights intermediate between those where longspurs (up to 9 cm) and Horned Larks and Lark Buntings (below 4 cm) foraged most successfully.

Different bird species respond to and forage in this vegetation in different ways, and of course the same species responds to vegetational differences by altering foraging behavior. Nevertheless, there can be very considerable dietary similarities among species that forage in different ways. Baldwin (1973) quantified these similarities for Lark Buntings and McCown's Longspurs in the shortgrass prairies in Colorado, where the former took 62% arthropods (40% grasshoppers) and 38% seeds, the latter 78% arthropods (27% grasshoppers) and 22% seeds. However, the bunting ate larger arthropods than the longspur, and its grasshoppers were twice the weight (and its weevils one and one-third the weight) of similar taxa in the longspur diet. Thus, despite similarities in dietary composition, there remain differences in the actual food items taken by these two species, and together with differences in foraging behavior and probably in the microsites searched with these different behaviors, these species utilize the shortgrass habitat in rather different ways. Despite differences in foraging behavior, size of diet items, and perhaps in the relative accessibility of various food items, given a general taxonomic similarity in diet short-grassland birds might be expected to respond in similar, rather than in different, ways to variable weather conditions year to year, especially if this weather variability affects food productivity rather than habitat structure. As far as I know, the data to test this notion are not yet available.

III. OTHER TEMPERATE GRASSLANDS

One approach to evaluate the extent to which grassland-bird communities depend on and respond to differences in the structure of their habitats is to compare birds in grasslands around the world. This can be done at various levels of resolution. Earlier, I compared censuses of grassland birds in a wide variety of habitats (Cody, 1966), with a view to supporting the idea that such habitats saturate ecologically with similar numbers of species, particularly with 3–4 passerines. Such crude comparisons are useful only if the effects of variations in census area and in vegetation structure and homogeneity are controlled and if historical effects and various local disturbances are minimal.

At another level of resolution, I compared the bird communities in two sites carefully matched for vegetation structure, in Chile and eastern Kansas, despite broad climatic differences between the sites (Cody, 1974b). Although the

Chilean grassland was cultivated and irrigated, the structural similarities between the two areas corresponded to striking similarities between their bird communities. These similarities extended beyond species' numbers to species' gross morphologies, behaviors, and overall utilization of the similar habitats (Table III).

A third approach might be to control for differences in climate and to compare bird species over habitat gradients between regions in which there is an overall similarity in weather and perhaps also in resource productivity and predictability. Such a comparison can be made between grassland birds in Mediterranean-climatic regions of the world (although such grasslands are a minor part of

TABLE III

Comparison of Vegetation Structure, Bird Communities, and Bird Morphologies in Two Matched Fields, Kansas and Central Chile[a]

Parameters	Kansas	Chile
Vegetation structure		
Mean height (m)	0.29	0.27
Vertical density (number of contacts)	5.10	6.23
Horizontal density (number of contacts /m)	8.68	9.26
Area under foliage profile	32.4	39.2
Bird community		
Number of species	3	3
Average between-species overlap in		
Habitat	0.63	0.60
Feeding height	0.78	0.89
Food/foraging behavior	0.18	0.21
Overall ecological overlap	0.53	0.56
Bird morphologies		
Eastern Meadowlark (*Sturnella magna;* Kansas)		
Red-breasted Meadowlark (*Sturnella militaris;* Chile)		
Body size (mm)	236	264
Bill length (mm)	32.1	33.3
Ratio bill depth/length	0.36	0.40
Grasshopper Sparrow (*Ammodramus savannarum;* Kansas)		
Yellow Grass Finch (*Sicalis luteola;* Chile)		
Body size (mm)	118	125
Bill length (mm)	6.5	7.1
Ratio bill depth/length	0.60	0.73
Horned Lark (*Eremeophila alpestris;* Kansas)		
Chilean Pipit (*Anthus correndera;* Chile)		
Body size (mm)	157	153
Bill length (mm)	11.2	13.0
Ratio bill depth/length	0.50	0.42

[a] From Cody (1974b).

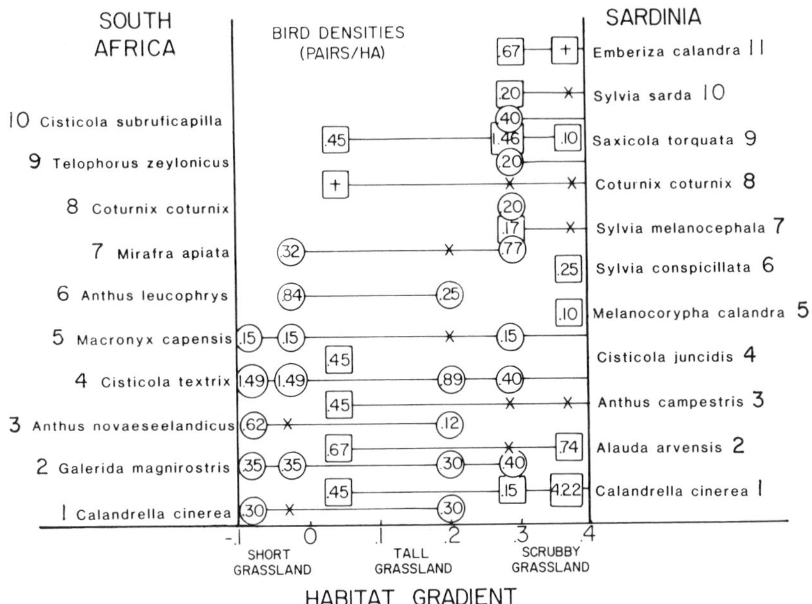

Fig. 6. Comparison of bird censuses in four South African and three Sardinian grasslands, each positioned on a habitat gradient from short to tall vegetation. Although species numbers per site and total species censused are similar, there are few similarities in the ways in which the closest ecological counterparts, in some cases conspecific or congeneric species, use the habitat gradient. See text for discussion.

Mediterranean-region vegetation and are often the product of disturbances such as fire and grazing). Two such regions, Sardinia and southern Africa, are compared in Fig. 6 (data from Cody, 1983). The habitat structure is quantified with a principal component measure of vegetation height and its "half-height," that at which half the vegetation density lies above and half below. With this habitat measure as abscissa, the densities and distribution of bird species on the low, grassy, and open end of the gradient can be compared.

From one point of view the two sets of census data appear to be very similar. For habitat $H < 0.33$, 11 species at 3 sites are found in Sardinia and 10 species at 4 sites in South Africa; all but a single species in each area are passerines, the exception being the Quail *Coturnix coturnix*. There are 3 lark species, 1 pipit, and 4 sylviine warblers in the Sardinian habitats and 3 larks, 3 pipits, and 1 sylviine warbler in the South African habitats. The censuses are completed by an emberizid bunting and a turdine chat in Sardinia and by a shrike in South Africa. At each census site, 5–7 species are found.

Beyond the numerical and taxonomic aspects of the censuses, there are few similarities in the way ecologically equivalent species are distributed over the grasslands. Related species are adjacent in Fig. 6, including 2 sets of conspecifics (lines 1 and 8), 2 sets of congenerics (lines 3 and 4), and 2 sets of confamilials (lines 2 and 10). On line 5 the Calandra Lark (*Melanocorypha calandra*) and the Cape Longspur (*Macronyx capensis*) are in different families, but the two are morphologically and behaviorally similar; they are the ''meadowlarks'' of these grasslands, and although the former is a lark and the latter a pipit, they are both large, chunky birds with pointed wings and stout, probing bills.

I consider that the birds of lines 1–5 and line 10 are reasonable morphological and ecological equivalents. But the distributions and densities of the species over this part of the habitat gradient are by no means well matched. Thus, although the same range of morphotypes, the same species numbers, and roughly the same overall densities occur in parallel in the two regions (similarities especially obvious when each grassland series is compared with other habitats on the gradient), there is no precise matching of morphotype with habitat structure. It appears that, while species numbers and their combined ecological range are governed at each site, the exact composition of the census is not. This implies that the presence of other species at the site must be an important factor beyond that of vegetation structure.

This point is emphasized by a third conspecific pair in the censuses, the Stonechat (*Saxicola torquata*). This bird occurs in shorter grasslands in Sardinia, reaching peak densities at $H = 0.25$ and occurring last at $H = 0.82$. The species also occurs on the South African gradient but only once, at $H = 0.75$, and not in the grasslands. There are many chatlike species in the South African habitats but no others in Sardinia; yet the stonechat's absence in South African grasslands cannot be attributable to the presence of chats there (there are none), but rather to more indirect or diffuse competition from the multitude of grassland larks and pipits and, perhaps, shrikes.

IV. TROPICAL GRASSLANDS

The increased climatic predictability of the tropics might be used to gauge the effect of this variable on grassland-bird communities, but, to date, few detailed studies have been published on the birds of tropical grasslands. The level of species packing of passerines in tropical grasslands was found by MacArthur *et al.* (1966) to be comparable to that in the temperate zone. This view is supported by the studies of Lill (1974), who found 3–5 species to be usual in various types of grassland in Guyana. For example, in short vegetation the Grassland Sparrow (*Myospiza humeralis*) coexists with the Chestnut-bellied Seedeater (*Sporophila castaneiventris*) and in open savannahs with the Wedge-tailed Grassfinch (*Emberizoides herbicola*) and the Grassland Yellow-finch (*Sicalis luteola*). Further, two Seed-finches could be found together (*Oryzoborus crassirostris* and *O. angolensis*), the former with Blueblack Grassquits (*Volatinia jacarina*) and three species of seedeaters (*Sporophila intermedia, S. lineola,* and *S. minuta*). Lill found slight differences in habitat use among species, but emphasized dietary differences and differences in foraging ecology related to the species' morphology.

A study of mixed-species feeding flocks in Costa Rican grasslands (Rubenstein *et al.*, 1977) also stresses the importance of subtle behavioral differences among species. The three species involved, Yellow-faced Grassquits (*Tiaris luteola*), Variable Seedeaters (*Sporophila aurita*), and White-collared Seedeaters (*S. torqueola*), are morphologically similar and feed on the ground, but may specialize to some extent on rarer but more energetically rewarding seeds, and in particular their foraging behavior is influenced by the proximity and identity of the nearest neighbor in the feeding flock. Interspecific interactions were also a feature of Lill's study (see earlier, this section).

In a study of the grassquit *Tiaris olivaceus* in mainland Costa Rica and on Jamaica, Pulliam (1973) found the bird in approximately the same number of habitats on the island but found it generally more rare, despite the reduced numbers of competitors there. The island birds were found to have the same means and variances, and the same sexual dimorphism, in bill measurements, but were considerably less stereotyped in seed preferences. These results emphasize again the behavioral rather than the morphological aspects of resource partitioning in grasslands and the fact that the link between bill morphology and diet is not a close one.

Perhaps the most detailed community-wide study conducted in tropical grasslands to date is that of Folse (1982) in Serengeti National Park, Tanzania. Folse conducted censuses year-round at five 1-km^2 sites from shortgrass to a grass-dominated open parkland. In this range of habitats he recorded 14 passerine and 3 nonpasserine species, numbers not unlike those in Sardinia and South Africa (see earlier, this section), especially in view of his much larger census areas. The shortgrass site A, with 4 passerine and 2 nonpasserine species, has a community composition remarkably like that in the Colorado shortgrass plains; larger species numbers, up to 9 passerines, were found in taller grass habitats.

All 18 of the grassland birds Folse censused varied significantly in density between habitats, and he concluded that habitat partioning was the principal means of niche segregation among species. Most species also varied seasonally between wet and dry seasons, with concomitant variations in grazing and fire. Birds respond to the amount of grass biomass present, but Folse was unable to attribute density changes to either resource limitation or to infrequent periodic bottlenecks in times of general resource superabundance. It appears that bird densities in this system might be governed by a complex of factors. Density shifts might follow regular (seasonal, climatic), periodic and unpredictable (fires), or intermediately predictable (grazing effects) shifts in food resource availability; such resource tracking may be further affected by the effects of a widely variable, local vegetation structure on foraging behavior and constrained by the effects of interactions with ecologically similar species.

V. ARCTIC–ALPINE BIRDS

Tundra habitats at high latitudes or at high elevations are occupied by similar sorts of birds—chats, longspurs, and buntings—as grasslands and are as repeatable in structure as grasslands from one part of the world to another. Table IV shows three censuses in such habitats, from 2700 m in the High Atlas of southern Morocco, from 3400 m in Rocky Mountain National Park, Colorado (Cody, unpublished data), and at 600 m in northern Iceland (Cody, 1963). The passerine birds number three species in each case, and habitat segregation among them is strong. In each case there are buntings (not closely related among sites), which are associated with the steepest slopes and rockier areas; they are slow searchers for insects and seeds around boulders and seeps. The other species have a far more active foraging behavior; Horned Larks and pipits are active searchers for insects in the grassier areas, and Wheatears pounce from boulder perches onto ground and low-flying insects and are characteristic of open slopes and plains.

At high latitudes or elevations productivity for tundra birds may be high, though of short duration, and selection to maximize its usefulness should be as

TABLE IV

Similar Bird Censuses from High-Elevation Alpine Sites in Colorado (Rocky Mountains) and Morocco (High Atlas Mountains) and from Arctic Tundra in Northern Iceland (Thorvalsdalur, Dalvik)

Colorado	Iceland	Morocco
	Slow-searching buntings, rockier sites:	
Brown-capped Rosy Finch (*Leucosticte arctoa*)	Snow Bunting (*Plectrophenax nivalis*)	Crimson-winged Finch (*Rhodopechys sanguinea*)
	Active foragers, grassier sites:	
Water Pipit[a] (*Anthus spinoletta*)	Meadow Pipit[a] (*Anthus pratensis*)	—
Horned lark[b] (*Eremophila alpestris*)	—	Horned Lark[b] (*Eremophila alpestris*)
—	Common Wheatear[c] (*Oenanthe oenanthe*)	Common Wheatear[c] (*Oenanthe oenanthe*)

[a] Qualifying foraging site: denser vegetation. Behavior: fast moving.
[b] Qualifying foraging site: more open vegetation. Behavior: slower moving.
[c] Qualifying foraging site: more open sites. Behavior: perch-and-pounce.

keen there as it is anywhere else. In Alaska, Lapland Longspurs (*Calcarius lapponicus*) select territories (average size 1.76 ha) in the spring at the time of snow melt, but before food resources can be assessed directly (Seastedt and MacLean, 1979). Nevertheless, there is a significant negative correlation between territory size and arthropod density later in the season, presumably because the birds can distinguish topographical variations which are correlated with their food production. Much earlier in the 1930s Tinbergen noted that Snow Buntings (*Plectrophenax nivalis*) in Greenland established territories before the snows were melted (see, e.g., Tinbergen, 1968), and, again, topographic variation might provide them with clues as to what sites will provide appropriate territories later in the season.

VI. OPEN-COUNTRY AND DESERT-EDGE SPECIES

In this section a brief mention will be made of habitat selection in bird species of dry, open country such as stony plains and dry steppe, habitats which are notable for their lack of vegetation cover and predominance of bare, often rocky ground. Chapter 7, this volume, treats habitat selection in shrub–steppe birds in more detail.

One of the predominant bird groups of the Chilean Andes is the ground-tyrants *Muscisaxicola*, chatlike tyrannid flycatchers that are common on the stony and sparsely vegetated higher slopes over a wide range of latitudes and elevations. In northern Chile there are five species, with widely overlapping ranges in breeding elevations; they cojointly cover over 4000 m, but no details are known about their breeding habitats. The species differ in size in a strikingly orderly fashion: *M. albifrons,* bill length 22.4 mm; *M. flavinucha,* 18.9 mm; *M. rufivertex,* 16.5 mm; *M. juninensis,* 15.4 mm; *M. maculirostris,* 13.3 mm; weights are not available, but weight ratios of adjacent species in the group must be close to 1.5.

The ranges of *Muscisaxicola* species from north to south in the Andes were summarized in Fig. 77 of Cody (1974a). By latitude 33°S four distinct sizes are represented, with *M. albifrons* dropping out and two additional species being added into existing size slots: *M. frontalis* is the same size as *M. flavinucha* and *M. albilora* is the same size as *M. rufivertex*. Further south, the larger sizes drop out, a smaller size (*M. macloviana,* bill length 12.7 mm) is added, and the number of distinct sizes declines to three and then to two.

The elevational ranges of the six *Muscisaxicola* species at latitude 33°S are given in Fig. 7. There is a great deal of interspecific overlap, with no apparent habitat differences except in the two smaller species, *M. maculirostris* and *M. alpina,* the former with the lowest and least sparsely vegetated breeding habitat and the only species to use perches in vegetation, and the latter with a high-elevation range in rockier areas. The foraging behavior of these birds is generally similar; all species employ the "perch-and-pounce" technique, in which rocks and small boulders provide elevated perches within a meter of the ground (usually <20 cm) and from which the surrounding area is scanned for ground-level arthropods. All do a minor amount of sallying after low-flying insects. The species do vary in the frequency with which they change perches and in the speed and distances they fly after prey, as illustrated in Fig. 7. Thus, body size is strongly coupled with foraging behavior, as it is also likely to be with prey size.

Where four or five *Muscisaxicola* species co-occur there is a great deal of

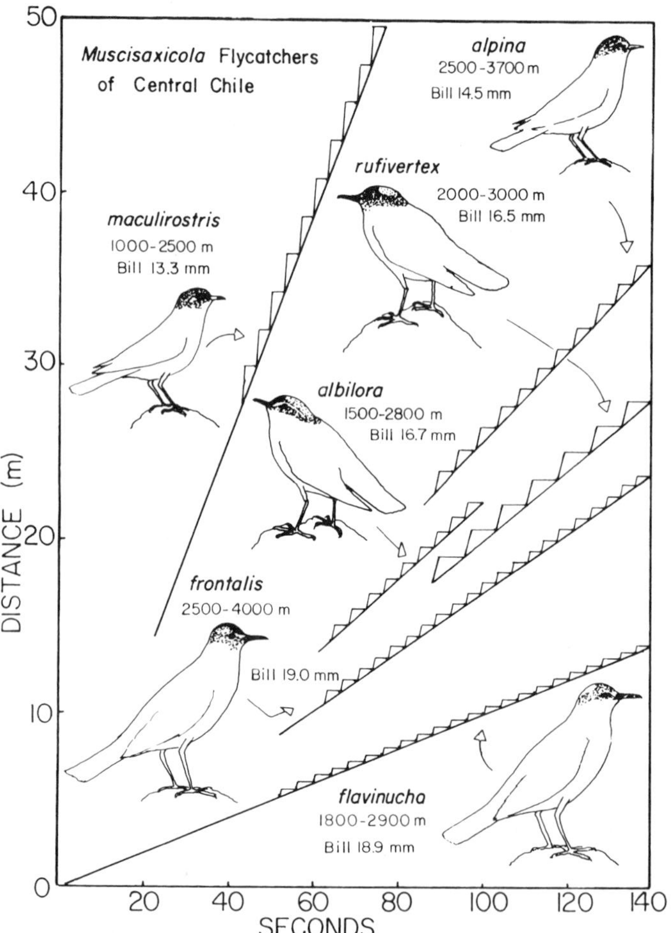

Fig. 7. The six species of *Muscisaxicola* of the central Chilean Andes show foraging behavior related to body–bill size. Similarly sized species either occupy different elevational zones (*M. maculirostris–M. alpina*) or are interspecifically agonistic (*M. frontalis–M. flavinucha, M. rufivertex–M. albilora*). From Cody (1970, 1974a).

interspecific interaction (Cody, 1974a; Smith and Vuilleumier, 1971). There are frequent threats and chases and interspecific territoriality between species of like size (*M. frontalis–M. flavinucha* and *M. rufivertex–M. albilora*); the intensity of direct behavioral interaction between species appears to be related to their similarity in size.

The *Oenanthe* wheatears of the Old World and Alaska are close ecological equivalents of the *Muscisaxicola* ground-tyrants and occupy tundra, deserts, stony heaths, and open ground from the arctic to South Africa. Unlike *Muscisaxicola,* species of *Oenanthe* are generally similar in size, as well as in foraging behavior, and appear to rely mainly on habitat segregation for their coexistence. I found 4 species in the High Atlas mountains of southern Morocco, where they occur in virtually all poorly vegetated sites; of 12 census sites with wheatears, 9 supported 1 species and 3 supported 2 species. *Oenanthe oenanthe* occurred only at the highest elevations, *Oe. deserti* mainly on the stony ergs south of the mountains on the edge of the Sahara, *Oe. leucura* broadly distributed in sparse, open scrub from sea level to 2200 m, and *Oe. hispanica* in somewhat thicker vegetation up to around 1500 m (M. L. Cody, unpublished data).

Wheatears are territorial in both breeding and winter seasons, and where several species occur together, they are interspecifically territorial. This was first reported by Stresemann (1950). A more recent study is that of Leisler *et al.* (1983), in which behavioral interactions and morphological differences among five species of *Oenanthe* were measured (see also Chapter 14, this volume). The work was conducted in Kenya, where *Oe. oenanthe. Oe. isabellina,* and *Oe. pleschanka* are winter visitors, and *Oe. lugens* and *Oe. pileata* are residents; resident and wintering chats in other genera were present. Males and females of all species occupied separate territories, which were defended both conspecifically and interspecifically. The resident species tended to be dominant in interspecific encounters, but this was equally influenced by body weight. Exactly as in *Muscisaxicola,* the smaller wheatears covered more ground in foraging than did the larger species, these feeding rates differing significantly for species with weight ratios around 2, but not significantly for species of smaller weight differences.

The most comprehensive study conducted to date in the genus is that of Cornwallis (1975), who worked in southwestern Iran, where 11 species breed (8 commonly) and 7 species overwinter (3 commonly). Among the breeding species, mean wing lengths range from 85 to 104 mm, bill lengths from 16.5 to 22.5 mm, and weights from 17 to 28 gm; thus, the species are all relatively similar in morphology. Cornwallis evaluated the extent of spatial overlap in the breeding season among the 9 more common species. Of the 36 pairwise species comparisons, 2 species pairs were completely allopatric and another 5 were largely separated by geographic range. Wheatears bred in steppe and subdesert, from sea level to nearly 4000 m. Eight of the species pairs in this geographic area maintained separate elevational ranges, and another eight pairs bred in largely disjunct ranges. Habitat is of course correlated to some extent with elevation, and habitat differences were found in 10 of 16 species pairs that segregate largely or entirely by elevation. Further, in another 10 species pairs, in which breeding ranges overlapped both in geography and in elevation, habitat differences were found to be significant.

Cornwallis felt that spatial segregation among species, by geographic range, elevational range, and/or breeding habitat, was the most important aspect of the coexistence of so many wheatear species in the region. Thus, at the highest elevations of greater than 2500 m in the montane zone, *Oenanthe oenanthe* and *Oe. xanthoprymna* occur, the former in flatter areas and the latter on steeper, rocky slopes. Five species, *Oe. finschii, Oe. isabellina, Oe. desertii, Oe lugens,* and *Oe. hispanica,* all breed at intermediate elevations around 2000 m. Of these, the first three breed in more level habitats, with *Oe. desertii* allopatric with the first two and *Oe. finschii* on stonier, less vegetated ground than the substeppe typical of *Oe. isabellina.* The eighth common breeder, *Oe. alboniger,* breeds around rocky gullies at low elevation.

Of the 36 species pairs (previously mentioned), 3 were not spatially separated: *Oenanthe lugens–Oe. finschii, Oe picata–Oe. alboniger,* and *Oe. picata–Oe. lugens.* Besides these species, 10 other species pairs exhibited marginal spatial overlap and were found in the same localities. Although there is a minor segregation of foraging behavior into either a ''perch-and-pounce'' mode or a ''dash-and-jab'' mode, the diets of species, especially at the same location, overlapped very considerably. Interspecific territoriality was recorded between species in all 3 pairs with extensive spatial overlap and in 5 of 10 pairs with marginal overlap. Thus, at the local level of territories, any residual potential overlap is eliminated by direct behavioral interactions among species. In integrating measurements on habitat structure and food availability, behavioral interactions, and biogeographic elements, this study is a model of how a diverse group of closely related species has come to co-occupy a limited geographic region.

VII. GRASSLAND BIRDS IN WINTER

A few grassland birds are resident (e.g., Eastern Meadowlarks in eastern Kansas), but most undertake local to long-distance migrations and overwinter in grasslands or other open habitats, such as deserts or desert edges at lower latitudes. Winter diversity in southern grasslands can be high; Emlen (1972) recorded 14 species in a grass–forb area in southern Texas, including such species as the Eastern Meadowlark, Savannah Sparrow, Vester Sparrow, Grasshopper Sparrow, Sprague's Pipit, and Mourning Dove.

Local overwintering populations may not be morphologically homogeneous, and Fretwell (1968, 1972) has discussed the extent to which different morphotypes in populations select different habitats and eat differently sized seeds. He presented evidence that Savannah Sparrows and Field Sparrows vary in size and that the larger individuals select winter habitats that produce larger seeds, and vice versa; individuals may also vary significantly in tarsus length with the habitat they occupy. The extent of such ecotypic variation and covariance of morphology with habitats is not known, but such relations could reflect a very strong selection for the correct habitat choice in both breeding and wintering seasons.

There appears to be a tendency for related species that occupy distinct breeding habitats to overwinter in different habitats. This segregation can be maintained in different populations of the same species; for example, different subspecies of the Oriental Greenfinch (*Carduelis sinica*) overwinter in different habitats around Kyoto (Nakamura, 1979). The emberizid buntings that partition Japanese grasslands by differences in habitat structure in the breeding season (e.g., Nakamura *et al.*, 1968) also segregate by habitat in winter. *Emberiza cioides* winters in pairs on creek banks and the edges of rice paddies, and *E. rustica*, on wetter surfaces in the fields, in the form of large flocks. *Emberiza spodocephala* is somewhat intermediate between these two, and *E. furcata* is found in grassy areas (Nakamura, 1973; Nakamura and Iijima, 1977).

Detailed work on the winter ecology of grassland birds has been conducted in Arizona by Pulliam and associates, where both resident and wintering species depend on seeds from late summer rains for sustenance (Pulliam and Brand, 1975), and their numbers are proportional to seed production (Pulliam and Parker, 1979; Dunning and Brown, 1982). There are thresholds of seed density to be surpassed before wintering finches can be supported; no granivorous birds occur where total seed weight is below 1 kg/ha. Seeds vary in size, and so do granivorous birds. Nine species treated in Pulliam (1983) had culmen lengths from 8.8 to 14.5 mm, and although the smallest species, Brewer's Sparrow (*Spizella breweri*), averaged smallest in mean seed size eaten, 1.04 mm, the largest species, Brown Towhee (*Pipilo fuscus*), had the largest mean seed size, 1.72 mm, and the rank correlation was 0.81, nevertheless he found a great deal of interspecific overlap in both types and sizes of seeds eaten.

Differences in habitat use are also important in these wintering finches. Pulliam and Mills (1977) showed that the degree to which different species ventured from cover into the grassland varied, with Vesper Sparrows foraging closest to cover, Chestnut-collared Longspurs furthest, and Savannah Sparrows and Grasshopper Sparrows at intermediate distances. This order is similar to that of these species in the vegetation density of their breeding habitats; it may reflect adaptations both to vegetation structure and to escape responses from predators.

Perhaps because of the formidable combination of variable local resource production in winter and variable numbers of a variety of granivorous birds of great mobility, each with variable population from their last breeding season and each with the ability to handle most of the variety of seed sizes, Pulliam was pessimistic about the degree to which wintering sparrows and their resources might be related in a deterministic fashion.

VIII. SYNTHESIS

Interspecific segregation by differences in vegetation structure is usual in sympatric grassland and open-country birds, and vegetation structure determines in turn the options in foraging behavior. In tallgrass prairies and in bare, open plains, these options might be minimal—illustrated by slow searching "buntings" ~15 gm and "wheatears" ~20 gm, respectively. In intermediate habitats, such as mixed-grass prairies and shortgrass plains, especially habitats that are more open, patchy, and variable than tallgrass prairies but more continuously vegetated than desert plains, a variety of foraging modes, and several bird species, can be viable simultaneously.

The food resources for grassland birds are the product of a variable climate and vary from year to year and from place to place, likely more so than does habitat structure. Evolutionarily, the appropriate response in grassland birds might be to maintain as generalized a morphology and behavior as possible, but such generalization will be constrained by interspecific competition and selection to minimize it by habitat and resource partitioning. Further, habitat structure and morphology constrain foraging behavior, and morphology is further constrained by winter diets that may be quite different from those in the breeding season. Yet, within such constraints grassland birds do seem to have maintained phenotypic plasticity in behavior and especially in diet, since most studies show broad diet overlaps among coexisting species and poor relationships between morphology and diet.

An ecological as opposed to evolutionary response to resource variability and unpredictability is resource tracking, where species maintain the behavioral flexibility to find and utilize sites where resources are sufficient. This behavioral flexibility has several components: (1) certain geographic areas might be used for breeding only in certain years, such as after unusually wet or unusually dry seasons; (2) certain habitats might be used only in certain circumstances, such as after recent fires; (3) territories can be located with respect to certain vegetational features and can vary widely in size with resource abundance; (4) diets can change, apparently opportunistically as different food taxa are available; (5) the presence in the territory of other species with similar ecological and especially food requirements can be regulated via direct behavioral interaction and result in partial or complete interspecific territoriality.

REFERENCES

Allaire, P. N., and Fisher, C. D. (1975). Feeding ecology of three sympatric sparrows in eastern Texas. *Auk* **92,** 260–269.

Baldwin, P. H. (1973). The feeding regime of granivorous birds in shortgrass prairie in Colorado, USA. *In* "Productivity, Population Dynamics and Systematics of Granivorous Birds" (S.C. Kendeigh and J. Pinkowski, eds.), Polish Sci. Publ. Warsaw.

Beason, R. C., and Franks E. C. (1974). Breeding behavior of the horned lark. *Auk* **91,** 65–74.

Blankespoor, G., and Krause, H. (1982). The breeding birds of Minnehaha County, South Dakota: Then (1907–1916) and now (1917–1975). *Am. Birds* **36,** 22–27.

Brownsmith, C. B. (1977). Foraging rates of starlings in two habitats. *Condor* **79,** 386–387.

Cink, C. (1974). Breeding bird census #100: Flood plain Tallgrass Prairie, Kansas. *Am. Birds* **28,** 1031–1032.

Cink, C. (1983). Winter bird census #41: Flood plain Tallgrass Prairie, Kansas. *Am. Birds* **37,** 40.

Cody, M. L. (1963). Ornithological studies. *In* "Edinburgh University Expedition, Iceland 1963, Final Report" (N. Stebbing, ed.), pp. 43–52. Edinburgh Univ. Edinburgh, Scotland.

Cody, M. L. (1966). The consistency of inter- and intra-specific continental bird species counts. *Am. Nat.* **100,** 371–376.

Cody, M. L. (1968). On the methods of resource division in grassland bird communities. *Am. Nat.* **102,** 107–147.

Cody, M. L. (1970). Chilean bird distribution. *Ecology* **51,** 455–464.

Cody, M. L. (1974a). "Competition and the Structure of Bird Communities." Monographs in Population Biology, Vol. 7. Princeton Univ. Press, Princeton, New Jersey.

Cody, M. L. (1974b). Optimization in ecology. *Science* **183,** 1156–1164.

Cody, M. L. (1983). Continental diversity patterns and convergent evolution in bird communities. *In* "Mediterranean-Type Ecosystems" (F. J. Kruger, D. T. Mitchel, and J. U. M. Jarvis, eds.), Vol. 43, pp. 357–402. Springer-Verlag, Berlin and New York.

Cornwallis, L. (1975). The comparative ecology of eleven species of wheatears (genus *Oenanthe*) in S. W. Iran. Ph.D. thesis. Oxford Univ., Oxford.

Crook, J. H. (1962). The adaptive significance of pair formation types in weaver birds. *Symp. Zool. Soc. London* **8,** 57–70.

Davis, J. (1974). Habitat preferences and competition of wintering juncos and golden-crowned sparrows. *Ecology* **54,** 174–180.

Dunning, J. B., Jr., and Brown, J. H. (1982). Summer rainfall and winter sparrow densities: A test of the food limitation hypothesis. *Auk* **99,** 123–129.

Emlen, J. T. (1972). Size and structure of a wintering avian community in southern Texas. *Ecology* **53,** 317–329.

Emlen, S. T., and Oring, L. W. (1977). Ecology, sexual selection and the evolution of mating systems. *Science* **197,** 215–223.

Evans, E. W. (1978). Nesting responses of Field Sparrows (*Spizella pusilla*) to plant succession on a Michigan old field. *Condor* **80,** 34–40.

Evans, F. C. (1964). The food of Vesper, Field and Chipping Sparrows nesting in an abandoned field in southeast Michigan. *Am. Midl. Nat.* **72,** 57–75.

Finzel, J. E. (1964). Avian populations of four herbaceous communities in southeastern Wyoming. *Condor* **66,** 496–510.

Folse, L. J., Jr. (1982). An analysis of avifauna–resource relationships on the Serengeti Plains. *Ecol. Monogr.* **52,** 111–127.

Fretwell, S. (1968). Habitat distribution and survival in the field sparrow (*Spizella pusilla*). *Bird Banding* **39,** 293–306.

Fretwell, S. (1972). "Populations in a Season Environment." Monographs in Population Biology, Vol. 5. Princeton Univ. Press, Princeton, New Jersey.

Gowthorpe, P. (1977). Territorialité et structures sociales d'une population de *Cisticola anonyma* (Muller) du Gabon: Un nouveau cas d'aide au nourrissage. *L'Oiseau et la Rev. Fran. de l'Orn.* **47,** 245–252.

Harmeson, J. P. (1974). Breeding ecology of the dickcissel. *Auk* **91,** 348–359.

Hebrard, J. J. (1978). Habitat selection in two species of *Spizella:* A concurrent laboratory and field study. *Auk* **95,** 404–410.

Hurley, R. J., and Franks, E. C. (1976). Changes in the breeding ranges of two grassland birds. *Auk* **93,** 108–115.

Johnson, D. H., and Springer, P. F. (1972). Breeding Bird Censuses #47–53: Mixed Prairie, North Dakota. *Am. Birds* **26,** 970–975.

Knodel, J. J. (1980). Breeding Bird Censuses #104–107: Annually burned and unburned Tallgrass Prairie, Kansas. *Am. Birds* **34,** 69–70.

Lack, D. L. (1935). Territory and polygamy in a bishop-bird *Euplectes hordacea hordacea* (Linn.). *Ibis* **77,** 817–836.

Leisler, B., Heine, G., and Siebenrock, K-H. (1983). Einnischung und interspezifische Territorialität überwinternder Steinschmätzer (*Oenanthe isabellina, O. oenanthe, O. pleschanka*) in Kenia. *J. Ornithol.* **124,** 393–413.

Lill, A. (1974). Behavior of the Grassland Sparrow and two species of seed-finch. *Auk* **91,** 35–43.

Lowther, P., and Seibel, D. E. (1982). Breeding Bird Census #115: Floodplain Tallgrass Prairie, Kansas. *Am. Birds* **36,** 80.

MacArthur, R. H., Recher, H., and Cody, M. L. (1966). On the relation between habitat selection and bird species diversity. *Am. Nat.* **100,** 319–332.

Maher, W. J. (1979). Nestling diets of prairie passerine birds at Matador, Saskatchewan, Canada. *Ibis* **121,** 437–452.

Martin, S. G. (1974). Adaptations for polygynous breeding in the Bobolink *Dolichonyx oryzivorus. Am. Zool.* **14,** 109–119.

Murray, B. G., Jr. (1969). A compartive study of the LeConte's and Sharp-tailed Sparrows. *Auk* **86,** 199–231.

Nakamura, H. (1979). Habitat differences among resident and migratory populations of Oriental Greenfinch *Carduelis sinica* groups in winter season. *J. Yamashina Inst. Ornithol.* **11,** 189–220.

Nakamura, T. (1973). Comparative ecology of emberizids wintering in central Honshu, especially of *Emberiza rustica* and *E. cioides. J. Yamashina Inst. Ornithol.* **7,** 139–159.

Nakamura, T., and Iijima, K. (1977). Buntings wintering in southern Kyushiu, especially the mode of dispersion in *Emberiza furcata* and *E. rustica. J. Yamashina Inst. Ornithol.* **9,** 1–19.

Nakamura, T., Yamaguchi, S., Iijima, K., and Kagawa, T. (1968). A comparative study on the habitat preference and home range of four species of *Emberiza* on peat grassland. *Misc. Rep. Yamashina Inst. Ornithol. Zool.* **5,4**(30), 313–336.

Orians, G. H. (1969). On the evolution of mating systems in birds and mammals. *Am. Nat.* **103,** 589–603.

Orians, G. H. (1978). "Some Adaptations of Marsh-nesting Blackbirds." Monographs in Population Biology, Vol. 14. Princeton Univ. Press, Princeton, New Jersey.

Orians, G. H., and Willson, M. F. (1964). Interspecific territories of birds. *Ecology* **45,** 736–745.

Pickwell, G. B. (1931). The Prairie Horned Lark. *St. Louis Acad. Sci. Trans.* **27,** 1–153.

Post, W., and Greenslaw, J. S. (1975). Seaside sparrow displays: Their function in social organization and habitat. *Auk* **92,** 461–492.

Pulliam, H. R. (1973). Comparative feeding ecology of a tropical grassland finch (*Tiaris olivaceus*). *Ecology* **54,** 284–299.

Pulliam, H. R. (1983). Ecological theory and the coexistence of sparrows. *Ecology* **64,** 45–52.

Pulliam, H. R., and Brand, M. R. (1975). The production and utilization of seeds in plains grasslands of southeastern Arizona. *Ecology* **56,** 1158–1166.

Pulliam, H. R., and Mills, G. S. (1977). The use of space by wintering sparrows. *Ecology* **58,** 1393–1399.

Pulliam, H. R., and Parker, T. A., III. (1979). Population regulation in sparrows. *Fortschr. Zool.* **25,** (2/3), 137–147.

Raitt, R. J., and Pimm, S. L. (1976). Dynamics of bird communities in the Chihuahuan Desert, New Mexico. *Condor* **78,** 427–442.

Risser, P. G., Birney, E. C., Blocker, H. D., May, S. W., Parton, W. J., and Wiens, J. A. (1981). "The True Prairie Ecosystem." *In* US/IBP Synth. Ser., Vol. 16, pp. 1–557. Hutchinson Ross, Stroudsburg, Pennsylvania.

Robins, J. D. (1971). A study of the Henslow's Sparrow in Michigan. *Wilson Bull.* **83,** 39–48.

Rotenberry, J. T. (1980). Dietary relationships among shrubsteppe passerine birds: Competition or opportunism in a variable environment? *Ecol. Monogr.* **50,** 93–110.

Rotenberry, J. T., and Wiens, J. A. (1980a). Temporal variation in habitat structure and shrubsteppe bird dynamics. *Oecologia* **47,** 1–9.

Rotenberry, J. T., and Wiens, J. A. (1980b). Habitat structure, patchiness, and avian communities in North American steppe vegetation: A multivariate approach. *Ecology* **61,** 1228–1250.

Rubenstein, D. I., Barnett, R. J., Ridgely, R. S., and Klopfer, P. H. (1977). Adaptive advantages of mixed-species feeding flocks among seed-eating finches in Costa Rica. *Ibis* **119,** 10–21.

Ryder, R. A. (1980). Breeding Bird Census #98: Shortgrass Plains, Colorado. *Am. Birds* **33,** 81.

Seastedt, T. R., and MacLean, S. F. (1979). Territory size and composition in relation to resource abundance in Lapland Longspurs breeding in arctic Alaska. *Auk* **96,** 131–142.

Shugart, H. H., Jr., and James, D. (1973). Ecological succession of breeding bird populations in northwestern Arkansas. *Auk* **90,** 62–77.

Smith, W. J., and Vuilleumier, F. (1971). Evolutionary relationships of some South American ground tyrants. *Bull. Mus. Comp. Zool.* **141,** 179–268.

Stepney, P. H. R., and Power, D. M. (1973). Analysis of the eastward breeding expansion of Brewer's Blackbird plus general aspects of avian expansions. *Wilson Bull.* **85,** 452–464.

Stresemann, E. (1950). Interspecific competition in chats. *Ibis* **92,** 148.

Tinbergen, N. (1968). "Curious Naturalists." Natural Hist. Library. Doubleday, Garden City, New York. (First published in 1958 by Basic Books.)

Weatherhead, P. J. (1979). Ecological aspects of monogamy in tundra-breeding Savannah Sparrows. *Auk* **96,** 391–401.

Weisbrod, A. R., and Stevens, W. F. (1974). The skylark in Washington. *Auk* **91,** 832–835.

Welsh, D. A. (1975). Savannah sparrow breeding and territoriality on a Nova Scotia dune beach. *Auk* **92,** 235–251.

Wiens, J. A. (1973). Interterritorial habitat variation in grasshopper and savannah sparrows. *Ecology* **54,** 877–884.

Wiens, J. A. (1974). Climatic instability and the "ecological saturation" of bird communities in North American grasslands. *Condor* **76,** 385–400.

Wiens, J. A., and Rotenberry, J. T. (1979). Diet niche relationships among North American grassland and shrubsteppe birds. *Oecologia* **42,** 253–292.

Wiens, J. A., and Rotenberry, J. T. (1980). Patterns of morphology and ecology in grassland and shrubsteppe bird populations. *Ecol. Monogr.* **50,** 287–308.

Wiens, J. A., and Rotenberry, J. T. (1981). Habitat associations and community structure of birds in shrubsteppe environments. *Ecol. Monogr.* **51,** 21–41.

Wittenberger, J. F. (1980). Vegetation structure, food supply, and polygyny in Bobolinks (*Dolichonyx oryzivorus*). *Ecology* **61,** 140–150.

Zimmerman, J. L. (1971). The territory and its density dependent effect in *Spiza americana*. *Auk* **88,** 591–612.

Zimmerman, J. L. (1982). Nesting success of dickcissels (*Spiza americana*) in preferred and non-preferred habitats. *Auk* **99,** 292–298.

Zimmerman, J. L., and Tatschl, J. L. (1975). Floodplain birds of Weston Bend, Missouri River. *Wilson Bull.* **87,** 198–206.

Chapter 7

Habitat Selection in Variable Environments: Shrub–Steppe Birds

JOHN A. WIENS
Department of Biology
University of New Mexico
Albuquerque, New Mexico

I. INTRODUCTION

The way in which an individual selects a habitat may have important effects on its access to food or mates, its vulnerability to predators, its exposure to climatic stresses, and so on. Selection might thus be expected to exert rather strong influences on habitat selection patterns. Given this, the patterns that we witness in nature should approach some optimum expression of habitat selection (Cody, 1974a).

This statement summarizes much of our thinking about avian habitat selection over the past several decades, which has been formalized in two related areas of

HABITAT SELECTION IN BIRDS

ISBN 0-12-178080-5

theory. On the one hand, Fretwell and Lucas (1969) have considered how individuals of a species should occupy habitats of varying quality or suitability under "ideal" conditions. This theory suggests that patterns of habitat occupancy should be closely linked to variations in individual fitness and that the best habitats should usually be fully saturated with individuals, unless a species is locally rare. On the other hand, species do not exist alone in habitats, but co-occur with other members of communities. The ideal pattern of habitat selection may therefore be modified by competitive interactions with other community members, producing patterns of habitat partitioning that facilitate the coexistence of ecologically similar species (Wiens, 1969; Cody, 1974b, 1978). If habitat or resource conditions vary through time, these variations may be closely tracked by changes in the habitat selection of the birds. Suitable ecological spacing between species is maintained.

These complementary views of habitat selection are founded on assumptions of relatively strong and continuous selection, of general equilibrium of communities in relation to resources, and of competition as a major guiding force in determining community structure and patterns of species' habitat occupancy. The veracity of these assumptions, however, has recently been challenged (e.g., Connell, 1980; Wiens, 1977, 1983a; Alley, 1982; Simberloff, 1983), especially as they may relate to variable environments. In erratically varying environments, so the argument goes, selection favoring optimal habitat selection may not always be intense; habitats may not always be fully saturated; resources may not always be limiting; and changes in habitats or resources may not always be closely tracked. In its extreme form, this "variable environment" view contrasts sharply with the more conventional "competitionist" view (Schoener, 1982), although in fact the differences may be more a matter of differing perspectives on environmental variation and its effects than strictly polarized alternatives (Wiens, 1983b; 1984b). The perspective one adopts, however, may influence the approach one takes to study habitat selection. Thus, a focus on competition as a major factor in structuring communities may produce an emphasis on documenting differences in habitat selection between co-occurring species (e.g., Noon, 1981; Cody, 1978; Diamond, 1978; Edington and Edington, 1983). A more nonequilibrium view, on the other hand, leads to a greater emphasis on documenting how individual species respond to habitat variation and is more likely to consider habitat selection as a process influenced by a multiplicity of factors (e.g., Sabo and Holmes, 1983; Karr and Freemark, 1983; Rice *et al.*, 1983). If interspecific interactions are important among these factors, their effects should be apparent.

The development of differing views of ecological communities leads to the crystallization of several questions I believe should guide our investigations of habitat selection. Is habitat selection by members of a species relatively constant over large geographic areas or between years? If habitat selection varies, does it represent a close tracking of resource levels? Do patterns of habitat selection of co-occurring species differ in the manners expected from species interactions? Do species, instead, respond to habitats in species-specific manners that are largely independent of interaction effects? Do patterns of habitat selection vary depending on the spatial scale on which they are viewed?

My objective in this chapter is not to provide answers to these questions; indeed, the research I will review has contributed much more to the development of alternative views and the formulation of such questions than to their resolution. Instead, I will synthesize some of the studies that John Rotenberry and I have conducted on habitat selection in breeding birds of shrub–steppe environments. I will adopt the multifactor approach noted above and will emphasize the habitat responses of two abundant and ecologically similar passerine species, rather than discuss overall community patterns. I will use this synthesis to focus attention on several aspects of avian habitat selection that may merit closer attention in future studies.

II. HABITAT SELECTION IN SHRUB–STEPPE BIRDS

A. The Setting and the Species

Our studies of the habitat relationships of breeding shrub–steppe birds have been conducted over the past decade in the northern portion of the Great Basin and adjacent areas, in Washington, Oregon, and northern Nevada. The region is topographically diverse: large fault blocks and elevated plateaus separate broad level valleys, many the dry remnants of Pleistocene lakes. The vegetation covering the plateaus and valleys is dominated by *Artemisia, Atriplex,* or *Sarcobatus* species among the shrubs and by various *Poa, Agropyron, Festuca,* and *Bromus* species among the grasses (Franklin and Dyrness, 1973). Large, unbroken expanses of shrub–steppe habitat are typical. Climatically, the region is typified by cold winters and hot summers, but diurnal variations may be great during the summer, with nighttime temperatures often falling below freezing. Most of the precipitation falls during winter and early spring. Annual precipitation usually averages less than 250 mm and is quite variable (Wiens, 1974; Rotenberry and Wiens, 1980a).

Breeding bird communities in these habitats generally contain 3–6 species: several finches, a lark, a thrasher, a meadowlark, and perhaps a flycatcher. With the exception of the flycatcher, all forage on the shrubs and grasses and on the ground and belong to a single ecological guild. Two species usually dominate the community: Brewer's Sparrow (*Spizella breweri*), which on average contributes 55% of the individuals present, and Sage Sparrow (*Amphispiza belli*), which accounts for 26% (Rotenberry and Wiens, 1980a). Sage Sparrows are larger than

Brewer's Sparrows (19 gm versus 12 gm), but both species place nests in similar situations, occupy broadly overlapping territories, and almost never interact behaviorally with members of the other species. I will focus on these two species in the following synthesis. Our methods and procedures for surveying the birds, measuring habitat features, and analyzing the relationships between these data sets are documented elsewhere (Rotenberry and Wiens, 1980b; Wiens and Rotenberry, 1981) and will not be repeated here.

B. Habitat Correlations at Different Spatial Scales

Our investigations of the habitat–occupancy patterns of these species have been conducted at several geographical scales, and it is instructive to consider these in a hierarchical fashion, from the broadest to the most localized.

1. General Biogeographic Patterns

The most inclusive scale incorporates the spectrum of habitat types from shrub–steppe through shortgrass prairies to midwestern tallgrass prairies (Rotenberry and Wiens, 1980b). Habitat variation at this "continental" scale was analyzed by subjecting the matrix of values of 22 habitat variables gathered in 26 samples to Principal Component Analysis (PCA). Three clear dimensions of variation in habitat structure emerged, the first corresponding to variation in horizontal heterogeneity or patchiness, the second portraying increasing vertical heterogeneity, and the third reflecting variation in the density and coverage of low forbs. Densities of most of the bird species varied significantly with these synthetic dimensions of habitat structure variation, and in particular several typical shrub–steppe bird species (including Sage and Brewer's Sparrows) were positively associated with the first two PCA axes, reflecting the horizontal patchiness and vertical structuring of the habitats they occupied. Both Sage and Brewer's Sparrows exhibited statistically significant correlations between population densities and several individual measures of habitat structure, although the correlation patterns were stronger for Sage Sparrows (9 out of 22 correlation tests were significant) than for Brewer's Sparrows (4 out of 22). Overall, then, at this scale of investigation, we could clearly identify assemblages of species that covaried in their distributions and abundances, and these variations were strongly associated with differences in the structural configurations of habitats.

2. Within-Shrub–Steppe Patterns

The continental analysis revealed clear patterns of bird–habitat associations, but it may legitimately be argued that these do not reflect true habitat selection but only circumstances of the biogeographic ranges of species. Sage Sparrows, for example, are distributionally restricted to the intermountain west, southern California, and northern Baja California and do not occur in the prairies east of the Rockies. They thus will be absent for distributional reasons from much of the space defined by the continental PCA and may exhibit strong correlations simply because they occupy one extreme of the habitat gradients defined at that scale (Wiens, 1981). We therefore used a data set derived from more intensive studies conducted on 14 sites within the shrub–steppe of the Pacific Northwest over a 3-year period to conduct a parallel analysis aimed at determining the patterns of bird–habitat correlations at this "regional," within-habitat-type scale (Wiens and Rotenberry, 1981).

At this scale of investigation, no clearly defined suites of species that covaried in their distributions and abundances emerged. Only 8 of 91 pairwise correlation tests between species were statistically significant, a proportion not different from that expected by chance (power test, Rotenberry and Wiens, 1985, see Toft and Shea, 1983)—the species appeared to vary independently of one another. Four of the 8 significant correlations, however, involved Brewer's Sparrows, whose densities varied inversely with those of Loggerhead Shrikes (*Lanius ludovicianus*) and Rock Wrens (*Salpinctes obsoletus*) and paralleled those of Sage Thrashers (*Oreoscoptes montanus*) and Gray Flycatchers (*Empidonax wrightii*). Our field observations suggest that the positive associations probably reflect common responses to habitat features, while the negative correlations may be influenced by predator–prey relationships (shrike) or clearly different habitat affinities (wren).

Relatively few correlations also emerged when we compared variations in the densities of each species to various measures of habitat structure. For the bird species that are most common and widespread through the shrub–steppe (the two sparrows and the Sage Thrasher), only 6 of the 60 tests were statistically significant (again, a proportion not differing from chance). For Sage Sparrows, only 1 of 20 tests indicated a significant relationship; for Brewer's Sparrows only 2 of 20. Similarly weak correlational patterns emerged from tests of bird densities against variation in PCA dimensions defined by variation in habitat structure; Sage Sparrows were not correlated with any of the three significant components, while Brewer's Sparrows were negatively correlated with the third axis, which portrayed variation from open, flat habitats to those with more uneven, rocky outcrops.

Thus, for these species considered at this scale, variation in habitat structure appeared to account for relatively few of the variations in their distributions and abundances. This conclusion is reinforced when we consider patterns of temporal variation in habitat features and bird populations (Rotenberry and Wiens, 1980a). We conducted our regional shrub–steppe studies over a 3-year period that included one extremely dry year (27% of normal winter–spring precipitation) followed by two abnormally wet years (136 and 126% of normal). This climatic variation produced clear and significant changes in habitat structuring: an increase in coverage of annuals and grasses, a decrease in the amount of bare, unvegetated ground, and decreases in both vertical and horizontal heterogeneity.

Despite these changes, however, and despite substantial annual variation in the densities of the bird species on individual plots, there was no overall pattern of response in bird distributions or abundances to the habitat variations.

Traditionally, studies of avian habitat selection have emphasized structural features of habitats, rather than aspects of their floristics or plant species composition, for a variety of reasons. In our regional studies, however, we also recorded the coverages of different shrub species on the plots, permitting an analysis of bird–floristics relationships (Wiens and Rotenberry, 1981). For the three common and widespread shrub–steppe bird species, correlations with the coverages of shrub species were higher, overall, than with structural features of the habitat; 26% (7 out of 26) of the tests were statistically significant. Sage Sparrows varied in density parallel with variations in sagebrush (*Artemisia tridentata*) coverage, and inversely with coverage of cottonthorn (*Tetradymia spinosa*) and greasewood (*Sarcobatus vermiculatus*). Brewer's Sparrows exhibited no positive correlations, but were negatively correlated with hopsage (*Atriplex spinosa*) and budsage (*Artemisia spinescens*) coverages. The analysis of temporal variations (Rotenberry and Wiens, 1980a) did little to elucidate these patterns further as coverages of the slowly growing, long-lived shrub species changed little in response to the precipitation variations over the 3-year period.

All these analyses were based on linear statistical models, as has been the tradition in studies of avian habitat correlations. There is little a priori justification, however, for expecting birds to respond to changes in habitat features in a linear fashion, either as individuals or as populations. We have recently extended our analysis of the regional shrub–steppe data, using nonlinear "response surface" or trend surface models (J. T. Rotenberry and J. A. Wiens, in preparation; see Haggett *et al.*, 1977, Meents *et al.*, 1983). Does the use of more complex models produce any greater insights into habitat selection of the birds? In general, it does, as the R^2 values for the response-surface regressions were consistently higher than those obtained in the earlier linear regressions. If we consider correlations between Brewer's Sparrow densities and the axes of the PCA based on vegetation structure, for example, the degree of correlation with the third axis increased (linear $R^2 = 33.8\%$, response surface $R^2 = 46.6\%$); slightly more of the variation was explained. For Sage Sparrows the R^2 values also increased (8.9–35.5%) but remained nonsignificant. The effects of using the nonlinear model with floristic data were more impressive. Sage Sparrows, which previously had shown a significant linear correlation with the axis of a PCA (based on shrub-species coverages) that reflected primarily sagebrush coverage ($R^2 = 51.8\%$), displayed an even closer correlation with the response-surface regression ($R^2 = 70.4\%$). Brewer's Sparrows in the linear analysis were negatively correlated with the axis reflecting coverages of several spinescent shrub species ($R^2 = 16.2\%$); this relationship was improved considerably, chiefly through the complex quadratic and cross-product terms of the response-surface model ($R^2 =$

67.7%). Partial correlation analyses indicated an absence of any strong effects between bird species, again suggesting a general independence between species in their responses to habitat features.

Several general conclusions emerge from our consideration of bird–habitat relationships at this regional scale. First, it is apparent that the scale of analysis does make a difference, as different patterns are expressed for species such as Sage and Brewer's Sparrows at the continental and regional scales. Second, within the shrub–steppe region, there are some clear relationships between variations in the distribution and abundance of bird species and habitat variables, but these are not strongly expressed, especially for the bird species that are most characteristic of the shrub–steppe. Third, the correlational patterns generally tend to be stronger with floristically defined habitat features than with structurally defined variables, suggesting that the habitat selection responses of the birds may be directed more toward features of habitats or associated resources that are closely coupled with the presence and coverage of certain shrub species than toward general aspects of the appearance or physiognomy of habitats. Finally, there is substantial evidence in our various analyses leading to the conclusion that the different bird species respond to habitat variations largely independently of one another (see Chapter 6, this volume, for similar results in grassland birds) and that direct, pairwise species interactions play little role in determining the distributions of species among habitats at this scale.

There are in these data, however, some intriguing suggestions of possible negative interactions between Sage Sparrows and congeneric Black-throated Sparrows (*Amphispiza bilineata*). Over the entire region, densities of the two species varied inversely ($p < 0.05$) (Wiens and Rotenberry, 1981). Where the two species co-occurred, habitats contained significantly more shrub cover, were of taller stature, and included more shrub species than where only Sage Sparrows were present. Black-throated Sparrows varied in density with coverage of *Atriplex* species and thus exhibited quite different shrub-species affinities from the *Artemisia*-associated Sage Sparrows; in the PCA analysis, the species loaded on orthogonal (and thus statistically independent) axes. To some extent these patterns may reflect the more southerly distribution of *Amphispiza bilineata,* which would restrict it to areas with a generally more diverse shrub flora than the areas in the northern Great Basin that are overwhelmingly dominated by species of *Artemisia*. The analysis of temporal variation (Rotenberry and Wiens, 1980a), however, suggests that this explanation may be too simple. Over the 3 years included in that analysis, densities of the two species showed significant inverse variations on the five sites on which both species occurred. These patterns are certainly consistent with the hypothesis that the two species do interact competitively with one another, but, unfortunately, they are also consistent with the suggestion that the species simply have different habitat affinities, independent of proximate interactions, and that the patterns are an expression of these species-specific habitat responses. More intensive observations, ideally including experimental manipulations, are required to identify the processes that may be influencing the distribution and abundance of these species in this region. On one site where both were reasonably abundant, their territories overlapped broadly, and we witnessed no overt behavioral interactions (either vocal or physical) between the species, which suggests that the expression of their interactions, if it occurs, may be subtle.

3. *Patterns within Plots*

Most investigations of avian habitat selection, especially those conducted within a community context, have involved broad comparisons between sampling areas; the studies summarized above are no exception. This procedure, however, may obscure important variation in habitat selection. In particular, by averaging habitat measures for entire study plots, it is implicitly assumed that the plots are internally homogeneous. One hopes, then, that the density variations of the birds reflect processes operating at a between-plot rather than a within-plot scale. It is possible, however, that two plots, one homogeneous and the other heterogeneous, might have similar mean values for habitat features averaged over the plot but might nonetheless support quite different densities of a species, depending on how individuals responded to the within-plot variance. Moreover, while density is clearly a better index of habitat response than simple species presence or absence at a plot, it is still a somewhat crude indicator of habitat quality as it might be viewed by individuals (Van Horne, 1983). To sharpen our focus on what may really be going on in habitat selection, then, we must examine what goes on *within* plots. Do individuals establish territories or select breeding habitats in nonrandom positions with respect to habitat variation within a plot?

To assess this, we mapped the positions of territories of breeding individuals within 9-ha plots at three sites. Measures of habitat features were taken at points located within the plot by a stratified random design, and average values for the features at each sampling period were then determined for subsets of the points occurring inside (Present) or outside (Absent) the territorial boundaries of individuals of a species in a plot. This procedure faces two possible complications. First, within-season changes in habitat features may be so dramatic as to render the mean values extremely dependent on just when the measurements are taken, as Whitmore (1979) has documented for grassland habitats. This is not likely to be a major problem in the shrub–steppe, however, as there is no seasonal burst of plant production that compares in magnitude to that in eastern grasslands. Second, there may be a seasonal sequence to territory establishment, such that early territories differ systematically in habitat features from those established later on (Wiens, 1973). We lack data to address this problem, but territory establishment by breeding shrub–steppe birds occurs over a much shorter time period than that in grasslands.

Our analysis of the present–absent habitat patterns within our study plots is only in an early stage, but some preliminary results may indicate the sorts of patterns that emerge. First, consider the responses of Sage Sparrows to variation in several habitat features within sagebrush-dominated plots at sites in eastern Washington (ALE), central Oregon (Cabin Lake), and north-central Nevada (Star Creek) (Fig. 1). Areas occupied by individual territories clearly contained a greater coverage of sagebrush than areas unoccupied by the birds at both ALE and Star Creek, while at Cabin Lake the birds established territories in areas of somewhat lower sagebrush coverage than average (Fig. 1A). This pattern might be explained by suggesting that the birds were more selective of sagebrush areas in plots in which overall sagebrush coverage was low, but this is not the case—overall sagebrush coverage was 24% at Star Creek, 21% at ALE, and only 18% at Cabin Lake. It is clear, however, that the birds were preferentially settling in areas of higher sagebrush coverage within the ALE and Star Creek plots. It is also apparent that the between-site variation in sagebrush coverage within occupied areas was considerably less than that in unoccupied areas.

Differences in the coverage of bare (unvegetated) ground between present and absent areas were less dramatic than for sagebrush (Fig. 1B), but nonetheless occurred. On average, bare-ground coverage was somewhat greater in occupied areas than in absent areas at ALE and Cabin Lake but less at Star Creek, where overall bare-ground coverage was greatest (50%, versus 39% at Cabin Lake and 27% at ALE). Again, coverage of bare ground varied less between sites in occupied areas than in unoccupied areas. Sage Sparrows were apparently unresponsive to variations in grass coverage within plots, as coverage within territories closely matched coverage in unoccupied portions of the plot, despite the considerable differences in overall grass coverage between the sites (Fig. 1C).

How do these patterns compare to those expressed by Brewer's Sparrows? Brewer's Sparrows did not occur on the sagebrush-dominated plots at ALE or Star Creek, so the comparison must be confined to Cabin Lake. At that site, there was no evidence of a preferential response to coverage of sagebrush or of bare ground, although coverages in Brewer's Sparrow territories were quite similar to those in Sage Sparrow territories, for which a selective response seemed evident (Fig. 2). Brewer's Sparrows, on the other hand, occupied areas with somewhat greater grass coverage than characterized areas outside of territories; again, the coverage in their territories was quite similar to that in Sage Sparrow territories, which displayed little difference between present and absent areas.

To a degree, these preliminary within-plot analyses support the earlier analyses conducted at the broader, regional scale. Sage Sparrows exhibited a clear association with sagebrush in that analysis, and this positive response is also apparent on the within-plot scale at two of the three sites. They failed to vary significantly with coverage of bare ground or grass in the regional analysis, and within-plot differences in coverages of these habitat features are either slight or

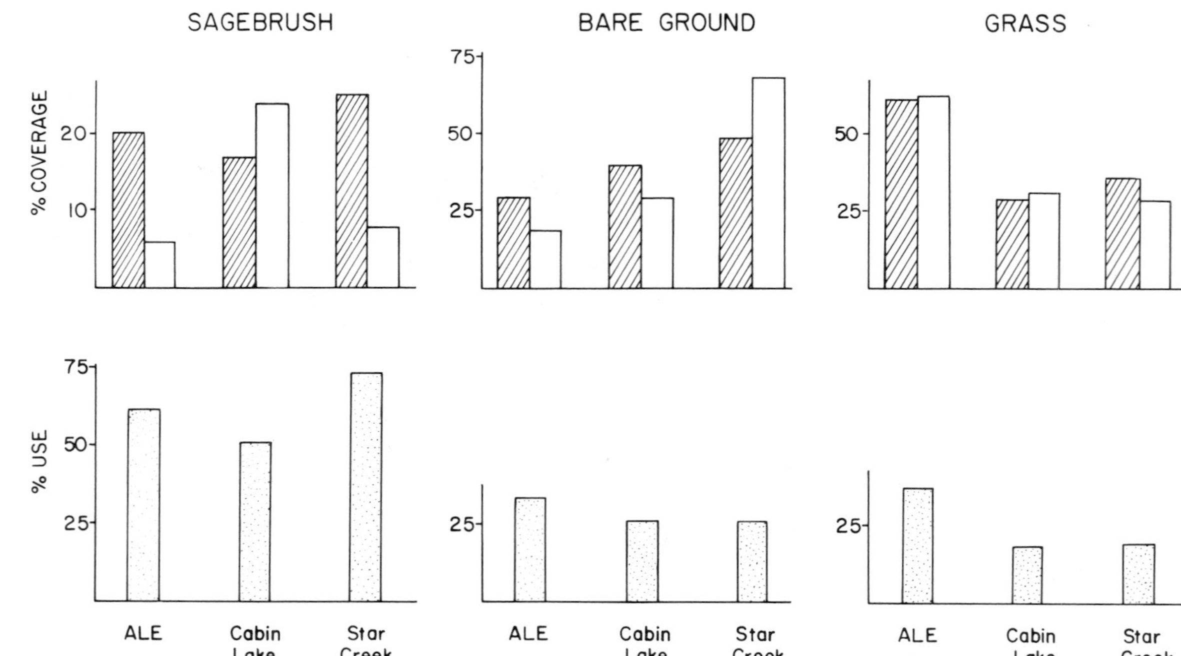

Fig. 1. (Above) Coverages of three components of shrub–steppe habitats in areas of plots inside [Present (hatched box)] versus outside [Absent (open box)] Sage Sparrow territories at three study sites: Note that coverage is scaled differently for the three features. (Below) Percentage of use in behavior of these habitat substrates by Sage Sparrows. See text.

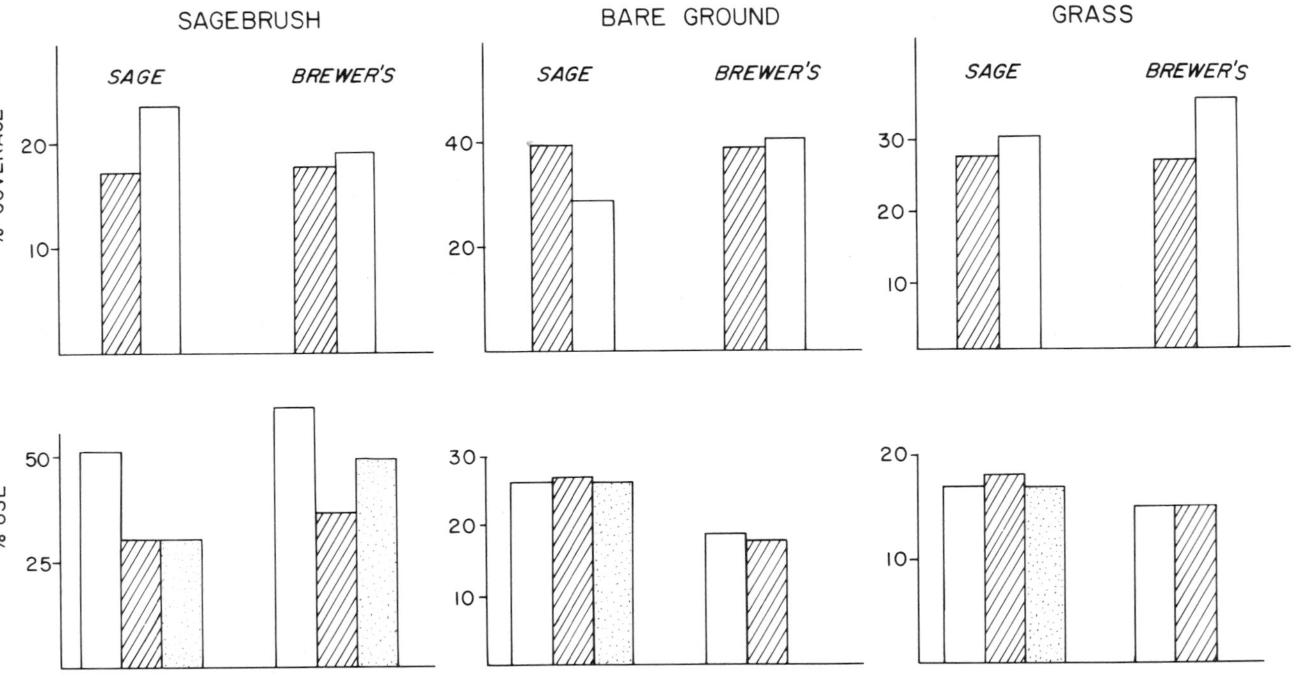

Fig. 2. (Above) Coverages of three habitat features in areas of a sagebrush-dominated plot at Cabin Lake that fell inside (hatched box) versus outside (open box) of territories of Sage and Brewer's Sparrows. (Below) Percentage of use of these habitat substrates in the total behavior of Sage and Brewer's Sparrows (open box) and in their foraging (hatched box) and singing (stippled box) behavior are considered separately. Coverages and percentage-use graphs are scaled differently for the different habitat features.

absent. On the one site where they occur, Brewer's Sparrows display the same general insensitivity to variations in these habitat features that characterized their regional distributions and abundances. The similarity in patterns of within-plot habitat occupancy for Sage and Brewer's Sparrows at Cabin Lake, however, is remarkable. It is not the sort of habitat differentiation that one might expect of species locked in intense competition with one another.

C. Patterns of Habitat Use

The above analyses are based on correlations, and no matter how complex or multifactorial such analyses become, they can at best indicate only statistical associations between birds and habitat features. In order to interpret such patterns in the context of habitat selection, we must assume that a correlation between density variations of a species or placement of territories in a plot and certain habitat features indicates a deeper functional relationship, that the correlation indicates something about how individual birds actually select and use habitats. Finding a significant correlation between territory placement and habitat variable x, for example, we presume that x is important as a cue in the habitat selection behavior or settling response of the individuals, that it is of some direct significance to the individuals as a resource, and/or that it is itself not an important habitat feature but is highly correlated with some other (unmeasured) feature that is important. Correlational approaches generally will not distinguish among these alternatives, although they may serve to sharpen hypotheses or suggest critical tests.

One way to begin to define the functional importance of habitat features to birds is through direct observation of their behavioral patterns of habitat use. If features of the habitat are used by individuals in a nonrandom fashion and if these patterns parallel patterns of apparent habitat selection revealed by various correlational procedures, our confidence in the importance of those habitat features to the birds and their role in habitat selection is strengthened. As part of our studies of shrub–steppe birds, we recorded the behavior of individuals and the habitat substrate on which the behavior occurred at 20-sec intervals throughout the breeding period over several years at several sites. These data are in a preliminary stage of analysis, and the details of our observational methods, analytical procedures, and results will be reported elsewhere (J. A. Wiens, J. T. Rotenberry, and B. Van Horne, in preparation). Some preliminary results that relate to the patterns of within-plot habitat occupancy presented above, however, can be described.

Consider first Sage Sparrow use of substrate at ALE, Cabin Lake, and Star Creek, averaged over all behaviors and over several years at the sagebrush-dominated plots at each site (Fig. 1). The percentage of all behavioral observations that occurred on sagebrush varied significantly among the three sites, with use being greatest at Star Creek and least at Cabin Lake (Fig. 1A). The rankings of sagebrush use for the sites were the same as those of sagebrush coverage within Sage Sparrow territories at the sites. Use of bare ground, on the other hand, was inversely related to the coverage of bare ground in territories; a greater percentage of behavioral observations occurred in association with bare ground where coverage was least, at ALE (Fig. 1B). The percentage of behavioral observations associated with grass paralleled grass coverage in occupied areas closely and was significantly greater at ALE than at the other two sites (Fig. 1C). Use of these different habitat features, then, varied among sites, generally in ways related to the variations in coverage of the features among sites. Sagebrush, however, was used substantially more than bare ground or grass. The correlations of Sage Sparrows with sagebrush coverage that emerged in the regional analysis and the nonrandom placement of territories with respect to sagebrush that was apparent in the within-plot analyses thus seem to correspond with a high degree of use of sagebrush by the birds in their behavior.

Comparisons of habitat use by Sage Sparrows with use by Brewer's Sparrows in the sagebrush-dominated plot at Cabin Lake are also revealing (Fig. 2). Areas occupied by Brewer's Sparrow territories contained the same amount of sagebrush cover as did Sage Sparrow territories, but Brewer's Sparrows centered more of their behavior on sagebrush than did Sage Sparrows, significantly so in 3 of 4 years. If the use of sagebrush is considered in terms of specific behaviors, however, the basis for this difference becomes clear. The two species used sagebrush for foraging to a very similar degree, but Brewer's Sparrows used it as a substrate in their singing behavior significantly more than did Sage Sparrows. Thus, despite the overall correlation of Sage Sparrows with sagebrush coverage and the absence of such a correlation for Brewer's Sparrows, the latter species actually used this substrate more, primarily for singing. Areas occupied by the two species were quite similar in coverage of bare ground, but a significantly greater proportion of Sage Sparrow behavior occurred there. Neither species, however, appeared to use bare-ground areas differentially for different behaviors (although Brewer's Sparrows never sang from the ground). Sage Sparrows also used grass cover somewhat more in their behavior than did Brewer's Sparrows, again despite the similarity of grass coverage within their territories. Overall, then, the patterns of similarity between the species that appeared in the within-plot analysis of habitat correlations breaks down somewhat when one considers the details of individual behavioral use of the habitat features, although not necessarily along the foraging dimensions expected if resource-based microhabitat partitioning via species interactions were important.

We have conducted another study of habitat use by shrub–steppe birds that provides a somewhat different perspective (J. T. Rotenberry and J. A. Wiens, in preparation). On two plots at our Fort Rock site, located a few kilometers from the Cabin Lake site in central Oregon, we measured 11 variables for each of 150 randomly selected "patches" of vegetation (contiguous associations of shrubs, understory herbs and grasses, and shrub detritus). We then used PCA to condense the variation in the vegetation variables into four major components, which collectively accounted for 73% of the variation in the original data set. We also quantified the behavioral patterns of patch use by Sage and Brewer's Sparrows during foraging bouts on the same plots by following foraging individuals and marking each vegetation patch used in a sequence and recording the length of time spent in each patch. We later measured the same 11 variables for these patches as were measured for the randomly selected patches. We then asked whether the distribution of the foraging-use patches in the habitat space defined by the PCA derived from the random patches differed significantly from chance expectations.

Both species clearly used patches in a nonrandom fashion, but in ways that were statistically indistinguishable between the species (Fig. 3). Both showed significantly high negative factor scores on Principal Component II on both study plots. This component contrasts patches dominated by sagebrush with those dominated by green rabbitbrush (*Chrysothamnus viscidiflorus*). Sage and Brewer's Sparrows thus showed a clear preference for foraging in sagebrush-

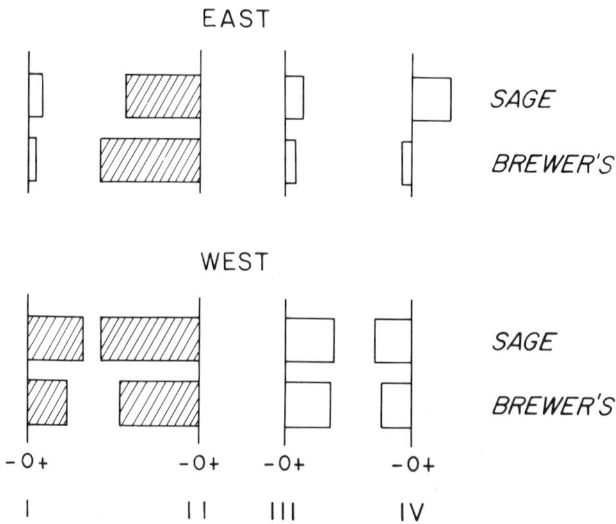

Fig. 3. Histograms reflecting the degree of positive (+) or negative (−) departure of features of habitat patches used by Sage and Brewer's Sparrows in their foraging on two plots (east and west), in relation to four dimensions of a PCA of attributes of randomly sampled habitat patches in each plot. Significant departures are hatched. Component I represents variation in size of patches [small (−)]; Component II reflects variation in contributions of sagebrush and of green rabbitbrush to patch vegetation [sagebrush (−)]; Component III contrasts compact, isolated patches (−) with more diffuse patches; and Component IV contrasts sagebrush coverage (−) with coverage of gray rabbitbrush.

dominated patches. The species also showed a significant departure from random patch use on Principal Component I, but only on the west plot. This component portrays variation in several aspects of shrub size. Although the two plots were quite similar vegetationally, patches on the east plot averaged somewhat larger than those on the west plot. Thus, where patches were relatively large (east), the birds used them in a nondiscriminating manner with respect to size, but where patches were on average smaller (west), use was selectively biased toward larger patches. This suggests that use of habitat components may be somewhat of a threshold process, the birds operating in a selective manner when habitat features fall below (or above) some threshold values but in a nonselective fashion on the other side of such thresholds.

D. An "Experiment" Testing Habitat Selection in Shrub–Steppe Birds

One way to determine the biological significance or reality of bird–habitat correlational patterns is through direct observations of the birds' behavior, as described above. Another way is to conduct experiments, manipulating habitat features in specific ways and then monitoring the responses of individuals or populations to the manipulations. If the correlations in fact reflect true causal linkages, the birds should respond, as they track the habitat change.

During our shrub–steppe studies, one such "experiment" was inadvertently conducted for us by the Oregon State Land Board and the Bureau of Land Management. Beginning in 1977, we had monitored habitat structure and floristic composition and breeding-bird densities at a site in south-central Oregon, Guano Valley. In the spring of 1980, an area of some 16 km^2 containing our study plot was aerially sprayed with 2,4-D, as part of a large-scale range improvement program in southern Oregon. The herbicide prompted a sudden burst of foliage growth by the shrubs, but at the end of the growing season, they died. The following fall the dead shrubs were broken down and removed in the process of drilling crested wheatgrass (*Agropyron cristatum*). The objective of such manipulations is to change a worthless shrub desert into a productive grassland, suitable for grazing at moderate stocking rates.

The experiment had obvious effects on habitat features. Overall height of the vegetation was reduced significantly between 1980 and 1981, the first posttreatment season. Sagebrush coverage was reduced from 27 to 4%, and grass coverage increased slightly (Fig. 4). In subsequent years, however, sagebrush coverage increased somewhat, primarily because the spraying was not completely successful and some regrowth from existing stumps occurred. In addition, the disturbance of the soil during the drilling operations permitted the establishment of some new sagebrush seedlings. Coverage increased to 12% in 1982 and 1983. Grass coverage increased dramatically, reflecting not only the establishment of

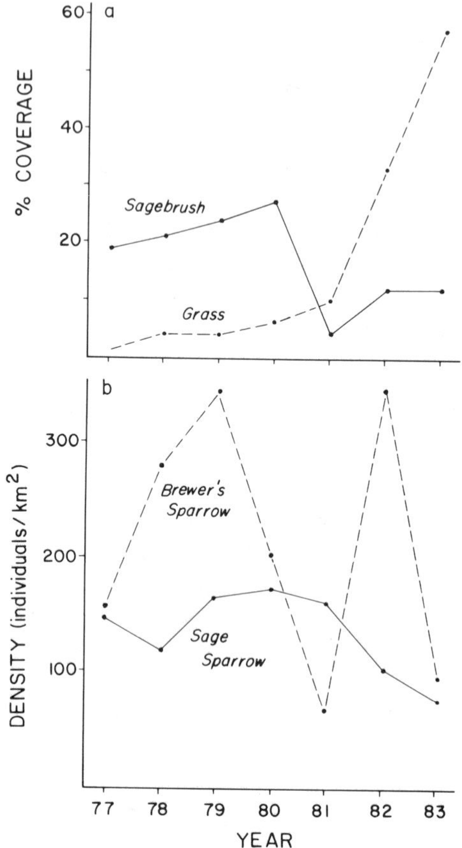

Fig. 4. Variations in coverages of sagebrush and grasses (a) and in densities of Sage and Brewer's Sparrows (b) at a shrub–steppe location (Guano Valley) prior to (1977–1979) and following (1981–1983) aerial herbicide spraying of the plot and planting of crested wheatgrass. See text.

the crested wheatgrass but also the invasion of cheatgrass (*Bromus tectorum*) following the soil disturbance (Fig. 4).

Given our previous correlational information on the habitat affinities of Sage and Brewer's Sparrows, we might have anticipated a rapid decrease in Sage Sparrow densities on the plot but no appreciable change in Brewer's Sparrow numbers. In fact, Sage Sparrow densities remained unchanged from previous levels in the summer following the manipulation, while Brewer's Sparrow densities decreased dramatically, continuing a trend evident on this plot the year before (Fig. 4). Had we stopped our observations then, we might have concluded that Brewer's Sparrows, rather than Sage Sparrows, were closely linked to sagebrush coverage and closely tracked this habitat perturbation. The 1982 and 1983 surveys, however, dispelled this conclusion. Brewer's Sparrow densities were high again in 1982 and dropped again in 1983, while Sage Sparrow densities decreased in 1982 and in 1983. The density variations of both species showed no concordance with density variations on other study plots in south-central Oregon during this time period.

What lessons can be drawn from this admittedly crude and unplanned experiment? Clearly, the species did not respond to this rather massive manipulation of their habitat in the rapid manner we might expect of a finely tuned habitat selection system, nor were the responses of the species consistent with the predictions of our correlational analyses. Why not? One possibility relates to the confounding influences of site tenacity. Our banding studies at Cabin Lake have indicated that adult Sage Sparrows (and to a lesser extent Brewer's Sparrows) that have previously bred on a plot have a strong likelihood of returning to the same location within a plot the following year; one male Sage Sparrow, for example, has returned to one of our plots for 6 successive years following his banding (at an unknown age), never shifting the center of his territory by more than 150 m between years. If individuals at Guano Valley were similarly site tenacious, many of the 1980 breeders might have returned to establish territories in the same locations in 1981, despite the major habitat changes [Hildén (1965) documented several instances of such scenarios]. Development of a clear response to the habitat manipulation might then require several years, depending on the turnover rate of adults in the local population. Site tenacity may thus have produced a major time lag in the response of populations to the habitat alterations.

III. DISCUSSION

When we consider the selection of habitats by birds, we do so through a human filter produced by our perceptions and preconceptions of nature. This leads us to consider patterns in a way that is either insensitive to variations in the scales on which the patterns are expressed or that imposes our own perception of scale upon that of the natural system, whether or not it is correct (Allen and Starr, 1982). This ignores the fact that patterns and processes in nature are sensitive to the scale on which they are viewed; as McCarty *et al.* (1956) and Karr and Freemark (1983) have observed for quite different disciplines, the scale on which systems are studied has a powerful influence on the conclusions that are reached. Thus, for example, the patterns and strengths of the relationships between shrub–steppe birds and habitat features that we have documented in our studies vary depending on the scale of investigation. At the most general, biogeographic, scale, we found clear and strong associations between the birds and several aspects of structure, most likely because that scale included a broad range of quite different habitat types (see also Haila *et al.*, 1980; Crawford *et al.*, 1981). As we reduced the scale, however, the tightness of the fit between birds and habitat became poorer, and some patterns vanished, while other patterns now emerged. It seems unlikely that any general biogeographical law (Blondel and Choisy, 1983) will resolve such scale difficulties.

Habitat selection, however, is a process that operates at the level of individual organisms, and whatever patterns of habitat associations emerge when we expand our scale to view populations, species, or communities must ultimately result from a summation of the responses of individuals to habitat features and their variations. It may be most appropriate, then, to investigate habitat selection at the scale of individuals. There is a basic paradox in this approach, however. If we study habitat relations at a broad spatial scale, we will not observe the true dynamics that produce any patterns we may discern, and thus we can only speculate on the processes that may have caused the patterns or on whether the patterns are in fact biologically real or simply epiphenomena of a particular, arbitrarily selected scale of study (Wiens, 1983a, 1985). By focusing on how individuals in local areas relate to habitats, it is more likely that the true dynamics of the system will be evident, but the data are more sensitive to both sampling and stochastic variation. Moreover, it is much more likely that events beyond the boundary of the study area will have important influences on what happens within the local area, especially for mobile organisms such as birds (May, 1981). By focusing too narrowly on the scale of individuals, we miss important events occurring over a larger scale, while at a larger scale we never observe the individual dynamics that may actually determine the patterns. There is no simple solution to this paradox, but hierarchical approaches, such as have been advocated by Allen and Starr (1982), hold some promise.

Adopting an approach focused on individuals does permit an identification of the factors that may influence habitat selection, and this may be helpful in understanding the habitat patterns we have found among shrub–steppe birds. My conception of the major influences on habitat selection is shown in Fig. 5. Individuals of a species may have an internal image or template of what constitutes a suitable or high-quality habitat, as a consequence of natural selection and/or prior experience and learning. This template may specify habitat requirements quite precisely. I have argued elsewhere (Wiens, 1984a), however, that selection may often produce a regional rather than a point solution to adaptive challenges, especially in variable environments—instead of a single optimum expression of habitat selection, there may be an optimality plateau, within which various expressions of habitat selection may be adaptively equivalent. Our observations of apparent thresholds in the response of Sage and Brewer's Sparrows to shrub size (Fig. 4) and the fact that such features as sagebrush coverage within

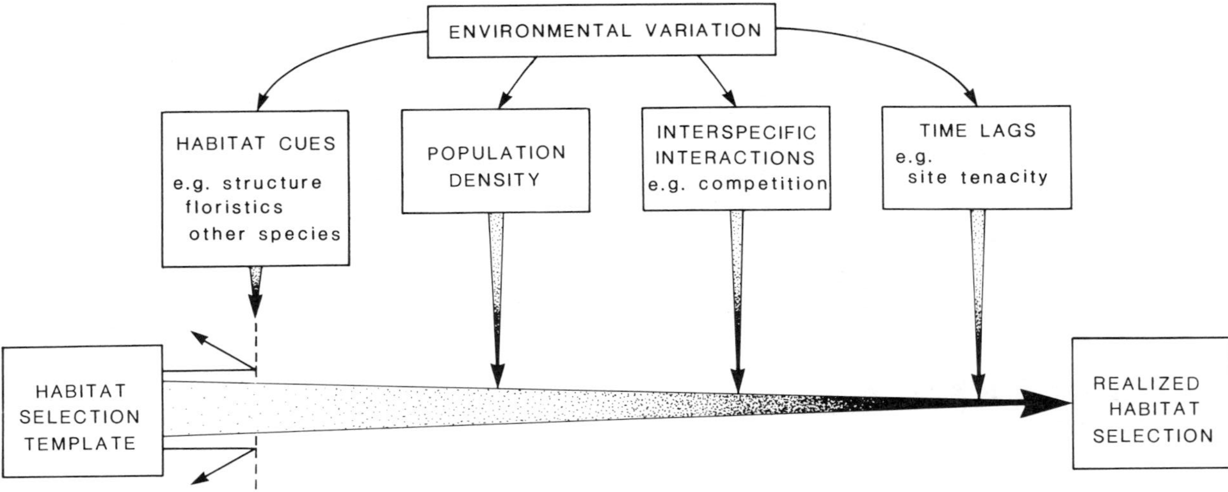

Fig. 5. A conceptualization of the influences of various factors on the expression of a basic habitat selection template of a bird as its realized habitat selection. See text for development.

territories vary, but do so less than coverage in areas not occupied (Fig. 1A), lend some support to this argument.

Whether individuals actually select a given habitat may depend on a process of matching habitat cues with their habitat selection template (Fig. 5). Svärdson (1949) and Hildén (1965) have characterized habitat selection as a two-stage process, in which individuals initially assess general features of the landscape and then respond to more specific features of habitats or microhabitats within the landscape in determining where to settle. In either case, various habitat features, such as the structural configuration or specific structural attributes of the habitat (Wiens, 1969) or the presence of other species (Slagsvold, 1980), may act as cues. Most studies have emphasized aspects of habitat structure (Roth, 1980), but several recent investigations have shown strong associations between birds and floristic composition of their habitats (Beals, 1970; Abbott *et al.,* 1977; Holmes and Robinson, 1981; Collins, 1983). Our correlational findings suggest that floristics may play some role in the habitat selection of shrub–steppe birds, and we have suggested that this may be effected through a linkage between plant species and arthropod herbivores (perhaps mediated by plant secondary chemistry), which in turn may influence the prey resource levels available on different plants to foraging birds (Wiens and Rotenberry, 1981); Holmes and Robinson (1981) and Collins (1983) have offered similar suggestions for deciduous-forest birds.

If a given habitat provides suitable cues to release a selection response, however, there is no certainty that it will in fact be occupied; a variety of other factors may intercede (Fig 5; see also Chapter 1, this volume). Svärdson (1949), for example, drew attention to the influences that the population density of a species in an area and in a given habitat could have on habitat selection by individuals and the breadth of habitat selection exhibited by a population, and Fretwell and Lucas (1969) formalized the argument. We do not have data to test such density dependence in habitat selection in the shrub–steppe systems, but the observation that none of our study areas has been fully packed with breeding territories during 8 years of study suggests that the populations are not fully saturating available habitats and that the density of populations present in an area may have rather little effect on habitat selection.

Svärdson (1949) also suggested that interactions *between* species would have a strong influence on habitat selection, through competitive exclusion. Thus, in communities containing close competitors, the range of habitat occupancy of a species might be restricted as the species partitioned available habitat. An alternative view, of course, is that the species occupy habitat independently of the densities of other species and that interspecific interactions do not influence habitat selection. While studies of avian habitat selection have traditionally emphasized the former view, there is increasing support from field studies in a variety of systems for the notion of species independence [e.g., Bond, 1957;

Haila *et al.*, 1980; Holmes and Robinson, 1981 (but see Chapter 9, this volume); Sabo and Holmes, 1983], although the studies of Rice *et al.* (1983) suggest that the importance of interactions versus independence may vary seasonally. Our investigations have generally supported the conclusion that breeding birds in the shrub–steppe respond to habitat variations largely independently of one another. This may not hold for all species in the community, however, as there is some evidence of negative interaction effects between Sage and Black-throated Sparrows where they co-occur. Unfortunately, correlative approaches can only suggest the possibilities of interactions or of independence, and our more detailed behavioral and experimental investigations have been conducted where Black-throated Sparrows do not occur.

A third factor that may cloud the expression of habitat preferences by individuals is the existence of time lags in responses to habitat conditions. The nature of the response of Sage Sparrows to the range improvement experiment at Guano Valley clearly indicates a lag of several years in response to a sudden, major change in both habitat structure and floristics. On another of our study plots, a rangefire one summer reduced sagebrush coverage from 8 to less than 1% and fostered a luxuriant growth of *Agropyron*. Sage Sparrow densities the following spring were lower than on a nearby control plot, but not as low as would have been expected on the basis of habitat structure (Rotenberry and Wiens, 1978). Best (1979) has documented lags within a breeding season by Field Sparrows (*Spizella pusilla*) in response to fire alterations of grassland habitat, and the data of Schroeder and Sturges (1975) on changes in Brewer's Sparrow densities following herbicidal spraying of shrub–steppe habitats indicate a lag of several years. In organisms that live and reproduce for several years, such as birds, these time lag effects may be commonplace.

Superimposed on the effects of habitat cues, population densities, interspecific interactions, and time lags on the expression of habitat selection by individuals are the consequences of environmental variations. If a given habitat suffers frequent disturbance, for example, it is unlikely that the realized habitat selection of most individuals in the population will correspond closely with the dictates of their habitat selection template, due to the influences of time lags. In a similar fashion, environmental variations may alter the absolute and relative abundances of individuals in populations, producing changes in the patterns and intensities of intraspecific and interspecific interactions to which an individual is exposed. Its habitat selection may be modified accordingly. Environmental variations are also likely to influence the various cues that may guide the habitat selection response of individuals. Some environmental variations (e.g., precipitation) may influence some habitat features more than others (e.g., grass versus shrub coverages in the shrub–steppe), and whether the birds respond to habitat variations may depend on whether the variable features are in fact relevant cues in the habitat selection process.

The interplay of these factors and the ways in which they are influenced by environmental variations may thus have fundamental effects on how closely the realized habitat selection of individuals in a population matches some pattern of evolved constraints. The consequence of this is that habitat selection is likely to be a dynamic process in time, with natural systems expressing considerable variance about the mean pattern predicted from any theoretical arguments. Habitat selection within a species is also likely to vary in space, partly because environmental variability may vary spatially, but also because each of the sets of factors influencing habitat selection (Fig. 5) itself varies geographically. The presence or form of various cues, for example, or the composition of the community of which the species is a member may change from place to place, altering the form of habitat selection in the species. Thus, despite the claims of Beals (1960) and of Noon *et al.* (1980) that habitat selection will usually vary little within a species, there are compelling reasons to expect such variation, as the studies of Collins (1983) and of Karr and Freemark (1983) have indeed shown.

These comments and observations may indicate why the scale of investigation plays such a critical role in the determination of patterns of habitat selection. If habitat selection is indeed affected by the factors shown in Fig. 5 and varies within a species in time and in space, any attempt to discern biologically real patterns of habitat selection that does not consider such factors and variations will end up averaging effects over some arbitrarily defined scale of space and time. The patterns that emerge may quite possibly be artifacts of averaging on that particular scale, and investigation at some other arbitrary scale would likely produce different results.

ACKNOWLEDGMENTS

John Rotenberry assisted in gathering much of the data on shrub–steppe birds and their habitats, and Beatrice Van Horne made important contributions to the analysis of the data and offered comments on a draft manuscript. Yevonn Ramsey prepared the figures. Our research has been supported by several grants from the National Science Foundation, most recently Grants DEB-8017445 and BSR-8307583.

REFERENCES

Abbott, I., Abbott, L. K., and Grant, P. R. (1977). Comparative ecology of Galapagos ground finches (*Geospiza* Gould):Evaluation of the importance of floristic diversity and interspecific competition. *Ecol. Monogr.* **47,** 151–184.

Allen, T. F. H., and Starr, T. B. (1982). "Hierarchy: Perspectives for Ecological Complexity." Univ. of Chicago Press, Chicago, Illinois.

Alley, T. R. (1982). Competition theory, evolution, and the concept of an ecological niche. *Acta Biotheor.* **31,** 165–179.

Beals, E. (1960). Forest bird communities in the Apostle Islands of Wisconsin. *Wilson Bull.* **72,** 156–181.

Beals, E. (1970). Birds of a *Euphorbia–Acacia* woodland in Ethiopia: Habitat and seasonal changes. *J. Anim. Ecol.* **39,** 277–297.

Best, L. B. (1979). Effects of fire on a Field Sparrow population. *Am. Midl. Nat.* **101,** 434–442.

Blondel, J., and Choisy, J. P. (1983). Biogéographie des peuplements d'oiseaux à différentes échelles de perception : De la théorie à la pratique. *Acta Oecologica Oecol. Gener.* **4,** 89–110.

Bond, R. R. (1957). Ecological distribution of breeding birds in the upland forests of southern Wisconsin. *Ecol. Monogr.* **27,** 351–384.

Cody, M. L. (1974a). Optimization in ecology. *Science* **183,** 1156–1164.

Cody, M. L. (1974b). "Competition and the Structure of Bird Communities." Princeton Univ. Press, Princeton, New Jersey.

Cody, M. L. (1978). Habitat selection and interspecific territoriality among the sylviid warblers of England and Sweden. *Ecol. Monogr.* **48,** 351–396.

Collins, S. L. (1983). Geographic variation in habitat structure of the Black-throated Green Warbler (*Dendroica virens*). *Auk* **100,** 382–389.

Connell, J. (1980). Diversity and coevolution of competitors, or the ghost of competition past. *Oikos* **35,** 131–138.

Crawford, H. S., Hooper, R. G., and Titterington, R. W. (1981). Songbird population response to silvicultural practices in central Appalachian hardwoods, *J. Wildl. Manage.* **45,** 680–692.

Diamond, J. M. (1978). Niche shifts and the rediscovery of interspecific competition. *Am. Sci.* **66,** 322–331.

Edington, J. M., and Edington, M. A. (1983). Habitat partitioning and antagonistic behaviour amongst the birds of a West African scrub and plantation plot. *Ibis* **125,** 74–89.

Franklin, J. F., and Dyrness, C. T. (1973). Natural vegetation of Oregon and Washington. *U.S.D.A. For. Serv. Gen. Tech. Rep. PNW* **8.**

Fretwell, S, D., and Lucas, H. L. (1969). On territorial behavior and other factors influencing habitat distribution in birds. I. Theoretical development. *Acta Biotheor.* **19,** 16–36.

Haggett, P., Cliff, A. D., and Frey, A. (1977). "Locational Analysis in Human Geography." Arnold, London.

Haila, Y., Järvinen, O., and Väisänen, R. (1980). Habitat distribution and species associations of land bird populations on the Åland Islands, SW Finland. *Ann. Zool. Fenn.* **17,** 87–106.

Hildén, O. (1965). Habitat selection in birds. *Ann. Zool. Fenn.* **2,** 53–75.

Holmes, R. T., and Robinson, S. K. (1981). Tree species preferences of foraging insectivorous birds in a northern hardwoods forest. *Oecologia* **48,** 31–35.

Karr, J. R., and Freemark, K. E. (1983). Habitat selection and environmental gradients: Dynamics in the "stable" tropics. *Ecology* **64,** 1481–1494.

McCarty, H. H., Hook, J. C., and Knos, D. S. (1956). The measurement of association in industrial geography. Iowa State Univ., Dept. Geography Report No. 1. Iowa State Univ. Press, Ames.

May, R. M. (1981). Modeling recolonization by neotropical migrants in habitats with changing patch structure, with notes on the age structure of populations. *In* "Forest Island Dynamics in Man-dominated Landscapes" (R. L. Burgess and D. M. Sharpe, eds.), pp. 207–213. Springer-Verlag, Berlin and New York.

Meents, J. K., Rice, J., Anderson, B. W., and Ohmart, R. D. (1983). Nonlinear relationships between birds and vegetation. *Ecology* **64,** 1022–1027.

Noon, B. R. (1981). The distribution of an avian guild along a temperate elevational gradient: The importance and expression of competition. *Ecol. Monogr.* **51,** 105–124.

Noon, B. R., Dawson, D. K., Inkley, D. B., Robbins, C., and Anderson, S. H. (1980). Consistency in habitat preference of forest bird species. *Trans. North Am. Wildl. Nat. Resour. Conf. 45th,* pp. 226–244.

Rice, J., Ohmart, R. D., and Anderson, B. W. (1983). Habitat selection attributes of an avian community: A discriminant analysis investigation. *Ecol. Monogr.* **53,** 263–290.

Rotenberry, J. T., and Wiens, J. A. (1978). Nongame bird communities in northwestern rangelands. Proceedings of the Workshop on Nongame Bird Habitat Management in Coniferous Forests of Western United States. *U.S.D.A. For. Serv. Gen. Tech. Rep. PNW* **64,** 32–46.

Rotenberry, J. T., and Wiens, J. A. (1980a). Temporal variation in habitat structure and shrubsteppe bird dynamics. *Oecologia* **47,** 1–9.

Rotenberry, J. T., and Wiens, J. A. (1980b). Habitat structure, patchiness, and avian communities in North American steppe vegetation: A multivariate analysis. *Ecology* **61,** 1228–1250.

Rotenberry, J. T., and Wiens, J. A. (1985). Statistical power analysis and community-wide patterns. *Am. Nat.* (in press).

Roth, R. R. (1980). Vegetation as a determinant in avian ecology. *Proc. Welder Wildl. Found. Symp., 1st,* pp. 162–174.

Sabo, S. R., and Holmes, R. T. (1983). Foraging niches and the structure of forest bird communities in contrasting montane habitats. *Condor* **85,** 121–138.

Schoener, T. W. (1982). The controversy over interspecific competition. *Am. Sci.* **70,** 586–595.

Schroeder, M. H., and Sturges, D. L. (1975). The effect on the Brewer's Sparrow of spraying big sagebrush. *J. Range Manage.* **28,** 294–297.

Simberloff, D. (1983). Competition theory, hypothesis-testing, and other community ecological buzzwords. *Am. Nat.* **122,** 626–635.

Slagsvold, T. (1980). Habitat selection in birds: On the presence of other bird species with special regard to *Turdus pilaris. J. Anim. Ecol.* **49,** 523–536.

Svärdson, G. (1949). Competition and habitat selection in birds. *Oikos* **1,** 157–174.

Toft, C. A., and Shea, P. J. (1983). Detecting community-wide patterns: Estimating power strengthens statistical inference. *Am. Nat.* **122,** 618–625.

Van Horne, B. (1983). Density as a misleading indicator of habitat quality. *J. Wildl. Manage.* **47,** 893–901.

Whitmore, R. C. (1979). Temporal variation in the selected habitats of a guild of grassland sparrows. *Wilson Bull.* **91,** 592–598.

Wiens, J. A. (1969). An approach to the study of ecological relationships among grassland birds. *Ornithol. Monogr.* **8,** 1–93.

Wiens, J. A. (1973). Interterritorial habitat variation in Grasshopper and Savannah sparrows. *Ecology* **54,** 877–884.

Wiens, J. A. (1974). Climatic instability and the "ecological saturation" of bird communities in North American grasslands. *Condor* **76,** 385–400.

Wiens, J. A. (1977). On competition and variable environments. *Am. Sci.* **65,** 590–597.

Wiens, J. A. (1981). Scale problems in avian censusing. *Stud. Avian Biol.* **6,** 513–521

Wiens, J. A. (1983a). Avian community ecology: An iconoclastic view. *In* "Perspectives in Ornithology" (A. H. Brush and G. A. Clark, Jr., eds.), pp. 355–403. Cambridge Univ. Press, London and New York.

Wiens, J. A. (1983b). Interspecific competition [Letter to the Editors]. *Am. Sci.* **71,** 234–235.

Wiens, J. A. (1984a). Resource systems, populations, and communities. *In* "A New Ecology: Novel Approaches to Interactive Systems" (P. W. Price, C. N. Slobodchikoff, and W. S. Gaud, eds.), pp. 397–436. Wiley and Sons, New York.

Wiens, J. A. (1984b). On understanding a nonequilibrium world: Myth and reality in community patterns and processes. *In* "Ecological Communities: Conceptual Issues and the Evidence'' (D. S. Strong, D. Simberloff, L. G. Abele, and A. B. Thistle, eds.), pp. 439–457. Princeton Univ. Press, Princeton, New Jersey.

Wiens, J. A. (1985). Vertebrate responses to environmental patchiness in arid and semi-arid ecosystems. *In* ''Natural Disturbance: The Patch Dynamics Perspective'' (S. T. A. Pickett and P. S. White, eds.). Academic Press, New York.

Wiens, J. A., and Rotenberry, J. T. (1981). Habitat associations and community structure of birds in shrubsteppe environments. *Ecol. Monogr.* **51,** 21–41.

Chapter 8

Habitat Selection in Temperate Marsh-Nesting Birds

JOANNA BURGER

Department of Biological Sciences
Bureau of Biological Research
Rutgers University
Piscataway, New Jersey

I. INTRODUCTION

A. Objectives

Aquatic habitats provide abundant food and sufficient nesting and brood-rearing cover for nesting birds, and water provides a barrier to many mammalian

HABITAT SELECTION IN BIRDS

ISBN 0-12-178080-5

predators. Many species of birds throughout the world nest in wetlands, including some loons, grebes, herons, ducks, hawks, rails, shorebirds, gulls, terns, skimmers, owls, and many passerines. These species represent a wide range of nest-site and brood requirements, temporal patterns of habitat use, and life history strategies. Birds in taxonomically unrelated families often compete for the same resources and use them at the same times. Yet spatial and temporal competition in marshes is usually only examined in closely related species (e.g., Wiens, 1965; Snelling, 1968; Bergman *et al.*, 1970; Faaborg, 1976; Burger and Shisler, 1978a; Orians, 1973; 1980), although some authors have examined competition between two or three species in unrelated families (e.g., Burger, 1973; Neuchterlein, 1981; Nudds, 1982). Yet no one has examined the relationships of many different species nesting in the same wetland areas. In this paper I examine habitat and nest-site selection in marsh-nesting species, concentrating on prairie marshes (central United States and Argentina) and tidal salt marshes (Atlantic coast). I compare the spatial and temporal overlap of the dominant species in these marshes and examine differential responses to the physical constraints in them. Habitat selection in marsh-nesting species involves a series of choices (habitat choice, territory selection, nest-site selection) leading to a specific nest-site choice, and I discuss and contrast these in different species.

B. Wetland Types

1. Definitions

Wetlands are lands where saturation with water is the dominant factor determining the nature of soil development and the types of plants and animals living there (Cowardin *et al.*, 1979). Wetlands may be transitional between aquatic and terrestrial ecosystems where the water table is near the surface. At least periodically, the land supports hydrophytes, plants that prefer a moist to inundated environment and are adapted to the oxygen deficiency of waterlogged soils (Ogaard *et al.*, 1981).

Cowarden *et al.* (1979) developed a classification system for wetlands and deep-water habitats, which includes five systems: marine, estuarine, riverine, lacustrine (nontidal water areas with less than 30% tree, shrub, or emergent vegetation cover), and palustrine (nontidal, with over 30% tree, shrub, or emergent vegetation cover). They further subdivided these systems into 55 different classes defined by bottom type and structure and major plant form, but only a few could be called marshes. It is difficult to define a marsh, but in this paper I use the term to mean any aquatic habitat with emergent vegetation. Usually marshes do not have trees, although trees may grow along the upland edges. Using the Cowardin *et al.* (1979) classification, marshes are called emergent wetlands and occur in estuarine, riverine, lacustrine, and palustrine systems.

2. *Types of Marshes*

Generally, marshes can be characterized by salinity, seasonality, and openness. Marshes range from tidal salt marshes to freshwater marshes and can be seasonal, semipermanent, or permanent. Some marshes are entirely covered with emergent vegetation, some have vegetation surrounding a large expanse of open water, and others have many channels and small ponds scattered throughout. Nontidal water movement and amount of edge area are determined by the degree of openness. For many species, edge is an important cue in nest-site selection.

In the New World temperate zone, large marshes are characteristic of the estuarine environment along the central and northeastern Atlantic coast of North America and of the prairie and pampas regions of central North America and southern South America. The freshwater marshes of the prairies and pampas were largely created by receding glaciers and are slowly filling in as succession proceeds toward climax grasslands. The prairie region of North America, however, is also dotted with small marshes and ponds, called prairie potholes (Stewart and Kantraud, 1971). Salt marshes of eastern North America date to the last glaciation, although they are continually expanding and growing with the accumulation of peat depositions (Teal and Teal, 1969). Generally, salt marshes are located behind barrier beach islands and on the mainland surrounding tidal bays.

There are many smaller marshes with emergent vegetation scattered throughout North and South America, and these are used extensively by nesting birds, particularly species that nest solitarily. The close proximity of small marshes to potential food sources found in rural, agricultural, or forest lands makes many of these marshes particularly attractive.

The classification of marshes on the basis of percentage of open water is essentially a physiognomic classification. That is, marshes can be classified by form and structure rather than by vegetation composition. As mentioned above, edge, interface between water and vegetation, is extremely important for nesting birds, and the amount of edge often relates directly to the number or placement of nests (Burger, 1974a). Merely knowing the percentage of open water often is not sufficient, because marshes can have equivalent amounts of open water but differ drastically in the amount of edge. These physiognomic classifications of marshes are exceedingly important, and I find making detailed sketches useful in capturing the essence of a marsh.

Since birds use the vegetation of marshes for nest placement, plant form and structure play key roles in habitat selection. Thus, the actual species of vegetation present may not be as important as the plant form. For example, in salt marshes along the east coast of North America. Red-winged Blackbirds (scientific names are listed in the Appendix) nest in both *Iva* and *Baccharis*. These shrubs have similar structure and strength, are about the same height, provide the same cover, and look very similar. Differences in form or structure, however, can be readily noted by birds. In prairie marshes, I found that Red-wings prefer to place nests in cattails (*Typha*) rather than in *Scirpus* reeds. Reeds provide less support and are easily blown in the wind. Clearly, species that nest on the water surface, such as grebes, are less affected by the structure of the vegetation, but are equally affected by the open water–vegetation pattern.

Diversity of plant structure types also leads to increases in diversity of nesting birds. A marsh composed entirely of monocot emergent plants has fewer nesting birds than one that contains monocot and dicot emergents plants, as well as low shrubs. The diversity of plant structures allows for niche diversification among species and results in greater species packing.

C. Marsh-Nesting Species

Birds that nest in wetlands belong to several orders: for example, loons (Gaviiformes), grebes (Podicipediformes), cormorants (Pelecaniformes), herons, bitterns, storks, ibises (Ciconiiformes), swans, geese and ducks (Anseriformes), hawks, ospreys and caracaras (Falconiformes), limpkins, cranes, rails, moorhens and coots (Gruiformes), oystercatchers, jacanas, avocets, stilts, gulls, terns (Charadriiformes), furnariids, flycatchers, wrens, blackbirds, swallows, warblers, and sparrows (Passeriformes). Although not all members of these groups nest in marshes, some species do. Further, even within a given species, not all members nest in marshes. For example, Double-crested Cormorants at Agassiz National Wildlife Refuge in Minnesota nest in marshes (in dead trees or other structures), while on Duck Island in Maine they nest on a dry, rocky substrate in a marine ecosystem.

Spatially, birds nesting in marshes may nest anywhere along the continuum from solitarily nesting (well spaced out from conspecifics) to densely colonial nesting (e.g., Common Terns may nest 0.5 m apart). Again, there is great diversity among birds in the species specificity of nest spacing. Some species usually nest solitarily (American Coot, some ducks), others semicolonially (Willet; Burger and Shisler, 1978b; Howe, 1982), and others almost always colonially (terns, egrets). In some species, some individuals nest solitarily, while other nest in colonies (Great Black-backed and Herring Gulls; Burger and Lesser, 1980). Even within a species, colonies may be sparse or dense, and within a colony, nest spacing often varies from dense to sparse groupings (e.g., Common Tern; Burger and Lesser, 1978).

D. Why Nest in Marshes?

In an evolutionary sense birds nest in marshes because they have higher fitness by nesting there than in other habitats. The obvious measure of fitness is total reproductive success (Lack, 1954; Williams, 1966; Ricklefs, 1977). Reproductive success over a lifetime is maximized by having ready access to mates, nest sites and materials, and food (Schoener, 1974) and having low rates of mortality of eggs and young due to predators or inclement weather. The habitat choice birds make, however, must be a compromise between maximizing acquisition of mates, nest sites, and food and minimizing losses due to predation and inclement weather.

Marshes provide an abundant and continuously renewed source of food. Freshwater marshes have frequent insect emergences (Orians, 1961), and salt marshes have periodic mosquito hatches, as well as renewed food resources, twice a day when the tide floods marshes, bringing in fish and invertebrates. Although high avian population densities may exist on marshes, competition for food is reduced because the food taken in any given time period does not affect what is available thereafter because of continual renewal by insect emergences and tidal flow.

Competition for mates should not be related to marsh nesting per se. Competition for nesting space, however, can be intense particularly for colonial marsh-nesting species (Lack, 1968; Burger, 1981), such as blackbirds (Orians, 1961, 1980), gulls (Montevecchi, 1978), and herons (Burger, 1978a,b).

Vegetation structure for available nesting is relatively simple, and vertical stratification is minimal (see Cody, 1968, 1974). Low sedges and grasses may be intermixed with taller emergents (reeds, cattails) and low shrubs, but, generally, there are no high bushes or trees. Thus, nesting birds have adapted to marshes by building floating nests, elevated nest platforms, or cup nests attached to several vegetation stems (see Section III,A,3).

All species of birds must contend with predation, and for many species, predation is the primary cause of egg and chick mortality (e.g., Patterson, 1965; Lack, 1968). Nesting in marshes over water, however, reduces the likelihood of mammalian predators such as foxes (*Vulpes*) and raccoons (*Procyon lotor*) that are responsible for high levels of nest and egg predation on dry land (Krunk, 1964; Southern and Southern, 1979; Greenwood, 1982). However, marsh-nesting species are still exposed to predation by mink (*Mustela vison;* Burger, 1974a; Burger and Lesser, 1978).

Although marsh-nesting birds avoid most mammalian predators, they are not protected from avian predators. Ricklefs (1969) noted that, among all temperate zone passerines, marsh-nesting birds suffer the highest nest mortality rate, mainly through predation. Predators include not only hawks, crows, owls, and Black-crowned Night Herons (Kruuk, 1964; Burger and Hahn, 1977; Montevecchi, 1977; Veen, 1977) that do not nest in marshes, but also some species that do nest in marshes, such as gulls (Burger, 1979a), marsh wrens (Picman, 1980), Common Grackles (Meanley and Webb, 1963), and Chimango Caracaras (Burger, 1974b). Frequent nest parasitism by species such as Redhead Ducks (Weller, 1959, 1968; Talent *et al.,* 1981) and Brown-headed Cowbirds (Shipley, 1979) reduces reproductive success of the host species. Unlike most dry-land nesting habitats, marshes are subject to periodic tidal flooding and unpredictable high storm tides (salt marshes) and to flash floods due to heavy rain and thunderstorms (freshwater marshes). In salt marshes the effects of high tides have been examined in gulls and terns (Montevecchi, 1978; Burger and Lesser, 1978; Burger, 1980), skimmers (Burger, 1982), and Clapper Rails (Andrews, 1977), where high tides can account for total egg and chick losses in any given year. Further, I have observed high wash-out rates in some years for other ground-nesting salt marsh species such as Willets, American Oystercatchers, Mallards, Gadwalls, and sparrows. In freshwater marshes thunderstorms can substantially raise water levels and flood out surface-nesting species, such as gulls (Burger, 1974a), terns (Bergman *et al.*, 1970), coots, and ducks.

Overall, the advantages of superabundant food and reduced mammalian predation presumably outweigh the disadvantages of competition for nest sites, avian predation, nest parasitism, and mortality due to flood tides. The habitat choice and nest-site selection of marsh-nesting species have evolved to maximize the advantages and minimize the disadvantages.

II. TYPES OF HABITAT SELECTION

A. Introduction

Habitat selection involves the choice of particular habitats among the available habitats (Partridge, 1978) and results in birds being nonrandomly distributed in space. Habitat selection can be viewed as involving a series of choices (Gochfeld, 1977): general habitat selection, territory selection, and nest-site selection. These three choices need not be made at the same time nor even by the same member of a pair.

B. General Habitat Selection

Habitat selection involves the choice of a major habitat type such as a field, forest, or marsh. Although habitat categories can be further refined, a bird must select a general habitat for its reproductive activities. Habitat selection may be genetic or learned or may vary from season to season for individuals or species (see Partridge, 1978; Buckley and Buckley, 1980). Habitat selection usually occurs at the beginning of the reproductive season when individuals return after migration. In temperate marsh-nesting species, this selection occurs anywhere from late February (Herring Gulls on salt marshes) to early June (Common Terns on salt marshes). In general, both males and females select a habitat.

C. Territory Acquisition

Territory acquisition involves the specific choice (usually by the male) of a territory, which is vigorously defended against all other males. Females may initially select a male, either by his qualities or those of his territory (Orians, 1969; Lenington, 1980; Yasukawa, 1981). In polygynous species, females also select a territory, usually within the territorial boundary of the male (Orians, 1961; Case and Hewitt, 1963). Usually in colonial marsh-nesting species, such as gulls, terns, and grebes, males select territories. Territory selection, particularly in species where the male must advertise for mates, should contain the necessary open water for displaying or should contain display perches.

Territory acquisition usually occurs shortly after habitat selection, except where species use a communal area for display and mate acquisition (some gulls, grebes, and ducks). Some marsh-nesting species select a new territory each year, but other species show a high degree of territory fidelity, returning to the same territory each year for several years (Tinbergen, 1956; Vermeer, 1963; Ferguson, 1981). In such species, territory selection may occur only once or twice in the lifetime of the bird.

D. Nest-Site Selection

Nest-site selection involves the specific choice of a site to build a nest and, in marsh-nesting species, usually occurs just prior to egg laying. Since marshes are subject to flooding, early construction of a nest might result in exposure to floods and require the pair to construct a nest several times.

At the time of nest-site selection the bird can choose only a site within its territory and must maximize the potential for cover and protection from sun, floods, and predators. Nest-site selection may involve a joint selection by both males and females (e.g., gulls; Tinbergen, 1960; Burger, 1974a), a choice among nests already constructed by males (e.g., coots), or a choice made entirely by females (most ducks, Red-winged Blackbirds, Case and Hewitt, 1963; Orians, 1980).

E. Variability in Choices

At all levels there is a great deal of variability in habitat selection, among and within species and among and within individuals from year to year. Such variability can best be explained by examining some marsh-nesting species.

Franklin's Gulls nest colonially in prairie regions of North America (Burger, 1974a). There is almost no variability in general habitat selection; colonies are in cattails and reeds with a high percentage of edge–open water interface. Gulls arrive on the breeding colony in mid-April, select territories within a week, and build nests within another week. Thus, all three types of habitat selection occur rapidly and before vegetation has changed.

Laughing Gulls nest in *Spartina* salt marshes along the northern Atlantic coast, but they also nest on dry land in Florida (Dinsmore and Schreiber, 1974) and Puerto Rico. On dry land they nest in vegetated sections or on open sand. Thus, there is variability in colony site (general habitat) selection. In salt marshes, they arrive in mid-April, select territories in late April, but do not build nests until early to mid-May, just prior to egg laying.

Herring Gulls have the greatest variability in general habitat selection, since they nest on rocky ledges, sandy or rocky islands, mowed grasslands, sand dunes, and salt marshes. In salt marshes, colony site (general habitat) selection can occur in late February, and territory acquisition occurs in early March. Nest-site selection does not occur until mid-April, just prior to egg laying. Thus, there is more variability in the timing of their habitat choices.

F. Proximate Cues in Habitat Selection

Proximate cues in general habitat selection may involve tradition (for species with fidelity; Bongiorno, 1970), the pressure of conspecifics or protector species (see Neuchterlein, 1981), and physical features. Marsh-nesting species can nest only when the physical environment is suitable. Thus, differences in water level and vegetation growth can influence suitability and thus determine the timing of nesting. Presumably physical characteristics and social factors, such as the presence of conspecifics, feature prominently in territory selection. The constraints of mate acquisition and territory defense make it essential for territories to be easily defended, yet allow the male to display to potential mates. Nest-site selection involves primarily physical features of the habitat which render the nest, eggs, and young protected from floods, heat stress, and predators. Avoidance of nest parasites and potential predators also plays a role in specific choices of nest sites.

I suggest that the factors influencing the three types of selection vary and that the most important factors used for one type of selection may not be important for the other types of selection. Figure 1 illustrates the types of factors and their relative importance to general habitat, territory, and nest-site selection. Although the relative weights might vary among habitats (marshes, forests, fields), these factors are all critical to avian habitat selection.

These three types of habitat selection during the reproductive season are separate events, although obviously some of the same factors enter into each choice. However, since habitat selection may occur only once in a bird's lifetime, territory selection may occur once or several times, and nest-site selection may occur once or twice in a season, the factors affecting these choices might vary.

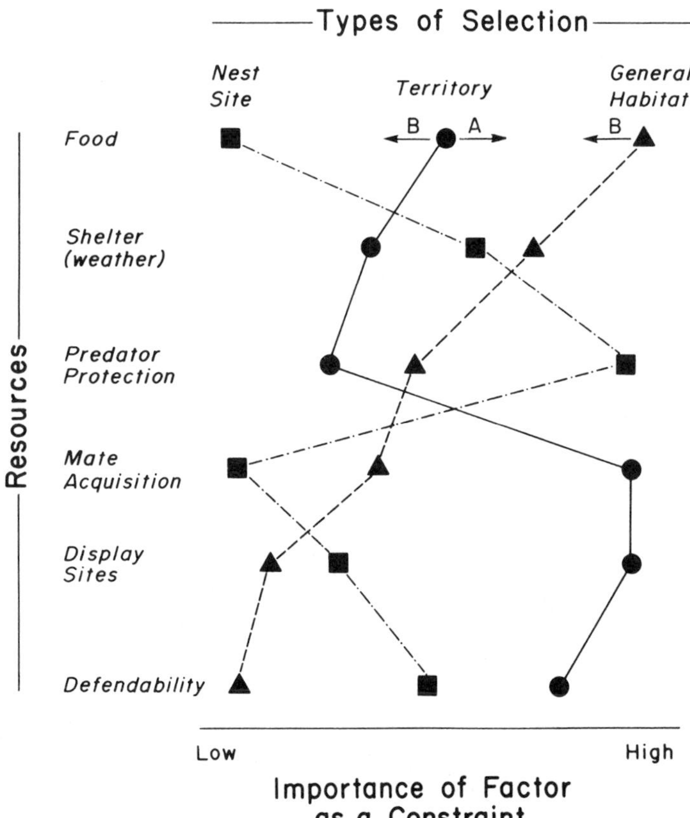

Fig. 1. Relative values of constraints influencing general habitat, territory, and nest-site selection in birds: A, birds that feed in their territories; and B, birds that do not feed in their territories.

By examining each type of selection separately, we can begin to understand the evolutionary selection factors affecting habitat choice. Many authors confuse and lump these three types of selection, but future research should analyze them separately.

III. NEST-SITE SELECTION IN MAJOR MARSH TYPES

A. Prairie Cattail Marsh

1. Habitat Description

In inland regions of North America there are large marshes with hundreds or thousands of hectares of cattail stands. The prairie marshes are exposed to violent climatic fluctuations which lead to different percentages of open water in different years (Stewart and Kantrud, 1972). These freshwater marshes often have large expanses of open water fringed with stands of cattails and tules (*Scirpus*) bisected by channels and creeks. The ecotone with upland areas is fringed with sedges (*Carex*) and grasses. Although shallow, these freshwater marshes are permanent and contain abundant submerged vegetation.

Freshwater marshes (persistent emergent wetlands; Cowarden *et al.*, 1979) are dominated by species that remain standing at least until the start of the next growing season. For example, cattail stems may break over the winter so that the dead stems are only half as tall as live stems, but they still provide support and cover to nesting birds arriving in March and April.

In the following discussion, the data were collected at Agassiz National Wildlife Refuge in northern Minnesota (1969–1972). The marsh contains over 2500 ha of freshwater marshes divided into sections by man-made dikes. Much of the refuge is managed with a periodic schedule of drawdowns and burning of emergent vegetation.

2. *Horizontal Species Stratification*

The dominant nesting species at Agassiz are grebes, Canada Geese, Black-crowned Night Herons, ducks, rails, coots, Franklin's Gulls, Black Terns, Forster's Terns, Long-billed Marsh Wrens, and Red-winged Blackbirds. Although the marsh is not complex in terms of horizontal stratification (Cody, 1974), differences in cattail density and dispersion and the presence of tules, grasses, and sedges in shallow water provide an opportunity for territories and nest sites to be stratified. Most species did not begin nest construction until the marsh vegetation began to grow in late April.

Figure 2 illustrates the horizontal stratification of the most common species nesting in Agassiz. In general, most species preferred to nest on edge areas adjacent to open water. Terns and grebes generally nested in areas with sparse cattails, although terns nested in sections entirely of sparse cattails, whereas grebes frequently nested in sparse sections of cattails or tules surrounded by areas of denser vegetation. Both grebes and terns build floating nests only loosely attached to vegetation stems. Terns occasionally nested on top of muskrat houses, but this was far less common than reported by Bergman *et al.* (1970).

The gulls, night herons, coots, ducks, marsh wrens, and blackbirds all nested in a wide range of habitats, from sparse cattails to dense cattails bordering open channels and pools. Franklin's Gulls selected sparse cattails adjacent to open water and switched colony sites when cattails became too dense. Red-winged Blackbirds and marsh wrens did not select the sparsest cattail areas, because they required several stems in close proximity for nest attachment. Black-crowned Night Herons also avoided nesting in very sparse cattails at the very edge of channels and usually nested 2–10 m in from the edges. In general, over 80% of

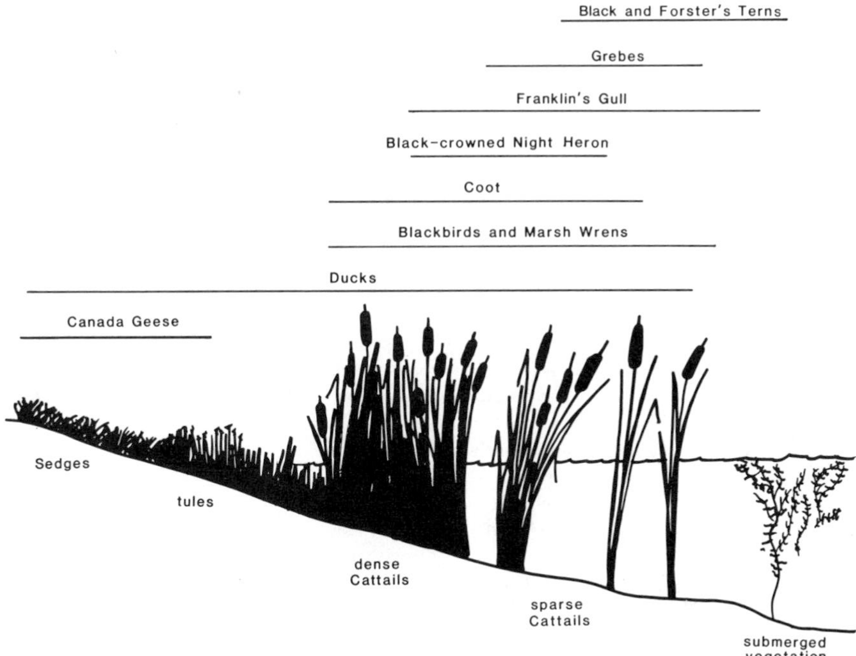

Fig. 2. Horizontal stratification of species in a prairie marsh in Minnesota. Lines indicate areas where each species breeds. See text for further explanation.

the individuals of each species nested within 30 m of open water, clearly avoiding dense stands of cattails (see Table I). For some (Franklin's Gull, Black-crowned Night Heron, coot, and Ruddy Duck) 90% of the nests were within 10 m of open water or channels. In addition to selecting edge sections of the marsh, coots frequently nested in small cattail islands of 1 to 2 m in diameter.

Soras and Virginia Rails and American Bitterns were less common as nesting species, but usually nested in the marsh edges in sedges, grasses, and cattails that were dry by the end of the summer. Canada Geese usually nested in forbs or sedges at marsh edges or on dikes. They also nested on artificial nest structures scattered throughout the marshes.

3. Vertical Stratification

Freshwater marshes in general have little vertical stratification compared to adjacent uplands where trees and shrubs grow. At Agassiz the vast majority of the marsh was homogeneous cattail stands with some *Scirpus* sections. The dead cattail stems from the previous year remain in the spring, slowly sinking into the water during early summer; new growth begins in late April–early May.

TABLE I

Nest Characteristics of Some Nesting Species at a Freshwater Marsh (Agassiz National Wildlife Refuge, Minnesota)

Species	N	Horizontal location	Nest height above water (cm)	Nest rested on water	Nest depth below water (cm)	Cattail[a] density (stems/4 m^2)
Western Grebe	32	Within 8 m of open water	14.2 ± 9.3	Yes	40.1 ± 13.2	50–95
American Coot	20	Within 10 m of open water	35.1 ± 8.6	Yes	6.1 ± 7.3	40–160
Forster's Tern	25	In sparse vegetation	12.1 ± 6.1	Yes	6.5 ± 4.3	20–60
Franklin's Gull[b]	100	Within 100 m of open water and 30 m of channels	14.7 ± 4.3	Yes	18.3 ± 4.2	40–115
Black-crowned Night Heron	20	2–10 m from channels	42.2 ± 6.1	Partially	None	60–110
Red-winged Blackbird	50	All over marsh, concentrated near channels and open water	62.2 ± 24.1	No	None	60–130

[a] Tule density for Grebes.
[b] Some data from Burger (1974a).

Even with little vertical stratification of the vegetation, birds place their nests at different levels. Black and Forster's Terns and Grebes build floating nests, usually attached to emergent or submergent vegetation. Although the tern nests are flimsy, grebes build substantial nests with material extending up to 45 cm below the water surface by the end of incubation (Table I). Grebe nests were built of wet, matted, dead vegetation in sections of the marsh with submerged and barely emergent dead *Scirpus* or cattail stems.

Franklin's Gulls, coots, and several ducks (Ruddy, Mallard, Blue-winged Teal) also built nests on the water surface. These nests were always attached to emergent vegetation, usually dead cattail stems from the previous year. Franklin's Gulls built flat nests with extended platforms; the ducks and coots built taller nests, and some had ramps leading up to them. Black-crowned Night Herons built nests attached to cattail stems that were partially touching water, but the main bulk was above water (Table I). Thus, they required sturdy, old cattail stems for nest attachment.

Red-winged Blackbirds and Long-billed Marsh Wrens both built cup-shaped nests attached to three to five cattail stems. The nests were well above the water level (Table I). At Agassiz, blackbird nests were at similar nest heights as reported for other cattail marshes (Bernstein and McLean, 1980). Wrens and blackbirds always built nests in old, dead cattail stems, presumably because they are stable and not growing. As has been reported in other studies, Redwings preferred to nest in cattails (Orians, 1980) and nested in sparser cattails than Marsh Wrens (Picman, 1980), apparently to partially avoid Marsh Wren predation. Picman (1980) found that Redwing nest success increased as the distance from Marsh Wren nests increased. Marsh Wren males build up to an average of 22 nests per male (Verner and Engelsen, 1970), and it would be interesting to compare site characteristics of nests that are actually used for egg laying with those of unused nests.

4. Social Factors Affecting Nest-Site Selection

Marsh-nesting birds ranged from solitary (American Bittern, Sora, and Virginia Rail) to colonial species (Franklin's Gull, Forster's Tern, and Black Tern). Some species (grebes) sometimes nested solitarily at Agissiz, but usually nested in groups of 3 to 20 nests. Partially this clumping of nests is a result of courtship activities early in the season. The grebes did not engage in group defense of nests, however, as did terns and gulls. Group antipredator behavior is one important advantage of colonial nesting (reviewed in Burger, 1981). Red-winged Blackbirds, although not usually thought of as colonial, tended to clump, and some areas of the marsh had many pairs, while other seemingly suitable areas had none.

Another social factor affecting nest choice is social parasitism (Neuchterlein, 1981). Some species nest in colonies of heterospecifics, deriving advantages from them. For example, at Agassiz, Black-crowned Night Herons always nested in colonies with Franklin's Gulls and shifted locations whenever the gulls shifted. The gulls often shifted sites a week or two after returning to the old site (see Burger, 1974a), and the night herons then shifted even though some had constructed nests in the old site. Black-crowned Night Herons left the colony silently whenever a predator approached, and they did not engage in any antipredator behavior. The gulls, however, mobbed and dive-bombed all predators in the colony. Several other species sometimes were found nesting in Franklin's Gull colonies (Double-crested Cormorants, Western Grebes), but they did not shift sites with the gulls.

5. Habitat Selection in Franklin's Gull

Franklin's Gulls arrive in mid-April at Agassiz and choose colony sites where the average vegetation cover ranges from 0 (open water) to 100% (very dense cattails) and where water dispersion ranges from no open water or channels to over 50% open water. Generally, Franklin's Gulls selected colony sites with between 40 and 50% vegetation cover, and all had open water (Table II).

In late April gulls begin to settle on territories, and males select sites with perches from which they can display to their mates and to nearby males. Gener-

TABLE II

Habitat Selection in Franklin's Gulls (Agassiz National Wildlife Refuge, Minnesota)

	Colony site selection		Territory selection[c]			Nest-site selection	
			Percent display perch		Percent mean vegetation cover[d]	Percent perch site used for nest site	Distance to vegetation[e]
Colony	Percent vegetation[a]	Vegetation distribution[b]	At water level	Above water level			
A	50	B	49	51	15 ± 12	49	0
B	45	C	70	30	12 ± 10	66	0
C	53	C	68	32	16 ± 12	68	0
D	51	B	35	65	21 ± 19	35	0
E	42	C	71	29	8 ± 11	70	0
F	48	C	63	37	13 ± 8	58	0
G	36	C	75	25	6 ± 10	72	0
Timing:	Mid-April		Late April–early May			Early May–mid-May	
Range available:	0–100	A–C	0–100	0–100	0–100	—	0–10 m

[a] From aerial surveys and aerial photographs.

[b] A, even throughout area; B, Open water areas with dense vegetation (>50% cover); C, Open water areas with sparse vegetation (<50% cover).

[c] Franklin's Gulls do not select territories without perches (based on sample of 100 nests observed from blinds over 3 years).

[d] Mean ± standard deviation, based on 100 nests in each colony.

[e] In all cases they attached nests to vegetation.

ally, over half of the males selected sites where there was a perch (a group of bent-over cattails) at water level (Table II). However, even though they selected colony areas with about 50% vegetation cover, they established territories in areas with only an average of 6 to 21% cover (Table II). Thus, colony and territory site selection differed.

Nest-site selection occurred in early to mid-May. All birds built nests attached to vegetation, and from 49 to 72% of the birds used the same perch site for a nest site (Table II). Nonetheless, from 28 to 51% of the birds selected another place in their territory for a nest site.

6. *Summary of Characteristics Used for Nest-Site Selection*

At Agassiz National Wildlife Refuge birds used social and physical features to select nest sites. Social factors include attraction to conspecifics (gulls, terns, Western Grebes), lower levels of attraction to conspecifics (other grebes, Red-winged Blackbirds), attraction to heterospecifics that provided antipredator behavior (Black-crowned Night Herons), avoidance of species that are egg predators (blackbirds avoiding marsh wrens), and nest parasites (other ducks avoiding Redheads).

Physical factors involved in nest-site selection were cattail density, cattail height, presence of strong stems from the previous year, and cattail density. To build nests, birds required some structure for attachment, either at water level or above it (passerines). Although most species attached nests to dead cattails from the previous year, nest construction did not usually begin until the new growth was 20–30 cm high. New growth provided cover as protection against predators and weather (sun, rain). Dense cattails between a nest and open-water areas provided a buffer against wave action caused by violent rain and thunderstorms. For water level–nesting species, proximity to open water was an important antipredator strategy, because adults frequently could not fly from nests located in cattails. I found dead adult gulls, terns, and ducks killed by mink while they were trying to swim rapidly through dense cattails.

B. Pampas Tule Marsh

1. Habitat Description

The ecological counterpart of North American prairie marshes in South America is the pampas tule marshes. Pampas marshes can also cover hundreds of hectares and are exposed to periodic violent rain and thunderstorms. These freshwater marshes are also classified as persistant emergent wetlands by Cowardin *et al.* (1979). Thus, they are dominated by species (cattails or tules) that remain standing all winter, providing cover and support for nesting birds in the spring.

Like freshwater marshes in the prairies of North America, the pampas marshes can range from almost pure stands of tules or cattails with little open water to shallow lakes fringed with a wide band of emergent vegetation. More commonly, pampas marshes contain intermixed areas of open water and emergent vegetation, broken by small pools and channels.

The data and observations presented below were taken in the alkaline marshes on San Jose Estancia, Murphy, Province of Sante Fe, Argentina (1972–1973). This marsh had an extensive expanse of open water (3 km wide) fringed by large areas of tule (*Scirpus californicus*) stands, with scattered areas of cattails. Tule density varied from sparse to dense.

2. Horizontal Species Stratification

The dominant species nesting at San Jose were Silver and Rolland's Grebes, Coots (*Fulica* spp.), Chimango Caracaras, herons and egrets, Brown-hooded Gulls, Brown and Yellow Marshbirds, Rush Tyrants, Rushbirds, and Scarlet-headed Blackbirds.

There was only limited horizontal stratification, because there were few differences in tule density and a limited number of sections with cattails and sedges. Nonetheless, the birds were horizontally stratified (Fig. 3), with some species usually nesting in the sparse tules (48–30 stems/80 cm^2; gulls and grebes), other species nesting in dense tules (40–120 stems/80 cm^2; passerines), and other species nesting only in the densest tules and cattails (120–180 stems/80 cm^2; herons, egrets, ibises; see Burger, 1979b).

Since the herons, egrets, and ibises actually built their nests above the water surface but supported by the vegetation, they could only nest in the densest tules and cattails. Nests were always constructed in dead material from the previous year. Similarly, the passerines that constructed nests in the tules above the water also nested in dense tules, so there were sufficient stems to securely attach a nest.

Coots nested in the widest range of habitats, from very sparse tules and cattails to dense vegetation. Chimango Caracaras nested in sparse to intermediate tules (25–45 tules/80 cm^2), usually near open-water areas. I found only two Southern Screamer nests (at the marsh edge) and three Maguari Stork nests (in dense tules near open water).

3. *Vertical Stratification*

As was the case in prairies marshes, birds at San Jose did not all nest at the same level. Grebes and gulls built floating nests on the water; coots and Caracaras built more substantial nests on the water; White-faced Ibises built nests on the water that were 12–18 cm high; Snowy Egret (25–35 cm) and Great Egret (35–65 cm) nests were built in the tules or cattails well above the water surface;

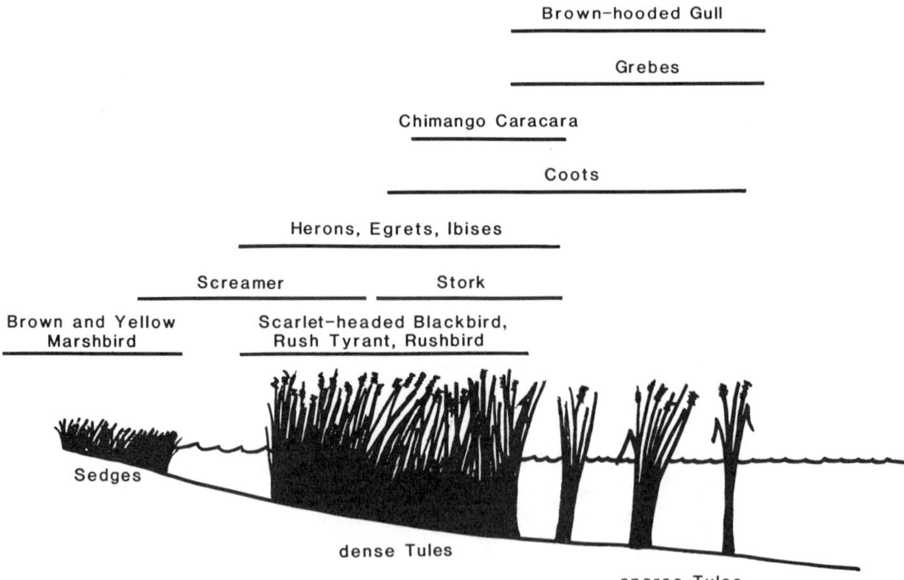

Fig. 3. Horizontal stratification of birds nesting in a tule marsh in Argentina. Lines indicate where each species breeds.

and the passerine nests were well above the water (>35 cm). The vertical nest placement influenced the horizontal placement since tules provide less structure than cattails, and denser tules were required to adequately support nests.

4. *Social Factors Affecting Nest-Site Selection*

At San Jose herons, egrets, ibises, and Rolland's and Silver Grebes were highly colonial. The herons, egrets, and ibises nested together in one colony, although some ibises attempted to nest in the colonies with Brown-hooded Gulls. I did not find any solitarily nesting grebes. Grebes rested in monospecific colonies, in mixed-species colonies with other grebes, and in mixed-species colonies of grebes and Brown-hooded Gulls (Burger, 1974c). Brown-hooded Gulls were also colonial, although 16 or 107 gulls nested solitarily (Burger, 1974b).

Grebes depart quickly from their nests and swim to open water whenever predators approach. One advantage of their nesting in gull colonies is that the gulls provide early warning of approaching predators and mob them (Burger, 1984). This was particularly important to the grebes, and grebes nesting in gull colonies suffered lower egg predation (1%) than those in colonies without nesting gulls (1–32%; Burger, 1984). The grebes' social parasitism of the gulls has a cost, however, because some gulls succeeded in usurping grebe nests, dumping the grebe eggs into the water.

Since Chimangos were predators of grebe and gull eggs and adults, grebes and gulls avoided nesting near nesting Chimangos. The Chimangos were loosely colonial, in that 3–12 nests would be clumped together in one section of the marsh.

5. *Summary of Characteristics Used for Nest-Site Selection*

Social factors such as conspecific attraction (gulls, chimangos, grebes) and attraction to closely related species (herons, egrets) affected nest-site selection for colonial species. Grebes tended to select gull colonies to nest in, because of the positive benefit of early predator warning (social parasitism), even though some grebe nests were taken over by the gulls. Wherever possible gulls and grebes avoided nesting close to Chimangos, because they were predators.

Physical parameters used in nest-site selection were distance to open water, vegetation density, presence of strong stems from the previous year, and presence of cattails (or tules).

C. *Spartina* Salt Marshes

1. *Habitat Description*

Salt marshes extend up the Atlantic coast from northern Florida to Maine, and they are also classified as persistent emergent wetlands (Cowarden *et al.*, 1979).

Salt marshes are covered by vegetation and subject to daily, periodic inundations by tides. Major vegetation is cordgrass (*Spartina alterniflora*), salt hay (*S. patens*), and shrubs (*Iva frutescens* and *Baccharis halimifolia*). Vegetation zonation is dependent on the frequency of tidal flooding as *Spartina alterniflora* is usually flooded daily, and areas where shrubs grow are usually flooded only on very high storm tides. With the advent of the Army Corps' deposition of dredge spoil on some salt-marsh islands, *Phragmites* now grow in the sandy sections of islands.

Natural salt marsh islands are bisected by many meandering creeks and channels. Salt marshes usually border bays and may be attached to either the mainland or barrier beach islands. Many salt marshes, however, are isolated islands in the middle of bays.

The data and observations reported below were collected on several salt marsh islands in Barnegat Bay (1975–1983). There are over 250 salt marsh islands in the bay, and there are bird colonies on over 30 of them.

2. *Horizontal Species Composition*

The most common species nesting in salt marshes are Clapper Rails, Mallards, Gadwalls, gulls, terns, Black Skimmers, herons, egrets, Red-winged Blackbirds, and Long-billed Marsh Wrens. Horizontal stratification occurs in salt marshes (Fig. 4), although most species occupy a broad range of habitats.

Clapper Rails nest closest to the incoming tide, along creeks and channels. Laughing Gulls and Forster's Terns nest in low areas of *Spartina alterniflora,* but select the highest spots in these areas. Although Common Terns and Black Skimmers also nest in the *Spartina alterniflora* zone, they nest on wrack strewn by high tides. Oystercatcher nests were also on wrack or *S. alterniflora* or were in high areas under bushes or *Phragmites.* Mallards and Gadwalls nest along a wider gradient, but they also select high spots. There are species differences, however, in that most Mallard nests (N = 110) were in *S. alterniflora* (85%) compared to *S. patens* (13%) or *Iva* (2%). Most Gadwall nests (N = 70) were in *S. alterniflora* (60%), and the rest were in *S. patens* (17%), *Iva* (9%), and *Phragmites* (14%). The few Shovellor nests were all in *S. patens*. Willets nested in a wide range of habitats, from high spots on *S. alterniflora* to under bushes and *Phragmites*. For Willet nests (N = 30), 30% were in *S. alterniflora,* 60% were in *S. patens,* and 10% were under bushes.

Herons, egrets, and ibises nested only in bushes or in *Phragmites* located on the highest sections of the salt marsh. Such areas were almost never exposed to flood tides. In 10 years of field work in these salt marshes, I have seen heron nests washed out in only 1 year. Red-winged Blackbirds and Marsh Wrens nested only in the high sections of the Marsh, whereas Seaside and Sharp-tailed Sparrows nested throughout the marsh, including the low areas with *Spartina alterniflora* and *S. patens*.

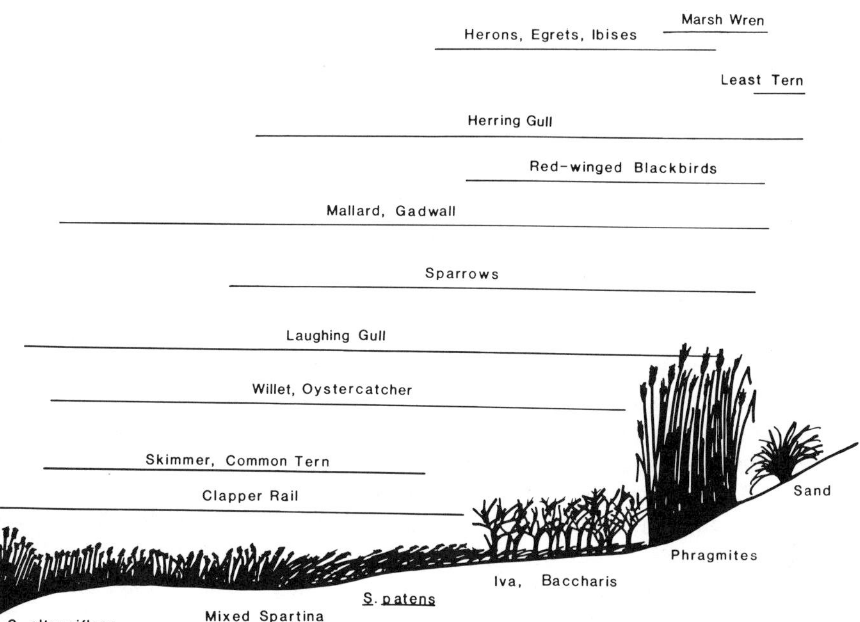

Fig. 4. Horizontal stratification of species in a salt marsh in New Jersey. Lines indicate areas where each species breeds.

3. Vertical Stratification

Salt marshes have more vertical stratification than the freshwater marshes described previously, because bushes are an integral part of salt marsh vegetation. Bushes are shaped differently from grasses and *Phragmites,* providing support and cover for nests.

Ground-nesting species include Clapper Rails, Willets, ducks, sparrows, and Herring Gulls. These species build nests on the ground in high places and build high nest structures to avoid tides (see Jackson 1983). Since there are daily high tides, birds can assess the relative height of the marsh by standing at the nest site for a few days or weeks prior to egg laying and by selecting *Spartina patens* or bushes.

Willets also nest on the ground in *Spartina* or under bushes. Nests (N = 30) were surrounded by live vegetation ($\bar{X}$ = 85 ± 18%). Vegetation cover over the nest was high ($\bar{X}$ = 54 ± 22%), it ranged from 5-85%. Even when Willets nested in *S. alterniflora*, they selected spots with higher vegetation ($\bar{X}$ = 38 ± 4 cm) compared to areas 1m away ($\bar{X}$ = 19 ± cm). In all cases the vegetation immediately surrounding the nest was taller than the adjacent vegetation. By pulling down a few of the taller grasses, the Willets built partially domed nests.

Some species, such as Laughing Gulls and Common Terns, nest either on the marsh itself or on wrack on top of the marsh (Montevecchi, 1978; Burger and Lesser, 1978; Burger, 1979a). When Common Terns and Laughing Gulls nest on the marsh ground, they build nests which are up to 18 cm high. Nest building increases markedly following a high tide (Burger, 1979c). Other species, such as American Oystercatcher and Black Skimmer, always nest on these high wracks, usually selecting the highest areas of the wrack.

Herons and egrets usually construct nests well above the ground in bushes or *Phragmites* (Burger, 1979b). Since the nests are elevated they must be supported by strong branches or stems. Thus, herons and egrets usually nest only in dense bushes or *Phragmites*. Glossy Ibises, Black-crowned Night Herons, and Snowy Egrets sometimes build nests on the ground, if nest sites are limited and competition from larger species is prevalent (Burger, 1979b).

Seaside and Sharp-tailed Sparrows nest on the ground in well-concealed nests. I found few nests, but these were in high sections of *Spartina alterniflora* (half were on spoil piles), were suspended in the grass, and were domed. Post (1974) studied Seaside Sparrows on Long Island and found that in natural, unditched marshes, nest sites are limiting, often forcing birds to nest quite close together. He reported that the requisite for a nest site is a high spot immune from flooding, yet low enough to be protected from aerial predators, rain, and wind. The mean height of 17 nests was 19.0 ± 4.5 cm (Post, 1974). Birds apparently selected nest sites by using the highest persistent old vegetation that was not flattened by winter tides. In ditched marshes, spoil piles provide abundant, suitably high nest sites, and sparrows are spaced out (Post, 1974).

4. Social Factors Affecting Nest-Site Selection

Social attraction among conspecifics results in skimmers, terns, gulls, herons, egrets, and ibises nesting in colonies. Herons, egrets, and ibises seem to be equally attracted to colonies with any one of these species. Thus, a single pair of Common Egrets may nest in a colony with other egrets and herons, but would not nest alone in a marsh. Within heronries, nest-site choice is severely affected by the other species, since there is intense competition for nest sites. Larger species nest higher, and smaller species nest lower (Burger, 1978b).

In salt marsh colonies, Black Skimmers always nest with Common Terns and are apparently deriving antipredator benefits from them. In 8 years of examining 5–12 Black Skimmer colonies a year, I found all Black Skimmers were in Common Tern colonies. Whenever an intruder comes close to a tern–skimmer colony, the skimmers leave immediately, whereas the terns mob and dive-bomb the intruder. Oystercatchers similarly nest near or in Common Tern and Laughing Gull colonies.

Willets are somewhat attracted by conspecifics, as they are loosely colonial in some sections of the marsh. Further, for the 30 Willet nests examined in 1982 and 1983, only 30% were outside of tern and gull colonies.

5. Temporal Factors Affecting Nest-Site Selection

Some salt marsh species are nonmigratory (Great Black-backed Gull) or migrate only short distances (some Herring Gulls, Clapper Rails), but most species migrate to the southern United States or South America. Even the nonmigratory individuals must return to their nest sites in the marshes.

In general, the species nesting in salt marshes build their nests only a few days before egg laying, and, in many, the nest is not completed until after egg laying (Common Tern, Laughing Gull, Clapper Rail). Overall, egg laying for most species begins from mid-April to late May, although there is variation among species (Fig. 5). In comparing the different species, I note that (1) there is no immediate difference in the egg-laying dates for colonial compared to solitary species, and (2) there is no relationship between the initiation date of egg laying and the total egg-laying period.

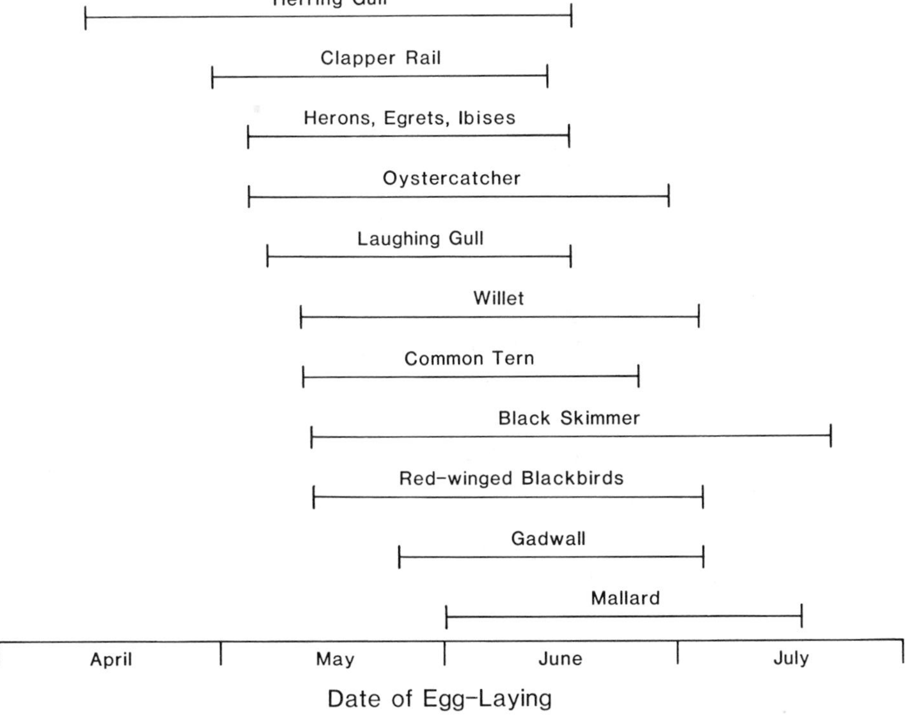

Fig. 5. Temporal patterns of nesting in salt marsh species in New Jersey.

Most salt marsh species will re-lay if their nests and eggs are destroyed by high tides early in the nesting season. Most species, however, move to another section of their territory (when only a few nests have been washed out), move to another part of the colony (when many have been washed out), or move to another island (when most or all have been washed out). Thus, even when flood tides have washed out nests, nest-site selection for the new clutch is influenced by social factors.

The initial dates of egg laying are also important in nest-site selection, because they influence competition among species. That is, since Herring Gulls nest earlier than Laughing Gulls, the Laughing Gulls have to compete directly and displace the Herring Gulls, or they must select sites not already occupied by them. Similarly, Gadwalls and Mallards cannot nest on the high spots already selected by Willets or Laughing Gulls (even though they might otherwise have selected these spots).

6. *Habitat Selection in Herring Gulls*

Herring Gulls, recent invaders into salt marshes (Burger, 1977), can be used to illustrate the three levels of habitat selection in a salt marsh–nesting species. The key factor in colony site selection seems to be the presence of bushes (either *Iva* or *Baccharis*). Herring Gulls clearly select the section of the island with the concentration of bushes (Table III). Further, all islands (n = 8) in Barnegat Bay newly occupied by nesting Herring Gulls have had some bushes.

Territory selection, occurring in early March, seems to involve selecting sections with open areas for display purposes (Table III). Nest-site selection occurs over a month later, and the birds select sites closer to vegetation than to their display site and with a higher percentage of cover (for protection of eggs). Many Herring Gulls make nest scrapes between territory acquisition and egg laying, and these sites frequently are used as nest sites. Nonetheless, the qualities used for display and nest sites differ and need to be examined separately in gulls and other species. Species that arrive on the breeding grounds very close to egg laying time (e.g., Common Terns) might be expected to have more concordance between the characteristics used in the three types of nesting-habitat selection.

7. *Summary of Characteristics Used for Nest-Site Selection*

Social factors such as strong attraction for conspecifics (gulls, terns), strong attraction for closely related species (herons, egrets, ibises), and weak attraction for conspecifics (Willets) influence nest-site selection. Oystercatchers and Black Skimmers nesting in Common Tern colonies illustrate an attraction for protector species.

The tidal constraints of salt marshes have shaped nest-site selection for all species nesting there. Species nest either on high spots in low sections of the marsh (rails, some Laughing Gulls, some Common Terns, Willets), on wrack (Skimmers, Oystercatchers), on high sections of the marsh (Herring Gulls, her-

TABLE III

Habitat Selection in Herring Gulls (Barnegat Bay, New Jersey)

Colony	Number of pairs in colony[a]	Colony site selection: Percent *Iva* on entire island	Colony site selection: Percent *Iva* in gull colony	Territory selection: Percent *Iva* within display site[b] (m)	Territory selection: Distance of site *Iva*[c] (m)	Nest-site selection: Percent *Iva*	Nest-site selection: Distance of nest to *Iva*[d] (m)	Percent display site-nest site
Clam Island	800	8	20	7 ± 11	3.1 ± 1.8	17 ± 8	1.1 ± 1.3	5
Carvel Island	80	40	22	12 ± 7	3.6 ± 1.4	13 ± 11	0.8 ± 0.6	9
Sandy Island[e]	56	1	15	3 ± 4	2.8 ± 0.8	14 ± 13	0.4 ± 0.5	4
Harvey Sedge	211	9	12	3 ± 5	4.3 ± 1.2	12 ± 6	1.2 ± 2.3	5
West Vole Sedge	65	4	7	2 ± 2	5.3 ± 2.1	15 ± 4	2.3 ± 3.1	8
High Bar	35	1	18	5 ± 4	4.1 ± 1.8	16 ± 11	0.6 ± 1.1	3
NW Lavelette	12	1	10	5 ± 6	3.3 ± 2.6	18 ± 4	0.2 ± 0.3	0
Timing:		Mid-February–early March		Early March		Mid-April–mid-May		
Range available:		0–80	—	0–100	0–30	0–100	0–25	

[a] 1982 Census.

[b] Display site taken as center of territory, determined from blinds. Based on 50–100 nests in each colony (randomly).

[c] For Sandy Island, distance was measured to *Iva* or *Solidago* since both provided cover for nests.

[d] Aerial surveys and aerial photography.

[e] Subisland B [Burger and Shisler (1978c)].

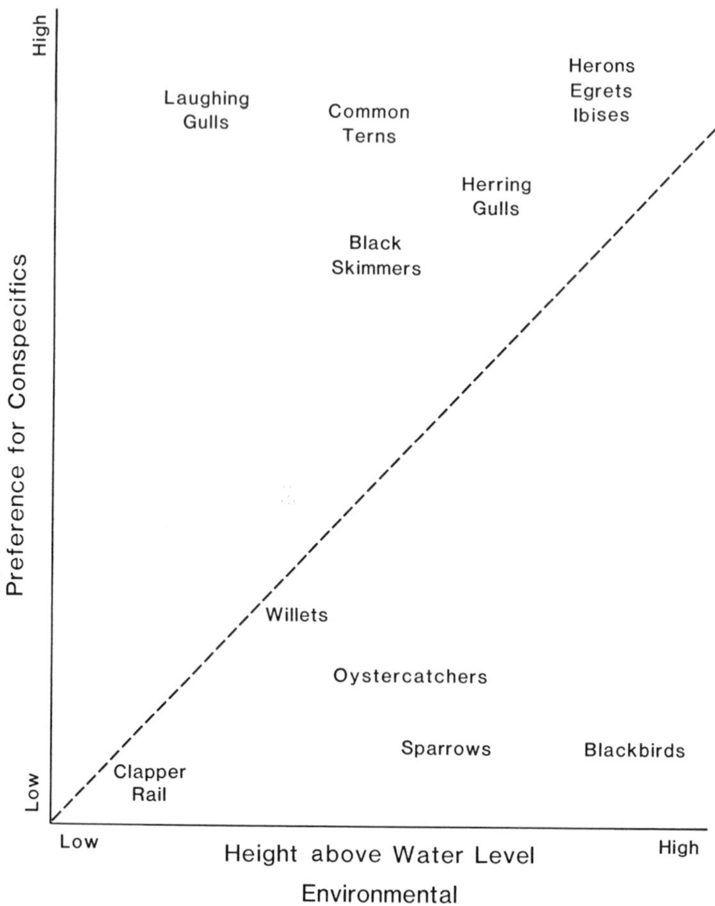

Fig. 6. Relationship of social factors (attraction to conspecifics) to one environmental factor (height above water level) in the nesting pattern of several species nesting in New Jersey salt marshes.

ons, egrets, ibises), or above ground (sparrows, blackbirds, herons, egrets). Indicators of marsh height (differences in vegetation species, height of vegetation, presence of wrack) provide cues used by the birds to select safe nesting spots (see Fig. 6).

D. Conclusions: Comparisons of Avian Nesting in Salt and Freshwater Marshes

Birds nesting in both freshwater and saltwater marshes are exposed to avian predators as well as aquatic mammalian predators. Thus, in regard to predation pressures, the marshes are similar. In physical appearances salt and freshwater marshes are also similar, in that they have large stands of monocot vegetation less than 3 m tall that are bisected by creeks and channels. But there the similarities end and the differences begin: (1) salt marsh vegetation is less structurally strong than the cattails and tules of freshwater marshes, (2) salt marsh grasses bend in the breeze and cannot be used for high nests, (3) salt marshes undergo periodic tidal inundations and frequent flooding.

Birds in both types of marshes adjust to a lack of vertical and horizontal complexity, as well as to the occurrence of frequent storms. But salt marsh species must also adapt to the daily tidal inundations. Most salt marsh species respond by selecting the highest spots possible to avoid flooding, while nesting on islands that are low enough so mammalian predators (rats, raccoons, foxes) cannot survive the flood tides of late winter. Thus, their choices of nest sites are a compromise between choosing high sites and being on low islands. Nest height is enhanced by nesting on wracks, building up nests, or suspending nests in the grass. For freshwater species the compromise is between nesting far enough from dry land to reduce mammalian predation and avoiding the periodic floods from heavy rains. Freshwater marshes are more stable with respect to nesting conditions, and this results in relative niche width being less in freshwater compared to salt marshes (compare Figs. 1 and 4).

APPENDIX

Taxonomic Listing

Common name	Genus and species
Silver Grebe	*Podiceps occipitalis*
Western Grebe	*Aechmorphorus occidentalis*
Silver Grebe	*Rollandia rolland*
Double-crested Cormorant	*Phalacrocorax auritus*
Common Egret	*Casmerodius albus*
Snowy Egret	*Leucophoyx thula*
Black-crowned Night Heron	*Nycticorax nycticorax*
American Bittern	*Botaurus lentiginosus*
Glossy Ibis	*Plegadis falcinellus*
White-faced Ibis	*Plegadis chihi*
Maguari Stork	*Euxenura maguari*
Southern Screamer	*Chauna torquata*
Canada Goose	*Branta canadensis*
Mallard	*Anas platyrhynchos*
Gadwall	*Anas strepera*
Blue-winged Teal	*Anas discors*

(*continued*)

Taxonomic Listing *(Continued)*

Common name	Genus and species
Shoveler	*Spatula clypeata*
Redhead	*Aythya americana*
Ruddy Duck	*Oxyura jamaicensis*
Chimango Caracara	*Milvago chimango*
Clapper Rail	*Rallus longirostris*
Virginia Rail	*Rallus limicola*
Sora	*Porzana carolina*
American Coot	*Fulica americana*
American Oystercatcher	*Haematopus palliatus*
Willet	*Catoptrophorus semipalmatus*
Great Black-backed Gull	*Larus marinus*
Herring Gull	*Larus argentatus*
Laughing Gull	*Larus atricilla*
Franklin's Gull	*Larus* pipixcan
Brown-hooded Gull	*Larus maculipennis*
Forster's Tern	*Sterna forsteri*
Common Tern	*Sterna hirundo*
Least Tern	*Sterna albifrons*
Black Tern	*Childonias niger*
Black Skimmer	*Rynchops niger*
Rushbird	*Phleocryptes melanops*
Many-colored Rush Tyrant	*Tachuris rubrigastra*
Long-billed Marsh Wren	*Cistothorus palustris*
Sharp-tailed Sparrow	*Ammospiza caudacuta*
Seaside Sparrow	*Ammospiza maritima*
Red-winged Blackbird	*Agelaius phoeniceus*
Scarlet-headed Blackbird	*Amblyramphus holloserice us*
Brown and Yellow Marshbird	*Pseudoleistes virescens*
Common Grackle	*Quiscalus quiscula*
Brown-headed Cowbird	*Molothrus ater*

ACKNOWLEDGMENTS

Over the years I have had many fruitful discussions concerning marsh-nesting birds with the following people, and I thank them now: M. Gochfeld, B. G. Murray, P. J. Regal, W. E. Southern, and H. B. Tordoff.

I also thank the following for funding some of the work described in this paper: American Association of University Women, Frank M. Chapman Fund, and Johanna and Charles Busch Fund.

REFERENCES

Andrews, H. F. (1977). Nest related behavior of the clapper rail (*Rallus longirostris*). Ph.D. thesis, Rutgers Univ., Newark, New Jersey.

Bergman, R. D., Swain, P., and Weller, M. W. (1970). A comparative study of nesting Forster's and Black Terns. *Wilson Bull.* **82,** 435–444.

Bernstein, N. P., and McLean, E. B. (1980). Nesting of red-winged blackbirds in cattails and common reed grass in Mentor Marsh. *Ohio Acad. Sci.* **80,** 14–19.

Bongiorno, S. F. (1970). Nest site selection by adult laughing gulls (*Larus atricilla*). *Anim. Behav.* **18,** 434–444.

Buckley, F. G., and Buckley, P. A. (1980). Habitat selection and marine birds. *In* "Behavior of Marine Animals," Vol. 4, Marine Birds (J. Burger, B. L. Olla, and H. E. Winn, eds.), pp. 69–112. Plenum, New York.

Burger, J. (1973). Competition for nest sites between Franklin's Gull and the American Coot. *Wilson Bull.* **85,** 449–451.

Burger, J. (1974a). Breeding adaptations of Franklin's Gull (*Larus pipixcan*) to a marsh habitat. *Anim. Behav.* **22,** 521–567.

Burger, J. (1974b). Breeding biology and ecology of Brown-hooded Gull in Argentina. *Auk* **91,** 601–603.

Burger, J. (1974c). Determinants of colony and nest-site selection in the Silver Grebe (*Podiceps occipitales*) and Rolland's grebe (*Rollandia rolland*). *Condor* **76,** 301–306.

Burger, J. (1977). Nesting behavior of Herring Gulls: Invasion into *Spartina* salt marsh areas of New Jersey. *Condor* **79,** 162–169.

Burger, J. (1978a). Competition between cattle egrets and native North American herons, egrets and ibises. *Condor* **80,** 15–23.

Burger, J. (1978b). The pattern and mechanisms of nesting in mixed species heronries. *In* "Wading Birds," Research Report 7, pp. 45–58. National Audubon Society, New York.

Burger, J. (1979a). Competition and predation: Herring gulls versus laughing gulls. *Condor* **81,** 269–277.

Burger, J. (1979b). Resource partitioning: Nest site selection in mixed species colonies of herons, egrets and ibises. *Am. Midl. Nat.* **101,** 191–210.

Burger, J. (1979c). Nest repair behavior in birds nesting in salt marshes. *J. Comp. Physiol. Psychol.* **11,** 189–199.

Burger, J. (1980). Nesting adaptations of herring gull (*Larus argentatus*) to salt marshes and storm tides. *Biol. Behav.* **5,** 147–162.

Burger, J. (1981). A model for the evolution of mixed-species colonies of Ciconiiformes. *Q. Rev. Biol.* **56,** 143–167.

Burger, J. (1982). The role of reproductive success in colony site selection and abandonment in black skimmers (*Rynchops niger*). *Auk* **99,** 109–115.

Burger, J. (1984). Grebes nesting in gull colonies: Protective associations and early warning. *Am. Nat.* **123,** 327–337.

Burger, J., and Hahn, C. (1977). Crow predation on Black-crowned night-heron eggs. *Wilson Bull.* **89,** 350–351.

Burger, J., and Lesser, F. (1978). Selection of colony sites and nest sites by common terns *Sterna hirundo* in Ocean County, New Jersey, *Ibis* **120,** 433–449.

Burger, J., and Lesser, F. (1980). Nest site selection in an expanding population of Herring Gulls. *Bird Banding* **51,** 270–280.

Burger, J., and Shisler, J. (1978a). Nest site selection and competitive interactions of Herring and Laughing Gulls in New Jersey. *Auk* **95,** 252–266.

Burger, J., and Shisler, J. (1978b). Nest site selection in Willets. *Wilson Bull.* **90,** 599–607.

Burger, J., and Shisler, J. (1978c). The effects of ditching a salt marsh on colony and nest site selection by herring gulls (*Larus argentatus*). *Am. Midl. Nat.* **100,** 54–63.

Case, N. A., and Hewitt, O. H. (1963). Nesting and productivity of the red-winged blackbird in relation to habitat. *Living Bird* **2,** 7–20.

Cody, M. L. (1968). On the methods of resource division in grassland bird communities. *Am. Nat.* **102,** 107–147.

Cody, M. L. (1974). ''Competition and the Structure of Bird Communities.'' Monographs in Population Biology, Vol 7, pp. 1–318. Princeton University Press, Princeton, New Jersey.

Cowardin, L. M., Carter, V., Giolet, F. C., and LaRoe, E. T. (1979). Classification of wetlands and deepwater habitats of the United States. Fish Wildl. Serv. Biol. Serv. Program, FWS/OBS-79/31.

Dinsmore, J. J., and Shreiber, R. W. (1974). Breeding and annual cycle of laughing gulls in Tampa Bay, Florida. *Wilson Bull.* **86,** 419–427.

Faaborg, J. (1976). Habitat selection and territorial behavior of the small grebes of North Dakota. *Wilson Bull.* **88,** 390–399.

Ferguson, R. S. (1981). Territorial attachment and mate fidelity by horned grebes. *Wilson Bull.* **93,** 560–561.

Gochfeld, M. (1977). Colony and nest site selection by black skimmers. *Proc. Colonial Waterbird Group* **1,** 78–90.

Greenwood, R. J. (1982). Nocturnal activity and foraging of prairie raccoons (*Procyon lotor*) in North Dakota. *Am. Midl. Nat.* **107,** 238–243.

Howe, M. A. (1982). Social organization in a nesting population of Eastern Willet (*Catoptrophorus semipalmatus*). *Auk* **99,** 88–102.

Jackson, J. A. (1983). Adaptive responses of nesting Clapper Rails to unusually high water. *Wilson Bull.* **95,** 308–309.

Krunk, A. (1964). Predator and anti-predator behavior of the black-headed gull (*Larus ridibundus* L.). *Behav. Suppl.* **11,** 1–129.

Lack, D. (1954). ''The Natural Regulation of Animal Numbers.'' Oxford Univ. Press (Clarendon), London and New York.

Lack, D. (1968). ''Ecological Adaptations for Breeding in Birds.'' Methuen, London.

Lenington, S. (1980). Female choice and polygyny in redwinged blackbirds. *Anim. Behav.* **28,** 347–361.

Meanley, B., and Webb, J. S. (1963). Nesting ecology and reproductive rate of the red-winged blackbird in tidal marshes of the upper Chesapeake Bay region. *Chesapeake Sci.* **4,** 90–100.

Montevecchi, W. A. (1977). Predation in a salt marsh laughing gull colony. *Auk* **94,** 583–585.

Montevecchi, W. A. (1978). Nest site selection and its survival value among laughing gulls. *Behav. Ecol. Sociobiol.* **4,** 143–161.

Nudds, T, D. (1982). Ecological separation of grebes and coots: Interference competition or microhabitat selection. *Wilson Bull.* **94,** 505–514.

Nuechterlein, G. L. (1981). Information parasites in mixed colonies of Western grebes and forster's terns. *Anim. Behav.* **29,** 985–989.

Ogaard, L. A., Leitch, J. A., and Clambey, G. K. (1981). Wetland vegetation of the Prairie Pothole region: Research methods and annotated bibliography. *North Dakota Res. Rept.* **85,** 1–50.

Orians, G. H. (1961). Social stimulation within blackbird colonies. *Condor* **63,** 330–337.

Orians, G. H. (1969). On the evolution of mating systems in birds and mammals. *Am. Nat.* **103,** 589–603.

Orians, G. H. (1973) The red-winged blackbird in tropical marshes. *Condor* **75,** 28–42.

Orians, G. H. (1980). ''Some Adaptations of Marsh-Nesting Blackbirds.'' Monographs in Population Biology, Vol. 14, pp. 1–295. Princeton University Press, Princeton, New Jersey.

Partridge, L. (1978). Habitat selection. *In* ''Behavioral Ecology an Evolutionary Approach'' (Krebs, J. R. and Davies, N. B., eds.) pp. 351–376. Sinauer Assoc., Sunderland, Massachusetts.

Patterson, I. J. (1965) Timing and spacing of broods in the Black-headed gull *Larus ridibundus*. *Ibis* **107,** 433–459.

Picman, J. (1980). Impact of marsh wrens on reproductive strategy of red-winged blackbirds. *Can. J. Zool.* **58,** 337–350.

Post, W. (1974). Functional analysis of space-related behavior in the Seaside Sparrow. *Ecology* **55,** 564–575.

Ricklefs, R, E. (1969) An analysis of nesting mortality in birds. *Smithson. Contrib. Zool.* **9,** 1–48.

Ricklefs, R. E. (1977). On the evolution of reproductive strategies in birds: Reproductive effort. *Am. Nat.* **111,** 453–478.

Schoener, T. W. (1974). Resource partitioning in ecological communities. *Science* **185,** 27–38.

Shipley, F. S. (1979). Predation on red-winged blackbird eggs and nestings. *Wilson Bull.* **91,** 426–433.

Snelling, J. C. (1968). Overlap in feeding habitats of redwinged Blackbirds and common grackles nesting in a cattail marsh. *Auk* **85,** 560–585.

Southern, L. K., and Southern, W. E. (1979). Absence of nocturnal predator defense mechanisms in breeding gulls. *Proc. Colonial Waterbird Group* **2,** 157–162.

Stewart, R. E., and Kantrud, H. A. (1971). Classification of natural ponds and lakes in the glaciated prairie region. *Bur. Sport Fish. Wildl. Resour. Publ.* **92,** 1–57.

Stewart, R. E., and Kantrud, H. A. (1972). Vegetation of prairie potholes, North Dakota in relation to quality of water and other environmental factors. *Geol. Surv. Prof. Pap. U.S.* **585-D,** 1–36.

Talent, L. G., Krapu, G. L., and Jarvis, R. L. (1981). Effects of redhead nest parasitism on mallards. *Wilson Bull.* **93,** 562–563.

Teal, J., and Teal, M. (1969). "Life and Death of the Salt Marsh." Ballantine Book, New York.

Tinbergen, N. (1956). On the functions of territory in gulls. *Ibis* **98,** 401–411.

Tinbergen, N. (1960). "The Herring Gulls World." Collins, London.

Veen, J. (1977). Functional and causal aspects of nest distribution in colonies of the sandwich tern *Sterna s. sandivicensis. Lath. Behav. Suppl.* **20,** 1–193.

Vermeer, K. (1963). The breeding ecology of the Glaucous-winged Gull *Larus glaucesceus* on Mandarte Island, *B.C. Occas. Pap. B. C. Prov. Mus.* **13,** 1–104.

Verner, J., and Engelsen, G. H. (1970). Territories, multiple nest building and polygyny in the Long-billed marsh wren. *Auk* **87,** 557–567.

Weller, M. W. (1959). Parasitic egg-laying in the redhead (*Aythya americana*) and other North American Anatidae. *Ecol. Monogr.* **29,** 333–365.

Weller, M. W. (1968). The breeding biology of the parasitic black-headed duck. *Living Bird* **7,** 169–207.

Wiens, J. A. (1965). Behavioral interactions of red-winged blackbirds and common grackles on a common breeding ground. *Auk* **82,** 356–374.

Williams, G. C. (1966). Natural selection, the costs of reproduction, and a refinement of Lark's principle. *Am. Nat.* **100,** 687–692.

Yasukawa, K. (1981). Male quality and female choice of mate in the Red-winged Blackbird (*Agelaius* phoeniceus). *Ecology* **62,** 922–929.

Chapter 9

Dispersion Patterns and Habitat Responses of Birds in Northern Hardwoods Forests

THOMAS W. SHERRY
RICHARD T. HOLMES

Department of Biological Sciences
Dartmouth College
Hanover, New Hampshire

I. INTRODUCTION

Much of the current theory and information concerning bird habitat selection and community structure derive from studies of temperate-forest birds (e.g., Lack, 1933; MacArthur, 1958; Hilden, 1965; James, 1971). These forests offer a

HABITAT SELECTION IN BIRDS

ISBN 0-12-178080-5

wide variety of habitats that differ in extent, physical structure, and availability to birds over both space and time. Most species breeding in temperate forests are migratory insectivores that take most of their prey from plant foliage or from the ground (MacArthur, 1959; Holmes *et al.*, 1979). Because of the seasonality of the temperate environment and the migratory habit, these species must reestablish residence annually and quickly in appropriate habitats. This has probably resulted in strong selective pressures on their patterns of habitat choice.

Many studies have been conducted on the associations of particular bird species with habitats along environmental gradients (e.g., Bond, 1957; James, 1971; Cody, 1975; Able and Noon, 1976; Smith, 1977), among seral stages (Johnston and Odum, 1956; Martin, 1960; Shugart and James, 1973; Kendeigh and Fawver, 1981; Smith, 1982; May, 1982), and in habitats with either similar or contrasting physical characteristics (e.g., Maurer and Whitmore, 1981; Sabo and Holmes, 1983; Landres and MacMahon, 1983). Most of these studies have demonstrated strong correlations between bird species and certain habitat features, with major emphasis on vegetation structure (MacArthur and MacArthur, 1961; James, 1971; Whitcomb *et al.*, 1981; James and Wamer, 1982). These largely correlative approaches have proven useful in describing and predicting certain bird–habitat associations, but they cannot explain the mechanisms—psychological, ecological, or evolutionary—by which birds select their habitats. The relevant questions are: Why is vegetation structure important to birds, and what selective forces in the environment shape patterns of habitat selection? On what spatial scale(s) does habitat selection operate? How do the morphology and behavior of a species affect its choice of habitat? Answers to these kinds of questions, we believe, are necessary for a more complete understanding of bird habitat use.

In this chapter, we present preliminary evidence from a northern hardwoods forest that bird species are distributed differently within what appears to be a "homogeneous" habitat and that their responses depend on subtle differences in vegetation structure and floristics, on the presence or absence of other bird species, and on the species' social systems. We also find that detection of habitat selectivity is greatly affected by the scale at which we measure dispersion. We conclude that habitat selection by forest birds is a complex process that is not adequately described or explained by just demonstrating a correlation between bird presence and vegetation parameters.

II. HUBBARD BROOK STUDY AREA

For more than a decade we have been studying bird habitat and foraging ecology in the Hubbard Brook Experimental Forest, West Thornton, New Hampshire, but, until recently, we had confined our activities to species and events occurring primarily on a single 10-ha study plot (e.g., Holmes and Sturges, 1975; Holmes *et al.*, 1979; Sherry, 1979; Robinson and Holmes, 1982). It became apparent, however, that some species were selectively choosing certain areas and avoiding others. To more fully document the dispersion patterns of these forest birds, we therefore expanded our study area to approximately 1 × 3 km. This area included our original 10-ha study plot (Fig. 1a). On subjective grounds, the vegetation of this larger area appears relatively uniform and homogeneous and represents, by criteria typically used by ornithologists, "one habitat type."

The second-growth northern hardwoods forest at Hubbard Brook is dominated by American beech (*Fagus grandifolia*), sugar maple (*Acer saccharum*), and yellow birch (*Betula allegheniensis*), with occasional white ash (*Fraxinus americanus*), red spruce (*Picea rubens*), and eastern hemlock (*Tsuga canadensis*). The understory consists of striped and mountain maple (*A. pensylvanicum* and *A. spicatum,* respectively), hobblebush (*Viburnum alnifolium*), ferns (mainly *Dryopteris spinulosa*), various herbs, low ground cover such as *Lycopodium* and the seedlings and saplings of sugar maple and beech. The forest was selectively logged in the early part of this century, and its vegetation and history have been described recently by Bormann and Likens (1979).

Our large study area is located on a south-facing slope; it ranges in elevation from 400 to 600 m above sea level and contains a series of shallow valleys or watersheds, drained by small first- to third-order streams. Canopy height averages 20–25 m, being slightly greater in stream valleys and less on ridges. More coniferous vegetation occurs in the cooler and moister areas along streams and also on the shallower soils on the ridges that separate the small watersheds. The vegetation in these regions is also dominated by beech, sugar maple, and yellow birch, but red spruce, balsam fir (*Abies balsamea*), eastern hemlock (*Tsuga canadensis*), red maple (*Acer rubrus*), and white birch (*Betula papyrifera*) occur at increased frequencies.

III. DISPERSION PATTERNS OF FOREST BIRDS

Ecologists often ask how individual organisms are distributed or dispersed in space. The answers often provide clues to mechanisms of interaction between these individuals and point to factors in the environment that influence their distribution. Organisms can be dispersed in three different ways: (1) randomly, where the location of individuals is independent of where other individuals are located; (2) patchily, where more individuals are close to each other ("clumped") than occurs at random; or (3) evenly, where individuals are more equidistant from each other than occurs randomly.

Typically, ecologists have studied the dispersion pattern of relatively sessile

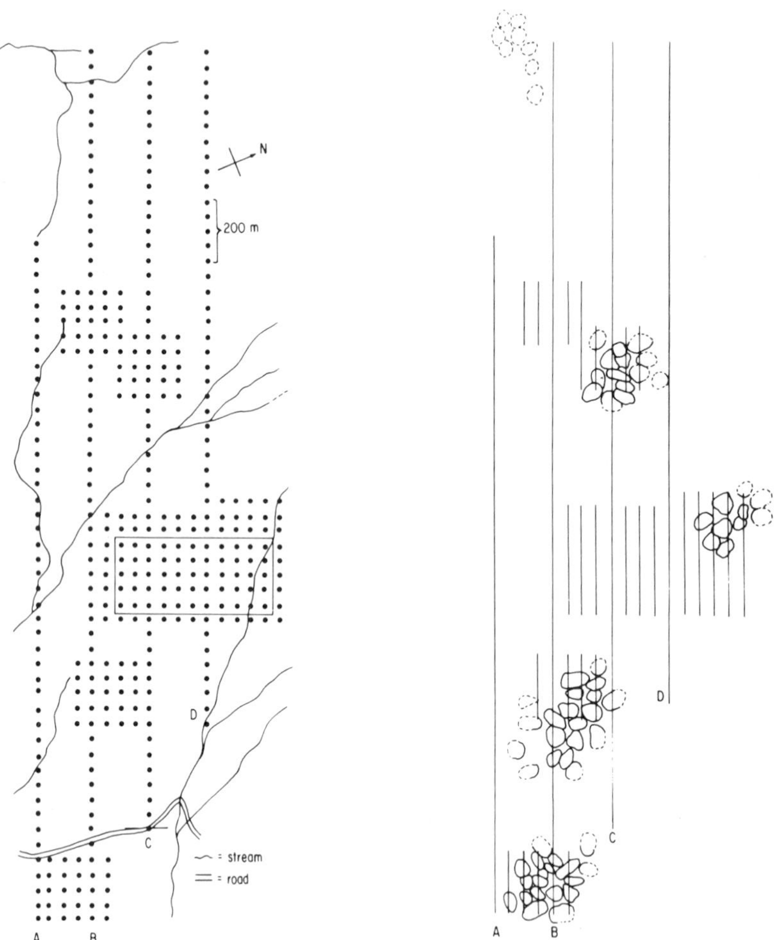

Fig. 1. Study area (a) and several examples of bird dispersion patterns (b,c,d) in the Hubbard Brook Experimental Forest, New Hampshire, June 1981. The four transect lines (A,B,C, and D) and

organisms such as plants, aphids on twigs, or granivorous insects in seeds. Forest birds, however, present special problems for analyses of dispersion, in part because they are mobile and often inconspicuous and because most defend territories (with invisible boundaries) over areas sometimes exceeding several hectares. In practice, it is difficult to define precisely the boundaries of avian territories within forests. Studies of American Redstarts (*Setophaga ruticilla*), for example, show that territories expand and contract with stage of the breeding

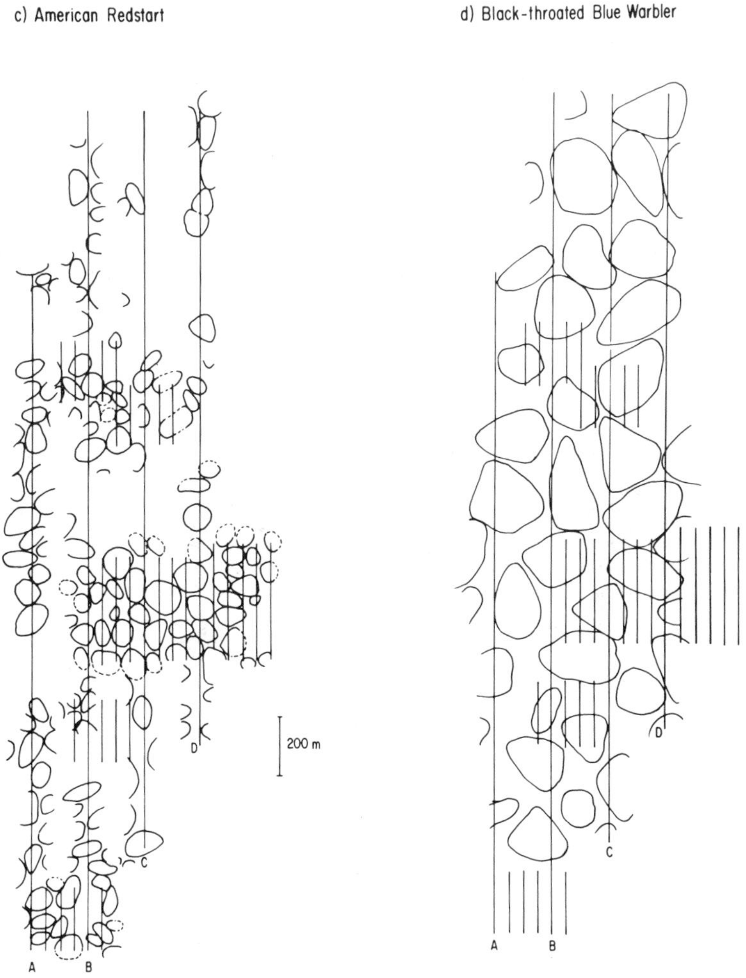

c) American Redstart

200 m

A B C D

the more extensive grids were marked at 50-m intervals with flagging. The rectangle at the center-right represents the original 10-ha study site described by Holmes and Sturges (1975). Dashed lines approximate territory boundaries that were not precisely mapped.

cycle (Yarrow, 1970) and shift in location as mates change or nest locations shift (Sturm, 1945). In designing a sampling scheme, we had to choose between sampling fewer territories in great detail or sampling more territories less precisely. We chose the latter, because we were concerned in the present study with the pattern of territory (and hence male bird) distribution over a relatively large area. We confine our analyses here to the seven most abundant bird species occurring in the Hubbard Brook Forest (Table I). These are all migratory pas-

TABLE I

Patchiness in Spatial Dispersion of Territories for Seven Abundant, Migratory Passerine Bird Species Breeding in the Hubbard Brook Experimental Forest, New Hampshire[a]

Species	Mean territory size[b]	1-ha Quadrats		4-ha Quadrats		16-ha Quadrats	
		L^c	M^d	L^c	M^d	L^c	M^d
Least Flycatcher	0.18 ± 0.001[e]	6.03	6.53	5.76	6.24	2.10	2.17
Red-eyed Vireo	1.10 ± 0.30	0.11	0.12	0.61	0.67	0.96	0.98
Black-throated Blue Warbler	3.60 ± 0.35	0.07	0.08	0.36	0.41	0.87	0.91
Black-throated Green Warbler	2.10 ± 0.25	0.07	0.08	0.87	0.96	0.78	0.83
Blackburnian Warbler	1.10 ± 0.19	1.00	1.10	0.04	0.05	1.21	1.30
Ovenbird	1.90 ± 0.22	0.31	0.37	0.44	0.49	0.85	0.88
American Redstart	0.43 ± 0.04[e]	0.59	0.60	1.04	1.08	1.02	1.04

[a] Results are shown for 50, 1-ha; 12, 4-ha; and 6, 16-ha quadrats (see text). Values of >1 indicate a clumped distribution; vlaues <1 indicate an even distribution.

[b] Ha ± 1 S.E. (R. T. Holmes and T. W. Sherry, unpublished data).

[c] Lloyd's index to "Patchiness" = $C = s^2/\lambda^2 + 1 - 1/\lambda$, where s^2 = variance, and λ = mean number of territories per quadrat (see Vandermeer, 1981).

[d] Morisita's index to "patchiness" = $(N/(N - 1))C$, where C = Lloyd's patchiness index (see footnote c), and N = total number of territories in all quadrats (see Vandermeer, 1981).

[e] Sherry (1979).

serines that arrive each spring in early to mid-May, establish territories, mate, reproduce, and then leave by August or September.

A. Methods of Measuring and Analyzing Bird Dispersion

We mapped the distribution of birds on a 10-km-long, 100-m-wide belt transect, comprising four parallel lines approximately equal in length and 200 m apart (see Fig. 1a). While moving along transect lines at a rate of 50 m every 6 min (cf., Holmes and Sturges, 1975), we recorded on a map the locations of all birds of both sexes seen or heard, especially long flights and all instances of countersinging by neighboring males. Four censuses of the entire 10-km transect were completed between 0630 and 1030 during June, the peak breeding period of these species. Most observations reported here were made in 1981. All observations were compiled on a transparent overlay of the belt transects. Estimated territory boundaries were drawn by one of us (T. W. S.) experienced with territory sizes in the study area (based on previous observations of individually marked birds of the species involved). For those species holding territories large enough to cross more than one transect line, territory boundaries between belts were interpolated from all observations, especially those of counter-singing males. This was facilitated by numerous observations of birds singing beyond the 50-m zone on each side of the transect line. For Least Flycatchers (*Empidonax minimus*), whose territories were the smallest of those studied, we gridded all additional areas between census lines where this species was found (Fig. 1a,b).

To characterize each species' dispersion pattern quantitatively, we calculated both Lloyd's and Morisita's "patchiness" indices (Vandermeer, 1981). Both indices are essentially variance : mean ratios of number of objects (territory centers in the present study) per quadrat. If territories are randomly distributed, the variance of the distribution equals the mean; if they are evenly distributed, the variance is less than the mean, because each quadrat will tend to have the same number of territories. If territory distribution is patchy, some quadrats will have many and others will have few, giving a variance : mean ratio greater than 1. Because measures of dispersion patterns are influenced by quadrat size (Vandermeer, 1981), we examined the dispersion pattern with five quadrat sizes (sample size in parentheses): 1 ha (50), 2 ha (35), 3 ha (25), 4 ha (12), and 16 ha (6). Quadrats were square, except for the 2-ha (100 × 200 m) and 3-ha ones (150 × 200 m), and were systematically superimposed on the study area with a distance between quadrats of at least the length of a quadrat's side, except for the very large 16-ha quadrats. Counts of territory occurrences were made once quadrat locations were superimposed onto territory maps of each species (e.g., Fig. 1). We considered a territory to be in a quadrat if its center was within or on the boundary line of that quadrat, and we defined the territory center as the center of the smallest circle that completely enclosed the territory. To test hypotheses, we compared observed counts of territories per quadrat with counts predicted by a poisson model using the chi-square statistic (Vandermeer, 1981).

Simberloff (1979) has noted that modified null (random) models are needed to analyze the spatial distribution of disks rather than points. This necessity arises because of constraints on the placement of rigid disks, whose centers in the case of disks with equal radii can lie no closer together than 1 diameter. Although Simberloff's model might appear appropriate for avian territories, we are not concerned here with the dispersion of territories, but rather with the dispersion of individual birds. Thus, we used the territory of a singing male bird to represent an "individual," and the presence or absence of such individuals within quadrats of varying sizes was used as a measure of dispersion. In essence, our methods quantify the extent to which territories either fill the environment (variance : mean ratio < 1) or aggregrate in restricted parts of the habitat (variance : mean ratio > 1). Clearly, the occurrence of territoriality in all species under study means that dispersion can never be truly "random", and randomness in the present study is discussed in a statistical sense only.

B. Dispersion Patterns of Forest Birds

Territory maps for three species (Fig. 1b,c,d) illustrate qualitatively the range in dispersion patterns of the common bird species at Hubbard Brook. Least Flycatchers (*Empidonax minimus*) have the smallest territories and, thus, potentially the greatest density of any species studied, but the territories were restricted exclusively to several patches dispersed widely across the study area (Fig. 1b), confirming earlier reports of flycatcher colonies (Davis, 1959). American Redstarts are more widely distributed than the flycatcher, but they too failed to use the entire area; some segments of the transects contained few redstart territories (Fig. 1c). In contrast, Black-throated Blue Warblers (*Dendroica caerulescens*) have much larger territories (Table I), which covered the study area more evenly. Of the other species, Ovenbirds (*Seiurus aurocapillus*) and Red-eyed Vireos (*Vireo olivaceus*) have territories dispersed widely, similar to the Black-throated Blue Warbler. Territories of Blackburnian Warblers (*Dendroica fusca*) and Black-throated Green Warblers (*Dendroica virens*) neither filled the habitat, nor clustered tightly, and represent an intermediate pattern. Qualitatively, territories of species were dispersed identically in 1982 and 1983, compared with 1981 patterns as illustrated in Fig. 1. Abundances of these species were similar over this 3-year period (T. W. Sherry and R. T. Holmes, unpublished data).

Morisita's and Lloyd's patchiness indices were consistent within a quadrat size for all seven species examined (Table I). Morisita's index gave higher patchiness values than Lloyd's, especially for species represented by relatively few territories. The seven species differed in their patterns of dispersion within a given quadrat size and also across quadrat sizes, indicating the importance of scale in considerations of spacing.

Least Flycatcher territories were clumped for all quadrat sizes, as indicated by patchiness values exceeding 1 (Table I). This pattern was statistically nonrandom for 1-ha quadrats ($\chi^2 = 5.9, p < 0.025$, 1 *df*) and for 4-ha quadrats ($\chi^2 = 5.74, p < 0.025$, 1 *df*). In both of these tests, more quadrats had zero territories than predicted by the poisson (random) model, and fewer quadrats had intermediate numbers, a pattern consistent with a clumped spatial distribution. We did not test statistically the dispersion patterns for 16-ha quadrats because (1) there were few of these quadrats (only 6), and (2) expected numbers of territories per quadrat were not small (values ranged from 2.33 to 13.67), in violation of poisson-model assumptions (Vandermeer, 1981).

Other than the Least Flycatcher, only the redstart and Blackburnian Warbler had patchiness indices greater than 1 (Table I), but in neither case was the tendency significantly different from random. For two out of three quadrat sizes, these species had patchiness indices close to 1, suggesting a random dispersion pattern. These species also illustrate how measures of dispersion may vary with different quadrat sizes. The redstart had overdispersed territories with 1-ha quadrats (but this trend was not statistically significant), while Blackburnian Warbler territories appeared overdispersed with 4-ha quadrats ($\chi^2 = 6.18$, $p < 0.01$, 1 *df*).

The redstart and Blackburnian Warbler had patchiness indices less than 1 for some quadrat sizes, and all other species had patchiness indices less than 1 for all quadrat sizes (Table I), indicating a tendency to be evenly spaced throughout the study area. We tested the null hypothesis that each species with patchiness index less than 1 was randomly distributed. For 1-ha quadrats, no species' index differed significantly from random. For 4-ha quadrats, the Blackburnian Warbler's index was nonrandom as shown above, and that of the Black-throated Blue Warbler approached statistical significance ($\chi^2 = 2.83$, $0.05 < p < 0.10$, 1 *df*). In each case, observed frequencies of quadrats with one territory exceeded the expected frequency.

We did no statistical tests for 16-ha quadrats, as mentioned earlier, but we did tests for 2-ha and 3-ha quadrats (T. W. Sherry and R. T. Holmes, unpublished data) because they are more comparable to the territory sizes of most species (Table I). Red-eyed Vireos had territories that were overdispersed with 2-ha quadrats ($\chi^2 = 3.31$, $0.05 < p < 0.10$, 1 *df*) and significantly overdispersed when 3-ha quadrats were used ($\chi^2 = 6.32$, $p < 0.025$, 1 *df*). Black-throated Blue Warblers had nearly nonrandom territories using 3-ha quadrats ($\chi^2 = 2.81$, $0.05 < p < 0.10$, 1 *df*). No other species had nonrandom dispersion patterns for these two quadrat sizes (the highly clumped Least Flycatcher was not tested). For both the vireo and the Black-throated Blue Warbler at these quadrat sizes, observed frequencies of one-territory quadrats exceeded predicted frequencies, indicating even dispersion.

An interesting result was the tendency for redstarts to be randomly dispersed, despite the fact that they are highly aggressive and territorial (Sturm, 1945; Ficken and Ficken, 1962) as all other overdispersed species and that they are also abundant (Fig. 1c). The most reasonable explanation, we think, is that the tendency to be territorial and overdispersed in some localized parts of the study area compensates for being patchily distributed on a larger scale.

Clearly, for this hypothesis to be reasonable, redstarts should be statistically overdispersed on some parts of our study area. We tested this prediction in 1982 within our intensive study area, which was expanded to 28 ha in 1981 (see Fig. 1a). Since we could not perform quadrat analysis of redstart dispersion in this small an area, we performed nearest-neighbor analysis. Our sampling of territory usage, moreover, was sufficiently intense (unlike that along the full belt transects) to justify greater confidence in the location of territory boundaries. Again, the center of a territory was by definition the center of the smallest circle completely enclosing the territory. About 25% of actual nearest-neighbor distances in this 28-ha area originally included reciprocal nearest-neighbor pairs. To avoid potential statistical problems associated with such pairs (Meagher and

Burdick, 1980), we used the next nearest neighbors in all such cases. The mean $\pm$ 1 *SE* distance to a nearest neighbor for redstart territory centers was 66.02 $\pm$ 2.72 m ($n = 41$). If the territory centers were ranodmly distributed, the mean $\pm$ 1 *SE* would be $1/(2\sqrt{P}) \pm \sqrt{(4 - \pi)/(4\pi PN)}$, where P is the density of points per area (44 territory centers per 28 ha = 0.0001571/m^2), and N is the number of points (Clark and Evans, 1954), giving 39.89 $\pm$ 3.14 m as the predicted values for randomly distributed territory centers. We could not reject the null hypothesis that the actual distances to the neighboring territories' centers were more or less variable than predicted distances ($F = 1.43$, $p > 0.05$, on 43 and 40 *df*). However, the mean of observed distances was not identical to the predicted mean ($t = 6.24$, $p << 0.001$). The difference was highly significant even when we used the mean nearest-neighbor distance (58.27 m) that included reciprocal nearest-neighbor pairs. We conclude that the significantly greater observed than predicted territory–territory distances were caused by the tendency for redstarts to spread throughout the 28-ha area as a result of territorial behavior. These results support our suggestion that the random distribution of redstart territories throughout the entire area encompassed by the 10-km transect was the result of counteracting tendencies for redstarts to avoid certain parts of the study area and to space themselves evenly within other, presumably preferred parts.

In summary, the seven most abundant birds in our study area were dispersed in diverse ways. The Least Flycatcher was clumped spatially, whereas Red-eyed Vireos, Black-throated Blue Warblers, and Ovenbirds tended to space themselves evenly across the entire study area. Territories of Blackburnian and Black-throated Green Warblers, as well as American Redstarts, tended to be randomly distributed at most scales (i.e., quadrat sizes), but our perception of the pattern often changed with quadrat size. Even a statistically random pattern may result from strongly nonrandom processes, as illustrated here by the American Redstart.

IV. VEGETATION PATTERNS AND AVIAN DISTRIBUTION

The variety of ways in which bird species were dispersed within our study area prompted us to ask whether vegetation heterogeneity in combination with habitat selectivity was important in producing these patterns. To test the hypothesis that individuals settled nonrandomly with respect to vegetation, we first characterized the vegetation along all transects. Next, we summarized the vegetation characteristics of that subset of plots occupied by each species and asked whether this subset was a random set of available plots.

A. Methods

We quantified vegetation characteristics on 200 contiguous 50-m-long plots centered along the 10 km of belt transect on which birds were censused (see

Section III). Trees were assessed on 10-m-wide plots along each 50-m segment of transect, while herbs and shrubs were assessed in 50-m-wide plots. The tree variables for each 50 × 50 plot were species dominance values, determined from diameter at breast height (*dbh*) of every live tree with *dbh* ≥ 10 cm. Several rare tree species that had frequency distributions of dominance values heavily biased by zero values (skewness coefficients ≥2) were eliminated, leaving 13 tree species for subsequent analyses. Tree basal area is a good index of available leaf area (Whittaker *et al.*, 1974), and previous studies demonstrate the importance of species-specific effects of plants on the foraging behavior of insectivorous birds (Holmes *et al.*, 1979; Holmes and Robinson, 1981).

Vegetation of the herbaceous (< 0.5 m) and shrub (0.5–5 m) layers was sampled differently. Percentage of coverage of herbaceous and shrub layers was estimated on a scale from 1 to 4, depending on whether the ground was 0–25%, 25–50%, 50–75%, or 75–100% covered by vegetation in that layer. To obtain floristic values for the herb and shrub strata, we ranked the most abundant species from 1 to 4 (by ground coverage). In total, the 2 ground-cover variables and 11 herb or shrub species had sufficiently nonskewed frequency distributions to be included in subsequent analyses.

To describe all available plots and to reduce the number of redundant variables, we performed separate principal components (PC) analyses for the tree data and for the herb–shrub data. Next, to locate each vegetation plot in the space defined by the new PC axes, we postmultiplied a standardized version (mean for each variable = 0, *SD* = 1) of the original data matrix (200 plots by 13 plant variables) by the factor–score coefficient matrix to get "standardized principal components" (McGee, 1978) for each vegetation plot. Then we selected for each bird species those 50-m segments along the transect lines covered by at least half a territory of that species (see Section III). For one pair of principal axes at a time, we then calculated a 95% confidence ellipse (Sokal and Rohlf, 1969; Cody, 1978; Collins, 1983a) for the mean (standardized) PC scores of each bird species. If the confidence ellipse for the all-species mean (i.e., the origin) of all available plots did not contain the mean for a particular species, we concluded that the species in question selected a habitat that was a statistically nonrandom sample of available habitats. We made no attempt to correct for outlying points, and the resultant increased variances make our conclusions about habitat selection conservative.

B. Vegetation Patterns and Bird Occurrences

Here we first describe patterns in the vegetation along the belt transect and then describe the distribution of birds with respect to vegetation patterns. For each bird species, we test the null hypothesis that its distribution is random with respect to vegetation parameters. Principal component analysis of the tree data produced four significant axes, which collectively explained 55% of the variation in original data (Table II). The first principal axis had a high positive correlation with sugar maple and strong negative correlations with red spruce, balsam fir, yellow birch, and red maple, suggesting that the PC I is a hardwood–mixed softwood axis. Beech and striped maple are positively loaded on the PC II axis, whereas white ash is negatively loaded on this axis. White ash, a midsuccessional species, reaches its elevational limit within our study area. It prefers richer and better-drained soils than beech, and beech (unlike ash) often spreads by root and stump sprouting, which may enable it to dominate large areas of forest, once established (C. Cogbill, personal communication). Thus, we interpret the second PC axis as one of soil quality and elevation. The third and fourth PC axes were much less easily interpreted and are not discussed, because the bird distributions were little affected by them (see Section IV,C).

Next, to show patterns in the vegetation across our study area, we plotted a continuous curve of standardized PC scores along lines representing the transects (Fig. 2). The resulting patterns show (1) a tendency for coniferous vegetation relative to deciduous to increase (i.e., negative scores on PC I), going from eastern to western portions of our study area, and (2) relatively high representation of white ash and less beech (i.e., negative PC scores on PC II) at both eastern and western ends (Fig. 2). To show this qualitatively, we can compare

TABLE II

Correlations (Loadings) of Tree Variables on First Four (Significant) Principal Component Axes[a]

Tree species	Principal component axes				Row
	I	II	III	IV	sum of squares
Sugar Maple	0.71				0.67
Beech		0.73			0.66
Yellow Birch	−0.56				0.38
White Ash		−0.63			0.66
Mountain Maple				0.69	0.59
Striped Maple		0.45	0.56		0.59
Balsam Fir	−0.63				0.45
Red Spruce	−0.68				0.49
Paper Birch				0.57	0.56
Red Maple	−0.46				0.37
Hemlock			0.70		0.61
% Variation	21.46	12.93	10.63	9.87	
Cumulative % variation	21.46	34.39	45.01	54.88	
Eigenvalue	2.36	1.42	1.17	1.09	6.04

[a] Only loadings ≥0.45 are reported (cf. Aspey and Blankenship, 1977).

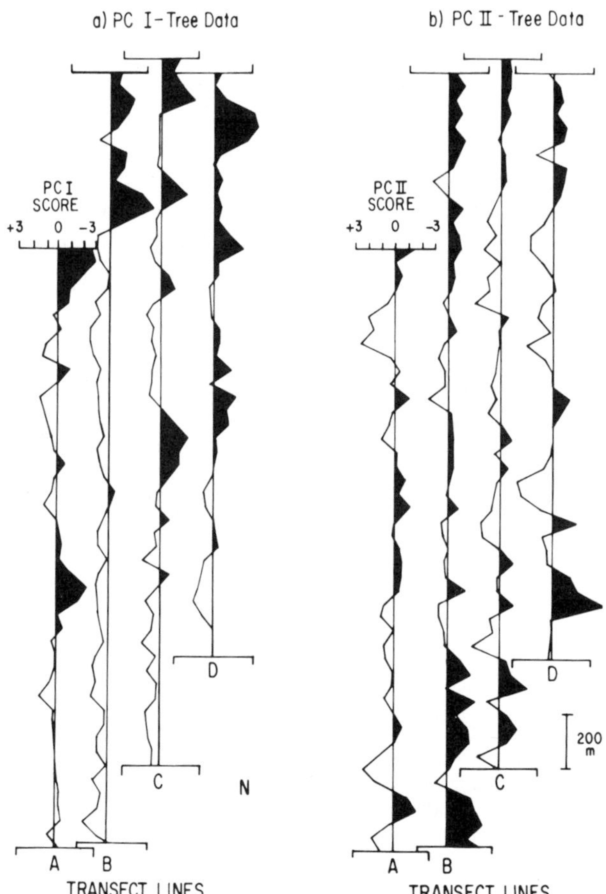

Fig. 2. Distribution of (a) principal components (PC) I and (b) PC II scores for tree variables along transects in the Hubbard Brook forest. Shaded areas (negative scores) on PC I reflect mixed hardwood (red maple, yellow birch)–conifer associations, whereas on PC II they represent the presence of white ash and better drained soils (see Table II and text). Transect lines conform with those in Fig. 1.

the settlement locations of particular bird species (e.g., Fig. 1) with the PC scores of the plots arrayed spatially as they occurred in the field (Fig. 2). The redstart and Least Flycatcher illustrate particularly clearly sparse settlement along segments of the transect where vegetation had more frequent coniferous elements.

To illustrate the distribution of all 200 available tree plots in PC space, we located each, using its standardized factor scores, in the space defined by PC axes I and II (Fig. 3a). Next, we plotted the 95% confidence ellipse for the bivariate mean of these factor scores (henceforth, the "mean ellipse") and the 95% confidence ellipse for the entire scatter of plots (henceforth, the "scatter ellipse"). The "origin" in PC space was by definition the mean PC score of all 200 plots. Then we plotted the 95% mean and scatter ellipses for PC scores of that subset of all plots in which each bird species occurred. The Black-throated Blue Warbler (Fig. 3b) was typical of all species, in that its scatter ellipse broadly overlapped that calculated for all 200 plots (compare Figs. 3a,b). Thus,

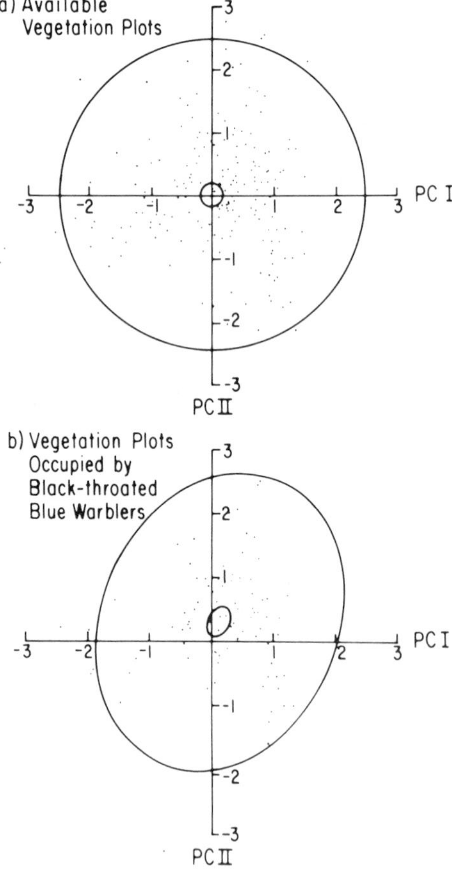

Fig. 3. Distribution of standardized PC scores (points) for tree vegetation plots in the space defined by the first two PC axes resulting from 95% confidence ellipses for the mean (small ellipse) and for all PC scores (large ellipse), superimposed on the bivariate scattergrams for (a) all 200 tree vegetation plots available, and (b) the subset of the 200 plots that were occupied by Black-throated Blue Warblers.

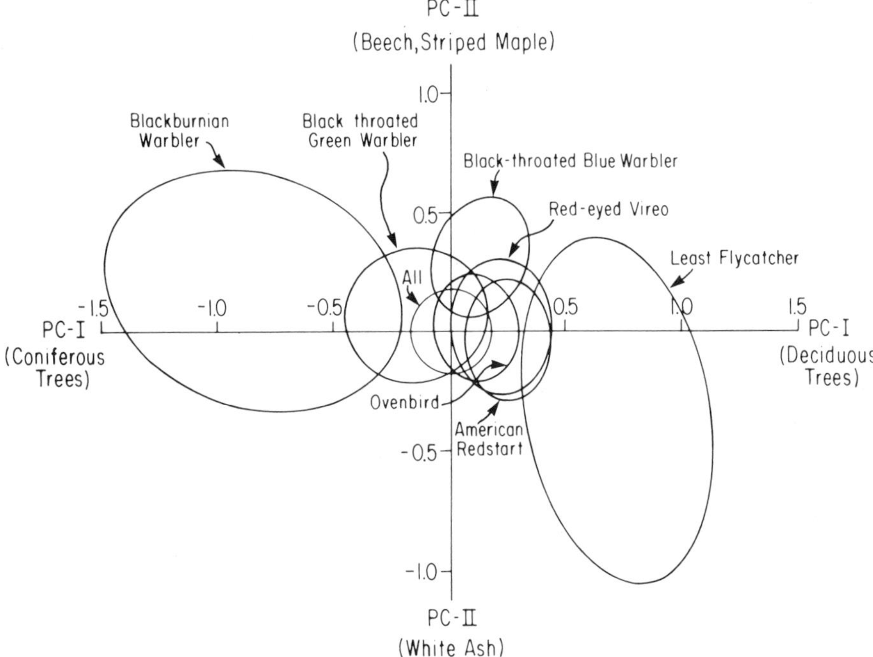

Fig. 4. Ninety-five percent confidence ellipses of the mean standardized PC scores (see Fig. 3) for all tree plots (circle labeled "All" centered at origin) and for subsets of all plots occupied by each of seven passerine bird species in the Hubbard Brook study area. Note expanded scale compared with that in Fig. 3.

all seven species were broadly distributed with respect to the components of vegetation measured in our study, and most, if not all, could be expected to have overlapping territories were it not for additional factors influencing their distribution patterns (see Section V).

Nonetheless, several species were not randomly distributed with respect to particular tree species. Specifically, the 95% mean ellipse for five bird species did not contain the origin, and some were not overlapping (Fig. 4). We interpret these results to mean that five species settled in habitats whose mean vegetation characteristics were unlikely (at the $p \leq 0.05$ level) to be described by random subsets of the available habitat and that those species whose mean ellipses do not overlap select different habitats. Least Flycatchers strongly preferred deciduous vegetation (positive PC scores for plots occupied), while the redstart and vireo did so less strongly. Blackburnian Warblers, by contrast, preferred areas with more coniferous vegetation. Black-throated Green Warblers had a tendency, although not statistically significant, to settle in areas with conifers. Ovenbirds showed a slight preference for deciduous vegetation but, like the vireo and

Black-throated Blue Warbler, were distributed evenly over much of our study area (Fig. 1, Table I). The mean ellipse of the Black-throated Blue Warbler showed little deviation along PC I, but was shifted significantly toward positive PC scores on the second axis. Thus, this warbler preferred areas with less white ash and more beech and striped maple. None of the seven bird species had mean ellipses which did not include the origin when plotted in the space defined by PC axes III and IV.

The method we used to quantify herb–shrub characteristics differed substantively from that used for trees, yet the results and conclusions from PC analysis of the two data sets are similar. Five statistically significant PC axes accounted for 68% of the variation in original variables (Table III). The positive beech and sugar maple loadings and negative spruce loadings, on PC I (Table III), define it as a deciduous–coniferous axis, just as PC I was based on tree data (Table II). PC II had positive loadings for total shrub as well as beech density and a negative loading for ferns. White ash was not included in the herb–shrub analysis, because it occurred only rarely in the near-ground vegetation.

The 95% mean ellipses for the 7 bird species, based on the herb–shrub data, are shown in Fig. 5. Again, Blackburnian Warblers and Least Flycatchers were

TABLE III

Correlations (Loadings) of Herb and Shrub Variables on First Five (Significant) Principal Component Axes[a]

Plant variable	Principal component axes I	II	III	IV	V	Row sum of squares
Herbaceous layer						
Beech	0.74					0.75
Sugar Maple	0.71					0.69
Striped Maple				−0.54		0.55
Viburnum	−0.70					0.65
Fern		−0.56				0.64
Lycopodium			−0.52		0.69	0.86
Density	−0.68					0.68
Shrub layer						
Density		0.71				0.68
Beech		0.71				0.77
Sugar Maple			0.80			0.78
Striped Maple	−0.57				0.47	0.61
Viburnum	−0.51					0.49
Red Spruce	−0.61					0.70
Percentage of variation	25.62	14.92	10.88	8.90	7.75	
Cumulative percentage of variation	25.62	40.54	51.42	60.32	68.07	
Eigenvalue	3.33	1.94	1.41	1.16	1.01	8.85

[a] Only loadings ≥0.45 are reported (cf. Aspey and Blankenship, 1977).

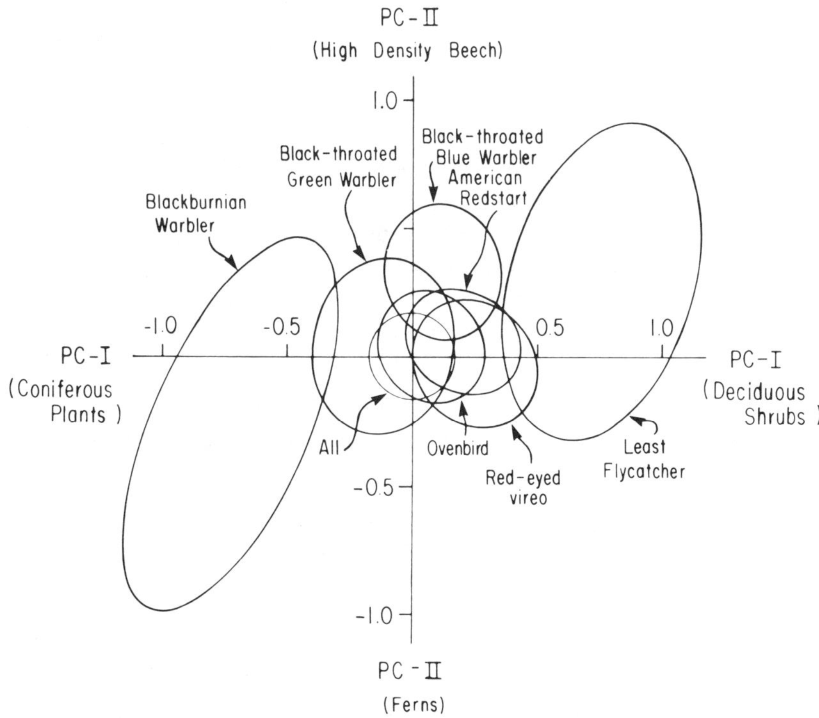

Fig. 5. Ninety-five percent confidence ellipses of the mean standardized PC scores (see Fig. 3) for all herb–shrub plots (circle labeled "All" centered at origin) and for subsets of all plots occupied by each of the seven passerine bird species in the Hubbard Brook study area. Note expanded scale compared with that in Fig. 3.

most distinct, the former occurring in areas with more coniferous elements, the latter in more deciduous sites. By its positive displacement on PC II, the Black-throated Blue Warbler, an understory species, showed a significant association with dense beech shrubs.

We conclude that our study area contained vegetative heterogeneity and that most bird species settled selectively with respect to floristic characteristics of the vegetation. The bird species showed strongest preference or aversion for deciduous versus coniferous plant species, and only Black-throated Blue Warblers showed preferences among deciduous tree species and in shrub coverage.

C. Habitat Selectivity

The patchy distribution of Least Flycatchers (Fig. 1, Table I), the avoidance of parts of the study area by redstarts (Fig 1.), and the coniferous associations of Blackburnian, and to some extent Black-throated Green Warblers, suggest that vegetation heterogeneity (Fig. 2) contributes to bird patchiness. To examine this relationship quantitatively, we asked whether those species with patchiest dispersion statistics (Lloyd's index, Table I) also tended to be more selective with respect to vegetation parameters (Fig. 4). As a measure of habitat selectivity for each species, we used the displacement (Euclidian distance) of the mean from the origin in PC 4-space (Table II). The relationship was consistently positive (r = 0.165, 0.45, and 0.74, when patchiness was quantified, with 1-ha, 4-ha, and 16-ha quadrats, respectively), but only statistically significant ($p < 0.05$) for patchiness calculated from 16-ha quadrats. These findings reinforce the importance of quadrat size to estimates of bird patchiness (see also James and Wamer, 1982). Census estimates of Least Flycatchers in the forest, for instance, would vary greatly depending on plot size and position.

D. Importance of Vegetation to Insectivorous Forest Birds

Our results demonstrate that dispersion patterns of forest birds result at least in part from species-specific preferences for, or avoidances of, certain vegetation characteristics. What is it about vegetation that is important? To what features of vegetation are these birds responding?

The fact that our vegetation measures were largely floristic suggests that plant species composition along the transects was important. In previous studies, we found that many of these same bird species preferentially select certain tree species for foraging (Holmes and Robinson, 1981; Robinson and Holmes, 1984). This choice seems to be influenced by the unique morphological and behavioral traits of individual bird species that allow them to differentially exploit arthropods among the various foliage structures exhibited by forest trees (Robinson and Holmes, 1982). Blackburnian and Black-throated Green Warblers primarily search for prey on plant surfaces near branches and rarely perceive or attack prey over long distances (Robinson and Holmes, 1982). Consequently, they forage mainly on plants like conifers and yellow birch that have foliage situated close to the branches where sitting arthropods are accessible to a perched bird. Least Flycatchers, in contrast, sit on perches within and below the canopy and scan the underside of leaves at considerable distances; they often make flying attacks on prey at distances of 1 m or more (Robinson and Holmes, 1982). For them, the openness and array of foliage around perch sites will affect their foraging success, and, hence, foliage structure may well influence their choice of habitat (but see later in this section). From this, one would expect habitat preferences of forest birds to be consistent through their breeding ranges, and there is some evidence for this (Noon *et al.*, 1980), although Collins (1983b) has found considerable variation for some parulids.

Those species discussed previously that have the strongest correlation with specific vegetation characteristics at Hubbard Brook also have the patchiest distribution and show the greatest habitat selectivity. The other species (Red-eyed Vireo, American Redstart, and Black-throated Blue Warbler) are more generalized in their habitat response (Fig. 4), search for prey on nearby and distant substrates (Robinson and Holmes, 1982), and are capable of using a variety of tree species that may differ in structure and/or insect abundance (Holmes and Robinson, 1981). They are also widely dispersed through eastern deciduous forests.

The main implication of these findings is that some bird species are present in the Hubbard Brook forest only where there is a particular kind or combination of vegetation types to which they are specifically adapted. For example, conifers are important for Blackburnian and to some extent Black-throated Green Warblers, while openness below and within the canopy of deciduous trees appears to be critical for Least Flycatchers. These species seem to be closely linked to these vegetation types, because of morphological and behavioral constraints that influence how they search for and capture insect prey. Foliage structure is also undoubtedly important in other ways, e.g., in providing nest sites, favorable microclimates, and places to hide from predators, yet, for forest birds, the opportunities and obstacles to foraging birds provided by different foliage configurations seem most influential (Holmes, 1981).

V. SOCIAL CONTEXT OF DISPERSION PATTERNS

The birds we have considered are each dispersed by agonistic behavior associated with territories on which mating, nesting, and feeding take place (i.e., "type A" territories; Wilson, 1975). The surprising result of the present study, given this homogeneity of social systems, is the diverse ways in which the birds are dispersed relative to each other and to vegetation parameters (Fig. 1). Quantitative studies of avian dispersion (Pruett-Jones and Pruett-Jones, 1982; Cooke *et al.*, 1983; Mumme *et al.*, 1983) are rare, and only a few (e.g., Rotenberry and Wiens, 1976) have contrasted variable dispersion patterns of sympatric bird species. Several kinds of evidence from our studies at Hubbard Brook suggest that bird distributions depend on the social context of individual interactions, both intra- and interspecific, and not just on responses to vegetation (discussed earlier).

Patterns of dispersion often indicate processes of population organization, and such is the case with our observations of the Least Flycatcher. Several observations suggest that strong aggregative behavior overrides vegetation characteristics to produce the highly clumped distribution of Least Flycatchers (Fig. 1). First, Least Flycatchers in our forest are always aggregated, suggesting a strong social bond. Second, we have sought in vain for consistent or measurable vegetation differences within occupied clumps of flycatchers versus adjacent unoccupied areas (T. W. Sherry and R. T. Holmes unpublished data, contra Sherry, 1979). Third, some Least Flycatcher aggregations have drifted in location between years, suggesting that suitable habitats are available in adjacent areas (T. W. Sherry and R. T. Holmes, unpublished data). The significantly clumped dispersion of territories, in addition to their relatively small size, suggests to us that proximity of neighbors (i.e., sociality) is an important, if not essential, dimension of Least Flycatcher ecology. Ardrey (1966) applied the term "noyau" for the often spatially discontinuous neighborhoods of animals that are aggressively territorial, yet at the same time mutually interacting and interdependent. Least Flycatchers seem in many respects to provide a good example of noyau social organization, but what are the advantages and disadvantages to Least Flycatchers of settling near other individuals of the same species?

Various studies have suggested that group living provides protection against predators, both directly and indirectly (Hamilton, 1971; Morse, 1977). In species like the Least Flycatcher that have small but clumped territories, the advantage may be for defense of nests from predators. Individuals could benefit directly from others that help to chase predators from an area or to warn of predators. Many migratory Tyrannidae and especially Least Flycatchers (Sherry, 1979) are aggressive and pugnacious toward other birds including predatory ones, a characteristic that likely contributes to relatively low nest losses (Murphy, 1983).

Protection of the food supply is another potential advantage of group membership. The small territories of Least Flycatchers compared with other Hubbard Brook species (Table I) have the potential consequence that an individual flycatcher can often benefit from neighbors that chase away broadly overlapping individuals of other species. Available evidence suggests that this advantage is important to Least Flycatchers. They are aggressive toward other small birds, particularly American Redstarts (Sherry, 1979), and redstarts have more similar diets, foraging heights, and foraging tactics compared with Least Flycatchers than with any other Hubbard Brook species (Sherry, 1979; Holmes *et al.*, 1979). Experimental removals of Least Flycatchers indicate that their behavior effectively reduces redstart abundance within flycatcher patches (see below; T. W. Sherry and R. T. Holmes, in preparation). Defense of resources interspecifically could be important to relatively specialized predators such as Least Flycatchers (Sherry, 1979), and their strong preference for deciduous vegetation (Figs. 4 and 5) may indicate a need for relatively productive sites whose insect resources might be greater or more predictable, on average. The Bell Miner (*Manorina melanophrys,* Meliphagidae) of Australia is strikingly similar to the Least Flycatcher in this regard. Intraspecifically territorial family groups of Bell Miners (just like mated pairs of Least Flycatchers) aggregate in food-rich areas and effectively defend the entire area against other species with overlapping feeding heights and food habits (Smith and Robertson, 1978; Loyn *et al.*, 1983).

Redstarts provide a second example of how social interactions, interspecific in this case, affect habitat use and dispersion. Specifically, redstarts were absent not only from relatively coniferous parts of the study area, as shown earlier in this section, but also from several highly deciduous, presumably satisfactory parts (Fig. 1c). These latter gaps in redstart distribution corresponded precisely with Least Flycatcher clumps (compare Figs. 1b,c and 2a), an observation we postulate is due to a direct species interaction. Sherry (1979) described several indirect lines of evidence that competition from Least Flycatchers influences redstart ecology. We tested the hypothesis directly by removing Least Flycatchers from two patches and simultaneously monitoring redstart abundance in the removal and control areas. In each case, redstart abundance increased in the experimental areas within two breeding seasons (T. W. Sherry and R. T. Holmes, in preparation), indicating that areas with Least Flycatchers were indeed suitable habitats for redstarts. It therefore appears that flycatchers restrict where redstarts will settle. Similarly, suitable habitat for Philadelphia Vireos (*Vireo philadelphicus*) is limited to areas with tall vegetation providing a refuge from the socially dominant Red-eyed Vireo (Robinson, 1981). Experimental demonstrations of interspecific competition in birds (Davis, 1973; Williams and Batzli, 1979; Dhondt and Eyckerman, 1980; Minot, 1981) are not unusual, and such demonstrations in other organisms are becoming increasingly common (Schoener, 1983).

Interspecific commensalism provides another class of interspecific social interactions that influence the dispersion and abundance of birds and may involve predators or food. An example of the former is the tendency of certain songbirds to aggregate where territories of *Turdus pilaris* help protect against nest predation (Slagsvold, 1980; see also Chapter 3, this volume). An example of the latter is the Ruby-throated Hummingbird (*Archilochus colubris*), which breeds in north-temperate forests (including Hubbard Brook) and depends on the activity of Yellow-bellied Sapsuckers (*Sphyrapicus varius*) for sap, an important food resource (Miller and Nero, 1983).

VI. SCALE OF HABITAT SELECTION

A common goal of habitat selection studies is to understand the distribution of species. We argue here that habitat selection by temperate zone–forest birds results from diverse agents operating on different spatial scales. Numerous studies, including the present one, have shown that dispersion patterns appear to vary with scale. At a large enough spatial scale, animals are rarely distributed at random, but, instead, tend to be aggregated (Taylor *et al.*, 1978; Taylor and Woiwod, 1982). It is not clear what biological processes are responsible for this pattern (Anderson *et al.*, 1982), but a variety of factors could be involved. In the present study, Least Flycatchers, and American Redstarts to a lesser extent, were patchy at the scale of our large, 1- × 3-km study area, in contrast to species such as the Red-eyed Vireo and Black-throated Blue Warbler, whose patchiness must be manifested (by the arguments of Taylor and colleagues) at a greater spatial scale.

Several processes that influence avian distribution and abundance operate locally, i.e., within areas the size of individual territories. Psychological processes, for example, affect habitat selection (Hilden, 1965) and are explicit in the "niche–gestalt" model of habitat selection (James, 1971; Whitmore, 1975; Collins *et al.*, 1982; James *et al.*, 1984). This paradigm stresses that individuals respond independently of other species to gross structural features of the habitat (e.g., vegetation height and density in various layers), presumably because these habitat features are correlated with such avian requisites as feeding and nesting sites. There is nothing explicit in discussions of the niche–gestalt that an individual bird need consider habitat characteristics beyond the limits of an area large enough for one or a few territories. Several common birds in our study area whose distribution is not overdispersed may respond directly to identifiable characteristics of the habitat. As indicated previously, the habitat selectivity and uneven distribution of the Blackburnian and Black-throated Green Warblers, probably results from positively correlated changes in the relative dominance of conifers and yellow birch, and from the heterogeneity of coniferous versus deciduous vegetation across our study area. The Black-capped Chickadee (*Parus atricapillus*), Brown Creeper (*Certhia familiaris*), and Solitary Vireo (*Vireo solitarius*) also tend to occur more frequently at Hubbard Brook where conifers are more abundant (R. T. Holmes and T. W. Sherry, unpublished data), perhaps because these latter species also depend directly on conifers and perhaps yellow birch for foraging substrates. Social interactions, finally, comprise an important class of influences on local avian habitat use. We have documented and discussed above several examples of intra- and interspecific interactions that influence the local distribution of forest birds. The literature contains numerous examples of interspecific territoriality and dominance hierarchies (see Wilson, 1975), two other kinds of social influences on local settlement or resource-use patterns.

To understand the distribution of birds in our study site, we found that it was not sufficient to know only the characteristics of the environment where one or a few territories could be located. Regional influences on the avifauna are also important. Studies of isolated forest tracts as "islands" have emphasized the importance of forest area to the distribution of forest birds (Whitcomb *et al.*, 1981; Ambuel and Temple, 1983). Warbler species such as the American Redstart and Ovenbird, for example, are particularly susceptible to local extinction within small patches of forest that typically remain in a fragmented landscape (Whitcomb *et al.*, 1981). Increased nest predation, nest parasitism, and competition from open-country species are some of the factors thought to decrease populations of forest-dwelling warblers and other migrants. The abundance of such migratory warblers as redstarts at Hubbard Brook, then, are probably related in part to the extensive forested areas found near our study site. Another important characteristic of the landscape is the proximity of nonforested habitats to a particular forested area. The temporary colonization of our study area by American Robins (*Turdus migratorius*) in the early 1970s was due in part to an irruption there of defoliating caterpillars (Holmes and Sturges, 1975), but probably also in part to the proximity of typical robin habitats, i.e., human residential areas, 7–8 km away. Most thrushes (*Hylocichla* and *Catharus* species) and Rose-breasted Grosbeaks (*Pheucticus ludovicianus*) in our study area move immediately after breeding to cutover vegetation, when such is available nearby. There they feed on fruits of *Prunus, Rubus,* and *Sambucus* species (R. T. Holmes, unpublished data), and the proximity of habitats with fruits probably increases postfledging survivorship of these forest birds. Finally, birds can also track food resources (Cody, 1981). At Hubbard Brook, for example, the abundance of many passerines increased coincident with or immediately following an outbreak of a defoliating insect (Holmes and Sturges, 1975), and, as mentioned above, the robin has occurred in our study area only in those years of high food abundance (Holmes and Sturges, 1975).

A discussion of the scale of habitat selection would not be complete without mention of geographic-scale processes and circumstances, many of which involve long, e.g., evolutionary, time scales. The abundance and, consequently, habitat-use patterns of migratory passerine birds is influenced by circumstances where these species winter (see Keast and Morton, 1980), often thousands of kilometers away. Several bird species temporarily disappeared from our study area in the mid-1970s as a consequence of unusually harsh winters in the southern United States, where these species winter (R. T. Holmes *et al.*, in preparation). Bark-foraging species in our study area are rare and thinly dispersed compared with many of the migrant insectivores (R. T. Holmes, unpublished data) and are derived evolutionarily from a different stock of birds than the migrants (Sabo and Holmes, 1983). In general, evidence is accomulating that to better understand avian distributions and habitat selection one must also consider historical and geographical processes (e.g., Cody, 1975; Rabenold, 1978; Vuilleumier and Simberloff, 1980; Sabo and Holmes, 1983).

VII. CONCLUDING REMARKS

By measuring the dispersion patterns of passerine birds breeding in a superficially homogeneous forest, we have shown that bird species differ markedly in their responses to the "same" habitat. We further show that habitat selection by forest birds is a result of diverse biological processes that operate on different spatial and temporal scales.

This approach to studying bird habitat selection differs in several respects from many other recent studies, most of which have utilized a multivariate correlational approach between the occurrence of individuals of a species and the surrounding habitat (mainly vegetation) (James, 1971; Collins *et al.*, 1982; and others). We consider these latter approaches to be useful first steps in generating hypotheses about bird–habitat relations and of value as a management tool. However, we do not find the correlational results to be very informative about the processes or dynamics of habitat selection. For instance, studies limited to correlational analysis do not take into account the fact that food abundance or scarcity can overshadow bird distribution–vegetation relationships or that birds may be absent in one season and present in the next due to a variety of reasons (see preceding sections).

For more complete understanding and analyses of habitat selection by forest birds, we believe studies must consider bird dispersion and habitat selection in a population and community context. Knowledge of vegetation patterns, foraging behavior, diet, intra- and interspecific social interactions, predation, and other mortality factors are all necessary to understand observed patterns of habitat selection. Furthermore, avian ecologists need to move beyond the correlational approach and start testing their findings through experimental manipulations in the field. We feel that carefully controlled alterations of vegetation structure or even of the presence of other species will provide important new insights into the mechanisms and procedures by which forest birds select and utilize their habitats.

ACKNOWLEDGMENTS

This research was supported by a grant from the National Science Foundation to Dartmouth College, and the fieldwork was conducted in the Hubbard Brook Experimental Forest, West Thornton, New Hampshire, which is operated by the Northeastern Forest Experiment Station, Forest Service, United States Department of Agriculture. We acknowledge the able field assistance of M. E. Cohn-Haft, G. K. Roderick, S. Sargent, J. D. Schell, E. J. Snyder, C. D. von Dohlen, C. J. Whelan, and F. W. Sturges. We thank F. W. Sturges, D. R. Peart, N. L. Rodenhouse, B. Steele, and C. J. Whelan for discussion and comments on the manuscript.

REFERENCES

Able, K. P., and Noon, B. R. (1976). Avian community structure along elevational gradients in the northeastern United States. *Oecologia* **26,** 275–294.

Ambuel, B., and Temple, S. A. (1983). Area-dependent changes in the bird communities and vegetation of southern Wisconsin forests. *Ecology* **64,** 1057–1068.

Anderson, R. M., Gordon, D. M., Crawley, M. J., and Hassell, M. P. (1982). Variability in the abundance of animal and plant species. *Nature* **296,** 245–248.

Ardrey, R. (1966), "The Territorial Imperative." Atheneum Press, New York.

Aspey, W. P., and Blankenship, J. E. (1977). Spiders and snails and statistical tales: Applicaton of multivariate analyses to diverse ethological data. *In* "Quantitative Methods in the Study of Animal Behavior" (B. A. Hazlett, ed.), pp. 75–120. Academic Press, New York.

Bond, R. R. (1957). Ecological distribution of birds in the upland forest of southwestern Wisconsin. *Ecol. Monogr.* **27,** 351–384.

Bormann, F. H., and Likens, G. E. (1979). "Patterns and Process in a Forested Ecosystem." Springer-Verlag, Berlin and New York.

Clark, P. J., and Evans, F. E. (1954). Distance to nearest neighbor as a measure of spatial relationships in populations. *Ecology* **35,** 445–453.

Cody, M. L. (1975). Towards a theory of continental species diversity: Bird distributions over Mediterranean habitat gradients. *In* "Ecology and Evolution of Communities" (M. L. Cody and J. M. Diamond, eds.), pp. 214–257. Harvard Univ. Press (Belknap), Cambridge, Massachusetts.

Cody, M. L. (1978). Habitat selection and interspecific territoriality among the sylviid warblers of England and Sweden. *Ecol. Monogr.* **48,** 351–396.

Cody, M. L. (1981). Habitat selection in birds: The roles of vegetation structure, competitors, and productivity. *BioScience* **31,** 107–113.

Collins, S. L. (1983a). Geographic variation in habitat structure for the wood warblers in Maine and Minnesota. *Oecologia* **59,** 246–252.

Collins, S. L. (1983b). Geographic variation in habitat structure of the Black-throated Green Warbler (*Dendroica virens*). *Auk* **100,** 382–389.

Collins, S. L., James, F. C., and Risser, P. G. (1982). Habitat relationships of wood warblers (Parulidae) in northern central Minnesota. *Oikos* **39,** 50–58.

Cooke, F., Findlay, C. S., Rockwell, R. F., and Abraham, K. F. (1983). Life history studies of the Lesser Snow Goose (*Anser caerulescens caerulescens*) II. Colony structure. *Behav. Ecol. Sociobiol.* **12,** 153–159.

Davis, D. E. (1959). Observations on territorial behavior of Least Flycatchers. *Wilson Bull.* **71,** 73–85.

Davis, J. (1973). Habitat preferences and competition of wintering juncos and golden-crowned sparrows. *Ecology* **54,** 174–180.

Dhondt, A. A., and Eyckerman, R. (1980). Competition and the regulation of numbers in Great and Blue Tits. *Ardea* **68,** 121–132.

Ficken, M. S., and Ficken, R. W. (1962). The comparative ethology of wood warblers: A review. *Living Bird* **1,** 103–122.

Hamilton, W. D. (1971). Geometry for the selfish herd. *J. Theor. Biol.* **50,** 175–180.

Hilden, O. (1965). Habitat selection in birds. *Ann. Zool. Fenn.* **2,** 53–75.

Holmes, R. T. (1981). Theoretical aspects of habitat use by birds. *In* "The Use of Multivariate Statistics in Studies of Wildlife Habitat" (D. E. Capen, ed.), pp. 33–37. *U.S.D.A. For. Serv. Gen. Tech. Rep. RM* 87. Rocky Mountain Forest and Range Experiment Station, Ft. Collins, Colorado.

Holmes, R. T., and Robinson, S. K. (1981). Tree species preferences of foraging insectivorous birds in a northern hardwoods forest. *Oecologia* **48,** 31–35.

Holmes, R. T., and Sturges, F. W. (1975). Bird community dynamics and energetics in a northern hardwoods ecosystem. *J. Anim. Ecol.* **44,** 175–200.

Holmes, R. T., Bonney, R. E., Jr., and Pacala, S. W. (1979). Guild structure of the Hubbard Brook bird community: A multivariate approach. *Ecology* **60,** 512–520.

James, F. C. (1971). Ordination of habitat relationships among breeding birds. *Wilson Bull.* **83,** 215–236.

James, F. C., and Wamer, N. O. (1982). Relationships between temperate forest bird communities and vegetation structure. *Ecology* **63,** 159–171.

James, F. C., Johnston, R. F., Wamer, N. O., Niemi, G. J., and Boecklin, W. J. (1984). The Grinnellian niche of the wood thrush. *Am. Nat.* **124,** 17–47.

Johnston, D. W., and Odum, E. P. (1956). Breeding bird populations in relation to plant succession on the piedmont of Georgia. *Ecology* **37,** 50–62.

Keast, A., and Morton, E. S. (1980), "Migrant Birds in the Neotropics: Ecology, Behavior, Distribution and Conservation." Random House Smithsonian Inst. Press, New York.

Kendeigh, S. C., and Fawver, B. J. (1981). Breeding bird populations in the Great Smoky Mountains, Tennessee and North Carolina. *Wilson Bull.* **93,** 218–242.

Lack, D. (1933). Habitat selection in birds, with special reference to the effects of afforestation on the Breckland avifauna. *J. Anim. Ecol.* **2,** 239–262.

Landres, P. B., and MacMahon, J. A. (1983). Community organization of arboreal birds in some oak woodlands of western North America. *Ecol. Monogr.* **53,** 183–208.

Loyn, R. H., Runnalls, R. G., and Forward, G. Y. (1983). Territorial Bell Miners and other birds affecting populations of insect prey. *Science* **221,** 1411–1413.

MacArthur, R. H. (1958). Population ecology of some warblers of northeastern coniferous forests. *Ecology* **39,** 599–619.

MacArthur, R. H. (1959). On the breeding distribution pattern of North American migrant birds. *Auk* **76,** 318–325.

MacArthur, R., and MacArthur, J. (1961). On bird species diversity. *Ecology* **42,** 594–598.

McGee, V. (1978). "The Multivariate Package of BASIC Programs," 3rd ed. Amos Tuck School of Business Administration, Dartmouth College, Hanover, New Hampshire.

Martin, N. D. (1960). An analysis of bird populations in relation to forest succession in Algonquin Provincial Park, Ontario. *Ecol. Monogr.* **41,** 126–140.

Maurer, B. A., and Whitmore, R. C. (1981). Foraging of five bird species in two forests with different vegetation structure. *Wilson Bull.* **93,** 478–490.

May, P. G. (1982). Secondary succession and breeding bird community structure: Patterns of resource utilization. *Oecologia* **55,** 208–216.

Meagler, T. R., and Burdick, D. S. (1980). The use of nearest neighbor frequency analysis in studies of association. *Ecology* **61,** 1253–1255.

Miller, R. W., and Nero, R. W. (1983). Hummingbird–sapsucker associations in northern climates. *Can. J. Zool.* **61,** 1540–1546.

Minot, E. O. (1981). Effects of interspecific competition for food in breeding Blue and Great Tits. *J. Anim. Ecol.* **50,** 375–386.

Morse, D. H. (1977). Feeding behavior and predator avoidance in heterospecific groups. *BioScience* **27,** 332–339.

Mumme, R. L., Koenig, W. D., and Pitelka, F. A. (1983). Are acorn woodpecker territories aggregated? *Ecology* **64,** 1305–1307.

Murphy, M. T. (1983). Ecological aspects of the reproductive biology of Eastern Kingbirds: Geographic comparisons. *Ecology* **64,** 914–928.

Noon, B. R., Dawson, D. K., Inkley, D. B., Robbins, C. S., and Anderson, S. H. (1980). Consistency in habitat preference of forest bird species. *Trans. North Am. Wildl. Nat. Resour. Conf.* **45,** 226–244.

Pruett-Jones, M. A., and Pruett-Jones, S. G. (1982). Spacing and distribution of bowers in Macgregor's Bowerbird (*Amblyornis macgregoriae*). *Behav. Ecol. Sociobiol.* **11,** 25–32.

Rabenold, K. N. (1978). Foraging strategies, diversity and seasonality in bird communities of Appalachian spruce–fir forests. *Ecol. Monogr.* **48,** 397–424.

Robinson, S. K. (1981). Social interactions and ecological relations of Philadelphia and Red-eyed Vireos in a New England forest. *Condor* **83,** 16–26.

Robinson, S. K., and Holmes, R. T. (1982). Foraging behavior of forest birds: The relationships among search tactics, diet and habitat structure. *Ecology* **63,** 1918–1931.

Robinson, S. K., and Holmes, R. T. (1984). Effects of plant species and foliage structure on the foraging behavior of forest birds. *Auk* **101,** 672–684.

Rotenberry, J. T., and Wiens, J. A. (1976). A method for estimating species dispersion from transect data. *Am. Midl. Nat.* **95,** 64–78.

Sabo, S. R., and Holmes, R. T. (1983). Foraging niches and the structure of forest bird communities in contrasting montane habitats. *Condor* **85,** 121–138.

Schoener, T. W. (1983). Field experiments on interspecific competition. *Am. Nat.* **122,** 240–285.

Sherry, T. W. (1979). Competitive interactions and adaptive strategies of American Redstarts and Least Flycatchers in a northern hardwoods forest. *Auk* **96,** 265–283.

Shugart, H. H., and James, D. (1973). Ecological succession of breeding bird populations in northwestern Arkansas. *Auk* **90,** 62–77.

Simberloff, D. (1979). Nearest neighbor assessments of spatial configurations of circles rather than points. *Ecology* **60,** 679–687.

Slagsvold, T. (1980). Habitat selection in birds: On the presence of other bird species with special regard to *Turdus pilaris*. *J. Anim. Ecol.* **49,** 523–536.

Smith, A. J., and Robertson, B. I. (1978). Social organization of bell miners. *Emu* **78,** 169–178.

Smith, K. G. (1977). Distribution of summer birds along a forest moisture gradient in an Ozark watershed. *Ecology* **58,** 810–819.

Smith, K. G. (1982). Drought-induced changes in avian community structure along a montane sere. *Ecology* **63,** 952–961.

Sokal, R. R., and Rohlf, F. J. (1969). "Biometry." Freeman, San Francisco, California.

Sturm, L. (1945). A study of the nesting activities of the American Redstart. *Auk* **62,** 189–206.

Taylor, L. R., and Woiwod, I. P. (1982). Comparative synoptic dynamics. I. Relationships between inter- and intra-specific spatial and temporal variance/mean population parameters. *J. Anim. Ecol.* **51,** 879–906.

Taylor, L. R., Woiwod, I. P., and Perry, J. N. (1978). The density dependence of spatial behaviour and the rarity of randomness. *J. Anim. Ecol.* **47,** 383–406.

Vandermeer, J. (1981). "Elementary Mathematical Ecology." Wiley, New York.

Vuilleumeir, F., and Simberloff, D. (1980). Ecology versus history as determinants of patch and insular distributions of high Andean birds. *Evol. Biol.* **12,** 235–378.

Whitcomb, R. F., Robins, O. S., Lynch, J. F., Bystrak, B. L., Klimkiewitz, M. K., and Bystrak, D. (1981). Effects of forest fragmentation on avifauna of the eastern deciduous forest. *In* "Forest Island Dynamics in Man-dominated Landscapes" (R. L. Burgess and D. M. Sharpe, eds.), pp. 125–205, Springer-Verlag, Berlin and New York.

Whitmore, R. C. (1975). Habitat ordination of passerine birds of the Virgin River Valley, southwestern Utah. *Wilson Bull.* **87,** 65–74.

Whittaker, R. H., Bormann, F. H., Likens, G. E., and Siccama, T. G. (1974). The Hubbard Brook ecosystem study: Forest biomass and production. *Ecol. Monogr.* **44,** 233–254.

Williams, J. B., and Batzli, G. O. (1979). Competition among bark-foraging birds in central Illinois: Experimental evidence. *Condor* **81,** 122–132.

Wilson, E. O. (1975). "Sociobiology: The New Synthesis." Harvard Univ. Press (Belknap), Cambridge, Massachusetts.

Yarrow, R. M. (1970). Changes in redstart breeding territory. *Auk* **87,** 359–361.

Chapter 10

Habitat Selection in Amazonian Birds

JOHN TERBORGH
Department of Biology
Princeton University
Princeton, New Jersey

I. INTRODUCTION

There can be no doubt that Amazonia harbors the most diverse bird communities in the world. One comes inescapably to this conclusion, whether comparisons with other parts of the world are made at a continent-wide scale (Amadon, 1973; Keast, 1972), at a regional scale (Haffer, 1978; Moreau, 1966; Medway and Wells, 1976), or at a local scale (Pearson, 1975; Karr, 1980; Fogden, 1972; Bell, 1982). Well over 1000 species are known to reside within the lowland forests of the Amazon basin, and more than 500 of these can occur within a single locality, encompassing no more than 10 km^2 (Donahue *et*

HABITAT SELECTION IN BIRDS

ISBN 0-12-178080-5

al., unpublished manuscript; Terborgh *et al.*, 1984). Within Amazonia it is clear that avian diversity is greatest in the west in a broad belt that runs along the base of the Andes from southeastern Colombia to northern Bolivia (Haffer, 1978).

The imposing richness of the Amazonian avifauna, indeed, of the Neotropical avifauna as a whole, has prompted a considerable body of speculation, both on the evolutionary mechanisms by which so many species might have originated (Haffer, 1969, 1974) and on the ecological mechanisms involved in accommodating such large numbers of species to the environment (Terborgh, 1980). In this essay on habitat selection, evolutionary–historical issues shall lie outside our purview, and the discussion shall be confined to questions about the contemporary ecology of the fauna. Can the extraordinary numbers of species be explained simply in relation to a more complex environment? Are habitat gradients in the tropics more finely subdivided (MacArthur *et al.*, 1966; Karr, 1971), or are the niches of species smaller (Klopfer and MacArthur, 1960) or of guilds larger (Schoener, 1971)? Later, I shall attempt to answer these questions, as well as some others that focus more directly on the mechanisms of habitat selection in Amazonian birds. How important are the structure of the vegetation, the availability of appropriate food resources, and the presence of competitors in limiting the number of habitats species occupy? These latter questions are more difficult and the data needed to answer them less plentiful, so my responses shall be rather tentative.

II. OVERVIEW

Our understanding of the processes of habitat selection in Amazonian birds is in an extremely primitive state. This is due, above all, to the formidable technical barriers in doing so simple and basic a thing as a breeding bird census. At this moment no census of an Amazonian bird community has ever been published, and without even so much as a census, what can one hope to say about habitat selection? (Fortunately, a census has recently been completed, and though it is still unpublished, we shall later have a preview of some of the results.)

As to why there is as yet no published census of an Amazonian bird community, I must embark on a short digression. Twenty years ago when I began to study birds in Peru, the state of ornithological knowledge there was roughly comparable to what it was in the United States during Audubon's time. There was no field guide nor even a general compilation of distributional information. Scores of species had never been illustrated, and totally unknown species were being discovered every year. Identification by sight was a chancy proposition at best; a few boldly marked or widespread species could be recognized with certainty, but field notes and even descriptions of birds in the hand, no matter how detailed, failed to discriminate among many of the others. There was no satisfactory alternative to collecting a reference specimen or pair of every species one wanted to identify with confidence. Each year one went home to the museum with a trunk full of specimens, and another increment of progress was achieved when names could finally be assigned to them. By this slow process, a working knowledge of the avifauna could gradually be acquired, but it was a knowledge inadequate to the task of doing precise work on habitat selection. There were simply too many obscure or cryptic species, and direct observation failed to reveal them all.

Nothing made this more clear than working with mist nets. Even after weeks of making observations in a tract of forest, nets would produce an embarrassing welter of new finds and would yield a picture of relative abundances that was totally at odds with one's direct experience. For this reason, ornithologists working in the tropics took to using mist nets with unreserved zeal. The results were obviously free of observer bias and, *ipso facto,* unassailable. Several studies of tropical bird communities relied on netting almost exclusively (Terborgh, 1971; Lovejoy, 1975; Fogden, 1972; Karr, 1971, 1980).

But, like most mechanical sampling techniques, netting has its limitations, rather severe ones. Acquired net shyness is a detriment to long-term studies, for example. I have found that the catchability of a population drops by 80–90% after a week of dawn to dusk netting. Even after a lapse of one year, experienced birds have about a 30% lower chance of being captured than naive birds (J. Terborgh, unpublished results). The long-term effects of net shyness are not the only problems with the technique. It is a very slow way to sample birds (per hour of investigator effort), and the sample sizes one accumulates are rather pathetically small relative to the diversity of the community. (A standard sample of 100 individuals, for example, may easily contain 40 species, for a mean of 2.5 individuals per species.) By far the most serious limitation for community studies, however, is the fact that even very large netted samples (i.e., >500 individuals) will generally not include more than 40% of the bird species in a tropical forest (Terborgh, 1977). It is obvious then that detailed analyses of community structure call for a better method.

A better method does exist; indeed, it has been routinely practiced by professionals and amateurs alike for over 40 years but mainly in temperate environments. This is the technique of spot mapping. A few attempts to spot map tropical forest bird communities have been published, but the results give an incomplete picture (MacArthur *et al.*, 1966; Orians, 1969). Numerous species known to be present simply did not appear in the census, and the resulting estimates of species diversity fell far short of reality (see Terborgh and Weske, 1969). These deficiencies were so apparent that further attempts to census tropical bird communities by spot mapping were not made. The impression became well entrenched among ornithologists that the method was inadequate to the task.

Having now gained considerable experience in spot mapping tropical bird communities, I am of the impression that the perceived inadequacies of the method were illusory; the real problem was with the investigators. The people who were studying tropical diversity at the time were in too much of a hurry. There were burning hypotheses to be tested, and quick results were desired. There was also the feeling that if one were armed with sharply focused questions, the answers could be obtained in a few weeks. The false assumptions behind this zealous attitude are now clearly apparent.

The first attempts to conduct spot-map censuses of tropical forest plots failed for two reasons. The first was that the investigators were unprepared. The vocalizations of many species were not known at the time, so only the most common and conspicuous birds could be censused. Second, the plot size used in these investigations was too small by at least an order of magnitude. These are key points, so it will be useful to expand on both of them.

A comprehensive knowledge of the vocal repertoires of all species in the community is a *sine qua non* for the successful application of the spot mapping method. In a few square kilometers of lowland forest in Panama or Costa Rica one can expect to find over 300 species of birds (Slud, 1960), and in western Amazonia, a similar area will contain over 500 species (Terborgh *et al.*, 1984). Obviously, no one can learn the songs, alternate songs, and call notes of several hundred species in a few weeks. Each sound must be tracked to its source, typically by making a tape recording and playing it back to attract the singer. Many species are rare, and some do not respond to playback. At our study site in southeastern Peru, where 530 bird species have been recorded, it took 10 years to acquire a comprehensive knowledge of bird vocalizations.

The second key point I wish to make pertains to the matter of scale. What range of plot sizes is appropriate for censusing tropical habitats? By North American standards, a plot of 5–10 ha is considered adequate, and one of 15 ha verges on the extravagant. Now, let us consider what this means in terms of the numbers of individuals of each species censused. A 10-ha plot of eastern deciduous forest can be expected to contain 10–15 pairs of the more common species and 2–10 pairs of most of the rest (Holmes and Sturges, 1975). Certainly, this is not overkill as an exercise in sampling. Nevertheless, for the sake of consistency, we can ask for a similar intensity of sampling in selecting an appropriate plot size for a tropical census. About the only information in the literature on the territory sizes of Amazonian birds is contained in Munn and Terborgh (1979). There it is pointed out that the common small insectivores of the forest understory (antwrens and ant vireos, tropical counterparts of warblers and vireos) possess territories of 6–8 ha. To include enough area to contain 10–15 pairs of these birds, one thus needs a plot of 100 ha. Not being aware of this 10-fold scaling difference, previous workers attempting to census tropical forest bird communities used plots that were barely of average size even by temperate standards, i.e., 1–5 ha (MacArthur *et al.*, 1966; Orians, 1969; Karr, 1971). In some extreme cases the plots used were smaller than the probable territory size of the most common species. When the community is examined on so small a scale, all but the most abundant species effectively become invisible or are passed off as "irregular species" (Karr, 1971). To avoid such misapprehensions, tropical habitats must be studied on a far greater scale than has heretofore been appreciated.

The previous commentary presents only one side of my overview of the primitive state of knowledge of the habitats of Amazonian birds—the ornithological side. The other side is botanical. To the ornithologist whose formative experience has been in the temperate zone, tropical vegetation is, at best, confusingly unfamiliar and, at worst, incomprehensible. Few ornithologists have made the effort to learn tropical botany; yet a simple drawing of analogies from one's temperate zone experience does not lead to an adequate understanding of tropical habitats. Ornithologists alone are not to blame for this deficiency, for the botanists have not been there to lead them. I know of not one single work dedicated to describing Amazonian plant formations. Such attempts as there have been to classify tropical vegetation are exceedingly rudimentary (Richards, 1952; Walter, 1973). The much used Holdridge (1967) system, for example, actually classifies climates, not vegetation. What is termed "Tropical Moist Forest" extends over more than a million square kilometers in Amazonia and includes a score or more of compositionally and structurally distinct vegetation formations.

In lower Amazonia the major types are termed "terra firme" (high ground) forest, igapo (seasonally flooded), and varzea (tidal swamps) (Black *et al.*, 1950; Richards, 1952; Lovejoy, 1972, 1975). Here at least are the beginnings of a useful classification, though one is still hampered by oversimplification.

In western Amazonia, where igapo and varzea are lacking, the botanical literature has even less to offer. In his introduction to the "Flora of Peru," the distinguished phytogeographer Weberbauer (1936) provides a 69-page account of the vegetation of the country, of which only 4 pages refer to lowland Amazonia. In this cursory passage, only two plant formations are mentioned, forest and "matorral" (literally shrubland, thicket), the latter term taking in the whole complex of successional and flood-disturbed vegetation in flood plains.

The most valient effort to date to define habitats in western Amazonia is by a group of ornithologists (Parker *et al.*, 1982), who, in frustration at the lack of guidance provided by their botanical colleagues, took matters into their own hands. The result is a simple scheme of six habitats, and from an ornithological perspective, this represents a major advance. The six are, in brief: high-ground forest, seasonally inundated forest, edge–second growth (man modified), river margins–beaches, riparian successional vegetation, and oxbow lakes. As we shall see, this is good so far as it goes, but a somewhat more detailed classification than this is needed to fill out a comprehensive picture.

I shall now proceed into the main body of the chapter. The first matter to occupy our attention will be the vegetation, and the question of how one is to recognize and define distinct habitats in Amazonia. I shall then continue with a description of the bird community at Cocha Cashu, a locality in southwestern Amazonia I have been studying in collaboration with students and colleagues for 11 years. Finally, in a more tentative vein, I shall try to isolate some of the mechanisms of habitat selection in this community.

III. VEGETATION TYPES IN WESTERN AMAZONIA

In this section I shall describe some of the vegetative features of western Amazonia. This will by no means be a comprehensive guide, for I am speaking strictly from my own limited experience. Furthermore, I shall only mention plant formations found in well-watered regions that would be classified as "Tropical Moist Forest" or "Tropical Wet Forest" under the Holdridge (1967) system. Those two zones cover much of the area under discussion, although there are scattered pockets or bands of dry forest, xeric scrub, and savannah within or adjacent to the more mesic forest types.

A. Landscape of Western Amazonia

Virtually all of western Amazonia east of the Andean foothills lies on relatively recent alluvial sediment, mostly derived from the Andes themselves. The land is flat or gently dissected by small streams. From the air the forest appears as a variegated and textured carpet that stretches in all directions to the horizon. The only features that catch the eye are major streams and rivers, with their associated sandbars, levees, backwaters, and oxbow lakes. Rivers have in fact created the whole landscape, so a classification of landforms and vegetation is most reasonably organized around the temporal and spatial relationships to fluvial activity.

As a first subdivision, it seems most natural to recognize the dichotomy between uplands and lowlands. By uplands, I refer to terrain that is elevated above the highest flood levels. Some 60–80% of western Amazonia consists of uplands, though very little of this great area lies more than 100 m above high water. The flora of these uplands is extremely diverse, the composition of the forest varying greatly from place to place in accordance with differences in soil type and rainfall (Gentry, 1982). Much of the area is covered by common soil types (sandy, silty, lateritic, etc.), but in certain regions there are distinctive or extreme soils that support sharply differentiated vegetation formations. For example, in the north (eastern Ecuador and northern Peru) there are pockets of sterile white sand soil, which, with the associated blackwater river systems, carry floristically distinctive forest types (Kinzey and Gentry, 1979). In the south (southern Peru, northern Bolivia) there are somewhat more fertile brown sand soils that in spots are heavily invaded by bamboo. At present, however, we know extremely little about either the flora or bird communities of the uplands, for a very simple reason. Western Amazonia, even now, is nearly roadless. The only practical channels of transportation are the rivers, and these typically lie in broad meander belts. From a random point on a river bank, the nearest upland may be several kilometers inland. Supply and transportation problems become acute away from rivers, and so neither local villagers nor biologists tend to stray far from their banks. Most of what we know thus relates to lowland, or floodplains.

Huge areas in Amazonia are subject to flooding, at least occasionally. Flooding may result directly from the overflow of rivers, or indirectly from the saturation of ground water tables. Of importance to the vegetation is the periodicity of flooding. The floodplain of the middle Amazon and its tributaries is continuously under water for nearly half the year. These areas are occupied by igapo forest. The farther upstream one goes, the less the water regime depends on averaging over the whole basin, and the more it depends on local precipitation. Since western Amazonia is basically a headwaters region, most of the rivers tend to rise and fall many times during a rainy season. The flooding thus tends to be of short duration, though of frequent occurrence. Under this regimen, one finds no counterpart of the igapo forest. Floodwaters normally drain away promptly from the levees immediately fronting on rivers, but further inland, often at the lateral fringes of the floodplains, the drainage is slow, giving rise to seasonal swamps.

Of even greater importance than the intensity and periodicity of flooding is the dynamic action of the rivers themselves. Nearly every river in upper Amazonia meanders. The dynamics of this process are surprisingly rapid. I have made measurements for the last 8 years on a bend of the Manu River near the field station at Cocha Cashu. At the point of greatest curvature, the bank on the outside of the bend loses approximately 20 m/year. At this rate the bend will progress a whole kilometer in just 50 years. Simple extrapolation suggests that the entire floodplain is swept out by the meandering river every 500–1000 years. Most of the vegetation in the floodplain is consequently undergoing primary succession, and rather little of it can be considered climax in the sense that the species composition is no longer changing in a progressive fashion.

As rivers erode their banks on the outside of bends, they create new land on the inside. A meander loop thus encloses a long tongue of fresh sediment, which is youngest distally and oldest toward the base. If one cuts a longitudinal transect from the tip of a tongue directly inland one passes through progressively older vegetation. In the Manu basin these tongues are typically 1–2 km in length. From the measurements mentioned above, one can thus surmise that the average loop develops for something like 100–200 years. Much of the vegetation in the meander belt is consequently not older than this.

B. Plant Succession in Meander Belts

The primary plant succession that occurs in meander loops is characterized by a sequence of progressively higher and longer-lived canopies. Open beaches are colonized in the dry season by a number of herbaceous annual species that are later buried or swept away by floods. The first perennial stage in the succession is monopolized by *Tessaria,* a fast growing composite tree that matures in 3–4 years. *Tessaria* invades the highest upper beaches, forming almost pure monotypic stands just riverward of the front levee. Along the Manu, these *Tessaria* stands are usually only 10–30 m wide, but may be 1–2 km in length, forming the outer vanguard of woody vegetation around the tips of meander loops. In 3 years the *Tessaria* reaches a height of 8–10 m. After that, vertical growth slows appreciably. In 3–5 years the stands are invaded by a tall cane (*Gynerium*), which spreads vegetatively from the next older belt of vegetation. Mixed stands of *Tessaria* and cane persist for several years more as *Cecropia* saplings grow up through them.

When *Cecropia* trees begin to overtop the *Tessaria* canopy, the latter dies and the succession enters what I term the *Cecropia–Gynerium* stage. This is somewhat longer lived, persisting for 5–10 years, and therefore covers a greater area than *Tessaria.* The cane, which reaches a maximum height of 10–12 m, eventually succumbs to increasing shading and is replaced by a more diverse understory of tree saplings and broad-leafed herbaceous plants (*Musaceae, Zingiberaceae, Marantaceae*).

Of the several tree species that are able to grow up through the cane and *Cecropia,* the most vigorous and prominent is a fig (*Ficus insipida*). These easily surpass the *Cecropia,* which attain heights of only 20–25 m. Unlike their forerunners in the succession, the figs are big long-lived trees, reaching, at the extreme, a diameter of 1.5 m and a height of around 40 m. For a time, figs tend to monopolize the canopy, but they are eventually joined by roughly equal numbers of two meliaceous trees, *Cedrela odorata* and *Guarea guidonia.* At this point the succession has reached what I term the *Ficus–Cedrela* stage. The understory is still rank with *Heliconia* and other broad-leafed herbs, but by now there is an irregular middle story composed of a number of small to medium-sized trees.

Viewed from the river, the vegetation on an extending meander tongue is conspicuously zoned. The crowns of the successive canopies, *Tessaria–Cecropia–Ficus insipida,* are fully visible, each above the previous level, like a chorus on risers. The boundaries between the zones are quite distinct, because the transitions between successive stages are of brief duration relative to the stages themselves.

We do not know yet how long the *Ficus–Cedrela* stage persists, but I would guess that it is within the range of 100–200 years. Because it is so much longer lived than the earlier stages, *Ficus–Cedrela* stands cover very large areas in the meander belt. Many of the stands are in fact recycled by the river before succession has gone to the next stage. Barring this fate, a stand does in time succeed to a much more diverse type of forest, for the *Ficus–Cedrela* canopy does not replace itself. Subsequent successional stages that are still on accreting land and subject to periodic inundation, I call simply "late successional forest."

Even more diverse, and presumably still older, forests occur within the meander belt on patches that have by chance escaped the ravages of the river for periods of several hundred years or more. Having persisted so long, these patches have built up through occasional increments of flood-borne sediment to the point that further inundations are quite infrequent (i.e., once in several to many years). The vegetation of these patches I term "high-ground forest." In contrast with upland forest, which occupies weathered soils altogether out of the floodplain, high-ground forest lies on deep, young alluvium in irregular disjunct patches within the floodplain. Presumably because of the optimal soil conditions, high-ground forests can attain great stature. Certainly, the most magnificent forests I have seen anywhere in Amazonia have been of this type. With frequent emergents reaching 50–60 m and a more nearly closed second stratum at 25–35 m, this is the classical multitiered tropical forest.

C. Lacustrine Succession

We have now reviewed the sequence of stages in riparian succession; of secondary importance is the succession that takes place in oxbow lakes. Every so often the river pinches off and short-cuts a meander loop, leaving a lake to occupy the former channel. Given time and enough sediment-laden floods, these lakes also yield to succession. Marshes at first encroach from the ends and sides, and later shrubs and, finally, trees invade the marsh. This process proceeds at a much slower pace than riparian succession, and some lakes survive for hundreds of years, as one can judge by examining the forest along their margins. Nevertheless, the inevitable floods and the attendant sedimentation eventually reduce these lakes to shallow depressions that no longer hold water through the dry season.

Another species of fig (*Ficus trigona*) comes to dominate the vegetation at this stage, forming nearly monotypic closed-canopy stands. Beneath the figs, which are only 15–20 m tall, are scattered smaller trees and a rank understory of 3-m *Heliconia*. I shall call this habitat "fig swamp." These swamps must persist for a long time, because the depressions they fill receive sediment at a very low rate. Eventually, however, the conditions must reach a stage at which other tree species overtop the figs, thereby transforming the stand into a type of late successional forest.

Fig swamps constitute a minor feature of Amazonian vegetation relative to swamps dominated by *Mauritia flexulosus,* a stately 25-m-tall fan-leafed palm. *Mauritia* swamps, called aguajales in Peru, occupy large areas in the Ucayali and Madre de Dios river basins. Unlike *Ficus trigona* swamps, which occur on oozy, unconsolidated muck, aguajales develop on firm sediment, but beyond this, I know little about how they form. They occupy seasonally or permanently inundated depressions, often well back from the main river channel.

This completes our catalogue of major vegetation types in western Amazonia. I do not claim that the list is complete, but it does include all the formations that will be mentioned in the ensuing discussion. One could easily define further subdivisions in the classification, especially in the later successional stages, but this would entail delving into floristic details that would be out of place in the present discussion. The scheme I have presented, imperfect as it may be, will serve nevertheless the purpose of facilitating a first-order overview of the match between birds and vegetation in a part of the world where neither are as yet well known.

IV. ORGANIZATION OF THE BIRD COMMUNITY AT COCHA CASHU

One of the few places in the western Amazonia where the habitat relationships of birds have been subjected to systematic examination is at Cocha Cashu Biological Station in the Manu National Park of southeastern Peru (11°51′S; 71°19′W). The station is located in the Manu River floodplain on the rim of an oxbow lake. All phases of both the riparian and lacustrine successions described above are represented in the immediate area (see Frontspiece). An extensive trail network provides access to about 5 km^2 of terrain surrounding the station, roughly half of which is occupied by high-ground forest and half by successional vegetation of various ages (Terborgh, 1983).

Between August 15 and November 15, 1982, five of us (C. Munn, T. Parker, N. Pierpont, S. Robinson, and J. Terborgh) undertook a census of the birds resident in a heterogeneous 110-ha plot. The plot included 82 ha of high-ground forest and 28 ha of successional and edge habitat (Table I). The results of the census, along with an account of the methods employed, will be published in full elsewhere. Here, I shall present summary data that bear on the issues of tropical diversity and habitat selection.

A. Species Diversity

First, let us consider simply the numbers of species found to be resident in each type of vegetation. By the term resident I do not imply necessarily that a species nested within the plot. In some cases nesting was confirmed, but more

TABLE I

Habitat Characteristics and Avian Diversity in the Seral Stages of Riparian Succession in Western Amazonia

	Habitat characteristics					
Seral stage	Stature (m)	Plant diversity	No. plant layers	Available resources	No. ha censused	No. resident birds
Open beach	0.5	Low	1	Insects, seeds (seasonally only, as subject to inundation for 7–9 mo/year)		2–5
Tessaria	8–10	1 Species	1	Insects, lizards (subject to intermittent flooding in wet season)	2	21
Gynerium–Cecropia	12–20	Low	2	Insects, fruit (seasonal), nectar (seasonal), few small vertebrates (seasonally inundated)	8	49
Fig swamp	15–20	Low	3	Insects, fruit (seasonal), nectar (year-round), small vertebrates, snails	5	72
Ficus–Cedrela	30–40	Low to intermediate	3	Insects, nectar (year-round), fruit (seasonal), seeds, small vertebrates	10	127
Late successional	35–50	Intermediate to high	4	Insects, nectar, fruit, large & small vertebrate prey, seeds, nuts	3	113[a]
High ground	35–55	Very high	5	Full spectrum of resources available on year-round basis	82	224

[a] Anomalously low species total due probably to the small area of this habitat in the census plot (3 ha).

often it was inferred. Apart from those species that probably nested within the plot, there were many we know did not nest there, but which were classified as residents nevertheless, because all or some part of the plot was included within a larger home range (e.g., swifts, parrots, fruit crows) or territory (large raptors). Such species comprise only a minor fraction (<10%) of the total for each habitat.

The entire 110-ha plot harbored at least 328 resident species, not including kingfishers, swallows, or other birds that forage predominantly in or over water. Of this total, 236, or 72% were found to occupy the high-ground forest. This number is radically higher than reported for censuses of mature tropical forest in Panama and Costa Rica (40 species, for a plot of unspecified size on Barro Colorado Island, Panama: MacArthur *et al.*, 1966; 56 species for a 2-ha plot in the former Panama Canal Zone: Karr, 1971; 64 species for a 1.2-ha plot at La Selva in Costa Rica: Orians, 1969). Some of the difference results from the fact that the avifauna of Amazonia is decidedly richer than that of the Central American lowlands; the rest is due to the inadequate methodology of the earlier studies, as explained above.

The richest forest bird communities of temperate North America contain only about 40 species; how is it possible to explain the presence of six times as many species in a tropical forest? In discussing their Panamanian results MacArthur *et al.* (1966) concluded that the additional species could be accounted for solely on the basis of the structurally more complex habitat if it were assumed that the Panamanian birds were recognizing five foliage layers versus the three that gave the best fit with temperate bird communities. Treated thus, results from the two regions could be scaled together on a plot of bird species diversity versus foliage height diversity. The greater diversity of tropical bird communities, by this view, was attributable to the presence of more foliage layers in the tropical habitats and a finer vertical partitioning of those layers by the resident birds. The Amazonian forest, however, is not apparently more complex structurally than the Panamanian forest, yet it contains six times as many species as MacArthur *et al.* reported in their census. Therefore, unless Amazonian birds are recognizing seven or eight vertical layers in the forest, the greater structural complexity of the habitat offers, at best, only a partial explanation of the elevated avian diversity.

The same conclusion was reached via different means by Terborgh and Weske (1969). They studied a series of man-made habitats in Peru and found that structurally simple vegetation (e.g., a coffee plantation consisting of coffee bushes shaded by *Inga adenophylla* trees) held many more birds than even the most complex temperate habitats. Clearly, structure alone is not the answer.

At about the same time, a further suggestion was offered by Orians (1969) and Karr (1971). They observed that Central American bird communities contained numbers of species that had no evident ecological counterparts in the temperate avifauna. Ant-following insectivores were one such group of species, and obligate frugivores were another. Meanwhile, Schoener (1971) noticed that large-billed insectivores were far more prominently represented in tropical than in temperate bird communities and proposed that this could be understood on the basis of his earlier finding that tropical insect samples contained many more large insects than comparable samples from temperate North America (Schoener and Janzen, 1968). To these suggestions must be added the earlier one of Klopfer and MacArthur (1960) that the elevated diversity of tropical bird communities could be understood if the species possessed smaller niches on the average than their temperate counterparts.

We now have three proposals to consider: (1) that tropical communities contain whole guilds of species that are not represented in counterpart temperate communities, (2) that tropical guilds may include species covering a broader range of sizes and, in particular, larger insectivores, and (3) that the niches of tropical birds are smaller than those of temperate birds.

These proposals lay neglected for nearly 10 years, perhaps because of the implicit difficulty of interpreting tropical–temperate comparisons. There are too many variables involved: latitude, climate, structure of the habitat, greatly differing diversities of trees, insects, etc., not to mention the very distinct historical and geographic origins of the respective avifaunas (Terborgh and Robinson, 1985). Even if tropical versus temperate comparisons are not fully interpretable, they may still be able to shed light on the three proposals.

I have reported on such a comparison (Terborgh, 1980). To control as well as possible for the inevitable structural differences between tropical and temperate forests, I selected as the temperate site an extensive virgin floodplain forest in South Carolina, lamentably, the only one of its kind in existence. The canopy of this forest averages 35 m tall, with emergents to 45 or 50 m. In stature the forest is thus not inferior to many tropical forests. It also rivals many tropical forests in structural complexity, having up to four distinct tiers of crowns (Terborgh, 1984a), as well as numerous lianes and a moderate load of epiphytes. Structurally simpler tropical sites I have studied have held well over 100 bird species, yet this South Carolina forest has only 40.

The tropical site was a square kilometer of high-ground forest near the Rio Llullapichis at 9° S latitude in central Peru. It held at least 207 bird species. A breakdown of both communities by guild revealed that every guild present in the temperate site was also represented in the tropical site, usually by many more species and that there were a number of additional, exclusively tropical, guilds. A partial list of these includes arboreal mast eaters (seed predators), terrestrial frugivores, insectivores that (1) follow army ants, (2) search vine tangles, and (3) glean dead leaves, and omnivores that feed on mixtures of fruit, insects, and nectar. Altogether, these additional guilds contributed 56 species, or 34%, of the 167 species by which the tropical site exceeded the temperate one. The proposal of Orians and Karr is thus upheld, but it clearly tells only part of the story.

A comparison of the largest and smallest birds in the 16 guilds the two communities held in common upholds Schoener's proposal as well. Although the largest species in several tropical guilds did exceed their temperate counterparts, in general, the differences were small so that an increased size range of tropical guilds could account for only 29 species, or 17% of the excess diversity.

By far the largest component of the excess tropical diversity, 82 species (49%), consisted of members of guilds that were common to both communities and that were within the size ranges circumscribed by their temperate counterparts. This component of the tropical excess was attributed to increased species packing.

At this point, because of the welter of variables potentially influencing the comparison, it becomes impossible to separate cause from effect. Is the species-packing really tighter in Amazonian guilds as Klopfer and MacArthur (1960) proposed? One's intuition says yes, because the average species is much less common and its ecological role much more narrowly circumscribed than in equivalent temperate guilds (Pearson, 1977b). Nevertheless, uncontrolled structural differences between the two sites, as well as other hidden variables, could mask another part of the story. In any case, there can be little doubt that each of the proposals we have considered contains an element of the answer to the riddle of tropical diversity and that none of them contains the whole answer. The riddle is thus akin to a jigsaw puzzle with many pieces, each of which must be carefully put in place after due consideration. Much the same can be said of the problem of habitat selection, as we shall see in the following section.

B. Habitat Selection

Having digressed at some length on the unrivalled avian diversity of Amazonia, I return now to the main theme of habitat selection. How do the 328 resident species of the Cocha Cashu study plot sort themselves into the available habitats (Table I)? The high-ground forest holds more species than any of the seral habitats, in fact, it holds as many as all the other habitats combined. Some of the extra species can be attributed to the much greater area of high-ground forest in the plot. The rest must be due to intrinsic difference in the environments offered by the habitats themselves.

Let us consider what these differences might be. A bird's habitat provides the following requisites of life: a suitable microclimate, appropriate foraging substrates, food resources, adequate nest sites, and cover. A habitat that fulfills these requisites may not be inhabitable, however, if it contains an excessive number of competitors, predators, or parasites.

We can now ask about the possible contributions of these factors to habitat selection in the Cocha Cashu bird community. Some can be eliminated from further consideration, because they show little variation between habitats. Cover for example, is abundantly available in most tropical vegetation, and so it is hard to see how it could be crucial to between-habitat discrimination. The same might be claimed for predators and parasites. Bird-eating hawks range through all habitats at Cocha Cashu, as do snakes, toucans, and monkeys, the principal nest predators. Exo- and endoparasites are potentially a factor, but we know nothing about them, much less how they might vary from one vegetation type to another. Brood parasites are rare (one species) and affect only about 1% of the species present. A factor that may eventually prove to be of much greater importance is the availability of a suitable microclimate, especially, for insectivores, the light environment (Pearson, 1971; Fitzpatrick, 1980). However, we are only on the threshold of understanding how microclimatic variation may influence species behavior and foraging efficiency.

The nesting habits of tropical birds are often narrowly stereotyped in ways that could easily lead to habitat specificity (Tomoff, 1974). Some species nest only over streams, others only in trees associated with certain species of stinging ants, others in the crowns of spiny palms, etc. (Koepcke, 1972). In this case it is legitimate to ask which came first in evolution, the habitat or nesting habit? If the nesting behavior evolved first, it could limit habitat occupancy. In certain extreme cases such as obligate cliff nesters, it is clear that the availability of nest sites is the limiting factor; but for most species this is not clear at all.

It is evident from the previous discussion that several of the factors that influence habitat quality cannot be evaluated further for lack of appropriate information. There remain in the list, however, three habitat quality factors that can be subject to some degree of analysis. These are between-habitat differences in foraging substrates, food resources, and competitors. I shall now take up each of these in turn, presenting evidence which suggests that all three play a role in explaining the observed patterns of habitat occupancy.

1. Role of Structure

An association between increasing structural complexity of the habitat, for example, along some environmental gradient, and increasing bird species diversity has been demonstrated in many studies, both in and out of the tropics (MacArthur and MacArthur, 1961; MacArthur *et al.*, 1966; Karr, 1971; Willson, 1974; Recher, 1969; Tomoff, 1974; Cody, 1975; Terborgh, 1977). The consistency of this result, as well as the ability to predict the presence or absence of particular bird species from structural measurements of the habitat (MacArthur *et al.*, 1962; Cody, 1978), has led to a widespread conviction that the structural qualities of a habitat are paramount in determining its occupancy by birds (James, 1971; Willson, 1974; Sabo, 1980).

My own feeling about this is that structure per se has been overemphasized in studies of avian habitats to the neglect of other, perhaps less easily measured factors that may be of equal or greater importance. Cody (1981) has sounded a similar precautionary note in a recent review. First, it is evident in the list of requisites discussed earlier that a habitat must meet multiple criteria of acceptability, of which structural adequacy is only one. (Since several criteria must be satisfied simultaneously, one is logically unjustified, without further evidence, in assuming that one criterion is more important than any other.) Second, in the association between avian diversity and structural complexity of the habitat, one does not find a cause and effect relationship, merely a seductive correlation. Third, there are reasons for thinking that some critical habitat features which are not so readily measured, in particular, the diversity of food resources, may be closely correlated with the structural complexity of vegetation. And fourth, to offer a counter example, it appears that in certain biogeographically isolated regions, the most diverse avian communities are not in the most structurally complex habitats (Patagonia: Vuilleumier, 1972; West Indies: Terborgh and

Faaborg, 1980; Tasmania: Ridpath and Moreau, 1966; Canary Islands: Chapter 3 this volume). All these points urge caution upon us in evaluating the role of structure in avian habitat selection.

In proceeding, I shall first demonstrate that in the community under study, avian diversity is correlated with the structural complexity of the habitat in the usual way. Then, I shall show that there are additional strong correlations with resources and competitors that should also be taken into account in reaching a final assessment.

Comparing the bird communities of the seral stages of riparian succession, we find that the number of resident species increases dramatically and probably monotonically with increasing age and stature of the vegetation (Table I). From the information provided in the table, it is apparent that the number of bird species per habitat is correlated with (1) the stature of the vegetation, (2) the number of superimposed strata of plants, (3) plant species diversity, (4) the variety of resources offered by the habitat, and (5) the temporal continuity of resource availability. Each of these intercorrelated variables could plausibly contribute to the trend in diversity, but in the absence of further information one is at a loss to say anything more. An independent role of structural complexity cannot be denied, but neither has one been conclusively demonstrated.

The situation in temperate habitats is somewhat simpler. Insectivores comprise a large majority of all bird species in the breeding season, although many of them migrate or alter their habitat use patterns at other times of year. A lack of year-round continuity in resource production thus may not carry the same consequences as it does in the tropics where many species occupy permanent territories.

TABLE II

Matrix of Number of Species in Common (Upper Right) and Percentage Similarity (Lower Left) for Resident Birds of the Habitats in Riparian Succession at Cocha Cashu

	Habitat[a]				
Habitat	HG	LS	F–C	G–C	T
HG	—	103	99	18	1
LS	61	—	80	24	0
F–C	56	67	—	32	2
G–C	13	30	36	—	9
T	<1	0	3	26	—

[a] HG, high ground; LS, late succession; F–C, *Ficus–Cedrela*; G–C, *Gynerium–Cecropia*; T, *Tessaria*.

TABLE III

Total Number of Resident Bird Species, Number of Habitat Exclusive Species, and Percentage of Exclusive Species in the Stages of Riparian Succession at Cocha Cashu, Manu National Park, Peru

	Successional stage				
	High-ground forest	Late successional forest	*Ficus–Cedrela* forest	*Gynerium–Cecropia* zone	*Tessaria*
Total number species	236	113	127	49	21
Number exclusive species	92	7	12	9	13
Percentage of exclusive species	39	6	9	18	62

To the extent that succession results in monotonic changes in several important habitat variables (Table I), it can be expected that the bird communities of adjacent stages will be more similar than those of more widely separated stages. This proves to be unambiguously true, as indicated by an orderly similarity matrix (Table II). Habitat occupancy patterns are thus highly contagious, and one finds few instances of disrupted or bimodal distributions along the successional gradient. This is reassuring, for it is consistent with one's intuitive judgment that the variables under consideration are indeed relevant ones.

Another way of addressing the issue of habitat specificity is to ask which habitats harbor the largest proportions of exclusive species (Table III). Here the results could not have been so easily predicted. Habitat-restricted species represent only a small fraction of the total in each habitat except at the ends of the gradient. This indicates that between-habitat turnover is not especially high, probably no higher than on a comparable temperate successional gradient. Between-habitat diversity (β-diversity) therefore does not make a major contribution to avian diversity in Amazonia, contrary to conclusions previously reached for Central America (MacArthur *et al.*, 1966).

Late successional forest and the *Ficus–Cedrela* zone are occupied mainly by species common to the high-ground forest; neither contains many exclusive species. It is the ends of the successional gradient that carry the most distinctive communities. Many more species occupy the high-ground forest than any other stage in the sequence, and it is these "extra" species that constitute the unique high forest element. This is not so at the other end of the gradient where, in spite of declining numbers of species, a much greater proportion of them are exclusive to a single stage. This is particularly true of the *Tessaria*. The canopy provided by the even-aged stands is extremely uniform, like that of a young pine plantation. In contrast, the canopy of a mature tropical forest is extremely irregular in both its vertical and horizontal dimensions. The difference is clear enough, but it does not directly account for the extreme differentiation of the *Tessaria* community. One would like to think that the structural distinctions were important in this case, but there are other major differences too, as we shall see in the next section.

2. *Role of Food Resources*

The availability of adequate food resources is so elementary an aspect of habitat selection that one wonders why it has not received more attention. Perhaps it is because temperate bird communities are composed mainly of insectivores and because the resources of such species are difficult to measure in ways that are relevant to their particular foraging habits. Tropical bird communities offer better opportunities for demonstrating the role of food resources in habitat selection because of their much greater trophic diversity.

In Table I, I indicated that the earliest stages in the successional sequence are deficient in certain classes of resources, a fact that is attributable to the very low plant diversity of these habitats. If one takes the bird community of the high-ground forest as a frame of reference, it can then be asked if the trophic structures of the bird communities of the early successional habitats depart from that of the high-ground forest as would be expected on the basis of the resources available (Table IV). The strongest case is that of the *Tessaria* stage. Given that the monotypic stands produce neither nectar nor fruit (the tiny seeds are wind dis-

TABLE IV

Representation (%) of Guilds in Habitats[a,b,c]

Guild	HG	F–C	LS	G–C	T	FS
Carr/rap/owl	28 (12)	19 (13)	20 (14)	6 (09)	4 (12)	11 (15)
Mast	17 (7)	8 (05)	8 (06)	6 (09)	2 (06)	2 (03)−
Frugivore	27 (11)	18 (12)	11 (08)	6 (09)	0 —	6 (08)
Nectarivore	7 (03)	10 (07)+	9 (06)+	5 (08)+	0 —	4 (06)+
Insectivore						
Terr	14 (06)	5 (03)−	4 (03)−	2 (03)−	0 —	2 (03)−
Bark	24 (10)	13 (07)	11 (08)	6 (09)	2 (06)	6 (08)
Foliage	21 (09)	12 (08)	13 (09)	8 (13)	4 (13)	12 (17)+
Sally (including night jars)	38 (16)	25 (17)	17 (12)	10 (16)	12 (38)++	4 (06)−
Other[d]	27 (11)	15 (10)	18 (13)+	7 (11)	3 (09)	7 (10)
Omnivore	39 (16)	24 (16)	31 (22)+	8 (13)	5 (16)	18 (25)+
	242	149	142	64	32	72

[a] Species totals greater than in Table I, because not restricted to census.

[b] HG, High ground; F–C, *Ficus–Cedrela*; LS, late succession; G–C, *Gynerium–Cecropia*; T, *Tessaria*; FS, fig swamp.

[c] Numbers in parentheses indicate percentages.

[d] Ant followers, vines, dead leaves, aerial.

persed), the absence of nectarivores and frugivores in the bird community is in accord with expectation, although the number of species involved is too small to permit a statistical confirmation of the result. Nectar and fruit are produced, at least seasonally, in the *Gynerium–Cecropia* and subsequent stages, all of which are occupied in good proportion by nectarivores and frugivores.

Some further patterns are apparent in Table IV. All habitats that have an understory composed of heliconias, gingers and marantads (most of which are hummingbird pollinated) contain an excess of hummingbirds [late succession (LS), *Ficus–Cedrela* (F–C), *Gynerium–Cecropia* (G–C), and fig swamp (FS)]. All habitats that are subject to seasonal inundation contain deficiencies of terrestrial insectivores (all but high-ground forest). The *Tessaria* harbors a great excess of sallying (including hovering) insectivores. This is perhaps attributable to the ascending branches of *Tessaria* and the upright attitude of the leaves, both of which could be impediments to hopping gleaners. Finally, there is an apparent excess of omnivores in the late successional forest and fig swamp. This is entirely a consequence of the fact that the organized mixed flocks of the canopy exploit both of these forest types, but not the *Ficus–Cedrela* zone (Munn, 1984). These flocks include a large number of omnivores (tanagers and honeycreepers). The absence of such flocks in the *Ficus–Cedrela* forest is quite striking, because the stature and vertical complexity of the canopy are conspicuously greater than in the fig swamp. In this instance the discrimination exercised by the birds seems clearly related to resource availability. The canopy of the *Ficus–Cedrela* zone is almost wholly occupied by three species of trees (see previous description of habitats), none of which produce fruit that is eaten by small birds (*Ficus insipida* is believed to be bat dispersed: Croat, 1978). The canopy of the fig swamp, in contrast, is dominated by *Ficus trigona,* a species that fruits year round and that is heavily utilized by birds. There are thus several features of the data that are much more easily explained by the presence or absence of resources in the respective habitats (or the occurrence of seasonal flooding) than by referring to their distinctive structural characteristics.

At this point it is appropriate to inject a note on the role of plant species diversity as a determinant of avian community diversity. Evidence from temperate North America on this issue is equivocal (MacArthur and MacArthur, 1961; Willson, 1974; James and Wamer, 1982). I would suggest that this is again a consequence of the fact that most North American birds are insectivores in the breeding season. Differentiation of the ecological roles of insectivores is based largely on differences in their search and capture behaviors, and these, in turn, relate to the parts of the vegetation to which the hunting insectivore directs its attention (Moermond, 1979; Fitzpatrick, 1980; Robinson and Holmes, 1982). It is thus among insectivores in particular that the structural diversity of the habitat is likely to be a major determinant of species diversity (Terborgh, 1977). It is not clear, however, that there is any direct or general coupling between plant species diversity and the structural diversity of the habitat. There may be up to a point, but the effect of plant diversity may saturate at such low values that it becomes lost in broader correlations.

Differentiation of the ecological roles of frugivores and nectarivores is based on very different features of the vegetation from those that are important to insectivores. Frugivores differ in their preferences for fruits of different size, hardness, nutritional quality, and mode of presentation by the plant (Diamond, 1973; Snow, 1981; Howe and Smallwood, 1982; Moermond and Denslow, 1983). All these distinctions are ones that do not relate directly to the structure of the habitat, but rather to the characteristics of different plant species. Much the same can be said of nectarivores whose roles are distinguished by the corolla length, sugar content, and degree of aggregation of the flowers they visit (Feinsinger, 1978). Thus, in the Andes, one finds about as many species of nectar feeding birds in the 3-m-tall elfin forest at 3,500 m elevation, where many plant species are bird pollinated, as in the 50-m-tall lowland forest, where most plants are insect pollinated (Terborgh, 1977). One can hardly doubt, therefore, that plant species diversity is a key element of habitat quality to frugivores and nectarivores, particularly so to those species that are year-round residents in a single type of vegetation (Stiles, 1978).

3. *Role of Interspecific Competition*

The presence of competitors may have either of two effects on a given species: (1) to reduce its abundance below what it would be otherwise or (2) to eliminate it altogether from certain habitats. Most attempts to demonstrate competition in natural communities have relied either on geographical comparisons of localities in which a presumed competitor is present in one and absent from another or on experimental introductions or removals (Diamond, 1970; Cody, 1978; Cole, 1983). Both of these approaches have yielded strongly positive results. Through comparisons of species-rich and species-poor mountain ranges in the Andes, for example, it has been shown that the elevational limits of up to two-thirds of the birds in the species-rich ranges are restricted by the presence of competitors (Terborgh and Weske, 1975; Terborgh, 1984b). Since a species' elevational range can be regarded as a facet of its habitat range, one can suspect that interspecific interactions may play an important role in determining the habitat utilization pattern of Amazonian birds. Evidence for this at present, however, is limited. I can offer just two sets of observations that are consistent with a competitive mechanism.

The first of these involves a geographical comparison. Between 1965 and 1972 I studied the distribution of birds in the Apurimac Valley of Peru (Terborgh, 1984b and references therein). I then moved the locus of my research to the Manu region to escape encroaching development. (The Apurimac Valley has subsequently become one of the world's two leading centers for the production of coca, the plant from which cocaine is extracted.)

A comparison of the two localities is of interest because the lowland avifauna of the Apurimac Valley included fewer than 300 resident landbirds, whereas over 400 occur in the Manu region. The relative poverty of the Apurimac fauna is probably attributable to the fact that the humid lowland habitat of the Valley is a disjunct patch that is isolated from the humid Amazonian lowlands far downstream by a pronounced rainshadow.

The habitat occupancy of the birds of the Apurimac Valley lowlands is documented in Terborgh and Weske (1969). Nearly all the same species occur in the Manu region, some 150 km to the east. The habitat distributions of the birds of this region are documented in Terborgh *et al.* (1984). Similarities and differences in the habitat utilization patterns of some 190 species can be judged by comparing data presented in the two publications. Allowing for some differences in the classification of habitats in the two localities and for the fact that the Apurimac Valley is disturbed while the Manu is pristine, it can be stated that the habitat occupancy patterns of the 190 species common to the two regions are similar if not identical in a great majority of cases. The comparison reveals only 12 unambiguous cases of habitat shifts (in either locality as judged by reference to the other, Table V). In 10 of the 12 cases (arguably 11, if not for a quirk of taxonomy) the shifts are associated with the presence–absence of a congener with which the species in question does not share its habitat. Nine of the 10 cases involve species that contract their habitat ranges in the Manu region in the presence of additional congeners there. The tenth case represents the reverse situation of a species with a contracted habitat range in the Apurimac Valley in the presence of a congener that is missing from the Manu fauna. In contrast, among 75 species that were alone in their genera in both localities, only two showed habitat shifts (the remaining 2 cases in Table V), and of 38 species that shared both localities with the same (one or more) congeners, none showed habitat shifts. These last two categories of species make up 60% of the total of 190 that the localities share in common, the remaining 40% consisting of species having different numbers of congeners in the two localities.

The results contain two improbable elements, first, that 10 of the 12 cases of habitat shift should occur in the 40% of the species pool that had differing numbers of congeners in the two localities ($p < 0.02$) and second, that in all 10 cases in which the presence–absence of a congener is associated with a habitat shift, the shift was a contraction of the habitat range where the additional congener was present ($p < 0.01$). One is thus drawn to the view that interspecific interactions are involved in most of these cases of habitat shift.

While it is of interest to our general goal of understanding the mechanisms of habitat selection in birds to observe that some species may exclude certain

TABLE V

Species Showing Pronounced Habitat Shifts in a Comparison of the Bird Faunas of the Apurimac Valley Lowlands and the Manu River Floodplain

Species	Habitat occupancy	
	Apurimac Valley	Manu
Crypturellus soui (Tinamidae)	Mature forest and plantations; *C. bartletti* absent	Restricted to edges; *C. bartletti* occupies middle and late successional stages
Buteo magnirostris (Accipitridae)	Occurred in mature forest as well as in all successional stages; *B. brachyurus* absent	Confined to river edges and early successional growth; *B. brachyurus* occupies mature forest
Leptotila rufaxilla (Columbidae)	Common in mature forest; absent from early successional vegetation occupied by *L. verreauxi*	Occupies full successional spectrum except *Tessaria*; *L. verreauxi* absent
Piaya cayana (Cuculidae)	Present throughout successional gradient including mature forest; *P. melanogaster* absent	Present throughout floodplain but replaced in upland forest by *P. melanogaster*
Trogon viridis (Trogonidae)	Present in mature forest as well as in successional stages, plantations; *T. melanurus* absent	Rare in *Ficus–Cedrela* forest; *T. melanurus* common in middle and late successional habitats
Monassa nigrifrons (Bucconidae)	Present in early through late successional habitats; *M. morphoeus* absent	Present throughout floodplain but replaced in upland forest by *M. morphoeus*
Phloeoceastes melandeucos (Picidae)	Present in forest interior as well as along edges, in plantations, etc.; *P. rubricollis* absent	Occurs along forest edges and in successional vegetation; *P. rubricollis* occupies forest interior
Antomolus dorsalis (Furnariidae)	Common in *Gynerium–Cecropia* stage; no bamboo in region; *A. rufipileatus* absent	Found only in bamboo; *A. rufipileatus* occupies *Gynerium–Cecropia* stage
Myrmotherula ornata (Formicariidae)	Dead-leaf forager in forest understory flocks; no bamboo in region; *M. leucophthalma* absent	Found only in bamboo where forages in dead leaves; *M. leucophthalma* is the dead leaf forager in forest understory flocks
Hypocnemis cantator (Formicariidae)	Common in mature forest where one of 21 Formicariid species	Confined to viney swamps and edges; 30 Formicariid species occupy mature forest

TABLE V (*Continued*)

Species	Habitat occupancy	
	Apurimac Valley	Manu
Machaeropterus pyrocephalus (Pipridae)	Present in all but one of 6 habitats studied, including mature forest; *P. coronata* absent	Rare and restricted to stream sides; *Pipra coronata* common in mature forest
Pipromorpha macconnelli (Tyrannidae)	Present in all habitats studied; *P. oleaginea* absent	Extremely rare in *Gynerium cecropia* stage; *P. oleaginea* occurs in all habitats

congeners from certain habitats in certain localities, it is of far greater interest to have an overall impression of how important interspecific interactions are to the habitat selection process. In short, are the cases cited above exceptional, or do they represent a common situation? Unfortunately, this broader question cannot as yet be answered in any decisive way, although results from the census at Cocha Cashu offer an outside estimate.

Spot maps of singing male birds reveal the occurrence of complementary habitat distributions in congener pairs in 35 genera, suggesting the possibility of habitat use being mediated by interspecific territoriality (Orians and Wilson, 1964; Cody and Walter, 1976). For many of the species pairs we found that congeneric neighbors patrolled abutting but non-overlapping territories, the common boundaries of which often, but not always, coincided with an obvious habitat boundary. Altogether about 80 species are involved, or roughly one-quarter of the 328 species that were recorded in the census. The figure of one-quarter is therefore an upper bound for the frequency with which congener interactions might be involved in controlling the habitat ranges of Amazonian birds. Further research is underway at this writing to establish a lower bound.

Diffuse competition [involving noncongeners and/or combinations of species (cf. Diamond, 1975)] may also be important in limiting habitat distributions, as seems to be the case in elevational distributions (Terborgh, 1984a), but appropriate tests of this possibility have yet to be devised.

V. SUMMARY

Research on the community ecology of Amazonian birds has only just begun, due to the heretofore daunting challenge of an unrivalled species diversity. Within a few square kilometers of floodplain habitat in western Amazonia it is possible to find over 500 birds. Until very recently, the vocalizations of many of the species were unknown, so it was impossible to take even the first step of conducting a census. One did not know the most elementary things, such as the approximate population densities of common species and sizes of their territories. Only within the last few years has our knowledge of vocalizations reached the point of permitting work at the community-wide level. The first comprehensive census of an Amazonian bird community has now been completed, and in this report I offer a preview of some of the results.

The first issue to occupy our attention is tropical diversity; why are there so many species, approximately six times the number that one finds in the richest forest communities in North America? This is a complex question to which there is no simple answer. Nevertheless, some understanding of the problem can be gained by examining in the light of today's evidence a number of hypotheses that were originally proposed more than a decade ago, that the increased diversity of tropical bird communities is due to: (1) greater structural complexity of the habitat, (2) the presence of entire guilds that are not represented in temperate bird communities, (3) larger guild niches reflecting broader underlying resource spectra, and (4) tighter species packing.

Censuses of some structurally simple plant formations in Amazonia (coffee and cacao plantations, certain stages of early successional vegetation, etc.) show conclusively that when the structure of the habitat is controlled, tropical bird communities are still far more diverse than their temperate counterparts. The greater structural complexity of a mature rainforest can doubtless account for some of its extra bird species, but not for all of them. Much the same can be said of the other proposed explanations of tropical diversity. New guilds (obligate frugivores, ant-followers, etc.) are present, but in total they account for only about a third of the additional species of a tropical forest. Niches also seem to be larger, because there are many large-billed and large-bodied insectivorous birds in the tropical forest (puffbirds, jacamars, motmots, oropendolas, etc.) and few or none of them in temperate forests. Nevertheless, such expanded guild niches seem to account for only another 17% (one-sixth) of the excess tropical diversity. The remaining 50% of the excess seems to be attributable to tighter species packing and to uncontrolled structural differences between tropical and temperate forests.

The main point to be made is that tropical diversity (or temperate unpoverishment) has many causes. It is clear that each of the hypotheses considered above offers a partial solution to the problem and that none of them provides a complete explanation. That is about all that can be said at present given the very crude nature of the comparative data currently available. For further progress it will be necessary to obtain better control over the relevant variables than has been achieved to date.

Lack of adequate control over variables also remains an impediment to better understanding of the process of habitat selection. Two levels of complexity confound the interpretation of empirical results. First, habitats must simultaneously satisfy multiple requirements—food, shelter, nesting sites, etc.,—so that identification of the proximal factors in a species' presence or absence is inherently difficult. Second, habitats are usually characterized in terms of their vegetation structure, but whenever structural gradients are examined in sufficient depth, it is found that several potentially relevant factors vary in parallel, occasioning severe ambiguities of interpretation.

The latter point is emphatically demonstrated in the case considered here, the natural (primary) successional gradient one finds in the meander loops of Amazonian rivers. The vertical complexity of the vegetation increases monotonically through succession from 1 to 5 plant layers, but accompanying this are concurrent changes in canopy height, plant diversity, the diversity of avian food resources produced by the habitat, and the continuity of resource availability through the annual cycle. It is thus not possible in our present state of ignorance to discriminate the quantitative roles of these factors as determinants of the pattern of avian habitat occupancy, though some suggestive hints can be obtained by looking at the guild structure of the communities. The essentiality of appropriate resources is exposed in the earliest successional stage. This is dominated by a single species of plant (*Tessaria integrifolia*) which provides neither fruit nor nectar in a form exploitable by birds. Accordingly, neither frugivores nor nectarivores are included among the 32 species resident in the habitat. From further qualitative associations of this type it is inferred that insectivores may respond to structural features of vegetation more or less independently of plant diversity, while frugivores and nectarivores may respond more directly to plant diversity through their need for a year-round supply of resources.

Habitat occupancy can also be affected by factors not directly associated with the vegetation, such as the presence of competitors, predators, and parasites. Competitors may prove to be particularly important in the highly diverse Amazonian avifauna, as spatially segregating congeners comprise about 25% of the species pool. A role of diffuse competition also seems likely, though positive evidence is so far lacking.

In sum, a variety of mechanisms are implicated in the pattern of habitat occupancy of Amazonian birds. One would like to evaluate these mechanisms, both as to functional detail and to quantitative importance, but at present this seems a distant goal.

REFERENCES

Amadon, D. (1973). Birds of the Congo and Amazon forests. A comparison. *In* "Tropical Forest Ecosystems in Africa and South America: A Comparative Review" (B. J. Meggers, E. S.

Hyensu, and W. D. Duckworth, eds.), pp. 267–277. Random House (Smithsonian Inst. Press), New York.

Bell, H. L. (1982). A bird community of lowland rainforest in New Guinea. I. Composition and density of the avifauna. *Emu* **82,** 24–41.

Black, G. A., Dobzhansky, Th., and Pavan, C. (1950). Some attempts to estimate species diversity and population density of trees in Amazonian forests. *Bot. Gaz.* (Chicago). **111,** 413–425.

Cody, M. L. (1975). Towards a theory of continental species diversities: Bird distribution over Mediterranean habitat gradients. *In* "Ecology and Evolution of Communities" (M. L. Cody and J. M. Diamond, eds.) pp. 214–257. Harvard Univ. Press (Belknap), Cambridge, Massachusetts.

Cody, M. L. (1978). Habitat selection and interspecific territoriality among the Sylviid warblers of England and Sweden. *Ecol. Monogr.* **48,** 351–396.

Cody, M. L. (1981). Habitat selection in birds: The roles of vegetation structure, competitors and productivity. *BioScience* **31,** 107–113.

Cody, M. L., and Walter, H. (1976). Habitat selection and interspecific interactions among Mediterranean sylviid warblers. *Oikos* **27,** 210–238.

Cole, B. J. (1983). Assembly of mangrove ant communities: Patterns of geographical distribution. *J. Anim. Ecol.* **52,** 339–347.

Croat, T. B. (1978). Flora of Barro Colorado Island. Stanford Univ. Press, Stanford, California.

Diamond, J. M. (1970). Ecological consequences of island colonization by southwest Pacific birds, I. Types of niche shifts. *Proc. Natl. Acad. Sci. U.S.A,* **67.** 529–536.

Diamond, J. M. (1973). Distributional ecology of New Guinea birds. *Science* **179,** 759–769.

Diamond, J. M. (1975). Assembly of species communities *In* "Ecology and Evolution of Communities" (M. L. Cody and J. M. Diamond, eds.), pp. 342–444. Harvard Univ. Press (Belknap) Cambridge, Massachusetts.

Feinsinger, P. (1978). Ecological interactions between plants and hummingbirds in a successional tropical community. *Ecol. Monogr.* **48,** 269–287.

Fitzpatrick, J. W. (1980). Foraging behavior of Neotropical tyrant flycatchers. *Condor* **82,** 43–57.

Fogden, M. P. L. (1972). The seasonality and population dynamics of equatorial forest birds in Sarawak. *Ibis* **114,** 307–343.

Gentry, A. (1982). Patterns of Neotropical plant species diversity. *Evol. Biol.* **15,** 1–84.

Haffer, J. (1969). Speciation in Amazonian forest birds. *Science* **165,** 131–137.

Haffer, J. (1974). Avian speciation in Tropical South America. *Publ. Nuttall Ornithol. Club.* **14,** 1–390.

Haffer, J. (1978). Distribution of Amazon forest birds. *Bonn Zool. Beitr.* **29,** 38–78.

Holdridge, L. R. (1967). Life zone ecology. *Occas. Pap. Trop. Sci. Cent.* (*San Jose, Costa Rica*).

Holmes, R. T., and Sturges, F. W. (1975). Bird community dynamics and energetics in a northern hardwoods ecosystem. *J. Anim. Ecol.* **44,** 175–200.

Howe, H. F., and Smallwood, J. (1982). Ecology of seed dispersal. *Annu. Rev. Ecol. Syst.* **13,** 201–228.

James, F. C. (1971). Ordinations of habitat relationships among breeding birds. *Wilson Bull.* **83,** 215–236.

James, F. C., and Wamer, N. O. (1982). Relationships between temperate forest bird communities and vegetation structure. *Ecology* **63,** 159–171.

Karr, J. R. (1971). Structure of avian communities in selected Panama and Illinois habitats. *Ecol. Monogr.* **41,** 207–233.

Karr, J. R. (1980). Geographical variation in the avifaunas of tropical forest undergrowth. *Auk* **97,** 283–298.

Keast, A. (1972). Faunal elements and evolutionary patterns: Some comparisons between the continental avifaunas of Africa, South America and Australia. *Proc. Int. Ornithol. Congr. 15th,* pp. 594–622.

Kinzey, W. G., and Gentry, A. H. (1979). Habitat utilization in two species of *Callicebus. In* "Primate Ecology: Problem Oriented Field Studies" (R. W. Sussman, ed.), pp. 89–100. Wiley and Sons, New York.

Klopfer, P. H., and MacArthur, R. H. (1960). Niche size and faunal diversity. *Am. Nat.* **94,** 293–300.

Koepcke, M. (1972). Über die Resistenzformen der Vogelnester in einem begrenzten Gebiet des tropischen Regenwaldes in Peru. *J. Ornithol.* **113,** 138–160.

Lovejoy, T. E. (1972). Bird species diversity and composition in Amazonian rain forests. *Am. Zool.* **12,** 711–712.

Lovejoy, T. E. (1975). Bird diversity and abundance in Amazon forest communities. *Living Bird* **13,** 127–191.

MacArthur, R. H., and MacArthur, J. (1961). On bird species diversity. *Ecology* **42,** 594–598.

MacArthur, R. H., MacArthur, J. W., and Preer, J. (1962). On bird species diversity II. Prediction of bird census from habitat measurements. *Am. Nat.* **96,** 167–174.

MacArthur, R. H., Recher, H., and Cody, M. L. (1966). On the relation between habitat selection and species diversity. *Am. Nat.* **100,** 319–332.

Medway, L., and Wells, D. R. (1976). "The birds of the Malay Peninsula," Vol. 5. Conclusion and survey of every species. Witherby, London.

Moermond, T. C. (1979). Habitat constraints on the behavior, morphology, and community structure of *Anolis* lizards. *Ecology* **60,** 152–164.

Moermond, T. C., and Denslow, J. S. (1983). Fruit choice in neotropical birds: Effects of fruit type and accessibility on selectivity. *J. Anim. Ecol.* **52,** 407–420.

Moreau, R. E. (1966). "The Bird Faunas of Africa and Its Islands." Academic Press, New York.

Munn, C. A. (1984). The behavioral ecology of mixed species understory and canopy flocks in Amazonian Peru. Ph.D. thesis, Princeton Univ., Princeton, New Jersey.

Munn, C. A., and Terborgh, J. W. (1979). Multi-species territoriality in Neotropical foraging flocks. *Condor* **81,** 338–347.

Orians, G. H. (1969). The number of bird species in some tropical forests. *Ecology* **50,** 783–801.

Orians, G. H., and Willson, M. F. (1964). Interspecific territories of birds. *Ecology* **45,** 736–745.

Parker, T. A., III, Parker, A., and Plenge, M. A. (1982). "An Annotated Checklist of Peruvian Birds." Buteo Books, Vermillion, South Dakota.

Pearson, D. L. (1971). Vertical stratification of birds in a tropical forest. *Condor* **77,** 453–466.

Pearson, D. L. (1975). The relation of foliage complexity to ecological diversity of three Amazonian bird communities. *Condor* **77,** 453–466.

Pearson, D. L. (1977a). A pantropical comparison of bird community structure on six lowland forest sites. *Condor* **79,** 232–244.

Pearson, D. L. (1977b). Ecological relationships of small antbirds in Amazonian bird communities. *Auk* **94,** 283–292.

Recher, H. F. (1969). Bird species diversity and habitat diversity in Australia and North America. *Am. Nat.* **103,** 75–80.

Richards, P. W. (1952). "The Tropical Rain Forest." Cambridge Univ. Press, London and New York.

Ridpath, M. G., and Moreau, R. E. (1966). The birds of Tasmania: Ecology and evolution. *Ibis* **108,** 348–393.

Robinson, S. K., and Holmes, R. T. (1982). Foraging behavior of forest birds: The relationships among search tactics, diet, and habitat structure. *Ecology* **63,** 1918–1931.

Sabo, S. R. (1980). Niche and habitat relations in subalpine bird communities of the White Mountains of New Hampshire. *Ecol. Monogr.* **50,** 241–259.

Schoener, T. W. (1971). Large-billed insectivorous birds: A precipitous diversity gradient. *Condor* **73,** 154–161.

Schoener, T. W., and Janzen, D. H. (1968). Notes on environmental determinants of tropical versus temperate insect size patterns. *Am. Nat.* **102,** 207–224.

Slud, P. (1960). The birds of finca "La Selva," Costa Rica: A tropical wet forest locality. *Bull. Am. Mus. Nat. Hist.* **121,** 49–148.

Snow, D. W. (1981). Tropical frugivorous birds and their food plants: A world survey. *Biotropica* **13,** 1–14.

Stiles, F. G. (1978). Temporal organization of flowering among the hummingbird food-plants of a tropical wet forest. *Biotropica* **10,** 194–210.

Terborgh, J. (1971). Distribution on environmental gradients: Theory and a preliminary interpretation of distributional patterns in the avifauna of the Cordillera Vilcabamba, Peru. *Ecology* **52,** 23–40.

Terborgh, J. (1977). Bird species diversity on an Andean elevational gradient. *Ecology* **58,** 1007–1019.

Terborgh, J. (1980). Causes of tropical species diversity. *Actis Congr. Int. Ornithol., 17th,* pp. 955–961.

Terborgh, J. (1983). Five New World primates: A study in comparative ecology. Princeton Univ. Press, Princeton, New Jersey.

Terborgh, J. (1984a). The vertical dimension of species diversity in temperate and tropical forests. *Am. Nat.* (in press).

Terborgh, J. (1984b). The role of ecotones in the distribution of Andean birds. *Ecology* (in press).

Terborgh, J., and Faaborg, J. (1980). Saturation of bird communities in the West Indies. Am. Nat. **116,** 178–195.

Terborgh, J., and Robinson, S. (1985). Guilds and their utility in ecology. *In* "Community Ecology: Pattern and Process" (J. Kikkawa, ed.). Blackwell, Oxford. (In press.)

Terborgh, J., and Weske, J. S. (1969). Colonization of secondary habitats by Peruvian birds. *Ecology* **50,** 765–782.

Terborgh, J., and Weske, J. S. (1975). The role of competition in the distribution of Andean birds. *Ecology* **56,** 562–576.

Terborgh, J. W., Fitzpatrick, J. W. and Emmons, L. H. (1984). Annotated checklist of bird and mammal species of Cocha Cashu Biological Station, Manu National Park, Peru. *Fieldiana, Zool.* (in press).

Tomoff, C. S. (1974). Avian species diversity in desert scrub. *Ecology* **55,** 396–403.

Vuilleumier, F. (1972). Bird species diversity in Patagonia (temperate South America). *Am. Nat.* **106,** 266–271.

Walter, H. (1973). "Vegetation of the Earth in Relation to Climate and the Eco-Physiological Conditions." Springer-Verlag, Berlin and New York.

Weberbauer, A. (1936). The phytogeography of the Peruvian Andes. *In* "The Flora of Peru" (J. F. MacBride, ed.), *Field Mus. Nat. Hist. Publ. Bot. Ser.,* Vol. 13, Part 1, no. 1, pp. 13–81.

Willson, M. F. (1974). Avian community organization and habitat structure. *Ecology* **55,** 1017–1029.

Part IV

A Variety of Approaches to Habitat Selection in Birds

Chapter 11

Habitat–Consumer Interactions in Frugivorous Birds

CARLOS M. HERRERA
Estación Biológica de Doñana
Seville, Spain

I. INTRODUCTION

Berrylike, fleshy fruits ("fruits" hereafter) are fed upon by a substantial fraction of the avifauna in many regions and habitat types (e.g., Rowan, 1970; Lein, 1972; Pearson, 1977). Geographical variation in the relative contribution of frugivores to local avifaunas is related to the proportional significance of fruit-producing species in the plant community. Frugivorous birds usually are absent or unimportant in grasslands, deserts, or scrublands on very nutrient-poor soils,

HABITAT SELECTION IN BIRDS

ISBN 0-12-178080-5

where the production of fleshy fruits tends to be uncommon among plants (Ellner and Shmida, 1981; Milewski and Bond, 1982; Herrera, 1984c). They acquire the greatest prominence in forested ecosystems or scrublands on fertile soils, where fruit-producing plants are widespread (Howe and Smallwood, 1982). The steep latitudinal gradient in fruit production rates from north temperate forests through Mediterranean scrublands to tropical forests (Table I) runs parallel to a pronounced variation in the relative contribution of frugivores to local or regional avifaunas (Lein, 1972; Morse, 1975; Pearson, 1977; Thompson and Willson, 1979; Herrera, 1984a). Karr (1971) estimates that 10–20% of the increased number of tropical-zone breeding bird species over temperate-zone birds in similar habitats is due to a substantially greater fruit availability. Variation among tropical forests in the abundance and diversity of frugivorous birds, particularly their relative scarcity in southeastern Asian dipterocarp forests, is also to be

TABLE I

Yearly Production of Fleshy Fruits in Several Regions and Habitat Types[a]

Source	Fruits/ha ($\times 10^3$)	Wet mass (kg/ha)	Dry mass (kg/ha)	Reference
Red maple forest (New Jersey)	16.3	2.7	1.3	Baird (1980)
Oak–hickory forest (New Jersey)	12.2	2.6	1.3	Baird (1980)
Floodplain forest (New Jersey)	74.0	22.6	7.4	Baird (1980)
Old pine plantation (Georgia)	—	32.9	—	Johnson and Landers (1978)
Mixed deciduous woodland (England)	14.9	1.8	0.5	Sorensen (1981)
Mixed deciduous forest (northern Spain)	45.2	10.5	2.1	J. Guitián (unpublished)
Mediterranean montane scrub (southern Spain)	65.6–433.2	11.9–104.5	6.1–40.1	Herrera (1984a)
Mediterranean lowland scrub (southern Spain)	1400.0	223.7	97.7	Herrera (1984a)
Pine forest (Texas)	—	—	64.4–129.9	Stransky and Halls (1980)
Tropical rainforest (Panama)	—	—	983.3	Leigh (1975)
Tropical wet forest (location unreported)	—	—	500	Hladik and Chivers, in Fleming (1979)
Old second growth forest (Guyana)	397.5	—	180.4	Charles Dominique *et al.* (1981)

[a] Figures obtained in most cases after recalculation of original data.

related to geographical variation in the significance of fruit-producing plants in the habitat (Fogden, 1972; Pearson, 1977; Karr, 1980).

Virtually any bird species, including raptors, ducks, gulls, or shrikes, may occasionally indulge in frugivory (e.g., Turček, 1961; Tutman, 1969; Ogawa, 1977; Thiollay, 1978; Grant, 1979; Lemke, 1979). It is, however, for those species that fruits play an important energetic or nutritional role that ecological implications of frugivory are most apparent. Detailed field studies assessing quantitatively the importance of fruit food for individual bird species are surprisingly scarce, but accumulating evidence is revealing that substantial frugivory may eventually prove to be more common than expected, particularly in non-tropical habitats (see, e.g., Brensing, 1977; Salomonson and Balda, 1977; Walsberg, 1977; Herrera and Jordano, 1981; Jordano and Herrera, 1981; Jordano, 1982; Debussche and Isenmann, 1983). Although frugivory tends to be most intense in some tropical and subtropical bird families (e.g., Coliidae, Pipridae, Dicaeidae, Trogonidae, Rhamphastidae, Cotingidae; Snow, 1981), there are also strongly frugivorous temperate representatives in the Mimidae, Turdidae, Sylviidae, and Bombycillidae (Putnam, 1949; Arvey, 1951; Havlin, 1977; Moore, 1978; Herrera, 1981d; Jordano and Herrera, 1981). This broad group of "strong" frugivores will be in the focus of the present contribution. It is beyond its scope to examine either the selective pressures involved in the evolution of avian frugivory or the set of adaptations enabling birds to subsist on fruit food (see, e.g., Snow, 1971, 1981; Morton, 1973; McKey, 1975; Foster, 1978; Herrera, 1984b, for an entry to the literature).

Relative to other well-studied groups such as insectivores or granivores, frugivorous birds have been much less frequently the subject of ecological studies. As a result, we know very little about their ecology, including the factors involved in their selection of habitat, and this will become apparent in the sections to follow. The aims of this chapter are to suggest a variety of factors which are presumably important in the choice of habitat by frugivorous birds and to highlight the peculiar mutualism-mediated relationship existing between seed-dispersing frugivores and the fruit-producing plant assemblage that enables these birds to promote, in part, the perpetuation and expansion of their preferred habitats.

II. FRUIT ABUNDANCE

Fruit supply is subject to strong spatiotemporal patterning. This is not, of course, a unique feature of this food type, but fruit conspicuousness renders these patterns much more readily discernible than those exhibited by cryptic avian food types such as foliage invertebrates or small seeds. On the other hand, the extremely patchy and erratic nature of fruit as a food supply for birds and the broad range of spatiotemporal scales on which fruit clumps occur have been repeatedly emphasized in the literature (e.g., Fogden, 1972; Karr, 1976; Bullock, 1978; Fleming, 1979), and there are reasons to suggest that fruit supply is much more unevenly distributed in time and space than other avian food types. This is presumably responsible for the broad within- and between-habitat utilization patterns ordinarily exhibited by species of frugivores (e.g., Rowan, 1967; Snow, 1962a; Brosset, 1981), as discussed in the following section.

A. Patterns in Time

A seasonal alternation of scarcity and superabundance seems to be an outstanding feature of fruit food in both tropical (Smythe, 1970; Foster, 1974; Frankie *et al.*, 1974; Crome, 1975; Alexandre, 1980; Lieberman, 1982) and nontropical (Sherburne, 1972; Thompson and Willson, 1979; Sorensen, 1981; Herrera, 1984a) habitats. Local fluctuations are least pronounced in tropical forests and increase in amplitude with increasing latitude.

Responses of tropical frugivores to seasonal fruit shortages range from temporal changes in food selection (Foster, 1977) to habitat shifts. The magnitude of the latter depends on the size of the spatial scale on which heterogeneity in seasonal fruit abundance patterns occurs. Resplendent Quetzals (*Pharomacrus moccino*) resident in Costa Rican cloud forest sequentially occupy four contiguous habitat types while closely tracking local abundance of their highly preferred lauraceous fruits (Wheelwright, 1983). *Dicaeus hirundinaceus* seasonally occupy all major Australian forest types, migrating in response to variations in the abundance of mistletoe berries, their main food (Keast, 1958). Long distance intratropical migration of *Vireo flavoviridis* and *Legatus leucophaius* in the Neotropics seems to have been selected for by seasonal changes in fruit abundance (Morton, 1977), and several species of Australian fruit pigeons regularly undertake migrations across lowland rainforest in response to local variations in fruit supply (Crome, 1975).

Increasing local seasonality in flowering and fruiting as one moves away from the tropics gives rise to long periods of extreme fruit scarcity occurring simultaneously over vast areas. This fact, by precluding the possibility of short-range habitat shifts, is responsible for the almost general absence of year-round frugivory among temperate birds. Most nontropical frugivores feed on insects at times of fruit shortage (e.g., Morton, 1973; Havlin, 1977), but habitat shifts in response to fruit supply may also occur. Altitudinal zonation of vegetation in the Guadalquivir Valley, southern Spain, provides an almost uninterrupted sequence of abundant fruit availability periods from early summer through late winter. *Sylvia atricapilla,* one of the most significant regional frugivores (Jordano and Herrera, 1981), tends to occupy at each season the altitudinal belts where fruits are most abundant (Fig. 1). This species inhabits in the course of a year virtually

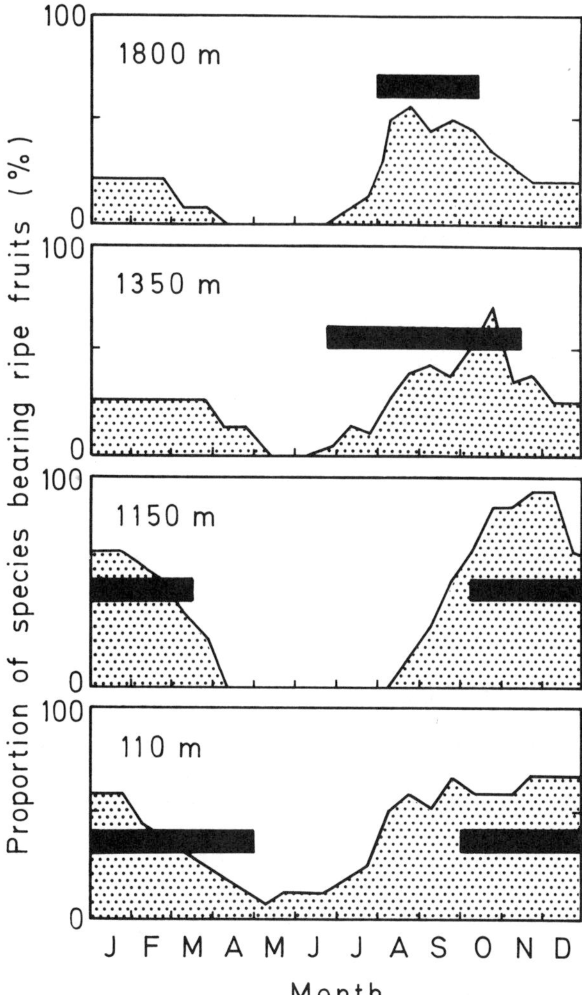

Fig. 1. Seasonal habitat occupancy (black bars) by *Sylvia atricapilla,* a strong frugivore, of four southern Spanish habitats occurring at different elevations and differing in fruiting phenology. Habitat occupancy during the period in which the bird is mainly insectivorous (April–May) is not shown, although it occurs at 1350- and 1150-m sites. From high to low elevation, habitats considered are pine (*Pinus nigra*) forest, pine forest–montane scrub ecotone, cool-climate montane scrub, and warm-climate lowland scrub.

all vegetational formations of the region, including habitats as disparate as successional deciduous scrub in mountains, highland pine forests, open juniper woodlands, and a variety of dense sclerophyllous scrublands (see also Bairlein, 1983).

Strong supra-annual variation in fruit abundance often occurs in montane and high-latitude plant communities (Laine, 1978; Herrera, 1984a). In contrast to periodical habitat shifts or regular migrations promoted by predictable seasonal rhythms, supra-annual fruiting patterns promote irregular displacements of frugivores over vast areas. *Turdus pilaris* and *Bombycilla garrulus* some years exhibit irruptive movements in northern and central Europe, apparently determined by extensive failure of *Sorbus* trees, their critical autumn–winter food (Ulfstrand, 1963). Provided some fruits are available, these irruptive birds may be found in any major habitat type (Bezzel, 1966; Tyrväinen, 1970, 1975; Kolunen and Vikberg, 1978).

Available information suggests that (1) species of frugivorous birds are often very catholic in their choice of major habitat types, (2) their ability to exploit fruits in a variety of habitats seems essential to the maintenance of frugivory, given the intense temporal fluctuations experienced by their food resource, and (3) gross habitat type selection is very often determined by fruit availability alone. I examine in subsequent sections the significance of other variables.

B. Patterns in Space

Horizontal distribution of fruit-producing plants, as well as their relative cover and species richness in the plant community, determine the patterns of spatial distribution of fruits within habitats. If differences in successional stages of contiguous vegetation patches exist, they can produce important horizontal heterogeneity in food supply for frugivores.

Fruit-bearing plants in temperate forests mostly are earlier successional shrubs whose fruit production intensity is very sensitive to shading (Halls, 1973; Baird, 1980). Fruits accordingly are concentrated in clearings and forest edges, and become scarce in the interior of mature forest stands dominated by nut- or cone-producing trees (e.g., Auclair and Cottam, 1971; Sherburne, 1972; Marks, 1974; A. J. Smith, 1975). Species of avian frugivores preferentially select these light gaps, relative to the forest interior (Willson *et al.*, 1982).

In Mediterranean scrublands, in contrast, fruit-bearing plants replace earlier successional non-fruit-producing shrubs and eventually dominate the vegetation in mature scrublands (Houssard *et al.*, 1980; Herrera, 1984c). This produces a very small scale patchiness in fruit distribution in late successional formations, where distribution of fruits becomes virtually continuous at times of peak availability (Herrera, 1984a). Small-sized frugivorous birds characteristic of these scrublands (*Erithacus rubecula, Sylvia atricapilla, S. melanocephala*) avoid early successional scrub and disturbed patches in favor of dense mature stands (Fernández, 1982).

Fruit abundance is predictably associated with light gaps in a Costa Rican lowland rainforest understory, and abundance of frugivoruous birds (mostly manakins, tyrant flycatchers, and tanagers) parallels fruit supply (Fig. 2). This pattern, however, does not seem to be general to all tropical rainforests. Understory frugivores in a Panamanian forest are not more abundant in treefall gaps than in nearby forest interior (Willson *et al.*, 1982), and no frugivorous species shows a preference for gaps (Schemske and Brokaw, 1981). Light gaps in these forests do not seem to produce more fruits than closed forest understory (Schemske and Brokaw, 1981). Further studies in tropical forests are needed to assess the generality of frugivores' responses to gap-related fruit abundance patterns.

The limited evidence available suggests that patchiness in successional stages of vegetation has important habitat implications for frugivorous birds. However, differences among plant community types in the distribution of the fruit-producing habit along the successional gradient (see, e.g., Opler *et al.*, 1980, Howe and Smallwood, 1982, Herrera, 1984a) preclude broad generalizations on the responses of frugivores to patchy habitats generated by asynchronous disturbances of the vegetation (White, 1979).

Patterns of habitat use by frugivores, particularly their daily displacements between distantly spaced fruit sources, are very poorly known. Both the isolation of fruiting patches and the food value of individual patches presumably have important effects on the time and energy budget of avian frugivores. One expects that the ability to exploit (on a daily basis) widely spaced and/or individually

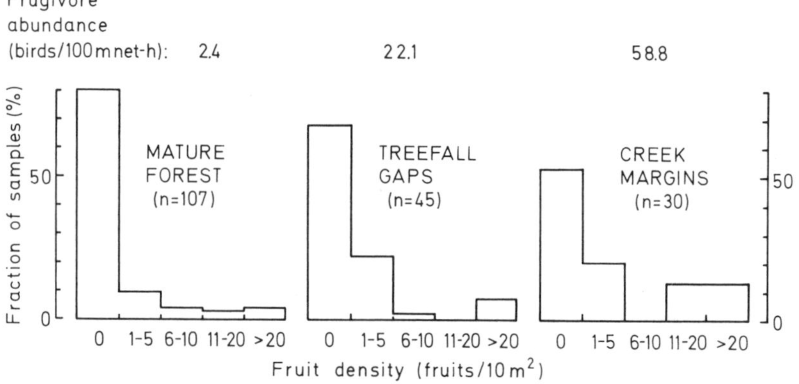

Fig. 2. Variation in fruit and frugivore abundance in the understory of a Costa Rican lowland rainforest (Finca La Selva, February 1980). Fruits were counted within 10 m^2 rectangular quadrat samples (*n*, number of samples examined in each habitat category. Frugivore abundance estimated by mist-net captures per netting effort. (Based on information supplied by F. G. Stiles.)

low-reward fruiting patches increases with body size, and the scanty field evidence available supports this expectation (Snow, 1962c; Walsberg, 1977; Davies, 1978; Davison, 1981). Interhabitat differences in average spacing and food value of fruiting patches shown previously will presumably affect differentially the selection of habitat by frugivores having different body sizes, but no field data are available to test this possibility.

Although isolated fruiting plants or widely spaced patches dominated by fruit-producing species often represent very dense concentrations of food for frugivores, territorial defense of these clumped resources is uncommon. As stated by Buskirk (1976): "Fruit cannot be banked over time and a suitable series of sequentially ripening fruit trees usually does not occur on a tract of land small enough to be defended by permanent territorial behavior. Defense of a superabundant fruit crop against trespassers would be energy consuming without yielding a realizable savings of food for future use." The few documented cases of feeding territories among frugivores (e.g., *Ilex, Juniperus;* Lederer, 1977a; Salomonson and Balda, 1977; Moore, 1978; Snow and Snow, 1984) involve plants having large crops of long-lasting fruits. This suggests that the temporary nature of most fruits actually renders its defense unprofitable. The importance of fruit crop size in determining territorial behavior has been shown by Lederer (1977b, 1981) for *Myadestes townsendi* feeding on juniper fruit. (See Buskirk, 1976; Karr, 1976; Walsberg, 1977; Moore, 1977, for further discussion on social systems of frugivores as they relate to the distribution in space of fruit supply).

III. FRUIT CHARACTERISTICS

Fruits are made up of seeds and pulp, and frugivorous birds handle them in a variety of ways. Some species feed on seeds alone, discarding the pulp (e.g., some finches, Newton, 1972). Others ingest the pulp and seeds together, cracking the latter in the bill or the gizzard (e.g., parrots, finches, gamebirds; Turček, 1961; Newton, 1972; Janzen, 1981), while some others feed only on the pulp and ignore the seeds (e.g., titmice, Sorensen, 1981). There is finally a fourth group, that of species ingesting whole fruits and later defecating or regurgitating the seeds intact (Snow, 1971). All these groups have often been pooled under the single denomination of "frugivores," but they greatly differ in important aspects of their feeding ecology (Snow, 1971, Herrera, 1984b). In particular, nutrient and energy yield should vary substantially according to the part(s) of fruit actually being used as food, since seeds are far more nutritious than fruit pulp for most species. Little is known about the food value of fruits to avian frugivores that digest seeds, and I concentrate for the remainder of the chapter on species ingesting whole fruits and using pulp alone as food. These are the most genuine frugivores, for they maintain a mutualistic relationship with their food plants and have been ultimately responsible for the evolution and maintenance of the fruit-producing habitat among plants (Snow, 1971; McKey, 1975).

A. Fruit Size

For birds that ingest whole fruits, gape width imposes an upper limit on the maximum size of fruit that can be ingested. Field observations confirm this intuitive assessment. Larger birds are able to ingest larger fruits. Small fruits are generally fed upon by more species than large ones, which can be handled effectively only by the limited subset of larger frugivores (Terborgh and Diamond, 1970; Leck, 1971; Diamond, 1973; Kantak, 1979; Herrera, 1981a). To the extent that average fruit cross diameter varies among habitat types, one should expect this variable to influence habitat selection by frugivores that ingest whole fruits.

The average fruit cross diameters of local fruit-producing species vary significantly with elevation in the Guadalquivir Valley, southern Spain. Plant species of Mediterranean scrublands tend to have smaller fruits than those of highland pine forests, and this elevational gradient is closely paralleled by variation in average gape width of local frugivorous species (Fig. 3). Scrubland frugivores

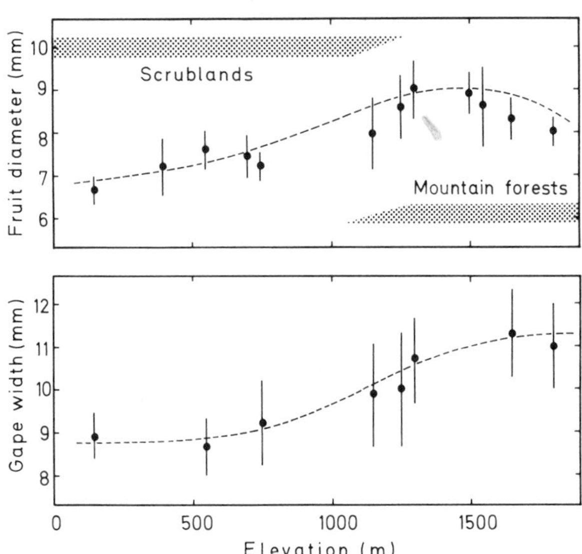

Fig. 3. Local averages for cross diameter of fruit species and gape width of frugivorous bird species experience parallel increases with elevation in the Guadalquivir Valley, southern Spain. Dominant frugivores in lower elevation scrublands are several small-sized (12–20 gm body weight) species, while larger-sized *Turdus* species (60–110 gm) predominate in highland forests. Vertical segments extend over ±1 *SE* of the mean.

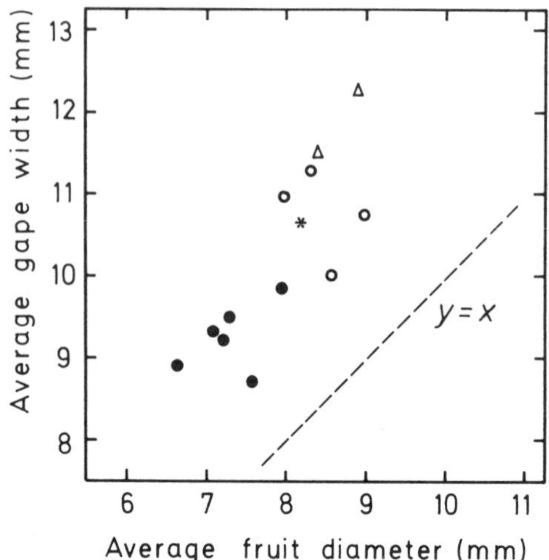

Fig. 4. Average gape width of avian frugivorous species is significantly related ($r = 0.801, p < 0.001$) to, and is about 2 mm larger than, average fruit cross diameter of local fruit-producing species in a sample of temperate, Mediterranean, and tropical habitats. Each symbol denotes a single locality: circles, southern Spanish habitats (filled circles, lowland scrub; open circles, highland pine forests); triangles, European temperate forests; asterisk, a Costa Rican highland locality (Cerro de la Muerte, 3100 m elevation). [Based on data in Sorensen (1981), information supplied by J. Guitián and P. Jordano, and author's unpublished data.]

are mostly small birds in the genera *Sylvia* and *Erithacus,* while large frugivores in the genus *Turdus* predominate numerically in highland forests (Jordano, 1982; C. M. Herrera, 1984a, unpublished data).

A plot of average fruit diameter against mean bird gape width reveals a significant linear relation between the two variables for southern Spanish locations (Fig. 4). Additional data from two temperate forests and a Neotropical highland locality fit remarkably well to the same general relation. Regardless of region and habitat type, therefore, the average gape width of local frugivores (feeding on whole fruits) runs roughly parallel to the average diameter of local fruit species, the two magnitudes differing by only 1–2.5 mm in the sample of localities considered (Fig. 4). Further data are needed to assess the generality of this relation, but the preliminary evidence presented strongly points to a significant role of fruit size alone in determining the composition of local assemblages of avian frugivores and, therefore, influencing habitat selection.

B. Fruit Quality

In marked contrast with other avian foods, the specific energetic and nutritional values of fruits vary dramatically among plant species, both among and within habitats. Lipid content of (dry) pulp may fall somewhere between 1 and 67%, protein between 1 and 25%, and minerals between 1 and 14% (White, 1974; Crome, 1975; Foster, 1978; Frost, 1980; C. M. Herrera, 1981a, unpublished data). Many ripe fruits contain toxic compounds or digestion inhibitors in the pulp whose nature and concentration vary greatly among species (Kear, 1958; Herrera, 1982b). Furthermore, the proportion of edible fruit pulp relative to seeds (which are not digested) is also very variable (Herrera, 1981a, 1982b), contributing to amplify interspecific variation in nutritional and energetic value of fruits to consumers. Interhabitat variation in the composition of fruiting plant assemblages leads to substantial variation in the nutritional configuration of food resources found by avian frugivores. Along the Guadalquivir Valley, local plant communities differ appreciably in the frequency distributions of protein and lipid content of fruit pulp (Fig. 5). Sclerophyllous scrublands under 1200 m elevation are characterized by the presence at each locality of several species having lipid content in excess of 20% dry weight of pulp, whereas high energy fruits are absent from highland pine forests. Although other factors are also involved (e.g., accessibility, secondary compounds in the pulp, palatability, pulp–seed weight ratio; Howe and Vande Kerckhove, 1980; Herrera, 1981c; Moermond and Denslow, 1983; Sorensen, 1983), nutrient and energy content of pulp are important in fruit choice by at least some species (Graber and Powers, 1981; Herrera, 1981b, 1984a), and one should expect interhabitat differences in the nutritional configuration of fruit resources to influence habitat selection by frugivores.

Small birds have higher metabolic rates and energy requirements relative to body weight than larger ones (Kendeigh *et al.*, 1977, Walsberg, 1980), a difference which is accentuated under low temperatures (Kendeigh, 1970). One should therefore expect to find in autumn–winter, when climatic conditions are most severe, a negative relation between body size and the proportional importance of energy-rich fruits in the diet. This relation does exist for southern Spanish frugivores (Fig. 6). Heavy fruit consumption by small birds in autumn–winter could be possible if an adequate supply of energy-rich fruits is available. In contrast, the lower relative energy requirements of large frugivores would enable them to subsist more easily on comparatively low-energy fruits. The size-related, differential reliance of these birds on energy-rich fruits (Fig. 6) could be adduced to explain their differential prominence in habitats at different elevations (Figs. 3 and 6), since the frequency of lipid-rich fruits varies with elevation in the region (Fig. 5).

High-energy fruits are produced by a minority of plant species in both tropical and nontropical habitats, and some frugivores relying on them for subsistence seem to be particularly sensitive to interhabitat variation in species composition of fruit supply. Geographical distribution and habitat type utilization of the African Palm-nut Vulture (*Gypohierax angolensis*), having a great dependence on the very oil-rich fruits of the palm *Elaeis guineensis* (Thiollay, 1978), are apparently dictated by the presence of this plant (Thomson and Moreau, 1957; Winterbottom, 1978). A close association exists also in the western Mediterranean Basin between *Sylvia melanocephala* and *Pistacia lentiscus,* whose very nutritious fruits are the basis of the bird's autumn–winter diet; the bird inhabits a variety of plant communities, greatly differing in physiognomy and composition, but presence of *P. lentiscus* seems indispensable in order to *S. melanocephala* to occupy a habitat in that season (C. M. Herrera 1984a, unpublished data). The association of Bellbirds (*Procnias*) with montane habitats in Central and South America has been interpreted as a consequence, in part, of their reliance on lauraceous fruits that are numerically more important in montane than lowland habitats (Snow, 1973).

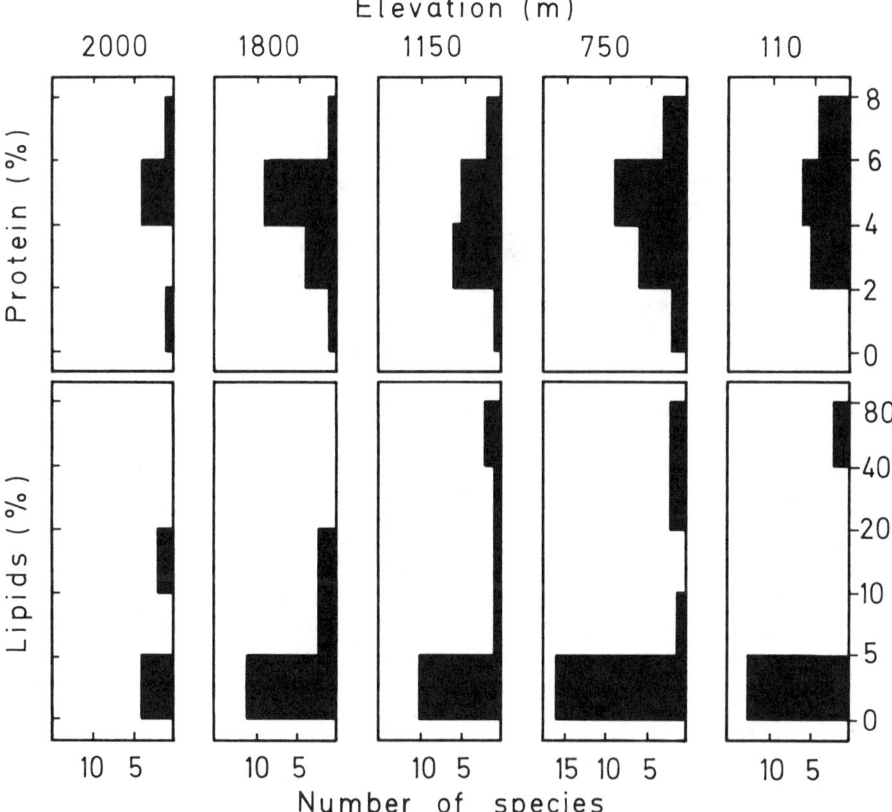

Fig. 5. Nutritional configuration of fruit species assemblages found by avian frugivores (as assessed by protein and lipid content of dry fruit pulp) varies markedly among plant communities occurring at various elevations in the Guadalquivir Valley, southern Spain. Note logarithmic scale used in the graphs for lipid content.

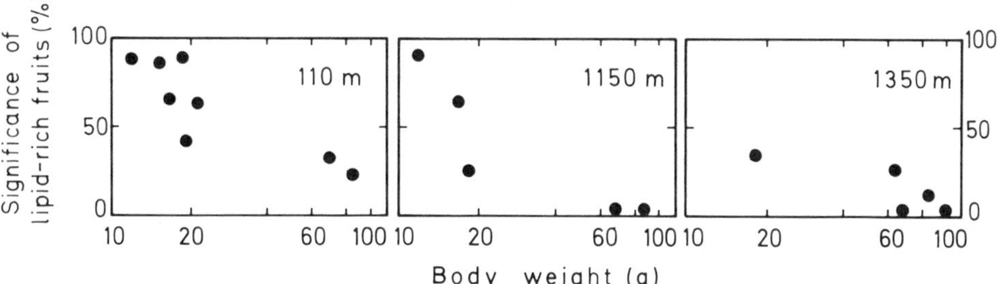

Fig. 6. In autumn–winter, the relative contribution (percentage of total fruits ingested) of lipid-rich fruits (lipid content of dry pulp greater than 20%) to the diet of southern Spanish avian frugivores declines with increasing body weight in three different habitat types (varying in vegetational composition and frequency of lipid-rich fruits; see Fig. 5). Dots represent individual bird species. [Based on data in Herrera (1984a and unpublished).]

Virtually nothing is known about the nutritional requirements of avian frugivores, but the limited information available seems to suggest that the strong nutritional imbalance characterizing the pulp of most fruit species forces many frugivores to have mixed-species diets in order to get a balanced input of energy, protein, vitamins, and minerals (Foster, 1978; Herrera, 1982a, 1984a). In southern Spain, single fecal samples of *Sylvia atricapilla* contain remains of up to nine fruit species, even though the birds face at times a virtually unlimited supply of highly energetic fruits. Other frugivores in the region and elsewhere behave similarly, ingesting a variety of fruit species over short time periods (Snow, 1977; Wheelwright, 1983; Herrera, 1984a). Although they are very poorly understood, nutritional constraints allow one to envisage subtle effects of interhabitat differences in nutritional configuration of fruit supply on habitat choice by frugivores.

We know next to nothing about the influence of secondary compounds in the pulp of ripe fruits on food selection by frugivores, but these do have some influence (Kear, 1958; Sherburne, 1972; Herrera, 1982a; Sorensen, 1983). The relative abundance of fruits defended (against fruit pests) chemically, by means of either digestion inhibitors like tannins or poisons like alkaloids, varies regionally and between habitat types (C. M. Herrera, unpublished data). Presumably, so does the chemical configuration of secondary compounds in fruits in different vegetation formations. It seems reasonable to expect these factors to have some significance in determining habitat choice by frugivores. Thus, the responses to habitat-specific chemical configuration of fruits presumably will differ among bird species as a function of somewhat species-specific attributes such as detoxification capacity (see Walker, 1983) and overall reliance on fruit for food. Blackbirds (*Turdus merula*) overwintering in southern Spanish montane pine forests strongly prefer the fruits of *Crataegus monogyna* over those of *Rosa canina,* apparently because of the much higher tannin concentration found in the pulp of the latter; the two species are virtually identical with regard to other pulp constituents (Herrera, 1985b). As a consequence, these birds tend to avoid *Rosa*-dominated forest understories in favor of *Crataegus*-dominated ones (C. M. Herrera, unpublished data).

IV. HABITAT–CONSUMER INTERACTION

A. Frugivores as Modifiers of Habitats

I have been considering so far the influence of habitats on frugivorous birds, but the influence of these consumers on their habitats deserves consideration also.

Through selectively dispersing the seeds of different plant genotypes in relation to their preferences, seed-dispersing frugivorous birds are able to exert directional selective pressures on fruit-producing plants. Since the birds and their food plants interact mutualistically, these pressures will result in the evolution of fruit traits that are favorable to the birds (e.g., conspicuousness, abundance). This, coupled with a similar selective potential by plants on their dispersal agents, forms the basis of plant–disperser coevolutionary processes as initially envisaged by Snow (1971) and McKey (1975). Subsequent studies have elaborated further on the notion of plant–disperser coevolution, and I omit here a more detailed presentation of these widely held coevolutionary notions (see Howe and Smallwood, 1982; Janzen, 1983, Herrera, 1985a, for reviews). It must be noted that, in addition to changes in individual plant species, plant–disperser coevolution may produce over the long-term an "improvement" of habitats for frugivores. The regular staggering of the fruiting seasons of coexisting plants as a way of alleviating competition for seed vectors has as a consequence an extended period of fruit availability (Snow, 1965; Smythe, 1970), with obvious benefits to birds.

In addition to bird–plant interactions taking place on an evolutionary time scale, other types of mutual influences occur between frugivorous birds and their food plants on an ecological time scale. To some degree, seed-dispersing birds potentially have an ability to shape their own habitats, an aspect which to date has not received more than anecdotal attention in avian ecological studies, although it is a unique feature of this group of birds. Plant ecologists have long recognized, however, the importance of avian seed dispersers in the dynamics of plant communities (Auclair and Cottam, 1971; Livingston, 1972; Marks, 1974; A. J. Smith, 1975; Harper, 1977).

In the Colorado Desert, berries of the mistletoe *Phoradendron californicum* are the Phainopepla's (*Phainopepla nitens*) major food during several months.

Resident birds are always associated with mistletoe, and local population densities usually reflect the local abundance of mistletoe berries (Walsberg, 1977). The Phainopeplas help in the distribution of the mistletoe, for they eat the berries and the seeds pass through the digestive tract and cling to the branches below, where they germinate (Crouch, 1943; Walsberg, 1975). High densities of mistletoe clumps are indicative of an extended residence of Phainopeplas in an area (Crouch, 1943). A similar relation presumably exists in Australia and southeastern Asia between some Dicaeidae and tropical mistletoes. These birds are heavily dependent on mistletoe berries for food, and they disseminate the seeds of their food plants (Docters van Leeuwen, 1954; Keast, 1958). The ability of some Dicaeidae to increase the carrying capacity of their environment and enhance favorable habitat attributes goes further than just spreading the seeds of their food plants; some species are pollinators of the same mistletoes which will later supply them with fruits (Docters van Leeuwen, 1954), and this pollinating activity obviously enhances fruit production to the birds' benefit. To the extent that mistletoes are a critical resource determining habitat selection by Phainopepla and Dicaeidae, the birds themselves are shaping their own habitats by introducing and spreading their critical food resource. In other words, birds are ultimately responsible for a habitat feature that promotes its utilization by the birds themselves.

The interaction between seed-dispersing frugivores and their habitats is vastly more complex than suggested by the previous simple examples. Most frugivores disperse the seeds of many plant species in the course of a season or over much shorter periods (Snow, 1962a, 1962b; Snow 1970, 1972; Snow and Snow, 1971; Frost, 1980; Cruz, 1981; Greenberg, 1981; Herrera, 1981b; Jordano and Herrera, 1981; Wheelwright, 1983), promoting simultaneously the recruitment of a broad plant assemblage rather than that of one or a few species. Birds ''assemble'' multispecies sets of seeds within themselves, thus in some sense restore sections of their habitats (in seed stage) in their guts and later in feces or regurgitations spread over their foraging areas. Nevertheless, many factors interact to determine the density, dispersion patterns, and composition of plant communities, and seed dispersal is only the first step in this process (Harper, 1977, Howe and Smallwood, 1982).

Regardless of these complications, however, the overall effects of multispecies seed dispersal by birds will generally be (1) an enhancement of the fruit-producing component of habitats and (2) a trend toward some ''improvement'' of habitat quality from the viewpoint of frugivores. Early successional shrubs of northern temperate forests occupy transient habitat patches created by treefall gaps and other disturbances and are later eliminated as succession proceeds (Auclair and Cottam, 1971; Marks, 1974; A. J. Smith, 1975). The preference of some frugivores for forest edges and clearings facilitates immigration of successional scrub seeds into newly opened patches; a few years later, the patch will become an important fruit source to birds, which will tend to select preferentially these spots over surrounding forest (Thompson and Willson, 1978; Willson *et al.*, 1982; McDonnell and Stiles, 1983). The species composition of seeds brought into the newly created gap by birds will be a function of the species composition of the fruit supply found by the birds in other gaps in the area but also of the differential preference of birds for the various fruit species (McDonnell and Stiles, 1983). Seeds of preferred species will be proportionally overrepresented in the incoming flux. It is likely that, other things being equal, these species will also tend to be overrepresented in the plant community that eventually dominates the patch and supplies further fruit to birds. The conclusion emerging from this example is that, to the extent that current species composition (and hence energetic, nutritional and chemical configuration of fruit supply) of fruiting patches in temperate forests has been shaped by specific fruit preferences of past frugivores, current habitat selection by these same seed-dispersing frugivores cannot be interpreted conventionally. Such frugivores play a far from passive role with respect to vegetation composition and structure.

The positive feedback which takes place between bird-dispersed plants and seed-dispersing birds on both ecological and evolutionary time scales raises the question: Are birds found where they are because the habitat attributes match their preferences, or do habitat attributes match birds' preferences to some degree because the birds have been there long enough to promote these attributes?

B. Southern Spanish Frugivores: A Case Example

As mentioned earlier, species of large-sized frugivores in the genus *Turdus* (mainly *T. torquatus, T. viscivorus, T. iliacus,* and *T. merula*) are predominant in undisturbed habitats in southern Spanish highlands in autumn–winter. The most extreme case is exemplified by Ring Ouzels (*T. torquatus*), which are largely confined to elevations above 1600 m where they feed almost entirely on the berries of Dwarf Junipers (*Juniperus communis nana*) (R. Zamora, unpublished data). In the highlands, plant communities are characterized by a dearth of lipid-rich fruits and the predominance of relatively large fruits. In contrast, small-sized frugivores predominate in lowland habitats, where average fruits are smaller and have higher lipid content. To what extent do differently sized frugivores select different habitats because of differences in the size and nutritional quality of fruits, as implied in earlier sections (the conventional look at habitat selection), and to what extent do size and quality of fruits differ among habitats because of differences over the long-term in the average body size of their complement of frugivorous species? This is the kind of circular reasoning that often emerges from analysis of coevolutionary processes and can be solved only by consideration of variables external to the plant–bird system (C. C. Smith, 1975).

Regardless of feeding habits, the proportional contribution of small species to bird communities tends to decline with increasing elevation (Affre, 1980; Thiollay, 1980; Lebreton and Broyer, 1981), probably as a consequence of size-related differential ability to cope with adverse thermal environments. This indicates, therefore, that the altitudinal segregation of differently sized frugivores in southern Spain is not necessarily a response to variation in some habitat attributes affecting its suitability to birds (fruit size and quality), as the correlative evidence alone would suggest. Independently of the plant–seed disperser interaction, fruiting plants living at different elevations have faced different size structures of avian disperser assemblages. Selection of small birds against large-fruited plants presumably has favored a decrease in fruit size in lowlands (Herrera, 1984a). In plant genera with species pairs whose members segregate altitudinally (e.g., *Pistacia, Daphne, Phillyrea, Lonicera*) the lowland species invariably has smaller fruits than its highland counterpart.

An analogous explanation may be adduced in relation to elevational variation in fruit quality. First, the small lowland frugivores have probably selected for increased energy content of fruits in their habitats more strongly than have large highland birds (Fig. 6, and Herrera, 1984a). Second, mild winters and evergreeness of plants have made possible the evolution in the lowlands of energy-rich winter fruits in response to selection by birds but not in the highlands, where greater climatic severity produces more deciduous plant species. Plant species having the most lipid-rich fruits in highland habitats are all evergreen.

Evolutionary processes do not seem to have contributed to the habitat–consumer matching in the case of Blackbirds (*Turdus merula*) overwintering in southern Spanish pine forests mentioned earlier. These birds prefer *Crataegus* fruits to those of *Rosa* and occupy preferentially forest understories dominated by the former species. They do, however, regularly ingest small numbers of *Rosa* fruits even where *Crataegus* is superabundant, apparently because these provide vitamins to the birds. As a result, birds tend to concentrate the few *Rosa* seeds they disperse under their preferred *Crataegus* plants, promoting a very close spatial association of both species permitted by the climbing habit of *Rosa* (Herrera, 1985b). It is relatively frequent to find pine forest understories having *Rosa* alone (avoided by *T. merula*), but habitats having *Crataegus* alone are extremely rare. It thus seems as if birds' behavior serves to add a *Rosa* component to *Crataegus*-dominated habitats. This has the consequence of birds finding their minor, albeit important, fruit food spatially close to their staple food species. The savings in locomotion costs lead perhaps to the observed preference for this habitat type (see Herrera, 1985b for further details on this complex interaction). To summarize, seed-dispersing birds have the ability to assemble in their guts the food species they like in the proportions they like, and some bird generations later, what they prefer may be found assembled again (in the way they like) in the form of adult, fruit-producing plants in bird-generated preferred habitats.

V. CONCLUDING REMARKS

In addition to illustrating the scant knowledge we have about frugivorous birds and their habitats, this chapter should have made apparent to the reader that, because of their mutualistic interaction with food resources, frugivores that disperse seeds are unique among birds in their relation with their habitats.

Frugivores are commonly less numerous in terms of species but often more abundant in terms of individuals than insectivores in bird communities (Snow and Snow, 1971; Karr, 1976; Brosset, 1981; Herrera, 1984a). This should be related to the fact that fruits can be abundant and tend to be conspicuous, but the number of ways they can be exploited efficiently is limited, thus affording little opportunity for specialization; the reverse is true of insects. These differences stem from the contrasting ecological relations fruits and insects maintain with their consumers (mutualism versus predation; Snow, 1971). Furthermore, as discussed earlier, strong spatiotemporal patterning of fruits will generally select against specialized habitat use. On the other hand, interspecific competition, a factor which has played a central role in the evolution of precise habitat selection among insectivores (Svärdson, 1949; Lack, 1971; Cody, 1978), is relatively unimportant in frugivorous species assemblages which most often exploit temporarily superabundant food (Fleming, 1979). And finally, seed-dispersing frugivores have a potential ability to make their habitats become, in ecological and evolutionary time scales, better "adjusted" to their preferences (although the extent to which this ability is realized in nature remains to be ascertained by future studies). For all these reasons, I contend that habitat selection by frugivores should not be analyzed using the usual ecological approaches that have proved useful with other birds maintaining nonmutualistic relations with their food resources. Some factors which may be relevant in studies of habitat selection by frugivores have been outlined previously. On the other hand, owing to their conspicuousness, properties of fruits are readily measurable (size, nutritional value, abundance) as compared to cryptic avian foods. Accordingly, the ultimate factors (see Hildén, 1965) determining habitat selection may be more easily assessed for frugivores than for other groups in which evidence based on quantification of proximate factors is the usual basis for habitat selection studies (e.g., Cody, 1968, 1978; James, 1971; Anderson and Shugart, 1974).

The implications of the habitat-shaping ability of avian seed dispersers transcend those related to the subject of habitat selection. One essential requirement for coevolutionary processes to take place is a reasonable degree of spatiotemporal coincidence of interacting counterparts, as nicely illustrated by highly coevolved host–parasite systems (Brooks, 1979; Waage, 1979). Coincidence will reinforce reciprocal selective pressures, hence promoting mutual adaptation. Despite the recent interest in plant–disperser coevolution (McKey, 1975; Howe and Estabrook, 1977), the implications of the ability of seed dispersers to shape their habitats in function of their preferences have been generally neglected in favor of the consideration of evolutionary processes occurring on longer time scales. The nonevolutionary component of the interaction between avian seed dispersers and their habitats may be seen as the first step toward coevolution. By differentially disseminating plant species with contrasting fruiting-related traits, birds are potentially able to shape plant communities which best meet their preferences; this has as a consequence a progressive reinforcement of their association with these habitats and an increase in the scope of evolutionary interactions through increased opportunities of spatial coincidence of species. Analogously, by differentially disseminating plant genotypes with different fruiting-related traits, birds are able to shape food species evolutionarily according to their preferences. This latter process has received the most attention from investigators, but I suspect that the nonevolutionary process of habitat–consumer interaction will eventually prove much more significant than expected and that some observed patterns of close plant–bird mutualistic congruency could be explained in these terms without invoking coevolution. Ecological and evolutionary interactions between seed dispersers and fruit plants occur simultaneously, and both processes tend naturally to produce patterns of mutual adjustment, so it may be difficult in practice to separate their effects. For this reason at least, nonevolutionary aspects should be considered in addition to evolutionary ones in studies of plant–disperser interaction at the community level.

ACKNOWLEDGMENTS

I thank Gary Stiles and the students of the OTS Course "Ecología de Poblaciones 1980" for obtaining data used in Fig. 2 and J. Guitián, P. Jordano, and R. Zamora for supplying useful unpublished information. Over the years, my field work in the Sierra de Cazorla has greatly influenced my thinking on the interaction between frugivores and their habitats and provided much of the data used here. I am greatly indebted to the Instituto para la Conservación de la Naturaleza, ICONA, for the invaluable logistical facilities that made my studies there both possible and enjoyable.

REFERENCES

Affre, G. (1980). Distribution altitudinale des oiseaux dans l'est des Pyrenees françaises. *L'Oiseau et R.F.O.* **50,** 1–22.

Alexandre, D. Y. (1980). Caractère saisonnier de la fructification dans une forêt hygrophile de Côte-d'Ivoire. *Terre et Vie* **34,** 335–359.

Anderson, S. H., and Shugart, H. H. (1974). Habitat selection of breeding birds in an East Tennessee deciduous forest. *Ecology* **55,** 828–837.

Arvey, M. D. (1951). Phylogeny of the waxwings and allied birds. *Univ. Kans. Publ. Mus. Nat. Hist.* **3,** 473–530.

Auclair, A. N., and Cottam, G. (1971). Dynamics of black cherry (*Prunus serotina* Erhr.) in southern Wisconsin oak forests. *Ecol. Monogr.* **41,** 153–177.

Baird, J. W. (1980). The selection and use of fruit by birds in an eastern forest. *Wilson Bill.* **92,** 63–73.

Bairlein, F. (1983). Habitat selection and associations of species in European Passerine birds during southward, post-breeding migrations. *Ornis Scand.* **14,** 239–245.

Bezzel, E. (1966). Die Invasion des Seidenschwanzes (*Bombycilla garrulus*) 1965/1966 in Bayern. *Anz. Ornithol. Ges. Bayern* **7,** 847–854.

Brensing, D. (1977). Nahrungsökologische Untersuchungen an Zugvögeln in einem südwestdeutschen Durchzugsgebiet während des Wegzuges. *Vogelwarte* **29,** 44–56.

Brooks, D. R. (1979). Testing the context and extent of host–parasite coevolution. *Syst. Zool.* **28,** 299–307.

Brosset, A. (1981). Evolution divergente des comportements chez deux bulbuls sympatriques (Pycnonotidae). *Alauda* **49,** 94–111.

Bullock, S. H. (1978). Plant abundance and distribution in relation to types of seed dispersal in chaparral. *Madroño* **25,** 104–105.

Buskirk, W. H. (1976). Social systems in a tropical forest avifauna. *Am. Nat.* **110,** 293–310.

Charles-Dominique, P., Atramentowicz, M., Charles-Dominique, M., Gerard, H. Hladik, H., Hladik, C. M., and Prevost, M. F. (1981). Les Mammiféres frugivores arboricoles nocturnes d'une forêt guyanaise: Inter-relations plantes-animaux. *Terre et Vie* **35,** 341–435.

Cody, M. L. (1968). On the methods of resource division in grassland bird communities. *Am. Nat.* **102,** 107–147.

Cody, M. L. (1978). Habitat selection and interspecific territoriality among the sylviid warblers of England and Sweden. *Ecol. Monogr.* **48,** 351–396.

Crome, F. H. J. (1975). The ecology of fruit pigeons in tropical Northern Queensland. *Aust. Wildl. Res.* **2,** 155–185.

Crouch, J. E. (1943). Distribution and habitat relationships of the Phainopepla. *Auk* **60,** 319–333.

Cruz, A. (1981). Bird activity and seed dispersal of a montane forest tree (*Dunalia arborescens*) in Jamaica. *Biotropica* **13,** 34–44.

Davies, S. J. J. F. (1978). The food of emus. *Aust. J. Ecol.* **3,** 411–422.

Davison, G. W. H. (1981). Diet and dispersion of the great argus *Argusianus argus, Ibis* **123,** 485–494.

Debussche, M., and Isenmann, P. (1983). La consommation des fruits chez quelques fauvettes méditerranéennes (*Sylvia melanocephala, S. cantillans, S. hortensis* et *S. undata*) dans la région de Montpellier (France). *Alauda* **51,** 302–308.

Diamond, J. M. (1973). Distributional ecology of New Guinea birds. *Science* **179,** 759–769.

Docters van Leeuwen, W. M. (1954). On the biology of some Javanese Loranthaceae and the role birds play in their life-history. *Beaufortia* **4,** 105–207.

Ellner, S., and Shmida, A. (1981). Why are adaptations for long-range seed dispersal rare in desert plants? *Oecologia* **51,** 133–144.

Fernández, V. (1982). Relaciones entre la estructura de la vegetación y las comunidades de pájaros en las Sierras de Algeciras. MS. thesis, Univ. Seville, Seville.

Fleming, T. H. (1979). Do tropical frugivores compete for food? *Am. Zool.* **19,** 1157–1172.

Fogden, M. P. L. (1972). The seasonality and population dynamics of equatorial forest birds in Sarawak. *Ibis* **114,** 307–343.

Foster, M. S. (1977). Ecological and nutritional effects of food scarcity on a tropical frugivorous bird and its fruit source. *Ecology* **58,** 73–85.

Foster, M. S. (1978). Total frugivory in tropical passerines: A reappraisal. *J. Trop. Ecol.* **19,** 131–154.

Foster, R. B. (1974). Seasonality of fruit production and seed fall in a tropical forest ecosystem in Panama. Ph.D. thesis, Univ. Microfilms Int., Univ. of Michigan, Ann Arbor, Michigan.

Frankie, G. W., Baker, H. G., and Opler, P. A. (1974). Comparative phenological studies of trees in tropical wet and dry forests in the lowlands of Costa Rica. *J. Ecol.* **62,** 881–919.

Frost, P. G. H. (1980). Fruit-frugivore interactions in a South African coastal dune forest. *In* "Acta XVII Congressus Internationalis Ornithologici" (R. Nohring, ed.), pp. 1179–1184. Deuts. Orn. Ges., Berlin.

Graber, J. W., and Powers, P. M. (1981). Dwarf sumac as winter bird food. *Am. Midl. Nat.* **105,** 410–412.

Grant, G. S. (1979). Ring-billed Gulls feeding on date fruits. *Condor* **81,** 432–433.

Greenberg, R. (1981). Frugivory in some migrant tropical forest wood warblers. *Biotropica* **13,** 215–223.

Halls, L. K. (1973). Flowering and fruiting of southern browse species. *U.S. For. Serv. Res. Pap.* SO–90, 1–10.

Harper, J. L. (1977). "The Population Biology of Plants." Academic Press, London and New York.

Havlin, J. (1977). Ein Vergleich der Nahrung des Stars und der Amsel. *Folia Zool.* **26,** 45–56.

Herrera, C. M. (1981a). Are tropical fruits more rewarding to dispersers than temperate ones? *Am. Nat.* **118,** 896–907.

Herrera, C. M. (1981b). Fruit food of Robins wintering in southern Spanish mediterranean scrubland. *Bird Study* **28,** 115–122.

Herrera, C. M. (1981c). Fruit variation and competition for dispersers in natural populations of *Smilax aspera*. *Oikos* **36,** 51–58.

Herrera, C. M. (1981d). Datos sobre la dieta frugívora del Mirlo (*Turdus merula*) en dos localidades del Sur de España. *Doñana Acta Vert.* **8,** 306–310.

Herrera, C. M. (1982a). Defense of ripe fruits from pests: Its significance in relation to plant–disperser interactions. *Am. Nat.* **120,** 218–241.

Herrera, C. M. (1982b). Seasonal variation on the quality of fruits and diffuse coevolution between plants and avian dispersers. *Ecology* **63,** 773–785.

Herrera, C. M. (1984a). A study of avian frugivores, bird-dispersed plants, and their interaction in Mediterranean scrublands. *Ecol. Monogr.* **54,** 1–23.

Herrera, C. M. (1984b). Adaptation to frugivory of Mediterranean avian seed dispersers. *Ecology* **65,** 609–617.

Herrera, C. M. (1984c). Tipos morfológicos y funcionales en plantas del matorral mediterráneo del sur de España. *Studia Oecologica,* (in press).

Herrera, C. M. (1985a). Determinants of plant–animal coevolution: The case of mutualistic vertebrate seed dispersal systems. *Oikos* **44** (in press).

Herrera, C. M. (1985b). Seed dispersal and fitness determinants in wild rose: Effects of hawthorn, birds, mice, and browsing ungulates. *Oecologia* **63** (in press).

Herrera, C. M., and Jordano, P. (1981). *Prunus mahaleb* and birds: The high-efficiency seed dispersal system of a temperate fruiting tree. *Ecol. Monogr.* **51,** 203–218.

Hildén, O. (1965). Habitat selection in birds. A review. *Ann. Zool. Fenn.* **2,** 53–75.

Houssard, C., Escarré, J., and Romane, F. (1980). Development of species diversity in some mediterranean plant communities. *Vegetatio* **43,** 59–72.

Howe, H. F., and Estabrook, G. F. (1977). On intraspecific competition for avian dispersers in tropical trees. *Am. Nat.* **111,** 817–832.

Howe, H. F., and Smallwood, J. (1982). Ecology of seed dispersal. *Annu. Rev. Ecol. Syst.* **13,** 201–228.

Howe, H. F., and Vande Kerckhove, G. A. (1980). Nutmeg dispersal by tropical birds. *Science* **210,** 925–927.

James, F. C. (1971). Ordinations of habitat relationships among breeding birds. *Wilson Bull.* **83,** 215–236.

Janzen, D. H. (1981). *Ficus ovalis* seed predation by an orange-chinned parakeet (*Brotogeris jugularis*) in Costa Rica. *Auk* **98,** 841–844.

Janzen, D. H. (1983). Dispersal of seeds by vertebrate guts. *In* "Coevolution" (D. J. Futuyma and M. Slatkin, eds.), pp. 232–262. Sinauer, Sunderland, Massachusetts.

Johnson, A. S., and Landers, J. L. (1978). Fruit production in slash pine plantations in Georgia. *J. Wildl. Manage.* **42,** 606–613.

Jordano, P. (1982). Migrant birds are the main seed dispersers of blackberries in southern Spain. *Oikos* **38,** 183–193.

Jordano, P., and Herrera, C. M. (1981). The frugivorous diet of blackcap populations *Sylvia atricapilla* wintering in southern Spain. *Ibis* **123,** 502–507.

Kantak, G. E. (1979). Observations on some fruit-eating birds in Mexico. *Auk* **96,** 183–186.

Karr, J. R. (1971). Structure of avian communities in selected Panama and Illinois habitats. *Ecol. Monogr.* **41,** 207–233.

Karr, J. R. (1976). Seasonality, resource availability, and community diversity in tropical bird communities. *Am. Nat.* **110,** 973–994.

Karr, J. R. (1980). Geographical variation in the avifaunas of tropical forest undergrowth. *Auk* **97,** 283–298.

Kear, J. (1958). Plant poisons in the diet of wild birds. *Bull. Br. Ornithol. Club.* **88,** 98–102.

Keast, A. (1958). The influence of ecology on variation in the Mistletoebird (*Dicaeum hirundinaceum*). *Emu* **58,** 195–206.

Kendeigh, S. C. (1970). Energy requirements for existence in relation to size of bird. *Condor* **72,** 60–65.

Kendeigh, S. C., Dol'nik, V. R., and Gavrilov, V. M. (1977). Avian energetics. *In* "Granivorous Birds in Ecosystems" (J. Pinowski and S. C. Kendeigh, eds.), pp. 127–204. Cambridge Univ. Press, London and New York.

Kolunen, H., and Vikberg, P. (1978). Irruptions of the waxwing *Bombycilla garrulus* in S. Finland. *Ornis Fenn.* **55,** 126–131. (In Finnish, with English summary).

Lack, D. (1971). "Ecological Isolation in Birds." Blackwell, Oxford.

Laine, K. (1978). Aspects of annual variation in the numbers of flowers, berries and seeds in the Kilpisjärvi area. *Kilpisjärvi Notes* **2,** 12–16. (In Finnish, with English summary).

Lebreton, P., and Broyer, J. (1981). Contribution à l'étude des relations avifaune/altitude. I. Au niveau de la région Rhône-Alpes. *L'Oiseau et R.F.O.* **51,** 265–285.

Leck, C. F. (1971). Overlap in the diet of some neotropical birds. *Living Bird* **10,** 89–106.

Lederer, R. J. (1977a). Winter territoriality and foraging behavior of the Townsend's Solitaire. *Am. Midl. Nat.* **97,** 101–109.

Lederer, R. J. (1977b). Winter feeding territories in the Townsend's Solitaire. *Bird Banding* **48,** 11–18.

Lederer, R. J. (1981). Facultative territoriality in Townsend's Solitaire (*Myadestes townsendi*). *Southwest. Nat.* **25,** 461–467.

Leigh, E. G. (1975). Structure and climate in tropical rain forest. *Annu. Rev. Ecol. Syst.* **6,** 67–86.

Lein, M. R. (1972). A trophic comparison of avifaunas. *Syst. Zool.* **21,** 135–150.

Lemke, T. O. (1979). Fruit-eating behavior of Swallow-tailed Kites (*Elanoides forficatus*) in Colombia. *Condor* **81,** 207–208.

Lieberman, D. (1982). Seasonality and phenology in a dry tropical forest in Ghana. *J. Ecol.* **70,** 791–806.

Livingston, R. B. (1972). Influence of birds, stones and soil on the establishment of pasture juniper, *Juniperus communis,* and red cedar, *J. virginiana* in New England pastures. *Ecology* **53,** 1141–1147.

McDonnell, M. J., and Stiles, E. W. (1983). The structural complexity of old vegetation and the recruitment of bird-dispersed plant species. *Oecologia* **56,** 109–116.

McKey, D. (1975). The ecology of coevolved seed dispersal systems. *In* "Coevolution of Animals and Plants" (L. E. Gilbert and P. H. Raven, eds.), pp. 159–191. Univ. of Texas Press, Austin.

Marks, P. L. (1974). The role of pin cherry (*Prunus pensylvanica* L.) in the maintenance of stability in northern hardwood ecosystems. *Ecol. Monogr.* **44,** 73–88.

Milewski, A. V., and Bond, W. J. (1982). Convergence of myrmecochory in mediterranean Australia and South Africa. *In* "Ant–Plant Interactions in Australia" (R. C. Buckley, ed.), pp. 89–98. Junk, The Hague.

Moermond, T. C., and Denslow, J. S. (1983). Fruit choice in neotropical birds: Effects of fruit type and accessibility on selectivity. *J. Anim. Ecol.* **52,** 407–420.

Moore, F. R. (1977). Flocking behaviour and territorial competitors. *Anim. Behav.* **25,** 1063–1065.

Moore, F. R. (1978). Interspecific aggression: Toward whom should a mockingbird be aggressive? *Behav. Ecol. Sociobiol.* **3,** 173–176.

Morse, D. H. (1975). Ecological aspects of adaptive radiation in birds. *Biol. Rev.* **50,** 167–214.

Morton, E. S. (1973). On the evolutionary advantages and disadvantages of fruit eating in tropical birds. *Am. Nat.* **107,** 8–22.

Morton, E. S. (1977). Intratropical migration in the yellow-green vireo and piratic flycatcher. *Auk* **94,** 97–106.

Newton, I. (1972). "Finches." Collins, London.

Ogawa, I. (1977). Pellet analysis of the bull-headed shrike *Lanius bucephalus* and the seasonal change of food habits. *Tori* **26,** 63–75. (In Japanese, with English summary).

Opler, P. A., Baker, H. G., and Frankie, G. W. (1980). Plant reproductive characteristics during secondary succession in neotropical lowland forest ecosystems. *Biotropica* **12,** 40–46.

Pearson, D. L. (1977). A pantropical comparison of bird community structure on six lowland forest sites. *Condor* **79,** 232–244.

Putnam, L. S. (1949). The life history of the Cedar Waxwing. *Wilson Bull.* **61,** 141–182.

Rowan, M. K. (1967). A study of the colies of southern Africa. *Ostrich* **38,** 63–115.

Rowan, M. K. (1970). The foods of South African birds. *Ostrich* **8,** Suppl., 343–356.

Salomonson, M. G., and Balda, R. P. (1977). Winter territoriality of Townsend's Solitaires (*Myadestes townsendi*) in a piñon–juniper–ponderosa pine ecotone. *Condor* **79,** 148–161.

Schemske, D. W., and N. Brokaw (1981). Treefalls and the distribution of understory birds in a tropical forest. *Ecology* **62,** 938–945.

Sherburne, J. A. (1972). Effects of seasonal changes in the abundance and chemistry of the fleshy fruits of northeastern woody shrubs on patterns of exploitation by frugivorous birds. Ph.D. thesis, Univ. Microfilms Int., Ann Arbor, Michigan.

Smith, A. J. (1975). Invasion and ecesis of bird-disseminated woody plants in temperate forest sere. *Ecology* **56,** 19–34.

Smith, C. C. (1975). The coevolution of plants and seed predators. *In* "Coevolution of Animals and Plants" (L. E. Gilbert and P. H. Raven, eds.), pp. 51–77. Univ. of Texas Press, Austin.

Smythe, N. (1970). Relationships between fruiting seasons and seed dispersal methods in a neotropical forest. *Am. Nat.* **104,** 25–35.

Snow, B. K. (1970). A field study of the bearded bellbird in Trinidad. *Ibis* **112,** 299–329.

Snow, B. K. (1972). A field study of the calfbird *Perissocephalus tricolor*. *Ibis* **114,** 139–162.

Snow, B. K. (1977). Territorial behavior and courtship of the male three-wattled bellbird. *Auk* **94,** 623–645.

Snow, B. K., and Snow, D. W. (1971). The feeding ecology of tanagers and honeycreepers in Trinidad. *Auk* **88,** 291–322.

Snow, B. K., and Snow, D. W. (1984). Long-term defence of fruit by mistle thrushes. *Ibis* **126,** 39–49.

Snow, D. W. (1962a). A field study of the black and white manakin, *Manacus manacus,* in Trinidad. *Zoologica N.Y.* **47,** 65–104.

Snow, D. W. (1962b). A field study of the golden-headed manakin, *Pipra erythrocephala,* in Trinidad, W. I. *Zoologica N.Y.* **47,** 183–198.

Snow, D. W. (1962c). The natural history of the oilbird, *Steatornis caripensis,* in Trinidad, W. I. Part 2. Population, breeding ecology and food. *Zoologica N.Y.* **47,** 199–221.

Snow, D. W. (1965). A possible selective factor in the evolution of fruiting seasons in tropical forest. *Oikos* **15,** 274–281.

Snow, D. W. (1971). Evolutionary aspects of fruit-eating by birds. *Ibis* **113,** 194–202.

Snow, D. W. (1973). Distribution, ecology and evolution of the bellbirds (*Procnias,* Cotingidae). *Bull. Br. Mus. (Nat. Hist.), Zool.* **25,** 369–391.

Snow, D. W. (1981). Tropical frugivorous birds and their food plants: A world survey. *Biotropica* **13,** 1–14.

Sorensen, A. (1981). Interactions between birds and fruits in a British woodland. *Oecologia* **50,** 242–249.

Sorensen, A. (1983). Taste aversion and frugivore preference. *Oecologia* **56,** 117–120.

Stransky, J. J., and Halls, L. K. (1980). Fruiting of woody plants affected by site preparation and prior land use. *J. Wildl. Manage.* **44,** 258–263.

Svärdson, G. (1949). Competition and habitat selection in birds. *Oikos* **1,** 157–174.

Terborgh, J., and Diamond, J. M. (1970). Niche overlap in feeding assemblages of New Guinea birds. *Wilson Bull.* **82,** 29–52.

Thiollay, J. M. (1978). Les rapaces d'une zone de contact savane-forêt en Côte-d'Ivoire: Spécialisations alimentaires. *Alauda* **47,** 147–170.

Thiollay, J. M. (1980). L'evolution des peuplements d'oiseaux le long d'un gradient altitudinal dans l'Himalaya central. *Terre et Vie* **34,** 199–269.

Thompson, J. N., and Willson, M. F. (1978). Disturbance and the dispersal of fleshy fruits. *Science* **200,** 1161–1163.

Thompson, J. N., and Willson, M. F. (1979). Evolution of temperate fruit/bird interactions: Phenological strategies. *Evolution* **33,** 973–982.

Thomson, A. L., and Moreau, R. E. (1957). Feeding habits of the palm-nut vulture *Gypohierax. Ibis* **99,** 608–613.

Turček, F. (1961). "Okologische beziehungen der Vögel und Gehölze." Slowak. Akad. Wissench., Bratislava.

Tutman, I. (1969). Beobachtungen an olivenfressenden Vögeln. *Vogelwelt* **90,** 1–8.

Tyrväinen, H. (1970). The mass occurrence of the fieldfare (*Turdus pilaris* L.) in the winter of 1964/65 in Finland. *Ann. Zool. Fenn.* **7,** 349–357.

Tyrväinen, H. (1975). The winter irruption of the fieldfare *Turdus pilaris* and the supply of rowanberries. *Ornis Fenn.* **52,** 23–31.

Ulfstrand, S. (1963). Ecological aspects of irruptive bird migration in northwestern Europe. *Proc. Int. Ornithol. Congr. 13th,* pp. 780–794.

Waage, J. K. (1979). The evolution of insect/vertebrate associations. *Biol. J. Linn. Soc.* **12,** 187–224.

Walker, C. H. (1983). Pesticides and birds-mechanisms of selective toxicity. *Agric. Ecosys. Environ.* **9,** 211–226.

Walsberg, G. E. (1975). Digestive adaptations of *Phainopepla nitens* associated with the eating of mistletoe berries. *Condor* **77,** 169–174.

Walsberg, G. E. (1977). Ecology and energetics of contrasting social systems in *Phainopepla nitens* (Aves: Ptilogonatidae). *Univ. Calif. Publ. Zool.* **108,** 1–63.

Walsberg, G. E. (1980). Energy expenditure in free-living birds: Patterns and diversity. *In* "Acta XVII Congressus Internationalis Ornithologici" (R. Nohring, ed.), pp. 300–305. Deuts. Orn. Ges., Berlin.

Wheelwright, N. T. (1983). Fruits and the ecology of Resplendent Quetzals. *Auk* **100,** 286–301.

White, P. S. (1979). Pattern, process, and natural disturbance in vegetation. *Bot. Rev.* **45,** 229–299.

White, S. C. (1974). Ecological aspects of growth and nutrition in tropical fruit-eating birds. Ph.D. thesis, Univ. Microfilms Int., Ann Arbor, Michigan.

Willson, M. F., Porter, E. A., and Condit, R. S. (1982). Avian frugivore activity in relation to forest light gaps. *Caribb. J. Sci.* **18,** 1–6.

Winterbottom, J. M. (1978). Birds. *In* "Biogeography and Ecology of Southern Africa" (M. J. A. Werger, ed.), Vol. 2, pp. 949–979. Junk, The Hague.

Chapter 12

Habitat Selection in Australian Honeyeaters, with Special Reference to Nectar Productivity

HUGH A. FORD

Department of Zoology
University of New England
Armidale, New South Wales
Australia

and

DAVID C. PATON

Department of Zoology
University of Adelaide
Adelaide, South Australia
Australia

HABITAT SELECTION IN BIRDS

ISBN 0-12-178080-5

I. INTRODUCTION

The honeyeaters (Meliphagidae) are a diverse passerine family, virtually confined to Australia, New Guinea, and the Pacific region. In Australia honeyeaters are the dominant family in most habitats, and often ten or more species coexist. They display remarkable adaptive radiation, filling niches occupied by members of a range of families on other continents (Keast, 1976). At one extreme there are specialized monospecific genera like the Painted Honeyeater (*Grantiella*) which feeds on mistletoe berries. At the other extreme is the large generalized genus *Meliphaga*. Although nectar is a major source of energy for many species, large quantities of insects, other carbohydrate foods, and some fruit are consumed (Pyke, 1980). There are excellent examples of closely related species occupying separate habitats (Ford and Paton, 1976a; Wykes, 1982). On the other hand, nectarivorous species may partition nectar sources along a gradient of nectar richness: larger species monopolizing the richest areas, smaller species more scattered sources (Ford and Paton, 1982). Many species are highly mobile, some showing regular migration and other more complex movements, which have been described as nomadic (Keast, 1968a). The more mobile species presumably select their habitats, often virtually on a day-to-day basis. Sedentary species, on the other hand, may be able to adapt to specific features of their habitat.

In this chapter we show that honeyeaters often occupy specific habitats and that this selection of habitat may be based on the production of nectar or other carbohydrate foods.

II. AUSTRALIAN HONEYEATERS

There are some 72 species of honeyeaters in Australia, with the greatest diversity in the northeast and central eastern coast. They occupy all habitats from stony and sandy deserts to tropical and temperate rainforests. Most species and the greatest densities are found in dry sclerophyll (*Eucalyptus*) forests and heath-

TABLE I

Main Genera of Australian Honeyeaters

Genus	Common name	Beak length (mm)	Weight (gm)	Number of species	Diet[a]
Meliphaga (incl. *Lichenostomus*)	Typical Honeyeaters	12–16	17–35	22	I > N
Melithreptus	Dark-headed Honeyeaters	12–15	12–20	6	I > N
Manorina	Miners	15–20	35–50	4	I > N
Phylidonyris	Yellow-winged Honeyeaters	16–22	14–20	5	N > I
Acanthorhynchus	Spinebills	20–25	10–14	2	N > I
Myzomela	Myzomelas	16	8	4	N > I
Anthochaera (incl. *Acanthagenys*)	Wattlebirds	20–24	45–150	4	N > I

[a] For diet I = insects, including honeydew, lerp, etc.; N = nectar.

land. It is not unusual for honeyeaters to represent 80% or more of all birds seen or captured in suitable habitats (e.g., Ford and Pursey, 1982).

Honeyeaters range in size from about 8 to 150 gm and have short to moderately long and decurved bills (Table I). Nectar is taken by all species, usually by perching on the plant or on the ground. Only the smaller species occasionally hover (see Pyke, 1981 for reasons). There is a positive correlation between beak length and the proportion of time spent foraging on nectar ($r = 0.56, p < 0.01$). Indeed, two guilds can be identified. The long-billed species depend largely on nectar, taking insects principally as a source of protein (Ford and Paton, 1976b, 1977; Recher and Abbott, 1970). Shorter-billed species take less nectar and more insects (Ford and Paton, 1976a). Honeyeaters are almost unique among birds in feeding extensively on and sometimes depending on other carbohydrate foods such as manna (a sugary exudate from damaged eucalypt leaves), honeydew (from various Homoptera), and lerp (the sugary protective coating of Psyllids) (Paton, 1980). These are important energy sources for long-billed species when nectar is scarce (Paton, 1982, 1985) and for short-billed species at all times of the year (Wykes, 1982).

Honeyeaters are major pollinators of many of the plants on whose nectar they depend, including species of Myrtaceae, Proteaceae, Epacridaceae, Loranthaceae, Fabaceae, and Myoporaceae (Ford *et al.*, 1979; Armstrong, 1979; Keighery, 1982). Some areas have plants in flower year-round, but in most there is a seasonal pattern of relative scarcity and abundance of nectar. This means that honeyeaters have to change their food or move around in search of flowers (Keast, 1968a).

Some examples of the main flower types are shown in Fig. 1. Simple cup-

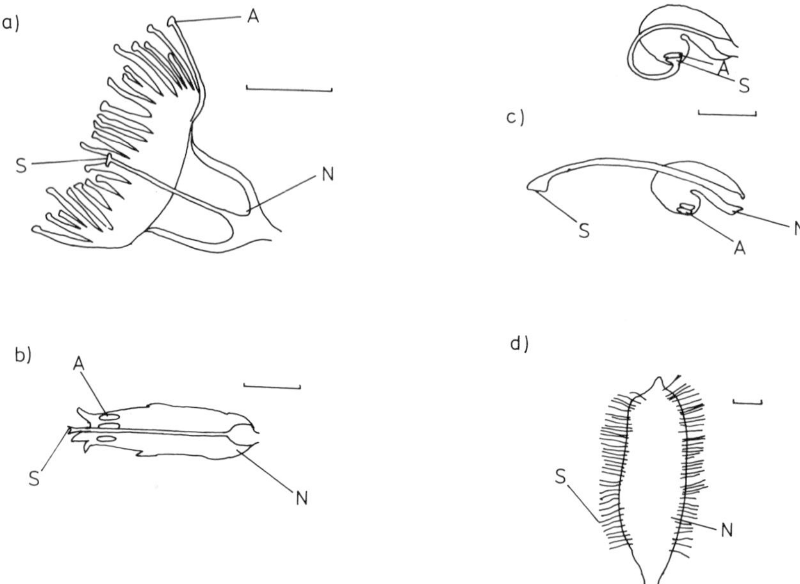

Fig. 1. Major floral types shown by bird-pollinated plants in Australia: (a) Cup-shaped, *Eucalyptus,* (b) tubular, *Astroloma,* (c) gullet-shaped, *Grevillea,* (d) brush inflorescence, *Banksia.* Scale-1 cm, A = anther, N = nectary, S = stigma.

shaped flowers (*Eucalyptus, Callistemon*) are usually the most common and are readily accessible to all honeyeaters and to other birds, mammals, and insects. Tubular flowers are found in the heaths (Epacridaceae), mistletoes (Loranthaceae), and *Eremophila* (Myoporaceae). These are accessible to most honeyeaters, though occasionally short-beaked species pierce the base of the corolla to obtain nectar. Gullet-shaped flowers are typical of the Proteaceae (*Grevillea, Banksia*) in which flowers are often arranged in large brushlike inflorescences. Nectar is accessible to most visitors in flowers of this type though stiff styles on some banksias may make visiting difficult for short-beaked species. At least in southern Australia there appear to be no cases of precise coadaptation between a single species of bird and one species of plant.

A. Habitat Separation in Short-Beaked Honeyeaters

The short-beaked genus *Meliphaga* (sometimes split into *Meliphaga* and *Lichenostomus*) has a total of 22 species. In South Australia the genus is neatly separated by habitat (Fig. 2) (Ford and Paton, 1976a). The only exceptions are the White-eared Honeyeater (*M. leucotis*), which is a bark-feeding specialist, and the Singing Honeyeater (*M. virescens*), which feeds mostly from shrubs and often takes fruit. All the others are generalized, taking nectar and insects from all substrates at all vegetation levels. Similar habitat separation is shown in New

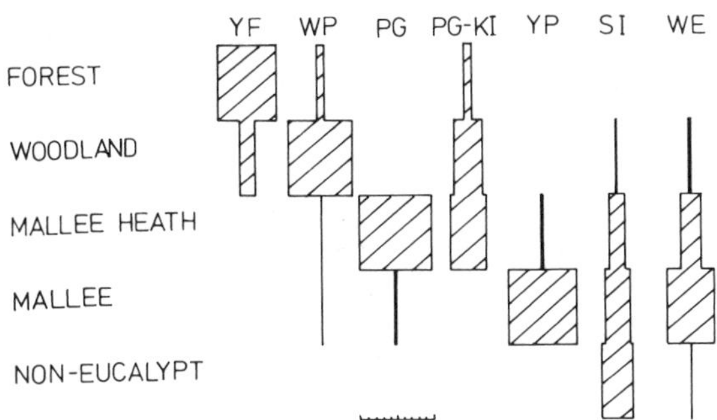

Fig. 2. Proportion of observations of five species of *Meliphaga* in each of six habitats. Forest, eucalypt forest with heathy understory; Woodland, grassy eucalypt woodland or riparian eucalypts; Mallee-heath, dense and many-stemmed eucalypts with heathy understory; Mallee, dense many-stemmed eucalypts with grassy or chenopod understory; Noneucalypt, *Acacia* or *Myoporum* woodland or chenopod shrubland. Habitats are arranged from wet to dry. YF, Yellow-faced; WP, White-plumed; PG, Purple-gaped on mainland; PG-KI, Purple-gaped on Kangaroo Island, YP, Yellow-plumed, SI, Singing; and WE, White-eared Honeyeaters.

South Wales (Recher and Abbott, 1970), though here there are extra species. The Lewin's Honeyeater (*M. lewinii*) occupies wet eucalyptus forest and rainforests, and the Fuscous Honeyeater (*M. fusca*) fits in between the White-plumed (*M. penicillata*) and the Yellow-faced (*M. chrysops*), in tall woodland (Ford and Bell, 1982). Ford and Paton (1976a) used only broad descriptions of habitat; more recently Wykes (1982) has quantified differences in habitats occupied by *Lichenostomus* (*Meliphaga*) in Victoria.

It is not clear how the differences in habitat displayed by most members of this genus arose. Once established, though, they are probably maintained by interspecific aggression. Honeyeaters are notoriously aggressive, and some, perhaps many, *Meliphaga* are cooperative breeders (Dow, 1980), making invasion of their territories especially difficult. Habitat choice is by no means fixed, as indicated by the expansion of habitat in the absence of potential competitors. The forest-dwelling Yellow-faced Honeyeater is absent from Kangaroo Island and southwestern Australia. The forests here are occupied by Purple-gaped (*M. cratitia*) and Yellow-plumed (*M. ornata*) Honeyeaters, respectively, which occupy mallee–heath and mallee in mainland eastern Australia (Fig. 2) (Slater, 1976). The Purple-gaped Honeyeater also inhabits woodland on Kangaroo Island in the absence of the White-plumed Honeyeater (Fig. 2).

The White-plumed Honeyeater's predilection for gums, especially *Eucalyptus camaldulensis,* is well known (e.g. Gannon, 1966). However, Green (1985) found that hand-reared birds showed no preference for native eucalypt leaves over exotic oak *Quercus* leaves. She suggested that the species' strong preference for eucalypts in the wild is because they find more food there (i.e., their preferences are learned).

Thus, it seems likely that *Meliphaga* honeyeaters select habitats where they can forage efficiently and where they will not be excluded by aggressive congeners.

The other two main genera of short-billed honeyeaters are *Melithreptus* and *Manorina*. Keast (1968b) has compared the habitats of *Melithreptus*, which differ but not as markedly as in *Meliphaga*. In the miners *Manorina*, there are representatives in wet forest (*Ma. melanophrys*), dry forest and woodland (*Ma. melanocephala*), and dry woodland (*Ma. flavigula*). These are very aggressive communal breeders (Dow, 1977, 1978), which exclude most other birds from their territories. Nevertheless, overlaps occasionally occur, and hybridization between *Ma. melanocephala* and *Ma. flavigula* has been reported (Dow, 1972). Perhaps similar mechanisms operate to separate members of this genus, as we have suggested for *Meliphaga*.

B. Habitat Separation in Long-Beaked Honeyeaters

In the long-beaked honeyeaters, genera tend to be well separated geographically (e.g., *Acanthorhynchus* spinebills, with species in the east and southwest) or to display considerable overlaps (*Anthochaera*, wattlebirds). *Phylidonyris* has the most species, several of which have been well studied (Recher, 1971, 1977; Paton, 1979). *Phylidonyris albifrons*, the White-fronted Honeyeater, is the arid habitat representative of the genus. In southern Australia the Tawny-crowned Honeyeater (*P. melanops*) occupies open heath, the Crescent Honeyeater (*P. pyrrhoptera*) sclerophyll forest, and the New Holland Honeyeater (*P. novaehollandiae*) occupies a wide range of forest and heath (Recher, 1971; Ford and Paton, 1977). The White-cheeked Honeyeater (*P. nigra*) is very similar morphologically to the New Holland. These two are generally separated geographically, though with appreciable overlap. The New Holland is more common south of Sydney, whereas the White-cheeked is more common to the north (Gannon, 1966). Gannon suggested that the White-cheeked prefers cooler and more humid sites. Recher (1977) found no obvious habitat difference between the two north of Sydney, though they bred at different times. He suggested that both nectar and insects were superabundant, alleviating competition between them. Both species occur in southwestern Australia, in different habitats (Recher, 1977). More interesting than habitat differences between long-beaked honeyeaters are differences between members of the guild based on nectar productivity. These are discussed in Section IV.

III. SEASONAL CHANGES IN ABUNDANCE OF NECTAR AND HONEYEATERS

Habitat selection implies an element of active choice of habitat rather than a passive presence in a habitat (Morse, 1980). This is hard to establish for more sedentary species, such as some species of *Meliphaga*. Where a species is more mobile, habitat selection can be inferred from a positive correlation between the species' abundance and the availability of food or some other critical resource. Nectar is a food that is easy to measure, and the abundance of honeyeaters has been related to the abundance of nectar, with varying degrees of quantification.

Keast (1968a) summarized the information on seasonal presence and absence of honeyeaters and concluded that they were highly dependent on the flowering of nectar-bearing plants. Some species, such as Yellow-faced and White-naped Honeyeaters (*Melithreptus lunatus*), show a regular long distance migration (Hindwood, 1956; Liddy, 1966). Interestingly, these are generalized rather than highly nectarivorous species. In most species though there are complex, chiefly local, movements which show some changes from one year to the next. In the first attempt to relate numbers of honeyeaters to abundance of flowers, Bell (1966) censused birds in heathland near Sydney. He found that the local abundance and distribution of honeyeaters were related to the flowering of their major food plants: *Banksia robur, B. spinulosa, B. marginata,* and *Lambertia formosa.* Over the past decade, more detailed studies have been carried out in a variety of habitats which are now summarized.

A. Woodland and Forest in South Australia

Ford (1983) counted birds in 16 sites near Adelaide over 12 months. He noted the intensity of flowering of the major nectar-bearing flowers in each site each month. Although this was only a crude measure of nectar abundance, it should allow sites and months to be ranked. Numbers of honeyeaters recorded each month were positively correlated with the flowering index for that month, when data from all sites were pooled (Fig. 3, $r_s = 0.84$, $p < 0.01$). Each of the eight common species was also positively correlated with flowering index ($p < 0.05$ in all cases). Birds and flowers both peaked in winter and were scarce in summer. As honeyeaters bred in late winter and spring (August–December), the winter peak was not simply caused by juveniles. In some species (White-plumed and Crescent Honeyeaters) increases were twofold; in others (Yellow-faced and New Holland Honeyeaters) they were about tenfold. Perhaps movements were local in the former and long distance in the latter group.

Eleven out of 16 sites showed a positive correlation between birds and flowers. Where individual species of birds were compared within sites they were positively correlated with flowering index in 43 out of 69 cases. When individual species of plants were compared with individual bird species, there were strong correlations for four plant species (not significant for 0–2 bird species), poor correlations for one species (4 significant, 3 not), and no correlations for three species. Interactions between plant species, e.g., a rather unproductive plant occurring with a very productive one with a different flowering season, clouded the overall pattern here.

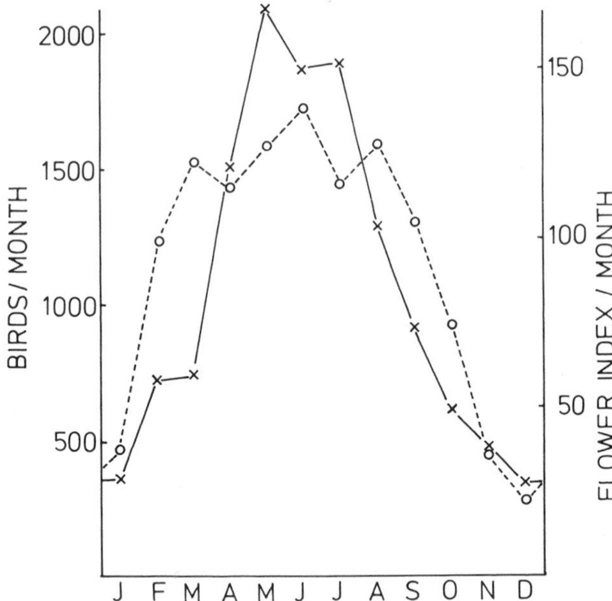

Fig. 3. Total numbers of honeyeaters of all species (x) and flowering index for all species (o), all sites combined, near Adelaide, South Australia. (Modified from Ford, 1983, reproduced with permission of the Australian Bird Study Association.)

On the whole then honeyeaters moved into sites as more flowers became available and out again as they became scarce. This was true equally for both short- and long-beaked honeyeaters. Nectar was depleted during the day for most of the year, except when flowers were most abundant. This suggests that although birds increased in response to an increase in nectar, they did not exploit the nectar fully.

B. Heathland in Western Australia

Collins and Briffa (1982) visited three sites every 3 months in heathland near Perth. One site had flowers and birds only in December; a second had the most birds and flowers in June and September. A third site had neither birds nor flowers in December but had both at other times of the year. Furthermore, some birds moved as much as 4 km between sites, indicating that they were actively selecting habitats where nectar was abundant.

C. Forest and Heathland in Victoria

Paton (1979) counted honeyeaters and measured the abundance of nectar in heathland near Cranbourne 40 km southeast of Melbourne and in sclerophyll

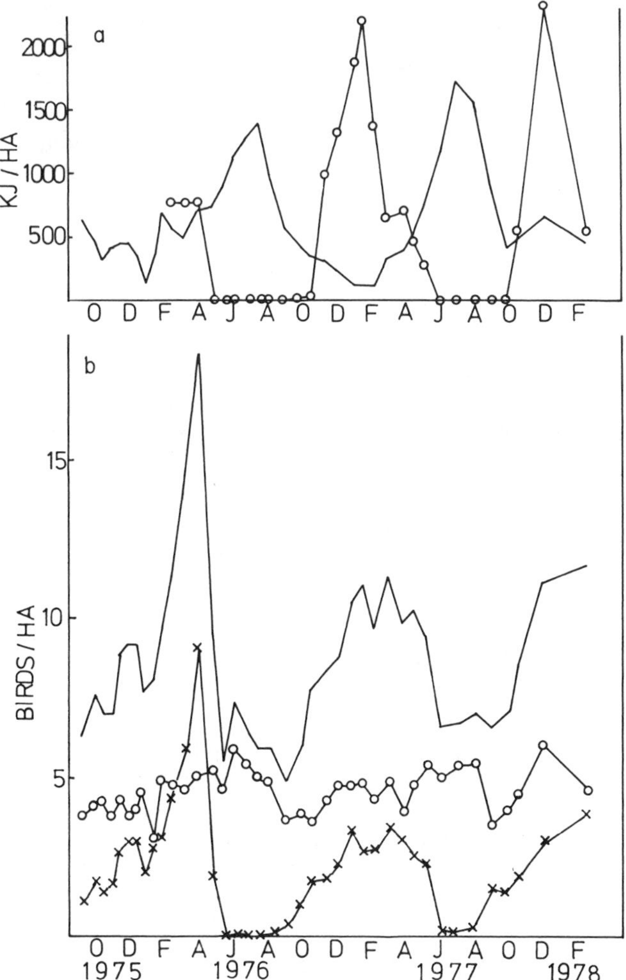

Fig. 4. (a) Productivity of nectar (———) and standing crop of manna (O—O) and (b) numbers of New Holland Honeyeaters (O), White-naped (x) and all honeyeaters, at Cranbourne, Victoria.

forest at Golton Vale 265 km northwest of Melbourne. Nectar production and honeyeater abundance at Cranbourne are shown on Fig. 4. New Holland Honeyeaters are positively correlated with nectar production ($r_s = 0.519, p < 0.01$). However, numbers of this species only vary about two-fold, whereas nectar varies tenfold. Clearly, New Holland Honeyeaters are primarily sedentary, and as nectar is depleted at all times of the year, they experience a relative shortage in summer. At this time of year they turn to other carbohydrate resources, especially manna, which peaks in midsummer (Paton, 1980, 1982).

For all honeyeaters combined and White-naped Honeyeaters, the only other common species, there is actually a negative correlation between abundance and nectar productivity ($r_s = -0.346$, $p < 0.05$, $r_s = -0.498$, $p < 0.01$). White-naped Honeyeaters move in as nectar becomes scarce. However, their abundance is strongly correlated with the standing crop of manna ($r_s = 0.809$, $p < 0.001$), as it is for all honeyeaters ($r_s = 0.784$, $p < 0.001$). Manna is the major food of White-naped Honeyeaters for most of the year (Paton, 1980). Since nectar was measured in terms of productivity, but manna was measured as a standing crop, they cannot legitimately be added. Perhaps a combination of the two plus other carbohydrate sources like honeydew would be the best predictor of honeyeater numbers.

The annual pattern of honeyeaters at Cranbourne is almost a mirror image of that for Adelaide (Fig. 3). The migratory White-naped and Yellow-faced Honeyeaters contribute largely to the peaks in summer and winter, respectively, in Victoria and South Australia. However, there is as yet no evidence that the same population migrates between the two areas.

At Golton Vale both honeyeaters and nectar peak in summer, and again the New Holland Honeyeater is the most abundant. However, the relationships are not significant ($r_s = +0.299$ for honeyeaters, $+0.438$ for New Holland, both $p > 0.05$, $n = 12$). The largely resident New Holland Honeyeaters switch to honeydew produced by psyllids and eriococcids under bark during the cooler months (Paton, 1980). The other birds, such as woodswallows *Artamus* also feed on nectar when it is most abundant.

In summary, although there is some relationship between nectar and honeyeaters, especially New Holland Honeyeaters, in Victoria this is considerably masked by the importance of other carbohydrates like manna and honeydew. For some species, e.g., White-naped Honeyeater, these other foods are more important than nectar.

D. Upland Heath in New South Wales

Ford and Pursey (1982) compared the abundance of honeyeaters and flowers at an upland site in northeastern New South Wales. Here the small Eastern Spinebill *Acanthorhynchus tenuirostris* was the commonest honeyeater, and *Banksia spinulosa* the major source of energy. The abundance of Spinebills was estimated by mist netting over 2 years and by censusing at 10 sites in 1 year (March–September). The numbers of productive inflorescences of *Banksia* in three 5×5 m plots were counted in 1 year, and the percentage of inflorescences that were productive at any one time was measured in the second year. In all cases the abundance of Spinebills was positively correlated with abundance of productive inflorescences ($r_s = 0.77$–0.83, $p < 0.01$). Spinebills increased in April as *B. spinulosa* started flowering and decreased in September as flowering finished.

No other honeyeaters were sufficiently common to examine correlations with flower abundance.

E. Heathland near Sydney

Pyke (1983) censused honeyeaters and measured productivity of nectar and biomass of insects at three sites near Sydney. There was no indication of any relationship between bird numbers and abundance of nectar or insects. Nectar was occasionally so abundant that it dripped from *Banksia* inflorescences. At other times it was totally inadequate to supply the birds' needs. There was no obvious alternative source of carbohydrate. As in the other sites New Holland Honeyeaters were most abundant, and a few were resident. Pyke (1983) concluded that most of the birds he saw were transient, moving through the sites without feeding. He had no idea where they had come from or where they were going.

Obviously, to understand the movements of honeyeaters one needs to know what is happening in a wide range of sites over a large area.

F. Summary

Studies that have looked for a relationship between productivity of nectar and numbers of honeyeaters have had varying success. There are several possible reasons why honeyeater abundance may not be closely correlated with the nectar supply.

Honeyeaters take other carbohydrate foods as sources of energy (Paton, 1980). The importance of these has only recently been appreciated, and unfortunately these are harder than nectar to measure. Consequently, no study has yet measured the productivity of all energy sources within a large area over a long period. One reason why Paton's data in Victoria (Section III,C) did not show strong relationships between birds and nectar was that manna and hemipteran exudates were abundant and used by birds when flower nectar was scarce.

A further possibility is that honeyeater populations are limited by protein rather than energy. Insects, not nectar, may determine their presence or absence (Carpenter, 1978). Pyke's (1983) data appear to discount this though. For New Holland Honeyeaters at least, demands for protein are remarkably low and can usually be satisfied in a short period. Collecting energy may take very much longer (Paton, 1982).

Most workers have assumed that honeyeaters have had almost exclusive use of the nectar sources. However other animals, including a variety of birds, such as silvereyes (*Zosterops*), lorikeets (Loriinae), and woodswallows, as well as insects and mammals, consume a portion of the available nectar for periods in the year. Quantities consumed by these animals may be hard to measure, particularly for the nocturnal visitors. Many of the birds and mammals opportunistically exploit superabundant nectar, though lorikeets and the Western Australian honeypossum (*Tarsipes rostratus*) feed year-round on nectar (Wooller *et al.*, 1983).

Insects, especially honeybees (*Apis mellifera*), may take large quantities of nectar in warm weather (e.g., Ford, 1979; Paton, 1979). Paton (1985) estimated that insects consumed 30–60% of the nectar produced by bird flowers during warmer months at Golton Vale, Victoria. However, it is not known if these insects merely take what the honeyeaters leave or whether they indirectly reduce densities of honeyeaters.

On occasion nectar may be unexploited by honeyeaters and other nectarivores, sometimes to the extent that it drips from inflorescences onto the ground (Carpenter, 1978). This may happen over an extensive area, suggesting a limit to the number of birds that are available to consume the nectar. As nectar production increases, there may be simply no birds to take advantage of it. Other animals such as rodents, marsupials, and bowerbirds may then exploit the nectar. Such superabundances seem to be particularly common in coastal New South Wales.

Another way in which nectar may become locally superabundant is when more profitable sources are available nearby. For instance it takes far longer to obtain sufficient nectar dispersed among many small tubular flowers, such as *Epacris,* than it does from a few *Banksia* inflorescences even though the former may offer a greater daily productivity per unit area (Paton, 1982).

IV. NECTAR PRODUCTIVITY, COMMUNITY STRUCTURE, AND HONEYEATER BEHAVIOR

A. Productivity and Honeyeater Ergonomics

All else being equal, one would expect birds to select habitats in which they can collect food most rapidly. Not only should this ensure that they obtain sufficient food to survive, but also that they may collect extra to breed, molt, or prepare themselves for migration or future shortages of food. In addition the less time they spend foraging, the shorter will be their exposure to predators. Thus, the overall productivity of nectar may be expected to influence honeyeaters, and, further, we should also expect standing crops of individual flowers or inflorescences to be important. One way in which birds can increase the standing crops of flowers on which they feed is by preventing other birds from visiting these flowers. Such territorial defense would be expected whenever nectar is not superabundant. However, as productivity declines, the area that needs to be defended increases, and defense becomes more costly or less successful. Honeyeaters differ in size and energy requirements, and the quantity of nectar they need for defense would vary accordingly. Furthermore, larger species tend to dominate smaller species (Ford, 1979).

If we consider both energy requirements of the birds and ease of defense of the flowers, we might expect the following pattern to occur as nectar sources change from poor (i.e., sparsely distributed) to rich (i.e., dense) in time or space. Poor sources can be economically exploited only by the small honeyeaters. As resources become richer progressively larger species can economically defend the nectar sources and exclude smaller species. Perhaps at an even higher level of nectar production all species could feed peaceably together with little interference, without depleting the nectar. Thus, changes in the productivity of nectar and how it is presented in flowers or inflorescences should influence not only the numbers of honeyeaters, but also which species are common and how they behave (Carpenter, 1978; Ford, 1979).

Nectar sources used by birds in Australia vary from banksia inflorescences where birds may take more than 1 ml (~5kJ) of nectar per minute to the tubular flowers of *Epacris* yielding less than 1 μl each. At the very best honeyeaters could collect only about 0.1 ml/min from *Epacris*. Honeyeaters differ in the rates at which they can probe flowers, larger species feeding faster than smaller ones. These faster feeding rates are insufficient to compensate for the increased energy requirements of the larger species. For example 110-gm Red Wattlebirds have energy requirements that are approximately five times those of an 11-gm Eastern Spinebill, yet they feed at best twice as fast. Thus, many of the less productive plants are uneconomical for the larger honeyeaters; wattlebirds, for instance, very rarely feed on *Epacris,* whereas spinebills often do. Banksias are readily visited by all honeyeaters, and they especially attract the large wattlebirds and aggressive New Holland Honeyeaters. Spinebills rarely visit banksias in South Australia and Victoria, probably because large species prevent them; in northeastern New South Wales Spinebills rely heavily on *Banksia* nectar (Ford and Pursey, 1982).

Thus, different honeyeaters appear to select different plant species and, consequently, the habitats where they are common. Probably all honeyeaters would prefer the same highly productive plant species and habitats, but some are denied access to these areas by larger, more aggressive species.

B. Sites of Differing Richness on Kangaroo Island

Ford and Paton (1982) chose three sites on Kangaroo Island, 100 km south of Adelaide, South Australia, that differed in plant species. They measured daily nectar production and assessed the relative abundance of different honeyeater species at each site. The poorest site had scattered tubular *Correa* flowers producing only 6 J of nectar/day. The next site had abundant *Adenanthos* flowers producing about 14 J of nectar/day, plus some *Banksia* inflorescences producing 8–10 kJ of nectar per day and a few *Eucalyptus cosmophylla* flowers. The richest site had abundant *E. cosmophylla* flowers which produced some 250 J of nectar per day.

The largest honeyeater, the Red Wattlebird *Anthochaera carunculata,* occupied only the richest site where it defended flowers. Many of the aggressive medium-sized New Holland Honeyeaters also fed here, and others defended territories in the intermediate site. The smaller Crescent Honeyeaters and Eastern Spinebills also fed here but were often chased. The Spinebill was most common in the poor site, where fewer Crescent and New Holland Honeyeaters occurred.

Although the larger species appeared to select sites where nectar production was high, this was not true of the smaller species. Because of the exclusion of the smaller by the larger species, had more sites been chosen, it is possible that the density of Eastern Spinebills would have shown a negative correlation with nectar production!

C. Temporal Changes in Nectar Production in Victoria

Just as one might expect larger species to occupy the best sites in an area at any one time, so might one expect them to move into areas as nectar production increases and to do so later than smaller species do. This is what Paton (1979) found in a site with *Banksia marginata* near Melbourne (Fig. 5). As the density of inflorescences increased, more New Holland Honeyeaters established territories. Later, Little Wattlebirds (*A. chrysoptera*) (65 gm versus New Holland, 20 gm) set up territories and excluded many New Holland Honeyeaters. As nectar production passed the peak the Wattlebird's territories expanded so that few New Holland Honeyeaters remained. Finally, the Wattlebirds departed, and the Honeyeaters returned. In this case changes in numbers of New Holland Honeyeaters were not in step with nectar production.

D. Response of Honeyeaters to Unpredictable Nectar Production

Where changes in nectar production are gradual and predictable the density of birds may change predictably, and territorial birds expand and contract their territories in response to changes in flower density. In some cases though, nectar production changes unpredictably from day to day. How do honeyeaters respond to this?

The upland site in New South Wales, discussed in Section III,D, has dense patches of *Banksia spinulosa* which flower in winter. In places nectar production per unit area may rival that of the best site on Kangaroo Island (Section IV,B). However, the most common species is the small Eastern Spinebill, which on Kangaroo Island is excluded to the poorer sites. Red Wattlebirds are present but show no inclination to be territorial. Some New Holland Honeyeaters do hold territories though; McFarland (1985) has plotted their territorial boundaries and followed nectar production. He found that overnight temperature determines the amount of nectar produced. When it falls below −3°C there is negligible nectar

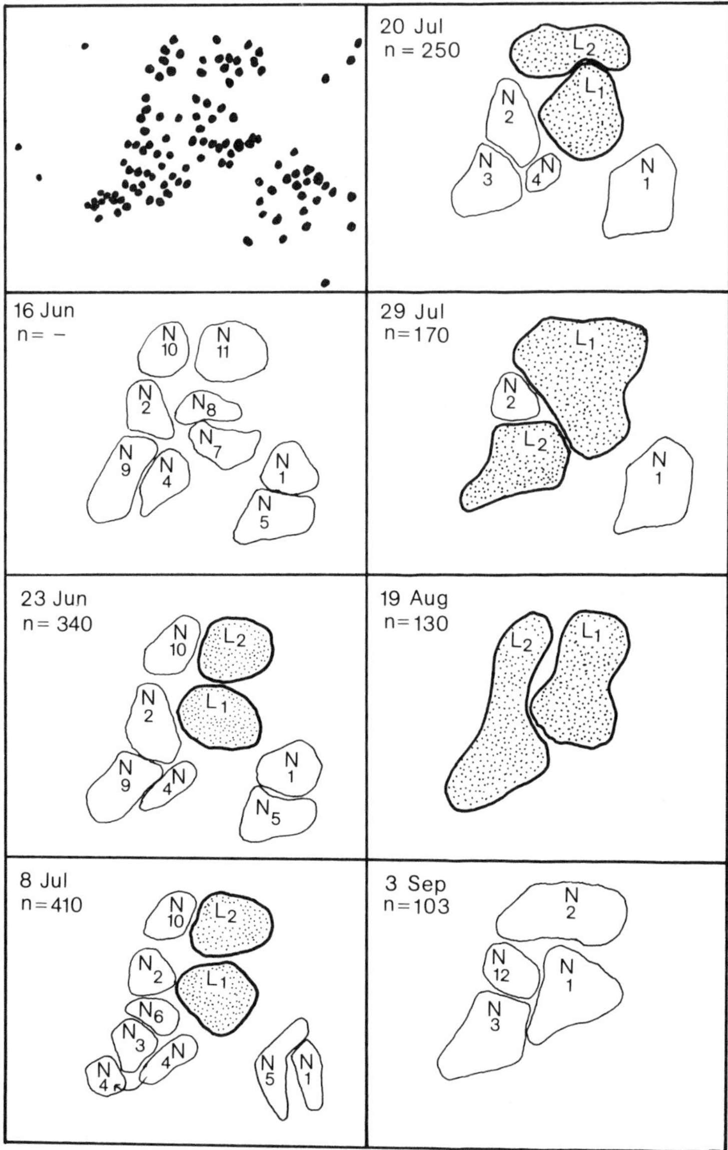

Fig. 5. Pattern of territories of New Holland Honeyeaters (N1, etc.) and Little Wattlebirds (L1, etc.) in *Banksia marginata* heathland at Cranbourne, Victoria, over 3 months (N = number of productive inflorescences in the site). Top left box indicates distribution of *B. marginata* plants in the site. (From Paton and Ford, 1983, reproduced with permission of Van Nostrand Reinhold Ltd.)

production. Daily productivity per territory varies from 1 to 230 kJ and, surprisingly, the New Holland Honeyeaters do not adjust their territorial boundaries to accommodate these changes. Days when nectar production in a territory exceeds the New Holland Honeyeaters' requirements of about 100 kJ/day outnumber days when little nectar is produced. There are rarely more than 3 "poor" days in a row; New Holland Honeyeaters probably put on fat on good days in order to survive the bad days. This rather precarious existence probably accounts for the scarcity of New Holland Honeyeaters and lack of territorial Red Wattlebirds.

Thus, if only average levels of nectar production are considered, these may be a poor guide to the numbers of individuals in an area and to the species that should be common.

V. DISCUSSION

A. Patterns of Habitat Selection Shown by Honeyeaters

Different populations of Australian honeyeaters show the whole spectrum of responses to changes in productivity of habitat, from sedentariness through local movements between neighboring habitats to long distance migration. To complicate matters further, different components of a population may be sedentary or mobile (Paton, 1979; hinted at in Ford and Pursey, 1982; Pyke, 1983).

Where populations are sedentary they often display a preference for a particular structural habitat, e.g., White-plumed Honeyeater for eucalypt woodland. Their lack of movement makes it hard to assess whether the habitats they occupy are selected, i.e., chosen in preference to other habitats, or whether the habitats result from relative survival, perhaps combined with chance and history. Sedentary populations may exploit a succession of flowering plants or, more often, turn to other foods when flower nectar is scarce. The *Meliphaga* species take many insects, whereas the New Holland Honeyeater turns to other carbohydrate foods (Paton, 1980, 1982).

There is frequently movement between neighboring habitats in response to local flowering peaks. Increases in birds correlated with increases in flowers have been shown in South and Western Australia and upland New South Wales (Ford, 1983; Collins and Briffa, 1982; Ford and Pursey, 1982), less successfully in Victoria (Paton, 1979, 1980, 1985), and not at all near Sydney (Pyke, 1983). Many factors prevent neat relationships between the abundance of birds and nectar. These include: exploitation of other foods, superabundance of nectar (i.e., insufficient birds), interspecific competition, use of nectar by other animals, and unpredictability of nectar sources. In addition, patterns within one site may be obscured by changes in nectar productivity in other nearby habitats (Pyke, 1983).

The long distance migrants (Yellow-faced and White-naped Honeyeaters) have been known for several decades (Hindwood, 1956). However, we still know little about the ultimate or proximate reasons for their movements. Even in these species there are often sedentary components of the population. Once the migrants have arrived at their destination they may still show local movements between neighboring habitats in response to flowering or abundance of other foods.

B. Comparison of Habitat Selection in Hummingbirds and Honeyeaters

Far more is known about the ecology of hummingbirds (Trochilidae) than is known about any other group of nectar-feeding birds. We do not wish to review this knowledge here, but merely seek to illustrate the parallels in habitat selection between hummingbirds and honeyeaters.

The situation is relatively simple in North America as here there are few species, most of which are long distance migrants. Species generally occupy different habitats for breeding but may overlap at other times. Perhaps the best known migrant is the Rufus Hummingbird (*Selasphorus rufus*), which breeds in the northwest, migrates along the Sierra Nevada and Cascade and Rocky mountains, and winters in southern Mexico (Phillips, 1975). It is an aggressive species, defending territories in a series of sites as flowers become abundant (e.g., Wyoming, Armitage, 1955; northwestern California, Gass *et al.*, 1976; southwestern California, Cody, 1968; Hixon *et al.*, 1983; Arizona, Kodric-Brown and Brown, 1978). What is more, in all sites territory size is rapidly adjusted as flower density changes so that the productivity of the territory remains about the same. This sensitivity to changes in productivity appears greater than that shown by honeyeaters, possibly because the small hummingbirds are less able to withstand periodic shortages of energy.

It is in the more complex communities of nectar-feeding birds in the Neotropics that one might expect hummingbirds to show the range of responses to nectar productivity that is shown by honeyeaters. Des Granges and Grant (1980) and Feinsinger (1980) summarized the movements of hummingbirds in sites in Mexico and in Costa Rica and Trinidad. All sites had resident species and local migrants. Residents from one habitat often became migrants into other habitats, suggesting sedentary and mobile components of the population. In other species apparently the whole population follows blooms among neighboring habitats. Long-distance migrants from temperate North America were most noticeable in Mexico.

Sufficient flowers were available year-round to sustain some residents, which were generalized and often territorial (e.g., *Panterpe* in Costa Rica, Wolf *et al.*, 1976 and *Amazilia* in Trinidad, Feinsinger and Swarm, 1982). Neither insects nor alternative sources of carbohydrate seem as important to hummingbirds as they are to honeyeaters. However, hummingbirds may become more insectivorous when they are sedentary and nectar is scarce (Pitelka, 1942; Young, 1971) or when they are breeding (Gass and Montgomerie, 1981). There have also been several recent reports of hummingbirds feeding on coccid honeydew (Edwards, 1982, and records therein). Perhaps future studies on hummingbirds should take more note of foods other than nectar.

Among the local migrant hummingbirds in the Neotropics there are specialists with very long and/or decurved beaks, adapted to flowers with morphologically similar corollas (e.g., various hermits and *Centropogon* and *Heliconia* spp., Wolf *et al.,* 1976; Stiles, 1975). There are no counterparts to these tightly coadapted birds and plants among the honeyeaters, at least not in southern Australia.

As might be expected there is a general relationship between number of flowers and abundance of hummingbirds (Carpenter, 1978). Gass and Montgomerie (1981) found that biomass of hummingbirds and available energy from nectar were positively correlated at sites in midlatitude alpine meadows and tropical lowlands. Stiles (1980) related abundance of flowers to numbers of hummingbirds and their annual cycle in a complex community in Costa Rican rainforest. Flowers are most abundant in the dry (March–April) and early wet seasons (June–August) and scarce in the late wet season (November–January). Stiles classed nine species of hummingbird as common or abundant in the first two seasons but only four species as such in the period of scarcity. Ranking the availability of flowers and abundance of birds there is a positive relationship for nonhermits and all species combined ($r_s = 0.79, 0.85, p < 0.01$). For hermits there are too many ties to accurately assess ranks, but these specialists appeared to respond less to the flushes of flowers. As they choose particular species of plants, the flowering of these and movements of the hermits may not coincide with the general pattern of nectar production.

Feinsinger and Swarm (1982) found no relationship between numbers of *Amazilia* (the dominant resident hummingbird) and flower abundance on Trinidad and Tobago ($r_s = -0.02, -0.20, p > 0.05$). However, numbers of all hummingbirds and all birds (including passerines) were correlated with flower abundance on Tobago ($r_s = 0.83, 0.81, p < 0.01$), whereas only the latter were correlated on Trinidad ($r_s = 0.26, 0.56, p > 0.05, p < 0.05$). Other hummingbirds and passerines moved into the study sites as flowers increased. Despite this, nectar was at times underexploited. To some extent the passerines (e.g., the Bananaquit *Coereba*) may have switched to other foods when nectar became scarce. There are similarities between the situation on Trinidad and that on Paton's Victorian sites. Both have a resident dominant species (*Amazilia tobaci* and *Phylidonyris novaehollandiae*) that responds to scarcity by diversifying its diet. There are nectar feeders whose numbers increase when nectar or other carbohydrate foods are abundant and more generalized species that turn to nectar at times of great abundance (honeycreepers and woodswallows).

In Mexico the migrant hummingbirds arrive in the dry season, coinciding with an abundance of cup-shaped flowers (Des Granges and Grant, 1980). The resident hummingbirds tend to visit tubular flowers, which were available year-round.

Where several hummingbird species coexist, they appear to choose different plant species or different patches of habitat. To some extent this is due to specialization, but the system may operate largely because there is a dominance hierarchy, with dominant species defending rich plants or patches and lower-ranking species either defending poorer sites or "trap lining" between dispersed nectar sources (Feinsinger, 1976; Wolf *et al.*, 1976; Des Granges and Grant, 1980). Not only size but also wing loading influences the position a species holds in the hierarchy (Feinsinger and Chaplin, 1975).

Overall, the Australian honeyeaters and Neotropical hummingbirds respond in similar ways to changes in productivity of flower nectar, their major energy source. Both groups are highly aggressive, frequently defending territories where nectar is sufficiently abundant. In both groups straightforward responses of particular species to nectar abundance may be complicated by interspecific competition and resource superabundance. Alternative foods seem to be more important for honeyeaters than for hummingbirds. On the other hand some hummingbird species show greater specialization for types or species of flowers than do honeyeaters. Some hummingbirds seem to live a more fugitive existence, flitting between temporary patches of abundance in neighboring habitats, than is shown by honeyeaters; possibly the small *Myzomela* honeyeaters behave in this fashion in tropical Australia and New Guinea.

ACKNOWLEDGMENTS

We thank David McFarland, who read the manuscript and provided some unpublished data. The manuscript was written while H. A. F. was on sabbatical leave in the Department of Biological Sciences, Stirling University, Scotland. He is grateful for the Department's hospitality and is especially grateful to Margaret Keilt for typing the manuscript.

REFERENCES

Armitage, K. B. (1955). Territorial behavior in fall migrant Rufous Hummingbirds. *Condor* **57,** 239–240.

Armstrong, J. A. (1979). Biotic pollination mechanisms in the Australian flora: A review. *N. Z. J. Bot.* **17,** 467–508.

Bell, H. L. (1966). A population study of heathland birds. *Emu* **65,** 295–304.

Carpenter, F. L. (1978). A spectrum of nectar-eater communities. *Am. Zool.* **18,** 809–819.

Cody, M. L. (1968). Interspecific territoriality among hummingbird species. *Condor* **70,** 270–271.

Collins, B. G., and Briffa, P. (1982). Seasonal variation of abundance and foraging of three species of Australian honeyeaters. *Aust. Wildl. Res.* **9,** 557–569.

Des Granges, J-L., and Grant, P. R. (1980). Migrant hummingbirds' accommodation into tropical communities. *In* "Migrant Birds in the Neotropics: Ecology, Behavior, Distribution and Conservation" (J. A. Keast and E. S. Morton, eds.), pp. 395–409. Random House (Smithsonian Inst. Press), New York.

Dow, D. D. (1972). Hybridization in the avian genus *Myzantha. Mem. Queensl. Mus.* **16,** 265–269.

Dow, D. D. (1977). Indiscriminate interspecific aggression leading to almost sole occupancy of space by a single species of bird. *Emu* **77,** 115–121.

Dow, D. D. (1978). Breeding biology and development of the young of *Manorina melanocephala,* a communally breeding honeyeater. *Emu* **78,** 207–222.

Dow, D. D. (1980). Communally breeding Australian birds with an analysis of distributional and environmental factors. *Emu* **80,** 121–140.

Edwards, E. P. (1982). Hummingbirds feeding on an excretion produced by scale insects. *Condor* **84,** 122.

Feinsinger, P. (1976). Organization of a tropical guild of nectarivorous birds. *Ecol. Monogr.* **46,** 257–291.

Feinsinger, P. (1980). Asynchronous migration patterns and the coexistence of tropical hummingbirds. *In* "Migrant Birds in the Neotropics: Ecology, Behavior, Distribution and Conservation" (J. A. Keast and E. S. Morton, eds.), pp. 411–419. Random House (Smithsonian Inst. Press), New York.

Feinsinger, P., and Chaplin, S. B. (1975). On the relationship between wing disc loading and foraging strategy in hummingbirds. *Am. Nat.* **109,** 217–224.

Feinsinger, P., and Swarm, L. A. (1982). "Ecological release", seasonal variation in food supply and the hummingbird *Amazilia tobaci* on Trinidad and Tobago. *Ecology* **63,** 1574–1587.

Ford, H. A. (1979). Interspecific competition in Australian honeyeaters—depletion of common resources. *Aust. J. Ecol.* **4,** 145–164.

Ford, H. A. (1983). Relation between number of honeyeaters and intensity of flowering near Adelaide, South Australia. *Corella* **7,** 25–31.

Ford, H. A., and Bell, H. L. (1982). Density of birds in eucalypt woodland affected to varying degrees by dieback. *Emu* **81,** 202–208.

Ford, H. A., and Paton, D. C. (1976a). Resource partitioning and competition in honeyeaters of the genus *Meliphaga. Aust. J. Ecol.* **1,** 281–287.

Ford, H. A., and Paton, D. C. (1976b). The value of insects and nectar to honeyeaters. *Emu* **76,** 83–84.

Ford, H. A., and Paton, D. C. (1977). The comparative ecology of ten species of honeyeaters in South Australia. *Aust. J. Ecol* **2,** 399–407.

Ford, H. A., and Paton, D. C. (1982). Partitioning of nectar sources in an Australian honeyeater community. *Aust. J. Ecol.* **7,** 149–159.

Ford, H. A., and Pursey, J. F. (1982). Status and feeding of the Eastern Spinebill *Acanthorhynchus tenuirostris* at New England National Park, north-eastern NSW. *Emu* **82,** 203–211.

Ford, H. A., Paton, D. C., and Forde, N. (1979). Birds as pollinators of Australian plants. *N. Z. J. Bot.* **17,** 509–519.

Gannon, G. R. (1966). The influence of habitat on the distribution of Australian birds. *Emu* **65,** 241–253.

Gass, C. L., and Montgomerie, R. D. (1981). Hummingbird foraging behavior: Decision-making and energy regulation. *In* "Foraging Behavior: Ecological, Ethological and Psychological Approaches" (A. C. Kamil and T. D. Sargent, eds.), pp. 159–199. Garland STPM Press, New York.

Gass, C. L., Angehar, G., and Centa, J. (1976). Regulation of food supply by feeding territoriality in the rufous hummingbird. *Can. J. Zool.* **54,** 2046–2054.

Green, R. (1985). Native and exotic birds in the suburban habitat. *In* "The Dynamic Partnership: Birds and Plants in Southern Australia" (H. A. Ford and D. C. Paton, eds.). Flora and Fauna Board of S. A., Adelaide (in press).

Hindwood, K. A. (1956). The migration of the White-naped and Yellow-faced Honeyeaters. *Emu* **56,** 421–425.

Hixon, M. A., Carpenter, F. L., and Paton, D. C. (1983). Territory area, flower density and time budgeting in hummingbirds: An experimental and theoretical analysis. *Am. Nat.* **122,** 366–391.

Keast, J. A. (1968a). Seasonal movements in the Australian honeyeaters (Meliphagidae) and their ecological significance. *Emu* **67,** 159–209.

Keast, J. A. (1968b). Competitive interactions and the evolution of ecological niches as illustrated by the Australian honeyeater genus *Melithreptus* (Meliphagidae). *Evolution* **22,** 762–784.

Keast, J. A. (1976). The origins of adaptive zone utilizations and adaptive radiations, as illustrated by the Australian Meliphagidae. *Proc. Int. Ornithol. Congr. 16th,* pp. 71–82.

Keighery, G. J. (1982). Bird pollinated plants in Western Australia and their breeding systems. *In* "Pollination and Evolution" (J. A. Armstrong, J. M. Powell, and A. J. Richards, eds.), pp. 77–90. Royal Botanic Gardens, Sydney.

Kodric-Brown, A., and Brown, J. H. (1978). Influence of economics, interspecific competition and sexual dimorphism on territoriality of migrant rufous hummingbirds. *Ecology* **59,** 285–296.

Liddy, J. (1966). Autumnal migration of Yellow-faced Honeyeaters. *Emu* **66,** 87–103.

McFarland, D. (1985). Community structure and territoriality in honeyeaters in an unpredictable environment. Ph.D. thesis, University of New England, Armidale, Australia.

Morse, D. H. (1980). "Behavioural mechanisms in ecology." Harvard Univ. Press., Cambridge, Massachusetts.

Paton, D. C. (1979). The behaviour and feeding ecology of the New Holland Honeyeater, *Phylidonyris novaehollandiae* in Victoria. Ph.D. thesis, Monash University, Melbourne, Australia.

Paton, D. C. (1980). The importance of manna, honeydew and lerp in the diets of honeyeaters. *Emu* **80,** 213–226.

Paton, D. C. (1982). The diet of the New Holland Honeyeater, *Phylidonyris novaehollandiae. Aust. J. Ecol.* **7,** 279–298.

Paton, D. C. (1985). Food supply, population structure and behaviour of New Holland Honeyeaters *Phylidonyris novaehollandiae* in woodland near Horsham, Victoria. *In* "Birds of the Eucalypt Forest and Woodlands: Ecology, Conversation and Management" (J. A. Keast, H. F. Recher, H. A. Ford, and D. Saunders, eds.). Surrey-Beattie, Sidney.

Paton, D. C., and Ford, H. A. (1983). The influence of plant characteristics and honeyeater size on levels of pollination in Australian plants. *In* "Handbook of Experimental Pollination Biology" (C. E. Jones and R. J. Little, eds.), pp. 235–248. Van Nostrand-Reinhold, Princeton, New Jersey.

Phillips, A. R. (1975). The migrations of Allen's and other hummingbirds. *Condor* **77,** 196–205.

Pitelka, F. A. (1942). Territoriality and related problems in North American hummingbirds. *Condor* **44,** 189–204.

Pyke, G. H. (1980). The foraging behaviour of Australian honeyeaters: A review and some comparisons with hummingbirds. *Aust. J. Ecol.* **5,** 343–370.

Pyke, G. H. (1981). Why hummingbirds hover and honeyeaters perch. *Anim. Behav.* **29,** 861–867.

Pyke, G. H. (1983). Relationships between honeyeater numbers and nectar production in heathlands in Sydney. *Aust. J. Ecol.* **8,** 217–234.

Recher, H. F. (1971). Sharing of habitat by three congeneric honeyeaters. *Emu* **71,** 147–152.

Recher, H. F. (1977). Ecology of co-existing White-cheeked and New Holland Honeyeaters. *Emu* **77,** 136–142.

Recher, H. F., and Abbott, I. J. (1970). Some differences in the use of habitat by White-eared and White-cheeked Honeyeaters. *Emu* **70,** 117–125.

Slater, P. (1976). "A Field Guide to Australian Birds: Passerines." Rigby, Adelaide, Australia.

Stiles, F. G. (1975). Ecology, flower phenology and hummingbird pollination of some Costa Rican *Heliconia* species. *Ecology* **56,** 285–301.

Stiles, F. G. (1980). The annual cycle in a tropical wet forest hummingbird community. *Ibis* **122,** 322–343.

Wolf, L. L., Stiles, F. G., and Hainsworth, F. R. (1976). Ecological organization of a tropical highland hummingbird community. *J. Anim. Ecol.* **45,** 349–379.

Wooller, R. D., Russell, E. M., Renfree, M. B., and Towers, P. A. (1983). A comparison of seasonal changes in the pollen loads of nectarivorous marsupials and birds. *Aust. Wildl. Res.* **10,** 311–317.

Wykes, B. (1982). Resource partitioning and the role of competition in structuring *Lichenostomus* honeyeater (and *Manorina melanophrys*) communities in southern Victoria. Ph.D. thesis, Monash University, Melbourne, Australia.

Young, A. M. (1971). Foraging for insects by a tropical hummingbird. *Condor* **73,** 36–45.

Chapter 13

Physiological Consequences of Microhabitat Selection

GLENN E. WALSBERG

Department of Zoology
Arizona State University
Tempe, Arizona

I. INTRODUCTION

Life depends on an essentially continuous exchange of mass and energy between an organism and its environment. This mass and energy transfer is critically sensitive to the local physical environment (the microclimate), which in turn is determined by the microhabitat an animal selects. Comprehending the physiological consequences of microhabitat selection therefore must be a central task of physiological ecology and one that presents notable challenges by requiring complex integrations of behavioral, physiological, and biophysical approaches.

HABITAT SELECTION IN BIRDS

ISBN 0-12-178080-5

Variation in energy transfer associated with microclimate selection directly affects the thermoregulatory demands with which a bird must cope. Though such demands may simply exceed the animal's capabilities to respond effectively and thus be lethal, probably more important are sublethal demands with which the animal can cope but which entail substantial power consumption. Current data indicate that maintenance metabolism (basal and thermostatic demands) accounts for approximately 40–60% of the total energy expended by free-living birds (Walsberg, 1983a). Variation in such maintenance demands is sensitive to the thermal stress imposed by the animal's microhabitat, and thus microhabitat selection can importantly influence the remaining fraction of the animal's total energy budget that can be allocated to elective activities, such as reproduction, resource defense, and social activities. Indeed, in at least some species temporal variation in the total energy budget is due primarily to changing thermostatic demands rather than changes in other components such as locomotor or synthetic activity (Walsberg, 1983a).

Mass exchange with the environment also is strongly controlled by microhabitat selection. Substances whose transfer is most obviously affected by environmental variables include respiratory gases and water vapor. The gaseous microclimate of birds has received extensive study only in the last 10 years; such analyses have dealt largely with determinants of egg dehydration during natural incubation.

The critical properties of microclimates include wind, radiation, air temperature, and humidity. These factors can interact in complex and variable manners that may appear initially puzzling. An example is the variable relation between coat color and solar heating in birds and mammals. Darker colors either may increase or decrease radiative heat gain depending on a suite of organismal and environmental properties, including wind velocity, coat structure, and coat optics (Walsberg, 1983b). Such potential variability substantially complicates studies of the physiological consequences of microsite selection. Studies of this subject traditionally have been divided between field observations of site selection and laboratory analyses of the physiological effects of various properties of the physical environment. The literature in avian thermobiology is replete with data describing the relation between air temperature and resting metabolic rate or evaporative water loss (for review, see Calder and King, 1974). The effects of variation in wind velocity and short-wave radiation have been studied less frequently, and some environmental factors, such as precipitation, have been ignored almost completely. Though laboratory studies often have entailed sophisticated analyses and have generated data of fundamental importance, the behavioral realism necessary to comprehend the ecological importance of such physiological effects usually is eliminated by the artificial setting. Observations of microhabitat selection by birds in nature also are available, though data necessary to confidently estimate the physiological consequences of such site selection commonly are lacking. The introduction of heat budget analyses to integrate field and laboratory observations of avian physiological ecology by workers such as Calder (1973a, 1973b) greatly improved our understanding of avian microclimate selection. In addition, theoretical and technical developments have substantially eased difficulties inherent in such integrations (e.g., Bakken and Gates, 1975; Bakken *et al.*, 1981; Robinson *et al.*, 1976; Walsberg and King, 1978a, 1978b, 1980).

II. GASEOUS ENVIRONMENT

A. Determinants and Consequences of Egg Dehydration

The avian egg respires through pores in the eggshell that also allow diffusive loss of water vapor. Consequently, eggs typically lose 9–18% of their initial mass in water during the course of incubation (Drent, 1975). The degree of egg dehydration is sensitive to the hydric microclimate to which the egg is exposed, and such dehydration is thought to importantly affect embryonic viability. Though high rates of evaporation from the egg may cause lethal desiccation, very low rates of evaporation also are deleterious. Evaporation of water from the egg allows formation of the internal air cell, which apparently is necessary in most bird species to allow transition between chorioallantoic respiration and pulmonary respiration (Romanoff and Romanoff, 1967; Visschedijk, 1968). In addition to this role in initiating pulmonary respiration, Ar and Rahn (1980) noted that such evaporative water loss may function importantly to maintain the embryo at constant hydration. Metabolic water production during development tends to increase the relative water content of the egg. Typical rates of evaporation loss balance this water production such that fractional hydration of the embryo tends to stay constant during incubation. Thus, maintenance of appropriate rates of vapor loss from eggs may play important roles in both embryonic respiration and osmoregulation. Evaluation of the significance of such effects is severely inhibited, however, by lack of information regarding embryonic responses to dehydration. Since sensitivity to dehydration varies markedly between adults of different species, it is reasonable to expect that similar variation may also characterize the avian embryo. Unfortunately, there are few available data to examine the relation between egg water loss and viability, and these are almost completely restricted to domestic species that have been isolated from natural selection for many generations.

Egg dehydration follows the relation

$$M_{H_2O} = \rho_{egg} - \rho_{nest}/r_{egg} + r_a$$

where $\dot{M}_{H_2O}$ is water loss (gm/sec), ρ_{egg} is the vapor density within the egg (gm/m^3), ρ_{nest} is the vapor density in the nest, r_{egg} is the resistance of the shell to vapor movement (sec/m), and r_a is the vapor resistance of the aerodynamic boundary layer adhering to the shell's outer surface (Walsberg, 1980). Microclimatic determinants of egg dehydration thus are represented by ρ_{nest} and r_a. However, the shell resistance (r_{egg}) is comparatively large and overwhelms effects due to variation in boundary layer resistances (Tracy and Sotherland, 1979). Microclimatic effects upon r_a, such as those due to wind velocity, are therefore unlikely to be important, and nest humidity thus represents the microclimatic factor of primary importance in determining egg dehydration. This key environmental factor is determined by a host of variables, however, including the evaporation of water from the adult's skin and the egg into the nest, the humidity

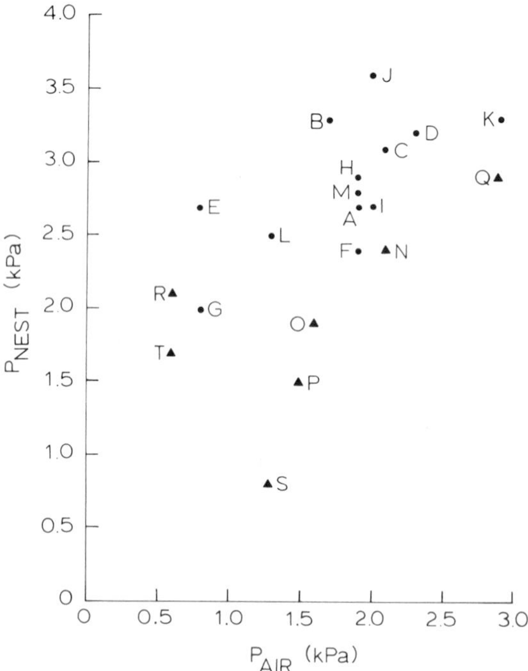

Fig. 1. Relation between general atmospheric humidity (P_{air}) and nest humidity (P_{nest}). Humidity is quantified as vapor pressure (kilopascals). Ground nesting species (circles) are: A. *Phasianus colchicus* (Rahn *et al.*, 1977), B. *Charadrius vociferous* (Grant, 1982), C. *Haematopus mexicanus* (Grant, 1982), D. *Recurvirostra americana* (Grant, 1982), E. *Larus glaucescens* (Morgan *et al.*, 1978), F. *Larus heermanni* (Rahn and Dawson, 1979), G. *Rissa tridactyla* (Morgan *et al.*, 1978), H. *Sterna fuscata* (Rahn *et al.*, 1976), I. *Sterna forsteri* (Grant, 1982), J. *Sterna maxima* (Vleck *et al.*, 1983), K. *Anous tenurostris* (Rahn *et al.*, 1976), L. *Columba livia* (Lomholt, 1976), M. *Chordeiles acutipennis* (Grant, 1982). Aerial nesting species (triangles) are: N. *Bulbulcus ibis* (Vleck *et al.*, 1983), O. *Casmerodius albus* (Vleck *et al.*, 1983), P. *Eudocimus albus* (Vleck *et al.*, 1983), Q. *Gygis alba* (Rahn *et al.*, 1976), R. *Phainopepla nitens* (Walsberg, 1983c), S. *Philetairus socius* (Bartholomew *et al.*, 1976), T. *Carpodacus mexicanus* (Walsberg, 1983c).

of the air surrounding the nest, and the nest structure (Walsberg, 1980). Considering this complex array of factors, it is not surprising that there is only a loose correlation between nest humidity and the humidity of the general atmosphere (t test, $p < 0.05$; $r^2 = 0.244$ (Fig. 1). However, it is clear that environmental properties are important in egg dehydration and there is increasing evidence that birds respond adaptively to such influences.

B. Altitudinal Effects on Egg Dehydration

Birds breed in habitats ranging up to 6500 m above sea level (Rahn, 1977). Because a change in barometric pressure produces an inverse change in the diffusivity of gases, water vapor tends to diffuse more rapidly from eggs held at higher elevations (Paganelli *et al.*, 1975; Paganelli, 1980). For example, standard atmospheric pressure at 4000 m altitude is reduced 61% below that at sea level. If other factors are held constant, this will increase the rate of vapor diffusion such that an egg that loses 16% of its initial mass in water during the course of incubation at sea level (a typical value; see Drent, 1970) will lose 26% of its mass when incubated at 4000 m.

Possible adaptive responses to such increased potential for desiccation at high altitude include increasing the initial water content of the egg, increasing the path length for vapor diffusion by increasing shell thickness, decreasing the number of pores in the shell, and changing the parental contribution to nest humidity. Little or no information is available describing changes in provisioning of the eggs with water or changing the parental contribution to nest humidity. Shell thickness has been examined in a variety of species and found not to vary with altitude (Carey, 1980). Adaptive shifts are seen, however, in shell porosity. In all species examined except *Pica pica,* shell conductance to water vapor varies inversely with altitude (Table I). These shifts apparently are produced by changes in shell porosity, since shell thickness is not correlated with altitude. In some cases, the reduction in shell porosity or conductance to water vapor is substantially greater than expected for the observed altitudinal change. The significance of this "overcompensation" is unclear. Possible explanations may include adaptations in response to (1) longer incubation periods at higher altitudes that would produce greater total egg water loss, (2) an increase in average egg temperature at higher altitudes that would increase egg vapor pressures and evaporation rates, (3) a decrease in vapor density at higher altitudes, or (4) difficulties at high altitude involving conservation of carbon dioxide or maintenance of acid–base balance by the embryo.

Almost all studies describing adaptive shifts in shell porosity are based upon analyses of interpopulation differences. It is not known to what degree such differences represent genetically fixed characteristics of local populations or phenotypic flexibility of individuals. For at least one species (*Gallus domes-*

TABLE I

Decrease in Water Vapor Conductance of Egg Shells (G_{H_2O}) Compared to Decrease in Barometric Pressure (P_B) Produced by Increased Altitude[a]

Species	P_B reduction (%)	G_{H_2O} reduction (%)	Reference
Gallus gallus	31	28	Rahn *et al.*, 1982
Gallus domesticus	37	32	Wangensteen *et al.*, 1974
Hirundo rustica	17	59	Packard *et al.*, 1977
Petrochelidon pyrrhonota	19	39	Sotherland *et al.*, 1980
Turdus migratorius	32	26	Carey, 1980
Agelaius phoenecius	32	24	Carey, 1980

[a] Computed as reduction compared to values characteristic of lowland habitat or population.

ticus), the data of Ledoux (1977) suggest that individuals possess such phenotypic flexibility and can adjust shell porosity in a manner appropriate for a particular altitude. The existence of such flexibility in wild species clearly is an important question, since it may affect the range of altitudes over which an individual has a high likelihood of successfully reproducing.

C. Nest Placement and Humidity

Nest location within a habitat can importantly affect humidity within the nest. Perhaps the most extreme case occurs in grebes, of which at least a few species build floating nests in which a pool of water accumulates around the eggs (Sotherland, 1979; Ackerman and Platter-Rieger, 1979). In *Podilymbus podiceps,* shell conductance to water vapor is about three times the normal values. This allows the embryo to experience a degree of dehydration typical of other species despite a very humid environment; 13% of initial egg mass is lost as water during the course of incubation (Ackerman and Platter-Rieger, 1979).

A second extreme in nest placement occurs in fossorial birds. Though few burrowing species have been studied, Birchard and Kilgore (1980) have compared the hydric environments occupied by cup-nesting Barn Swallows (*Hirundo rustica*) and burrowing Bank Swallows (*Riparia riparia*). Humidity averages substantially higher in Bank Swallow nests (2.7 kPa versus 1.2 kPa). Apparently as an adaptation to the higher nest humidity, Bank Swallows produce eggshells that are 43% more permeable to water vapor than Barn Swallow eggshells.

Less extreme variation in nest placement can also have significant effects upon humidity within the nest cup. Figure 1 illustrates that nests built on the ground tend to contain more humid air than do nests built aboveground. For these data, general atmospheric humidity does not differ significantly between ground and aerial nests (ANOVA, $p > 0.30$). In spite of this similar range of atmospheric humidities, vapor pressure within ground nests averages 48% higher than in aerial nests ($\bar{x} \pm SD = 2.87 \pm 0.443$ kPa in ground nests and 1.88 ± 0.739 kPa in aerial nests; the difference is statistically significant using analysis of variance [$p < 0.01$]). The cause of this difference is not clear, but it may reflect (1) the lower permeability to water vapor of typical ground nests compared to fibrous aerial nests (Walsberg, 1980, 1983c; Vleck *et al.*, 1983), (2) the soil acting as a water source for the nest cup (Vleck *et al.*, 1983), or (3) decreased wind velocity and convective loss of water vapor from nests near the ground (Vleck *et al.*, 1983). In 10 species studied by Vleck *et al.* (1983), those using aerial nests apparently compensate for decreased nest humidity by producing eggshells with a relatively low conductance to water vapor ($\bar{x} = 78\%$ of the value predicted by the equation of Hoyt, 1980). As with altitudinal adjustments, it is not known whether such variation in shell porosity represents phenotypic plasticity or a relatively fixed character of a population.

D. Adult Regulation of Nest Humidity

In contrast to the increasing evidence that egg dehydration is commonly regulated by adjustments in shell porosity, the limited data that are available do not support the hypothesis put forth by several workers that the adult bird behaviorally regulates nest humidity by varying ventilation of the nest (Rahn *et al.*, 1977; Morgan *et al.*, 1978; Ar and Rahn, 1980). As an adult incubates the clutch, water vapor diffusing from skin and eggs accumulates in the nest. These workers suggested that by varying the degree or frequency at which the adult allows this humid air to be flushed from the nest and replaced with drier air, the parent might regulate nest humidity and thus egg dehydration. This is an important possibility, since it would produce a particularly complex regulatory problem for the parent: it must simultaneously and independently regulate nest humidity and egg temperature. This could substantially limit the variety and timing of parental behavior. However, the few tests of this hypothesis that are available suggest that parent birds probably do not regulate nest humidity. In a previous analysis (Walsberg, 1980), I constructed a mathematical model that subsumed factors affecting egg water loss in nature: (1) the vapor pressure of the egg and the adult's skin, which are sources of water vapor to the nest, and (2) the resistances of the shell, the nest, and the parent's integument to vapor movement. Simulations using this model and limited empirical data indicated that regulation of egg water loss during incubation could be effectively accomplished by variation in shell porosity and that behavioral regulation of nest humidity is unlikely to be necessary for adjustment of egg dehydration. This suggested that the parent's behavior is largely released from selective pressures related to regulation of nest humidity. Later field tests of such predictions were conducted by artificially varying nest humidity during natural incubation in *Phainopepla nitens* and *Carpodacus mexicanus* (Walsberg, 1983c). No change was detected in ventilation of the nest by the parent birds, suggesting that nest humidity is not regulated. The species with which this test was conducted, however, construct fibrous nests that are quite permeable to water vapor. It remains possible that behavioral regulation of nest humidity may occur in those species that build substantially less permeable nests (e.g., clay nests?).

E. Oxygen and Carbon Dioxide

Variation in oxygen and carbon dioxide concentrations are likely to be important only in the case of birds occupying enclosed areas, such as nests or roosts within cavities. For example, it seems probable that gas tensions might deviate greatly from the open atmosphere when large numbers of birds roost communally in tree cavities (e.g., about 150 *Sitta pygmaea* roosting in a single cavity; Knorr, 1957). Unfortunately, data are not available to describe the conditions such birds experience.

Variation in the gaseous microclimate may also be important for the eggs and young contained in nests, as well as for parents attending cavity nests. For open nests, gas concentrations have been measured only in domestic fowl (Table II). In this species, CO_2 content of the air surrounding the nest is closely correlated with the CO_2 production of the developing embryo and rises from 0.05% early in incubation to 0.9% just prior to hatching (Burke, 1925; Walsberg, 1980). This peak CO_2 concentration occurs, however, during the period in which the embryo

TABLE II

Fractional Concentrations of Oxygen and Carbon Dioxide in the Nest

Species	Phase	F_{CO_2} (%)	F_{O_2} (%)	Reference
Alectura lathami	Incubation	8.16	13.2[a]	Seymour and Ackerman, 1980
Leipoa ocellata	Incubation	3.82	16.6[a]	Seymour and Ackerman, 1980
Gallus domesticus	Incubation	0.037–1.27	19.6–20.9[b]	Lamson and Edmond, 1914
		0.09–1.40	19.5–20.8[b]	Burke, 1925
Merops apiaster	Nestlings	1.2–6.5	15.1–20.5[c]	White *et al.*, 1978
Petrochelidon pyrrhonotta	Nest construction	0.10	20.8[b]	Withers, 1977
	Incubation	0.20	20.7[b]	
	Nestlings	0.32	20.6[b]	

[a] Fractional concentration calculated from partial pressure assuming barometric pressure = 101 kPa.

[b] F_{O_2} estimated from F_{CO_2} by assuming R.Q. = 0.73 (Drent, 1975) and that gas transport occurs primarily by convection. In this case, an increase in CO_2 concentration is paralleled by a similar decrease in O_2 concentration. Smaller reductions in O_2 concentration would occur if gas transport occurs primarily by diffusion.

[c] Calculated from author's data by assuming F_{CO_2} = 0.03% and F_{O_2} = 20.9% in the general atmosphere.

is least sensitive to such stress; data summarized by Lundy (1969) indicate that peak values occurring in the nest are unlikely to have significant effects upon embryonic viability. These data for CO_2 accumulation can be used to estimate O_2 depletion in the nest assuming that gas transport occurs primarily by convection and that the embryonic respiratory quotient is 0.73 (Drent, 1975). Such calculations indicate that O_2 concentration declines insignificantly to about 20% in the nests of domestic fowl (Table II). Since chickens lay relatively large clutches, these deviations in gas concentrations from that in open air probably are large compared to that of other open-nesting species. This suggests that accumulation of CO_2 or depletion of O_2 in the nest probably is not an important problem for typical species that build open, fibrous nests.

For species occupying enclosed nests, limited data hint that the gaseous microclimate is an important function of nest architecture. The Cliff Swallow (*Petrochelidon pyrrhonota*) builds a gourd-shaped nest of mud that apparently is well ventilated, since oxygen concentration declines to only about 20.6% during the nestling period, while carbon dioxide concentration rises to 0.3% (Table II). More extreme values are seen in the European Bee-eater (*Merops apiaster*), which digs a nest tunnel averaging 1.0 m long into soil (White *et al.*, 1978). The metabolism of birds, invertebrate animals, and bacteria substantially modify gas concentrations within the nest, with extreme values reaching 6.5% CO_2, 15.1% O_2, and 0.07% NH_3. Nest ventilation is facilitated by the pistonlike action of the adults moving through the tunnel and turbulent transport produced by winds inducing eddy formation at the outer tunnel entrance. Thus, extreme deviations in gas concentration occur near dawn on windless nights (White *et al.*, 1978). The physiological consequences for this species of exposure to the unusual nest atmosphere have not been studied, though labored and rapid breathing by nestlings was observed on some occasions. Other vertebrates without special adaptations would be severely stressed by the high concentrations of carbon dioxide and ammonia which *Merops* routinely tolerates by unknown mechanisms.

The most extreme gaseous environments occupied by any bird species probably are in the nests built by species in the family Megapodiidae. Various birds in this group incubate their eggs in nests that vary from excavations in volcanically heated sand to huge mounds of decaying forest litter and soil (e.g., 3.6 metric tons of material in nests of the Brush Turkey (*Alectura lathami*) studied by Baltin [1969]). Gas relations have been studied extensively in only two species, the Brush Turkey (Seymour and Rahn, 1978; Seymour and Ackerman, 1980) and the Mallee Fowl (*Leipoa ocellata*) (Seymour and Ackerman, 1980). Both species build large mounds of decaying vegetation in which the clutch is buried. Aerobic decomposition of the plant material generates heat, and egg temperature averages 34° (*Leipoa*) or 37° (*Alectura*) (Frith, 1956a, 1956b; Seymour and Rahn, 1978; Seymour and Ackerman, 1980). This egg temperature is sensed by the male, which periodically digs through the mound to the clutch and apparently tests the temperature with its head and beak. Egg temperature is adjusted by the male by either removing or adding material over the eggs. This regular construction of tunnels to the eggs probably is critical not only for thermoregulation but also for ventilation of the nest and maintenance of an appropriate gas composition (Baltin, 1969). Even with such activity, the restriction in gas diffusion produced by the large mound size and organic decomposition produces a notably hypoxic and hypercapnic environment (Table II). Such an atmosphere should substantially decrease diffusion gradients driving the embryo's exchange of oxygen and carbon dioxide across the shell. Apparently as an adaptation to such reductions in the driving gradients, the shell's resistance to gas exchange is decreased. This is accomplished by reducing the thickness of the shell, which reduces diffusion resistance about 26% in *Leipoa* and 65% in *Alectura* compared to that expected for a typical egg of appropriate mass and incubation period (Seymour and Ackerman, 1980). Thus, use of this striking extreme in nest environment critically depends on adaptations of embryonic physiology, interacting with behavioral adaptations of the adult, that maintain the nest within the limits of the embryo's tolerance.

III. THERMAL ENVIRONMENT

A. Nocturnal Roost-Site Selection

Birds typically are diurnal and restrict their foraging to daylight hours. Thus, the daily fast occurs at night when energy demands for thermostasis may be greatest. This coincidence of high thermoregulatory energy demands with fasting is most striking in nontropical areas during the winter, when nights are relatively long and cold. During such periods, birds become critically dependent on their comparatively limited capacities for energy storage. Though birds may store large fractions of their body mass as fat, avian power consumption occurs at such high rates that even extensive fat stores may represent at most but a few day's supply of energy for normal activity (Walsberg, 1983a). For example, the data for 10 passerine species reviewed by King (1972) indicate that under normal winter conditions these species commence their nocturnal fast with reserves sufficient for one night plus a portion of the following day. It is not unexpected that substantial mortality has been observed at nocturnal roosts during inclement weather [e.g., 4% mortality during a single night in the communal roost of blackbirds studied by Odum and Pitelka (1939)]. This suggests the existence of strong selective pressures for birds to select nocturnal roost sites that minimize thermoregulatory stresses. This might be accomplished by selecting sites characterized by increased air temperature or shelter from wind, precipitation, and radiative heat loss to the heat sink presented by a clear night sky. Most analyses have dealt primarily with the radiative and convective environment.

TABLE III

Energetic Savings Due to Winter Roost-Site Selection

Species	Roost site	Energy savings (%)[a]	Reference
Myiopsitta monachus	Nest	3.7[b]	Caccamise and Weathers, 1977
Pica pica	Fir grove	8.3	Mugaas and King, 1981
Turdus migratorius	Fir grove	4.4	Walsberg and King, 1980
Sturnus vulgaris	Pine woods	12–38	Kelty and Lustick, 1977
	Ficus grove	13–29[b]	Yom-Tov *et al.*, 1977
Passer domesticus	Nest box	13[b,c]	Kendeigh, 1961
Philetairus socius	Nest	43[b]	White *et al.*, 1975

[a] Energy savings reported as reduction in nocturnal power consumption compared to that expected if bird roosted in open.

[b] Value probably is an underestimate, since it accounts for effect of increased air temperature but not that of changes in wind velocity or radiation.

[c] Maximum estimate occurring at $T_A = -30°C$.

Table III summarizes estimates of the energetic savings accrued by use of sheltered roost sites. Data for three species occupying cavities or domed nests indicate that nocturnal power consumption is reduced up to 43% compared to the power consumption expected for a bird roosting in an exposed site. These values undoubtedly are underestimates, since authors accounted only for the effect of increased air temperature within the cavity. Increased long-wave radiation impinging on the bird and shelter from wind were not quantified, though for other species that roost inside of dense vegetation such shelter from wind and excessive radiation loss is estimated to reduce nocturnal energy requirements 4–38%. For birds roosting in cavities, a major source of energy savings is an increase in air temperature within the cavity resulting from the retention of metabolically produced heat (Kendeigh, 1961; White *et al.*, 1975; Caccamise and Weathers, 1977). Available data suggest that such heating of the local environment is unimportant for species not roosting in cavities, even in the case of communal roosts in which millions of birds may aggregate in a dense woodland (Kelty and Lustick, 1977). The possibility remains unexplored, however, that individuals roosting in very dense vegetation may significantly warm their local surroundings.

Complete or nearly complete heat budgets are available for several species that roost in vegetation rather than enclosed cavities (Kelty and Lustick, 1977; Mugaas and King, 1980; Walsberg and King, 1980). These studies indicate that the major source of energy savings accrues from reduced convective heat loss due to shelter from wind. Compared to values recorded in unsheltered locations, average wind velocity is reduced 72% at roosts used by wintering *Turdus mi-*

gratorius (Walsberg and King, 1980) and reduced from 84% to at least 98% at mixed species roosts occupied by *Sturnus vulgaris, Agelaius phoeniceus,* and *Molothrus ater* (Francis, 1976; Kelty and Lustick, 1977).

Several authors have suggested that a major advantage of nocturnal shelter seeking may be improvement of radiation balance (e.g., Moore, 1945). This would occur because vegetation typically exhibits thermal emissivities (ϵ) of 0.95–0.99 (Gates and Tantraporn, 1952), while dry air can have much lower emissivities (e.g., for dry air at 0°C, $\epsilon = 0.72$ [Campbell, 1977]). Thus, a roosting bird might increase by 30–40% the amount of thermal radiation it receives from the upper hemisphere of its radiative environment if the animal is exposed to a dense vegetation canopy rather than the clear night sky. Compared to shielding from forced convection, however, the limited data that are available suggest that such radiation effects are of minor importance. The relative importance of variation in wind velocity, the radiative environment over the bird, and the air temperature can be evaluated by calculating the standard operative temperature (T_{ES}) of a roosting bird's environment under a variety of conditions. Standard operative temperature is a thermal index that allows direct comparison of widely varying microclimates. It represents the temperature of a standard metabolic chamber that produces a degree of thermal stress on the bird equivalent to that produced by a natural environment (Bakken, 1976). Here, standard conditions within the metabolic chamber consist of still air with a black-body radiative environment (i.e., all radiation emitted from chamber walls at air temperature and $\epsilon = 1.0$). Modifying the formula of Bakken (1976) to incorporate thermal resistances rather than thermal conductances yields the equation

$$T_{ES} = T_B - (r_{TS}/r_T)\,(T_B - T_E)$$

Here, T_B is body temperature and is assumed to equal 40°C (Calder and King, 1974), r_T is the total thermal resistance between animal and environment under natural conditions, and r_{TS} is the total thermal resistance between animal and environment under standard metabolic chamber conditions. Both r_T and r_{TS} consist of the sum of r_B and r_E, where r_B is the resistance to heat transfer from the animal's core to its outer surface and r_E is the resistance to heat transfer between the animal's surface and the environment. This latter resistance incorporates effects such as forced convection and can be calculated according to the equations and assumptions of Walsberg and King (1978a).

The final term, T_E, represents the operative temperature of the natural environment (Bakken, 1976; Robinson *et al.*, 1976; Campbell, 1977). This is a single temperature index that subsumes all factors determining the net thermal gradient operating on the animal. It is calculated as

$$T_E = T_A + (r_E/\rho C_p)\,(R_{ABS} - \epsilon\sigma T_A^{\,4})$$

Here, ρC_p is a constant equal to 1200 J/m^3 – °K, T_A is air temperature, R_{ABS} is the radiation absorbed by the animal, ϵ is the emissivity of the animal's surface (here assumed to equal 0.98), and σ is the Stefan–Boltzmann constant (5.67×10^{-8} W/m^2 – ^{0}K^4).

These equations can be used to examine the sensitivity of a bird's net thermal stress to variation in wind and overhead cover. Appropriate data are available for the White-crowned Sparrow (*Zonotrichia leucophrys*) and *Calidris* spp. sandpipers. For *Zonotrichia,* the characteristic dimension for estimates of convection is estimated as 0.035 m and body resistance (r_B; sec/m) is estimated using the equation of Robinson *et al.* (1976)

$$r_B = 336 - 36u^{1/2}$$

where u is wind velocity (m/sec). This equation was computed from empirical data demonstrating a decrease in r_B with an increase in wind velocity, probably due to wind penetrating into the plumage and disrupting coat insulation. This effect should not be confused with the convective cooling that occurs at the coat surface due to disruption of the adherent atmospheric boundary layer. Traditional heat budget analyses usually account only for the latter effects of wind; data with which to estimate effects of wind penetration into plumage insulation are sparse. The effects of such penetration can be estimated by comparing results predicted using the equation of Robinson *et al.* (1976) for r_B as a function of wind velocity with an estimate derived by assuming wind does not disrupt coat insulation. For this estimate of r_B without wind penetration, the equation of Robinson *et al.* (1976) predicts r_B = 336 sec/m at 0°C and still air.

For simulations of *Calidris* thermal relations (Chappell, 1980), the characteristic dimension is assumed to equal 0.051 m and body resistance is computed as

$$r_B = 347 - 25.2\ u^{1/2}$$

The effect of wind penetration into the plumage will be examined by comparison of results with those predicted by assuming r_B = 347 sec/m, the value predicted by the above equation if wind velocity is zero. For simulations involving both species, r_E is assumed to equal 100 sec/m in a standard environment chamber (Robinson *et al.*, 1976; Mahoney and King, 1977).

For these analyses, air temperature is held at 0°C and the bird is assumed to be either completely exposed to the night sky or completely shielded by a vegetation canopy. In the case of complete exposure, the upper hemisphere of the radiative environment consists of air at ϵ = 0.72 (Campbell, 1977). For the case of complete cover, the upper hemisphere consists of vegetation at air temperature and ϵ = 0.98 (Gates and Tantraporn, 1952). In both cases, the bird is assumed to

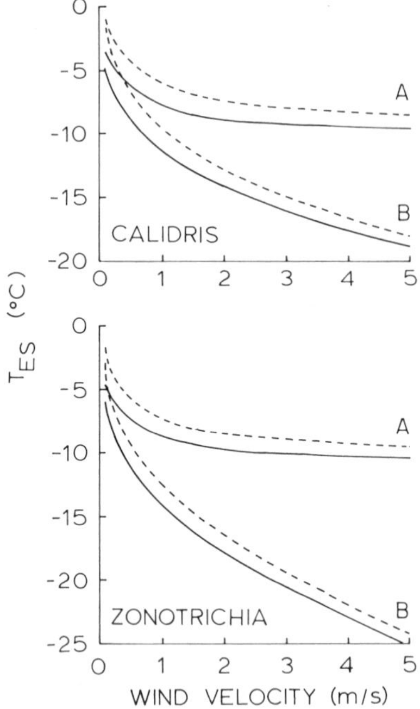

Fig. 2. Estimated sensitivity of standard operative temperature (T_{ES}) at nocturnal roost sites to variation in wind velocity and the radiative environment. The radiative environment above the bird is assumed to result from either complete exposure to a clear night sky (no overhead cover) or complete shielding from the night sky by vegetation cover. Results are estimated for both the case (A) in which wind acts only on the plumage surface and the more realistic case (B), in which wind penetrates into the plumage and disrupts coat insulation.

roost above vegetation so that the lower radiative hemisphere consists of vegetation at air temperature and $\epsilon = 0.98$.

Calculated values of standard operative temperature for *Zonotrichia* and *Calidris* exposed to this range of conditions are shown in Fig. 2. For both species, the effect of overhead cover is remarkably small compared to wind effects. Changing the radiative environment such that the bird is shielded by vegetation completely rather than being exposed to the cold night sky increases heat gain to the animal equivalent to only a 1–2°C increase in air temperature. This minor effect produced by the maximum possible change in the nocturnal radiative environment contrasts to the major effects produced by slight variations in wind velocity. Using the realistic case that accounts for wind penetration into the coat, an increase in wind velocity from 0.1 m/sec to 3.0 m/sec produces an increase in cold stress equivalent to an 11–19°C decline in air temperature in these species. This is about 10 times the maximum effect of changes in overhead cover. This strongly suggests that the primary basis of nocturnal roost-site selection in vegetation cover is reduction in wind velocity, rather than shielding from the night sky or exploiting small local variations in air temperature.

These simulations also demonstrate the importance of wind penetrating into the coat, rather than simply acting at the coat surface. This is particularly striking since the thermal benefits of the increased insulation depth produced by feather fluffing may be countered by greater wind penetration into the less dense plumage. Such wind penetration often has been ignored (e.g., Porter and Gates, 1969; Walsberg and King, 1980), though it clearly can be a major source of heat loss and deserves intensive study.

Use of particular microhabitats as nocturnal roosts may substantially reduce nocturnal energy expenditure but may also entail a substantial energy cost in daily travel to and from the roost site. Gyllin *et al.* (1977) and Yom-Tov *et al.* (1977) have calculated that in some cases this cost of transport may be greater than the savings accrued by selection of particular nocturnal roost sites. That use of such roost sites may increase a bird's total daily energy expenditure does not demonstrate, however, that energy conservation is not a major factor favoring shelter seeking. Such use of sheltered microclimates reduces thermostatic demands during periods in which the bird must rely on limited endogenous reserves and in which inclement weather could exceed the bird's capacity for heat generation. In such conditions, minimizing total daily power consumption may be unimportant compared to minimizing nocturnal energy requirements and thus facilitating survival during this stressful period.

B. Nest-Site Selection

Probably the most common example of microhabitat selection in birds is placement of nests in dense vegetation. Besides rendering the nest less conspicuous to predators, such placement may shield the nest from wind, excess nocturnal radiation loss, or excess diurnal heat gain from solar radiation. Site selection for thermal advantage has been described in a variety of species and often takes the form of locating nests on thermally favorable sides of trees (Balda and Bateman, 1973) or placement immediately beneath a sheltering structure (Calder, 1973b). It is notable, however, that other expressions of nest-site selection may be equally effective yet considerably more subtle. An example is seen in the Warbling Vireo (*Vireo gilvus;* Walsberg, 1981). Special adaptations in nest placement are not immediately apparent in this species. However, analyses using hemispherical photographs of the vegetation–sky mosaic over the bird revealed that nests are placed with regard to the woodland canopy such that exposure to direct solar radiation is reduced 47% during hot afternoon hours compared to the cool morning period. It is quite possible that similar types of site selection are widespread in birds yet remain undescribed because of their inconspicuous nature.

Effects of nest placement upon nocturnal radiation balance can be substantial.

For the skin surface area exposed above the nest, this reduction in radiative heat loss is equal to 180–220% of basal metabolism (BM) in *Calypte anna* (Calder, 1974), 192% of BM in *Selasphorus platycercus* (Calder, 1973b; BM estimated from equation of Aschoff and Pohl, 1970), and 49% of BM in *Promerops cafer* (Burger *et al.* 1976; BM calculated on surface-area basis). Savings due to reduced convective heat loss have been estimated less frequently but can be substantial. For example, *Promerops cafer* prefers to nest about 100 cm above ground in dense vegetation (Burger *et al.*, 1976). Shifting its nest to a region nearer the top of the vegetation where wind velocities are higher would increase convective heat loss from the bird's upper surface by an amount equal to 18% of BM (calculated using data of Burger *et al.* [1976] on a surface-area basis assuming nest height shifted to 200 cm).

However, such values calculated on an area-specific basis for the bird's upper surface exaggerate effects of site selection upon the bird's total heat budget. This is because a large fraction of the surface area is enclosed by the nest and substantially insulated from such effects. For example, Smith *et al.* (1974) estimated that the overhead vegetation canopy reduced radiative heat loss by an amount equal to only about 3% of whole-body energy expenditure. In the Willow Flycatcher (*Empidonax traillii*), placing the nest in a sheltered microhabitat rather than in the open reduces the bird's total power consumption by only 5% during the coldest period (0100–0400) (Walsberg and King, 1978b); this savings is split approximately equally between decreased radiative and convective losses.

It is striking that these savings due to microhabitat selection may be small compared to those produced by the addition of an insulating nest. For example, Walsberg and King (1978b) estimated for the Willow Flycatcher that the nest insulation reduces power consumption 19%, or almost four times the savings produced by use of a sheltered nest site. Data on nest insulation are accumulating (Skowron and Kern, 1980), but all extant data were collected in still air conditions and ignore the possible effects of wind penetration. Such penetration of wind into fibrous nests may disrupt nest insulation in addition to simply removing heat from the nest's outer surface. If the effects of wind penetration into fur or feathers (see Section III,A) gives an indication of effects in nests, then such penetration may vary widely and may be of major importance in the thermal relations of nesting birds.

To what degree do the thermal consequences of nest-site selection affect the parent's survival and production of offspring? Few data are available to address this critical point. In some species, workers have concluded that variation in site selection has no apparent effect on parental survival or reproduction (e.g., Verbeek, 1981). Other workers, however, have found important effects. The most complete data set is for Cactus Wrens (*Campylorhynchus brunneicapillum*), which build an enclosed, retort-shaped nest with a single side entrance. This nest is placed in spinescent vegetation with the opening oriented according to season and prevailing wind direction (Ricklefs and Hainsworth, 1969). Early in the breeding season during cool weather, this desert species orients the nest entrance away from prevailing winds and thus reduces convection within the nest cavity. Later in the breeding season during hot weather, nests are built such that the entrance faces the prevailing wind, thus increasing ventilation and cooling of the nests (Ricklefs and Hainsworth, 1969). Austin (1976) examined the reproductive success of 86 nests as a function of orientation. He studied individuals nesting late in the breeding season when nests tend to be oriented toward prevailing winds. In his population, a majority (54%) of the nests were oriented within a 100° angular field that included the direction of the prevailing wind. A minority of nests were "incorrectly" oriented over the remaining 260° of compass directions. Reproductive success was significantly greater in nests "correctly" oriented within the 100° angular field; 64% of eggs laid in such nests produced fledged young as opposed to only 40% of eggs laid in nests oriented "incorrectly."

The precise mechanism whereby changes in thermal stress affect reproductive success has not been clearly demonstrated in any species. However, the types of effects possible are seen in Zerba and Morton's (1983) study of White-crowned Sparrows (*Zonotrichia leucophrys*). These workers compared parental behavior during incubation at two nests tended consecutively by the same female; one nest was placed in the open and had essentially no protective canopy, while the second nest was placed under a dense vegetation canopy. Even in the 3000-m altitude site occupied by these birds, solar radiation at midday could produce lethal egg temperatures. Requirements for shielding the eggs from insolation restricted the female to the exposed nest for extended periods of time. Compared to the period in which she was attending the more sheltered nest, time available for activities such as foraging was significantly reduced. Not studied, but perhaps more important, were affects upon parental behavior during the nestling period when the female's foraging requirements probably peak (Walsberg, 1983a).

Finally, it is notable that most analyses of the microclimatic effects of nest-site selection have dealt with nocturnal cold stress. This distinct emphasis upon nocturnal effects is unfortunate, since diurnal stresses such as those from insolation can be of major importance (Zerba and Morton, 1983; Salzman, 1982). Indeed, it is striking that while most studies documenting effects of site selection upon reproduction and survival have dealt with problems of diurnal stresses, the microclimate at such times has received comparatively little attention.

C. Activity Away from the Nest or Roost

Because of ease of analysis, almost all studies quantifying the physiological consequences of microsite selection in birds have dealt with occupation of a

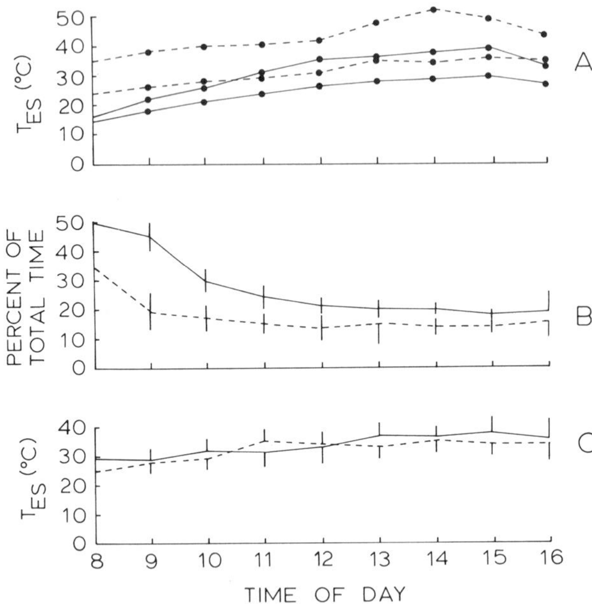

Fig. 3. Preliminary data describing microclimate selection by *Phainopepla* males in two contrasting habitats: Sierra Ancha (- - -) and Star Ranch (——). Values plotted are $\bar{x} \pm$ SD, $n = 10$. For simplicity of presentation, only mean values are given in section A. (A) Range of standard operative temperatures (T_{ES}) found in *Phainopepla* microhabitats. Two sets of data are given for each habitat, representing the maximum and minimum T_{ES} existing at a particular time. (B) Percentage of time spent by Phainopeplas exposed to full sunlight (warmest microclimate). (C) Average T_{ES} actually experienced by Phainopeplas during hourly period, weighted by fraction of time spent in each microclimate.

single site, either the nest or a nocturnal roost. The physiological consequences of variation in daily activity movements such as foraging are almost unknown, though adaptive shifts in microsite occupation have been suggested by a variety of analyses that correlated changes in activity with climatic variation (e.g., Grubb, 1975, 1977, 1978). Few data are available, however, to describe the physiological consequences of such effects. Two species for which data exist to allow quantifying such physiological correlates are the Phainopepla (*Phainopepla nitens*) and the White-crowned Sparrows.

Figure 3 shows preliminary data from comparative studies of microclimate selection in Phainopeplas breeding in two contrasting sites (G. E. Walsberg, unpublished data). Phainopeplas were studied during breeding in the Sierra Ancha of central Arizona at an elevation of 5100 m and at the Starr Ranch Audubon Sanctuary, located at about 300 m elevation and about 17 km east of the Pacific Ocean in Orange County, California. In both sites Phainopeplas occupy riparian woodlands bordered by chaparral and typically complete egg laying during the first week of June. However, the climate of the California site is ameliorated by an influx of cool and moist air from the Pacific Ocean, while the Arizona site receives no such buffering and is distinctly hotter. These differences were quantified using taxidermic mounts to measure operative environmental temperature (T_E) (Walsberg, 1982). Mounts consisted of a hollow copper cast of the bird's body covered by the animal's integument. When allowed to equilibrate with a particular microclimate, such models exhibit an internal temperature very near T_E for the particular species and environment studied (Bakken, 1976; Walsberg, 1982). In each field site, six such models were placed in each of three distinctive microclimates. During midday hours, the first microclimate class consisted of exposed perch sites located on top of vegetation, where at least 75% of the bird's surface directed toward the sun actually received direct insolation. The second site consisted of semishaded situations within the vegetation canopy in which 25–75% of the bird's appropriate surface received direct insolation. The third site consisted of well-shaded sites with less than 25% exposure to the solar beam. Wind velocity was simultaneously measured in each of these sites using the techniques of Walsberg (1982). The effects of different convection regimes on the bird's resistance to heat transfer was determined using the technique of Chappell and Bartholomew (1981), in which taxidermic models were filled with Wood's alloy and heated by internal resistance wires while held in a wind tunnel. This allows changes in total thermal resistance of the animal–environment system to be estimated as a function of wind velocity. These data allow calculation of standard operative temperature (T_{ES}) for Phainopeplas in their various natural microclimates. Here, T_{ES} is defined as in Section III,A: the temperature of a black-body chamber with still air that produces the same degree of thermal stress on the bird as its natural microclimate. Values presented are for males only; temperatures for females typically are 0.5–1.0°C lower. Simultaneously with measurements of T_E and wind velocity, the activity of individual Phainopeplas was timed with stopwatches 30 consecutive minutes during each hour to determine the amount of time spent in each of the three microclimate classes. Such simultaneous determinations of microclimate and activity were made for 10 males 5–7 days after their eggs hatched. Data were collected for a particular male on only 1 day. Weighting each microclimate's T_{ES} value by the amount of time spent in that microclimate by the bird allows calculation of the average T_{ES} experienced by the adult as it proceeds in its daily activity.

Preliminary results from this analysis demonstrate that T_{ES} typically is 5–13°C higher in midday at the Sierra Ancha site than at the Starr Ranch site. This apparently is due primarily to less intense solar radiation and lower air temperatures at Starr Ranch. Thus, the most exposed microclimate in the Sierra Ancha is characterized by T_{ES} values up to 52°C, which would produce substantial heat stress on Phainopeplas (Walsberg, 1977). Phainopeplas apparently adapt to such potential stress by shifting their activity to cooler microclimates. For example, Phainopeplas in the Sierra Ancha spend only 17% of midday hours exposed to full sunlight; this is 0.62 times the value for the birds in the more mesic Starr Ranch site (Fig. 3; the difference is statistically significant; ANOVA; $p < 0.05$). The result of such shifts in microclimate use during midday hours is that the birds experience average values of T_{ES} that are indistinguishable in the two sites: 32.3°C in the Sierra Ancha and 33.4°C at Starr Ranch. At no time during this 8-hr period does analysis of variance reveal a significant difference in T_{ES} ($p > 0.10$). Thus, these preliminary data suggest that thermoregulatory demands in this species are effectively regulated by changes in microsite use.

DeWoskin (1980) used meteorological measurements and operative environmental temperature theory to quantify the thermal significance of microclimate selection in foraging White-crowned Sparrows. Microhabitats occupied by these birds ranged from open fields to dense hedges. These sites varied sufficiently in thermal characteristics to produce differences of up to 20% in a sparrow's metabolic rate. Such thermal properties, however, apparently are not important determinants of microhabitat use. For example, flocks typically moved into dense cover during midday, though occupation of such shaded sites substantially increased cold stress on the animal. This movement into denser cover during midday at the expense of thermal benefits suggests that nonthermal factors are primary determinants of such microsite selection. Empirical evidence to critically evaluate alternative selective pressures in these species is lacking, but possible factors include changes in cost–benefit relations of foraging (DeWoskin, 1980) or protection from diurnal predators.

These analyses of microhabitat selection by *Phainopepla* and *Zonotrichia* are hardly a sufficient data base for useful generalizations regarding microsite use during daily activity. At a minimum, however, such studies demonstrate that recently developed biophysical techniques are effective means for estimating the physiological consequences of microhabitat selection even in the case of daily movements through complex natural environments.

IV. CONCLUDING COMMENTS

Information describing the physiological consequences of microhabitat selection has expanded rapidly in the last 10 years. Technical and conceptual advances in integrating field and laboratory analyses have stimulated a major thrust toward placing relatively well-known physiological mechanisms in their proper ecological context. This integrative approach is inhibited, however, by major gaps that remain in understanding basic physiological responses of birds to important environmental stresses. Two examples that suffice for illustration are the effects of precipitation and the interaction of cold stress and exercise-induced thermogenesis.

Perhaps the most striking gap in our knowledge of the relations of birds to a single environmental property deals with effects of precipitation. While the effects of variation in other physical phenomena may be comparatively subtle, effects of precipitation often are dramatic; several workers have observed mass mortality after cold rains (e.g., Odum and Pitelka, 1939; Kessler *et al.*, 1967). Unfortunately, the physiological ecology of such wetting is virtually unstudied, though Lustick and Adams (1977) examined its thermal consequences in starlings (*Sturnus vulgaris*). The lower critical temperature of completely wet starlings (summer-acclimated) was elevated from 22.5°C to 40°C by wetting. Maximum power consumption in these drenched birds occurred at only 25°C at a rate expected to be reached at about −40°C in dry birds. (This value was calculated using the regression equation of Lustick and Adams [1977] and is only a rough approximation since it requires extrapolation below the range of their measurements.) Thus, these data suggest that complete wetting increases heat drain from the bird equivalent to that produced by about a 17–65°C decrease in air temperature. No summer-acclimated bird survived more than 1 hr when drenched at air temperatures below 10°C.

Such data clearly requires further development but does indicate that even rare episodes of rain may present major selective pressures influencing shelter seeking. It is notable that selection of sites that provide overhead cover during roosting or nesting has generally been attributed to improvement of radiation balance but may serve much more important functions of deflecting rain. This possibility is reinforced by the comparatively slight advantages that simulations reported in Section III,A suggest are produced by such improvements in radiation balance. Major problems that await future workers include analyzing the physical bases of water resistance in birds, the physiological effects of variation in duration and intensity of wetting, and the extent to which adaptations to avoid wetting have coerced the evolution of shelter seeking.

A second critical gap lies in comprehending the relationship between heat produced as a by-product of activity and a bird's thermostatic requirement. Though it might be expected that muscular production of heat associated with activity reduces the need for a bird in a cold climate to use shivering thermogenesis, the relevant empirical data are ambiguous (Walsberg, 1983a). Results in some cases indicate a partial or complete substitution of the waste heat of activity for thermostatic demands and in other cases indicate an addition of exercise thermogenesis to cold-induced thermogenesis. The mode of this interaction may be of major ecological importance, because if the waste heat of activity substitutes for shivering thermogenesis there may be little difference in the power consumption of an active bird and one that is immobile but shivering. This would significantly affect the cost–benefit relations of activities such as territorial defense and foraging, since in cold conditions these may entail little or no net energy cost compared to a less active bird.

ACKNOWLEDGMENTS

This analysis was supported in part by National Science Foundation grant DEB 80–04266.

REFERENCES

Ackerman, R. A., and Platter-Rieger, M. (1979). Water loss by pied-billed grebe (*Podilymbus podiceps*) eggs. *Am. Zool.* **19,** 921.

Ar, A., and Rahn, H. (1980). Water in the avian egg: Overall budget of incubation. *Am. Zool.* **20,** 373–384.

Aschoff, J., and Pohl, H. (1970). Rhythmic variations in energy metabolism. *Fed. Proc., Fed. Am. Soc. Exp. Biol.* **29,** 1541–1552.

Austin, G. T. (1976). Behavioral adaptations of the Verdin to the desert. *Auk* **93,** 245–262.

Bakken, G. S. (1976). A heat transfer analysis of animals: Unifying concepts and the application of metabolism chamber data to field ecology. *J. Theor. Biol.* **60,** 337–384.

Bakken, G. S., and Gates, D. M. (1975). Heat transfer analysis of animals: Some implications for field ecology, physiology, and evolution. *In* "Perspectives in Biophysical Ecology" (D. M. Gates and R. B. Schmerl, eds.), pp. 255–290. Springer-Verlag, Berlin and New York.

Bakken, G. S., Buttemer, W. A., Dawson, W. R., and Gates, D. M. (1981). Heated taxidermic mounts: A means of measuring the standard operative temperature affecting small animals. *Ecology* **62,** 311–318.

Balda, R. P., and Bateman, G. C. (1973). The breeding biology of the Piñon Jay. *Living Bird* **11,** 5–42.

Baltin, S. (1969). Zur Biologie und Ethologie des Talegalla-Huhns (*Alectura lathami* Gray) unter besonderer Berücksichtigung des Verhalteus während der Brutperiode. *Z. Tierpsychol.* **26,** 524–572.

Bartholomew, G. A., White, F. N., and Howell, T. R. (1976). The thermal significance of the nest of the Sociable Weaver, *Philetairus socius:* Summer observations. *Ibis* **118,** 402–410.

Birchard, G. F., and Kilgore, D. L. (1980). Conductance of water vapor in eggs of burrowing and nonburrowing birds: Implications for gas exchange. *Physiol. Zool.* **53,** 284–292.

Burger, A. E., Siegfried, W. R., and Frost, P. G. H. (1976). Nest-site selection in the Cape Sugarbird. *Zool. Afr.* **11,** 127–158.

Burke, E. (1925). A study of incubation. *Bull. Univ. Montana Agric. Exp. Station* **178,** 1–43.

Caccamise, D. F., and Weathers, W. W. (1977). Winter-nest microclimate of Monk Parakeets. *Wilson Bull.* **89,** 346–349.

Calder, W. A. (1973a). An estimate of the heat balance of a nesting hummingbird in a chilling climate. *Comp. Biochem. Physiol.* **46,** 291–300.

Calder, W. A. (1973b). Microhabitat selection during nesting of hummingbirds in the Rocky Mountains. *Ecology* **54,** 127–134.

Calder, W. A. (1974). The thermal and radiant environment of a winter hummingbird nest. *Condor* **76,** 268–273.

Calder, W. A., and King, J. R. (1974). Thermal and caloric relations of birds. *In* "Avian Biology" (D. S. Farner and J. R. King, eds.), Vol. 4, pp. 259–413. Academic Press, London and New York.

Campbell, G. S. (1977). "An Introduction to Environmental Biophysics." Springer-Verlag, Berlin and New York.

Carey, C. (1980). Adaptation of the avian egg to high altitude. *Am. Zool.* **20,** 449–459.

Chappell, M.A. (1980). Thermal energetics of chicks of arctic-breeding shorebirds. *Comp. Biochem. Physiol.* **65,** 311–317.

Chappell, M. A., and Bartholomew, G. A. (1981). Standard operative temperatures and thermal energetics of the antelope ground squirrel. *Ammospermophilus leucurus. Physiol. Zool.* **54,** 81–93.

DeWoskin, R. (1980). Heat exchange influence on foraging behavior in *Zonotrichia* flocks. *Ecology* **61,** 30–36.

Drent, R. H. (1970). Functional aspects of incubation in the Herring Gull. *Behaviour,* Suppl. **17,** 1–132.

Drent, R. H. (1975). Incubation. *In* "Avian Biology" (D. S. Farner and J. R. King, eds.), pp. 333–420, Academic Press, London and New York.

Francis, W. J. (1976). Micrometeorology of a blackbird roost. *J. Wildl. Manage.* **40,** 132–136.

Frith, H. J. (1956a). Breeding habits in the family Megapodiidae. *Ibis* **98,** 620–640.

Frith, H. J. (1956b). Temperature regulation in the nesting mounds of the mallee fowl, *Leipoa oscellata* Gould. *C.S.I.R.O. Wildl. Res.* **1,** 79–95.

Gates, D. M., and Tantraporn, W. (1952). The reflectivity of deciduous trees and herbaceous plants in the infrared to 25 microns. *Science* **115,** 613–616.

Grant, G. S. (1982). Avian incubation: Egg temperature, nest humidity, and behavioral thermoregulation in a hot environment. *Ornithol. Monogr.* **30,** 1–75.

Grubb, T. C., Jr. (1975). Weather dependent foraging behavior of some birds wintering in a deciduous woodland. *Condor* **77,** 175–182.

Grubb, T. C., Jr. (1977). Weather dependent foraging behavior of some birds wintering in a deciduous woodland: Horizontal adjustments. *Condor* **79,** 271–274.

Grubb, T. C., Jr. (1978). Weather dependent foraging rates of wintering woodland birds. *Auk* **95,** 370–376.

Gyllin, R., Kallander, H., and Sylven, M. (1977). The microclimate explanation of town centre roosts of Jackdaws (*Corvus monedula*). *Ibis* **119,** 358–361.

Hoyt, D. F. (1980). Adaptation of avian eggs to incubation period: Variability around allometric regressions is correlated with time. *Am. Zool.* **20,** 417–426.

Kelty, M. P., and Lustick, S. I. (1977). Energetics of the starling (*Sturnus vulgaris*) in a pine woods. *Ecology* **58,** 1181–1185.

Kendeigh, S. C. (1961). Energy of birds conserved by roosting in cavities. *Wilson Bull.* **73,** 140–147.

Kessler, F., Glitz, M. L., and Buritt, H. F. (1967). High mortality of a population of cowbirds wintering at Columbus, Ohio. *Ohio J. Sci.* **67,** 48.

King, J. R. (1972). Adaptive periodic fat storage in birds. *Proc. Int. Ornithol. Congr. 15th,* pp. 200–217.

Knorr, O. A. (1957). Communal roosting of the Pygmy Nuthatch. *Condor* **59,** 398.

Lamson, G. H., and Edmond, H. D. (1914). Carbon dioxide in incubation. *Bull. Storrs Agric. Exp. Station* **76,** 219–257.

Ledoux, T. (1977). The inverse relationship between high altitude and total pore area in the eggshell of the chicken. *Fed. Proc., Fed. Am. Soc. Exp. Biol.* **36,** 534.

Lomholt, J. P. (1976). Relationship of weight loss to ambient humidity of birds' eggs during incubation. *J. Comp. Physiol.* **105,** 189–196.

Lundy, H. (1969). A review of the effects of temperature, humidity, turning and gaseous environment in the incubator on the hatchability of the hen's egg. *In* "The Fertility and Hatchability of the Hen's Egg" (T. C. Carter and B. H. Freeman, eds.), pp. 143–175. Oliver and Boyd, Edinburgh.

Lustick, S., and Adams, J. (1977). Seasonal variation in the effects of wetting on the energetics and survival of starlings (*Sturnus vulgaris*). *Comp. Biochem. Physiol.* **56,** 173–177.

Mahoney, S. A., and King, J. R. (1977). The use of the equivalent black-body temperature in the thermal energetics of small birds. *J. Thermal Biol.* **2,** 115–120.

Moore, A. D. (1945). Winter night habits of birds. *Wilson Bull.* **57,** 253–260.

Morgan, K. R., Paganelli, C. V., and Rahn, H. (1978). Egg water loss and nest humidity during incubation in two Alaskan gulls. *Condor* **80,** 272–275.

Mugaas, J. M., and King, J. R. (1980). Annual variation of daily energy expenditure by the Black-billed Magpie: A study of thermal and behavioral energetics. *Stud. Avian Biol.* **5,** 1–78.

Odum, E. P., and Pitelka, F. A. (1939). Storm mortality in a winter starling roost. *Auk* **56,** 451.

Packard, G. C., Sotherland, P. R., and Packard, M. J. (1977). Adaptive reduction in permeability of avian eggshells to water vapour at high altitudes. *Nature* **266,** 255–256.

Paganelli, C. V. (1980). The physics of gas exchange across the avian eggshell. *Am. Zool.* **20,** 329–338.

Paganelli, C. V., Ar, A., Rahn, H., and Wangensteen, O. D. (1975). Diffusion in the gas phase: Effects of ambient pressure and gas composition. *Respir. Physiol.* **25,** 247–258.

Porter, W. P., and Gates, D. M. (1969). Thermodynamic equilibria of animals with environment. *Ecol. Monogr.* **39,** 227–244.

Rahn, H. (1977). Adaptation of the avian embryo to altitude: The role of gas diffusion through the egg shell. *In* "Respiratory Adaptations, Capillary Exchange, and Reflex Mechanisms" (A. S. Paintal and P. Gill-Kumar, eds.), pp. 94–105. Vallabhbai Patel Chest Institute, Univ. of Delhi, Delhi.

Rahn, H., and Dawson, W. R. (1979). Incubation water loss in eggs of Heermann's and Western Gulls. *Physiol. Zool.* **52,** 441–460.

Rahn, H., Paganelli, C. V., Nisbet, J. C. T., and Whittow, G. C. (1976). Regulation of incubation water loss in eggs of seven species of terns. *Physiol. Zool.* **49,** 245–259.

Rahn, H., Ackerman, R. A., and Paganelli, C. V. (1977). Humidity in the avian nest and egg water loss during incubation. *Physiol. Zool.* **50,** 269–283.

Rahn, H., Ledoux, T., Paganelli, C. V., and Smith, A. H. (1982). Changes in eggshell conductance after transfer of hens from an altitude of 3,800 m to 1,200 m. *J. Appl. Physiol.* **53,** 1429–1431.

Ricklefs, R. E., and Hainsworth, F. R. (1969). Temperature regulation in nestling Cactus Wrens: The nest environment. *Condor* **71,** 32–37.

Robinson, D. E., Campbell, G. S., and King, J. R. (1976). An evaluation of heat exchange in small birds. *J. Comp. Physiol.* **105,** 153–166.

Romanoff, A. L., and Romanoff, A. J. (1967). "Biochemistry of the avian embryo." Wiley, New York.

Salzman, A. G. (1982). The selective importance of heat stress in gull nest location. *Ecology* **63,** 742–751.

Seymour, R. S., and Ackerman, A. A. (1980). Adaptations to underground nesting in birds and reptiles. *Am. Zool.* **20,** 437–447.

Seymour, R. S., and Rahn, H. (1978). Gas conductance in the eggshell of the mound-building Brush Turkey. *In* "Respiratory Function in Birds, Adults and Embryonic" (J. Piiper, ed.), pp. 243–246. Springer-Verlag, Berlin and New York.

Skowron, C., and Kern, M. (1980). The insulation in nests of selected North American songbirds. *Auk* **97,** 816–824.

Smith, W. K., Roberts, S. W., and Miller, P. C. (1974). Calculating the nocturnal energy expenditure of an incubating Anna's Hummingbird. *Condor* **76,** 176–183.

Sotherland, P. R. (1979). Adaptation of avian eggs to conditions above-ground: An alternative to reducing porosity. *Am. Zool.* **19,** 921.

Sotherland, P. R., Packard, G. C., Taigen, T. L., and Boardmand, T. J. (1980). An altitudinal cline in conductance of Cliff Swallow (*Petrochelidon pyrrhonota*) eggs to water vapor. *Auk* **97,** 177–185.

Tracy, C. R., and Sotherland, P. R. (1979). Boundary layers of bird eggs: Do they ever constitute a significant barrier to water loss? *Physiol. Zool.* **52,** 63–66.

Verbeek, N. A. M. (1981). Nesting success and orientation of Water Pipet *Anthus spinoletta* nests. *Ornis Scand.* **12,** 37–39.

Visschedijk, A. H. J. (1968). The air space and embryonic respiration. 2. The times of pipping and hatching as influenced by an artificially changed permeability of the shell over the air space. *Br. Poultry Sci.* **9,** 185–186.

Vleck, C. M., Vleck, D., Rahn, H., and Paganelli, C. V. (1983). Nest microclimate, water-vapor conductance, and water loss in heron and tern eggs. *Auk* **100,** 76–83.

Walsberg, G. E. (1977). Ecology and energetics of contrasting social systems in *Phainopepla nitens* (Aves: Ptilogonatidae). *Univ. Calif. Publ. Zool.* **108,** 1–63.

Walsberg, G. E. (1980). The gaseous microclimate of the avian nest during incubation. *Am. Zool.* **20,** 363–372.

Walsberg, G. E. (1981). Nest-site selection and the radiative environment of the Warbling Vireo. *Condor* **83,** 86–88.

Walsberg, G. E. (1982). Coat color, solar heat gain, and conspicuousness in the Phainopepla. *Auk* **99,** 495–502.

Walsberg, G. E. (1983a). Avian ecological energetics. *In* ''Avian Biology'' (D. S. Farner, J. R. King, and K. Parkes, eds.), Vol. 7, pp. 161–220. Academic Press, London and New York.

Walsberg, G. E. (1983b). Coat color and solar heat gain in animals. *BioScience* **33,** 88–91.

Walsberg, G. E. (1983c). A test for regulation of nest humidity in two bird species. *Physiol. Zool.* **56,** 231–235.

Walsberg, G. E., and King, J. R. (1978a). The heat budget of incubating Mountain White-crowned Sparrows (*Zonotrichia leucophrys oriantha*) in Oregon. *Physiol. Zool.* **51,** 92–103.

Walsberg, G. E., and King, J. R. (1978b). The energetic consequences of incubation for two passerine species. *Auk* **95,** 644–655.

Walsberg, G. E., and King, J. R. (1980). The thermoregulatory significance of the winter roosts selected by robins in eastern Washington. *Wilson Bull.* **92,** 33–39.

Wangensteen, O. D., Rahn, H., Burton, R. R., and Smith, A. H. (1974). Respiratory gas exchange of high altitude adapted chick embryos. *Resp. Physiol.* **21,** 67–70.

White, F. N., Bartholomew, G. A., and Howell, T. R. (1975). The thermal significance of the nest of the Sociable Weaver *Philetairus socius:* Winter observations. *Ibis* **117,** 171–179.

White, F. N., Bartholomew, G. A., and Kinney, J. L. (1978). Physiological and ecological correlates of tunnel nesting in the European Bee-eater, *Merops apiaster. Physiol. Zool.* **51,** 140–154.

Withers, P. C. (1977). Energetic aspects of reproduction by the Cliff Swallow. *Auk* **94,** 718–725.

Yom-Tov, Y., Imber, A., and Otterman, J. (1977). The microclimate of winter roosts of the starling *Sturnus vulgaris. Ibis* **119,** 366–368.

Zerba, E., and Morton, M. L. (1983). The rhythm of incubation from egg laying to hatching in Mountain White-crowned Sparrows. *Ornis Scand.* **14,** 188–197.

Chapter 14

Morphological Aspects of Habitat Selection in Birds

HANS WINKLER

Institut für Limnologie der Österreichischen
Akademie der Wissenschaften
Mondsee, Austria

and

BERND LEISLER

Max-Planck-Institut für Verhaltensphysiologie
Vogelwarte Radolfzell,
Radolfzell-Möggingen, Federal Republic of Germany

I. INTRODUCTION

For all animals the quality of the habitat where they carry out their activities is of vital importance and most species have an influence on their selection of habitat. Birds, without doubt, belong to the most mobile group of organisms. Yet, many species are confined to very specific habitats. Very similar species, however, may occupy distinctly different habitats, a fact which led not only to

HABITAT SELECTION IN BIRDS

ISBN 0-12-178080-5

the discovery of new species but also to the intensive study of habitat selection in birds, an important field of research in ornithology.

Many modern studies of avian habitat selection are related to questions of community structure, competition, and other synecological concepts (Cody, 1981). Ecomorphology centers more on the features of an organism. Together with other disciplines, such as ethology, which concentrate on the immediate interaction between the individual and the environment, it can provide data for a more autecological interpretation (Leisler and Winkler, 1985). A better understanding of such general problems as convergence and composition of species assemblages (e.g., Karr and James, 1975; Leisler, 1977b; Bairlein, 1981) should emerge.

It is easy to demonstrate that morphology is closely associated with habitat selection. Body size is one important feature that certainly determines habitat selection. Within the bustards (Otidae), for example, large and small birds often select different habitats in the same region [the Great Bustard (*Otis tarda*) and the Little Bustard (*Tetrax tetrax*), and the large Kori Bustard (*Choriotis kori*) and some smaller species (*Lophotis ruficrista, Eupodotis senegalensis, Lissotis melanogaster*)]. The vegetations of these regions have structural characteristics related to body sizes and negotiable only by birds of a specific size. The smaller-sized Reed Warbler (*Acrocephalus scirpaceus*) occurs in denser reeds than the larger Great Reed Warbler (*A. arundinaceus*). Even more subtle size differences may effectively determine the utilization of breeding holes and the interspecific competition in hole-nesting birds (Löhrl, 1970, 1977).

Body size together with its physiological consequences determines diving abilities and, as a consequence, habitat selection in cormorants Phalacrocoraciidae (Winkler, 1983). In tits (Paridae) lighter species are found more often in the periphery of trees (Perrins, 1979), and, similarly, small body size is likely an adaptation for specialized foliage insectivory in parulid warblers (Greenberg, 1979). But lighter weight is not the only prerequisite for utilizing thinner twigs. When the heavier species develop better clinging abilities they may be found on lighter supports, as is the case of the Great Spotted Woodpecker (*Picoides major*) and the lighter Middle Spotted Woodpecker (*P. medius;* Winkler, 1973; Jenni, 1983). In shorebirds, the length of the leg is related to wading depth (Baker, 1979).

Examples of this sort can be listed almost endlessly. Commonly, many of these relationships have been investigated only in the laboratory, and traditional laboratory work is not sufficient. A complete understanding can be attained only if behavioral and ecological studies in the field are included (Bock, 1977). In addition problems arising with some results may be avoided if shape is studied in addition to the size of morphological characters alone.

Besides interspecific variation there is a lot of intraspecific variation which should have consequences for habitat selection. Phenotypic nongeographic intraspecific variation can originate from sexual dimorphism, ontogenetic variation, and temporal variation (Thorpe, 1976). Ontogeny in birds is rather rapid. Functionally important structures are either fully developed when needed (e.g., the hind limb of the reed warbler; Impekoven, 1962) or grown adequately with use (e.g., the wing of gallinaceous birds; Glutz, Bauer, and Bezzel, 1973).Molt is an important periodical event which has many ecological implications (Stresemann and Stresemann, 1966); habitat selection is affected mainly in birds which shed large parts of their plumage simultaneously (Haukioja, 1971). For instance, surface-feeding ducks hide in dense vegetation when molting their wing feathers, whereas diving ducks prefer open and deep water (Bezzel, 1964). Often sexual dimorphism in size is linked with differences in habitat selection (Morrison, 1982; Tiainen, 1982). Other examples of sexual morphological differences will be discussed later in some detail.

Size differences may also influence dominance relationships, which in turn determine the quality of the habitat occupied (e.g., Ficken and Ficken, 1967). Intrasexual variation in morphology, mainly size, is also known to be related to habitat quality (Lundberg *et al.*, 1981; Catchpole *et al.*, 1985). An interesting example of intralocality variation is the correlation between plumage color and the local habitat partitioning of some larks (Hoesch and Niethammer, 1940). However, it is still unclear whether and how the hypothesis that niche width and intraspecific variation are correlated (Van Valen, 1965) and the concept of ecotypic variation (Fretwell, 1969; Banks, 1970) are relevant for the problem of habitat selection.

In this chapter we concentrate mainly on the analysis of interspecies variation within close taxonomic units and discuss some cases where morphological differences between the sexes pertain to habitat selection.

II. MATERIAL AND METHODS

Most of our own data comes from work done over the past 10 years. Morphological characters of sylviine warblers are (1) external (e.g., bill length, wing length, tarsus length, tail length), (2) skeletal characters, and, most importantly, (3) functionally relevant characters as foot span, measures related to wing shape, rictal bristle length, and bill curvature. All together 53 characters, divided into 3 functional complexes have been measured (see Leisler and Winkler, 1985 for a complete list).

Habitat measurements for various species were taken from Leisler (1981), Cody and Walter (1976), Cody (1978), and Zbinden and Blondel (1981). Migration distances of European sylviid warblers were taken from Berthold (1973, 1979) and Curry-Lindahl (1981). Measurements on woodpeckers include bill length, wing length, tail length, and toes. All were taken from individuals of the subspecies studied in the field and collected in the same season. Field methods were those of Winkler (1979a).

The multivariate statistical methods (Morrison, 1976) we employed are Principal Component Analysis (PCA) of the correlations (rather than the covariances) between the log-transformed morphological characters and Discriminant Analysis (DA) of habitat measures of *Acrocephalus* territories (Leisler, 1981). Multiple Regression Analysis was used only to follow the method of Albrecht (1979) to demonstrate the relationship between one factor and two morphological compounds extracted with PCA.

III. RESULTS AND DISCUSSION

A. Fine Structure of Morphology and Habitat Relationships

1. Morphology, Vegetation Structure, and Habitat Selection

That vegetation structure is associated with habitat selection has been known for a long time (e.g., Hildén, 1965; James, 1971; Karr and James, 1975; Albers, 1978; review in Cody, 1981). Still, the role of morphology remained obscure in most cases. We shall illustrate this with one well-studied example (Leisler, 1981; Leisler and Winkler, 1985). In six species of Old World warblers (Sylviinae) of the genus *Acrocephalus* (marsh warblers), we examined both morphology (48 characters) and vegetation structure (12 characters) in the breeding habitats.

Discriminant analysis of the habitats resulted in a good separation on the first component, which was related to vegetation density in the lowest layers and in that above 1.5 m, to vegetation height, to the number of emergent elements, and to water depth. The first morphological axis as extracted with PCA mainly described features of the wing and was not correlated with habitat utilization. The second component, however, related the hind limb morphology to habitat structure, i.e., the first component of the DA (Fig. 1).

The morphological axis (ordinate) mainly expresses the ability to cling to vertical stems in the fashion of these warblers: the body is held more or less horizontal, and the feet are far apart. Long legs which fold up like a jackknife are required for this type of locomotion (Winkler and Bock, 1976). Species living in tall, erect vegetation where water is deep (*Acrocephalus arundinaceus, A. scirpaceus, A. melanopogon*) have better clinging abilities than species inhabiting vegetation on the drier side of the succession between water and land, with much undergrowth and fewer emergent elements (*A. schoenobaenus, A. paludicola*). How morphology acts in this case is still open to speculation. The differences in foot and leg structure could affect climbing behavior. The efficiency of pedal locomotion in turn might determine the success in competitive interactions with other species. Reed Warblers (*A. scirpaceus*) and Sedge Warblers (*A. schoeno-*

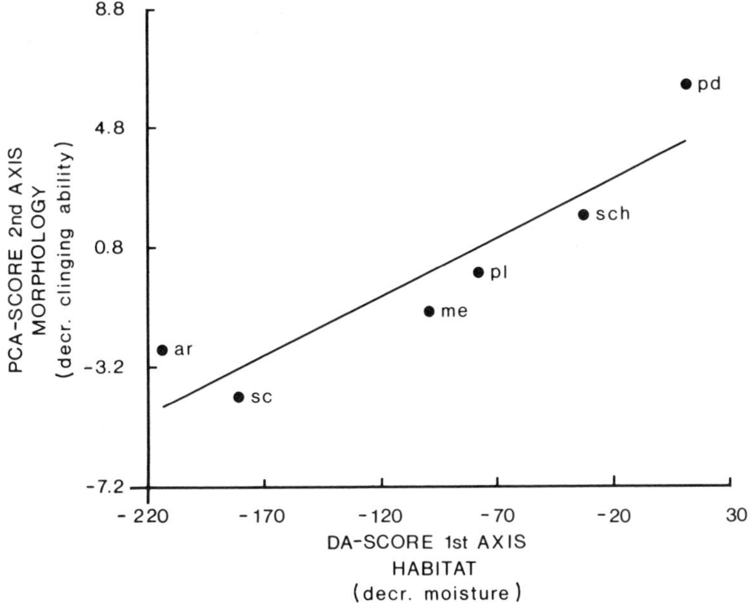

Fig. 1. Correlation between the first component of discriminant analysis of 12 breeding habitat characters and the second component of a PCA of 48 morphological characters in the genus *Acrocephalus. Acrocephalus paludicola* (pd), *A. schoenobaenus* (sch), *A. palustris* (pl), *A. melanopogon* (me), *A. arundinaceus* (ar), *A. scirpaceus* (sc), (r = 0.915).

baenus) are possible competitors. The Sedge Warbler arrives earlier from the wintering grounds and reacts aggressively even to the song of the later-arriving Reed Warbler, which remains neutral to the other species' song (Catchpole, 1978). Habitat utilization of these species overlaps (Leisler, 1981). Reed Warblers actively displace Sedge Warblers in habitats rich in reed (Svensson, 1978). In such habitats, marginally suitable for sedge warblers, Reed Warblers are better adapted morphologically. This may influence the outcome of such interspecific encounters.

2. *Interaction between Morphology and Substrate*

More insight can be expected when observations and experiments are conducted to ascertain how the actual interfacing between morphology and habitat structure takes place. Goldcrests (*Regulus regulus*) show a significant preference for spruce, whereas Firecrests (*R. ignicapillus*) show no particular preference (Fig. 2A). Leisler and Thaler (1982) found striking differences between the two species in the structure of the soles of the feet. Pads, papillae, and folds on the sole form a set of features which fit the substrate by augmenting friction and surefootedness (Rüggeberg, 1960; Lennerstedt, 1974). Only the top of the papillae in the pads of passerine species are free to penetrate the rugosities in the bark of branches and twigs (Lennerstedt, 1975). Figure 2B shows a medial view of the foot of the two species. In the Goldcrest the papillae of the pads are separated by deep furrows which are capable of gripping individual needles (Fig. 2C). Hence, Goldcrests have fewer difficulties in certain postures on coniferous twigs than do Firecrests. Some difficult situations can be handled only by the Goldcrest (B. Leisler and E. Thaler, in preparation).

How can the interaction between morphology and substrate lead to habitat selection? There are two possible ways to relate morphological endowment to the selection of the proper habitat. One way would be that evolution fostered a parallel behavioral adaptation such that an innate preference matches morphologically determined capabilities exactly. The other would be that the animal has to discover by trial and error which structures it can negotiate most efficiently with its physical constitution. Of course, both mechanisms can work at the same time. To understand the role of either morphology or ethology as pacemakers in evolution and for the notion that intraspecific variation in morphology is less than in behavior (Curio, 1977), it is of some theoretical interest to determine the relative importance of these two phenotypic aspects.

Some studies have shown that birds raised without information about their natural habitat or under uniform conditions select the species-specific habitat when presented with the appropriate choice (Partridge, 1974, 1979). Coal Tits (*Parus ater*) and Blue Tits (*Parus caeruleus*) reared with no previous experience of vegetation showed different preferences for their respective habitats: conifers

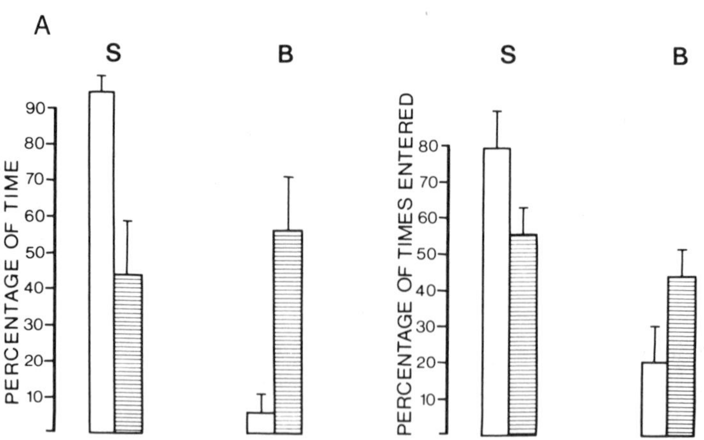

Fig. 2. (A) Selection of Goldcrests (*Regulus regulus*, white bars) and Firecrests (*R. ignicapillus*, dark bars) of spruce (S) and beech (B) in experimental chambers. Differences are significant at the 0.5% level ($n_1 = 11$ and $n_2 = 11$). (B) Feet of *R. ignicapillus* (above) and *R. regulus* (below). Note the well-developed pads in *R. regulus*. (C) Grip (left foot) of Goldcrest *R. regulus* on spruce needles. After a photograph by K. Wüstenberg.

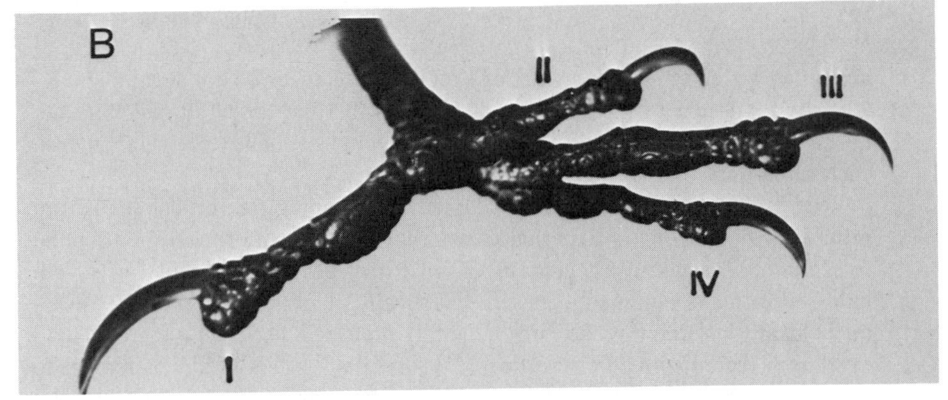
B
II
III
IV
I

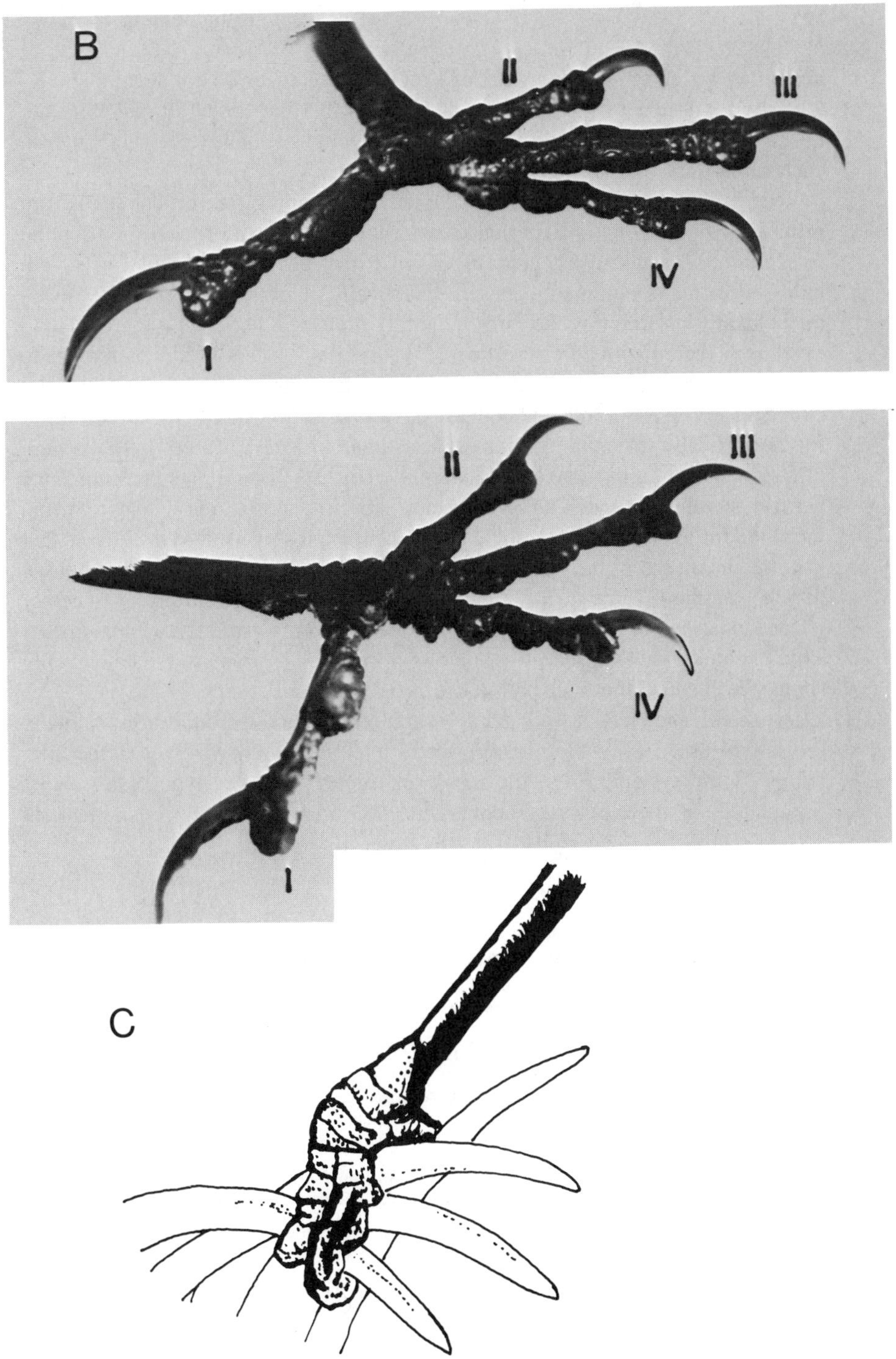

II
III
IV
I

C

and broad-leafed trees. These species-specific preferences do not develop gradually with exposure to the branches. It appears that the observed matching between behavior and foot morphology developed independently of one another in each bird.

Reed Warblers, extremely well adapted for climbing on vertical stems, are influenced by both early experience and short-term experience in their habitat selection. W. Ley (in preparation) raised experimental groups in cages with either horizontally mounted reed stems, vertically mounted ones, or both. Subsequent testing showed that the birds generally preferred vertical structures. However, birds reared with purely vertical structures hardly showed any preference in the test situation. This can be explained by the fact that novelty of the horizontal structures also plays a role and complicates the situation. In all cases, preference for the vertically structured habitat increased within 3 days (Table I). The system of habitat selection in these birds apparently consists of an innate preference for vertical structures, matching the species' specific morphology, and a strong incentive for learning. The learning component expresses itself in the effects that novelty has on the bird's behavior. The experience with their own morphology may be the main force in these changes. A counteracting reinforcing scheme, which consisted in presenting food on the horizontal structures in a further experimental group raised in a purely horizontally structured habitat, only slightly influenced the later preference for vertical structures.

In general, the basis for habitat selection remains an open question for many species (Morse, 1980). Morphology seems to be intimately involved in the fine-tuning of the system. From the results presented here the hypothesis can be formulated that there is a crude congruence between morphological endowment

TABLE I

Mean Percentages of Time Spent by Hand-Raised Reed Warblers (*Acrocephalus scirpaceus*) in Experimental Chambers Furnished with Vertical Reed Stems[a]

Rearing condition	*n*	Day of experiment		
		First	Second	Third
Horizontal elements	26	62.2	65.4	72.3
Vertical elements	7	46.6	47.3	52.7
Mixed elements	10	53.6	61.4	67.9

[a] An alternative chamber provided horizontal elements. Differences between rearing conditions are significant at the 5% level; differences between successive experimental days are significant at the 0.1% level; interactions are not significant. Data from Ley (in preparation).

and innate predilections. Fine-tuning of habitat choice is achieved by a special learning disposition which enables the bird to develop locomotor skills that well fit its physical capabilities and to find the correct habitat in which to fully exercise them. Clearly, there is much need for further research.

B. How Does Morphology Regulate Habitat Selection?

1. Historical Aspect

When in the course of evolution strong correlations between characters emerge, overall morphological variation is channeled into distinct directions. Once such preferred directions are established, for many possible reasons (e.g., internal epigenetic constraints, mechanical principles, correlations among selection pressures), it seems that within closer taxonomic units this limits the adaptive answers that are available to meet the challenges of selective forces. In sylviine warblers the bill appears to be a structure which is free to vary quite unrestrained and therefore can evolve into various shapes. In contrast, the hind limb forms a good example of strong intercharacter correlations.

In a study of the morphometry of 25 sylviine species, the variation of 15 characters was analyzed (Leisler and Winkler, 1985). PCA extracted two major axes of variation. The first was associated with foot size, development of the hind claw, and pelvis width. These characters are associated with clinging abilities; species scoring high on this component have poorer clinging abilities than those scoring low (Fig. 3). The second component is correlated positively with length of the middle claw and tarsus length and negatively with length of the inner toe. Functionally, this corresponds to decreasing walking ability.

In the plane defined by these two axes of variation, three clusters appeared. One cluster (upper left in Fig. 3) was composed of species with feet adapted for clinging, another (upper right, Fig. 3) was formed by species with a typical perching foot, and the third (middle low) encompassed species which do a lot of walking. The corresponding habitats are reed, trees, and dense ground vegetation, respectively. Within these clusters only slight modifications seem to be feasible. In the aquatic warbler (*Acrocephalus paludicola*), which occupies habitats similar to those inhabited by the grass warblers (*Locustella*), some tendencies to develop a walking foot are recognizable (Fig. 3). On the other hand, Savi's Warbler (*L. luscinioides*) living in reed shows an inclination to develop a clinging foot. Within the given range of possible morphological variation and covariation, neither species can fully exploit the possibilities offered by the intermediate habitat. Abilities evolved under a definite set of habitat conditions are, of course, not restricted to these specific habitats. Chiffchaffs (*Phylloscopus collybita*) are arboreal birds. They often cling on very thin pendent twigs (e.g., willows) and therefore possess some morphological characteristics placing them

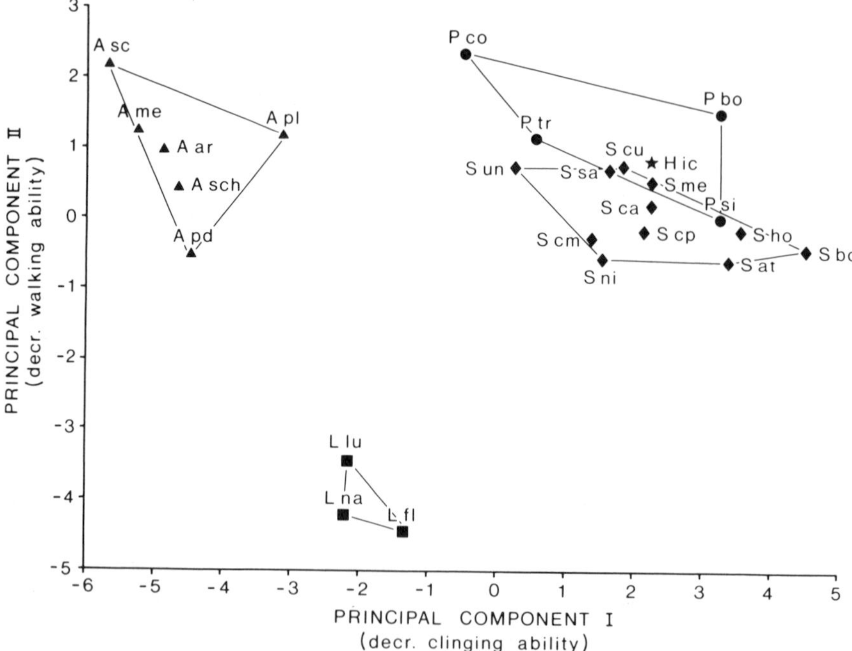

Fig. 3. PCA of the hind limb (15 characters) of 25 sylviid warblers. See text for further explanation. *Acrocephalus scripaceus* (A sc), *A. melanopogon* (A me), *A. palustris* (A pl), *A. arundinaceus* (A ar), *A. schoenobaenus* (A sch), *A. paludicola* (A pd); *Phylloscopus collybita* (P co), *P. bonelli* (P bo), *P. trochilus* (P tr), *P. sibilatrix* (P si); *Hippolais icterina* (H ic); *Sylvia undata* (S un), *S. sarda* (S sa), *S. curruca* (S cu), *S. melanocephala* (S me), *S. cantillans* (S ca), *S. conspicillata* (S cp), *S. hortensis* (S ho), *S. communis* (S cm), *S. nisoria* (S ni), *S. atricapilla* (S at), *S. borin* (S bo); *Locustella luscinioides* (L lu), *L. naevia* (L na), *L. fluviatilis* (L fl).

nearer to *Acrocephalus*. It is thus not surprising that this is the only species of its group which is able to use the rich food supply of reed belts in the fall (Bairlein, 1981). The other arboreal species having only a small perching foot (*P. bonelli, P. sibilatrix*) cannot resort to this option. Morphology in cases like these is therefore a very important aspect for the understanding of habitat selection. Character states and intercharacter correlations, evolved in specific environments, can pose quite stringent limitations as to what habitats a species can adapt, but also may constitute preadaptations for very different habitats that share some intrinsic physical properties with the original habitat.

2. *Natural Experiments*

The problem in characterizing the influence of morphology on habitat selection and in the assessment of its relative importance is that in the course of evolution habitats have also shaped morphological responses. To disentangle this network of mutual and reciprocal relationships, it is necessary to find instances where other factors are also involved and to determine morphology in such a way that habitat-based morphological relationships are excluded or are at least of minor importance. Relationships of the form factor $\Rightarrow$ morphology $\Rightarrow$ habitat should allow some insight into the role of morphology, a role more difficult to evaluate in morphology $\rightleftharpoons$ habitat relationships. Experimental manipulation would be advantageous but is difficult and moreover has seldom been done. Riehm (1970), for instance, shortened the tail of Long-Tailed Tits (*Aegithalos caudatus*) to a length similar to the Blue Tit and found that postures and movements on twigs which involve the tail subsequently became rarer. Some insight can also be gained from observations of injured birds in the field; these and other examples could serve as experiments carried out by nature.

A strong external factor not directly related to habitat use is migration. The need for long-distance flights is a strong selective force and results in typical morphological adaptations, which in turn limit morphological variation. Some degrees of freedom for adaptive responses to habitat requirements are lost; thus habitat selection has to adapt to morphology rather than the other way around. The 25 species of sylviine warblers may support this theory. Migration distances of the species studied morphologically were taken and related to morphology by the method of Albrecht (1979). Forty-eight morphological characters were analyzed with PCA. The first two components of morphometric variation (characters included all major complexes: feeding apparatus, hind limb, forelimb, and tail) were well correlated with migratory distance (Fig. 4). This correlation is almost exclusively a correlation of the second morphological component (ordinate) and migration distance. In other words, adaptations with respect to habitat are restricted only along the axis specified by the second principal component, not along the first axis (abscissa). The first axis mainly defines the ability to use either pedal locomotion (positive side, Fig. 4) or short, straight flights to cover intrahabitat distances. The strong influence of migration, on the other hand, precludes the evolution of highly maneuverable wings in the long-distance migrants. Habitats which would require such wings can be exploited only by species which are not forced to migrate far. The genus *Sylvia* presumably evolved in Mediterranean shrub habitats (see Chapter 17, this volume). Short-distance migration and predominantly aerial locomotion of the ancestors allowed only limited radiation of the descendants to the north (see, for example, Chapter 3, this volume). In the subtropics and tropics, radiation of round-winged species is unlimited; species-rich genera such as *Cisticola, Prinia, Bradypterus, Cettia,* and *Megalurus* have been able to evolve (Leisler, 1977a). Long-distance migration and the exploitation of the forest canopy seem to be compatible. The radiation of *Phylloscopus* warblers in the north, presumably centered in central Asia, would follow this pattern (Gaston, 1974). The restrictions posed by the need for

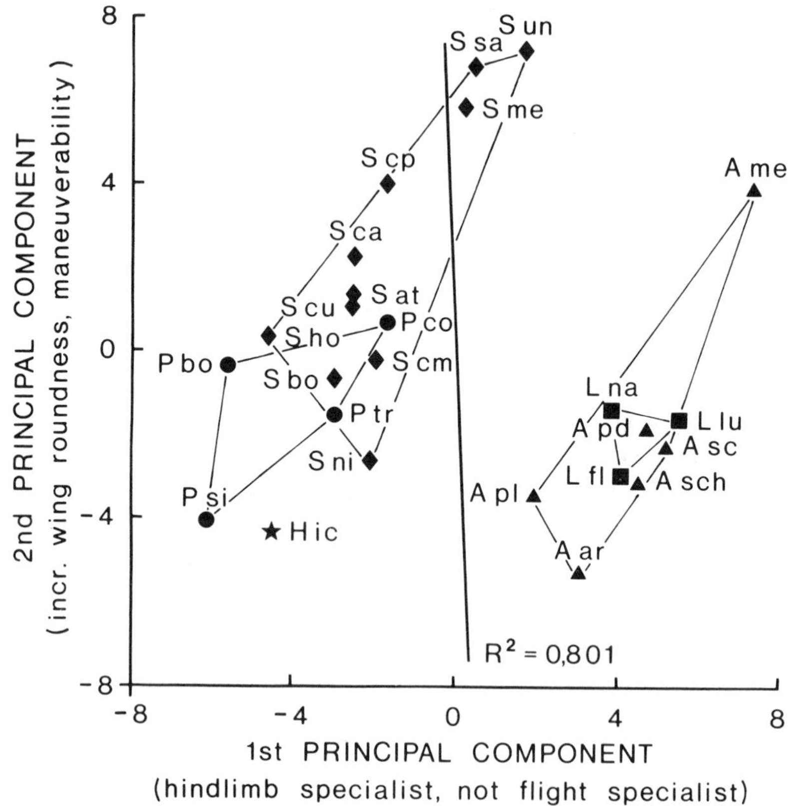

Fig. 4. PCA of total morphology (48 characters) of 25 sylviid warblers. Line denotes the direction of the variation of migratory distance. Species designations as in Fig. 3.

migration are the main reason for the fact that migratory species are more uniform than sedentary ones (Cox, 1968).

There are other factors, not immediately related to habitat selection, which do not restrict morphological variation but enhance it. Morphological sexual dimorphism, in principle, enlarges the intraspecific variation. In many cases differences in habitat selection accompany morphological differences between the sexes (e.g., Selander, 1966; Morrison, 1982; Tiainen, 1982). We shall return to this later. Only when factors are found that are uncoupled from habitat selection and that promote the differentiation of the sexes can the rationale of this chapter be met. Hence, we discuss a case where it is certain that the causal chain runs: sexual selection ⇒ sexual dimorphism ⇒ habitat selection. The ruff (*Philomachus pugnax*), a highly dimorphic species in which males court in leks, was studied by Parz-Gollner (1983) in an important migration area. Sexes showed significant differences in microhabitat selection (Fig. 5). Even though wading

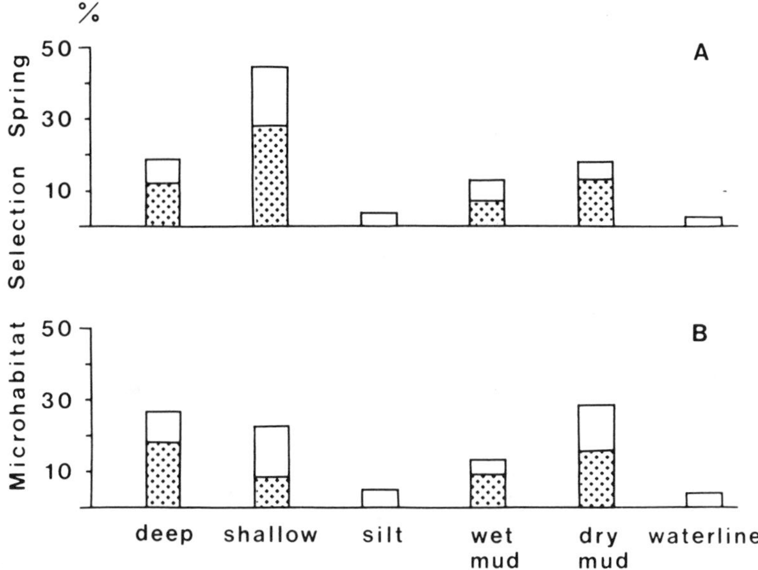

Fig. 5. Habitat selection of Ruff *Philomachus pugnax* in spring. Dark parts of bars indicate portion of habitats with vegetation. Data from Parz-Gollner (1983). (A) Female birds (n = 770). (B) Male birds (n = 1160).

depth was measured relative to features of the bird (tarsus), the larger males showed a clear preference for deeper water. In consequence, habitat utilization is broader in males than in females. Males display on leks; therefore, sexual selection favors larger size, which in turn allows the males access to habitat categories not open to females. Here ecological competition between the sexes does not seem to have the same role as in other shorebirds. Individual and sexual morphological differences have been shown to influence habitat choice significantly in some species (e.g., Puttick, 1981; Harrington, 1982), but competition as the main driving force could not be ruled out, although no overt competition occurred at the study sites (Puttick, 1981).

C. Consequences for Ecological Research

Examining the interface between morphology and habitat or ecology in general produces a better understanding of the relevant ecological structures. When ecologists study questions of habitat selection, they often choose features of the habitat according to criteria of methodological pertinence rather than according to biological relevance for the species involved. This may lead to questionable results, especially for the discussion of community structure and competition.

A simple and popular measure is vegetation height. This feature is quite easy

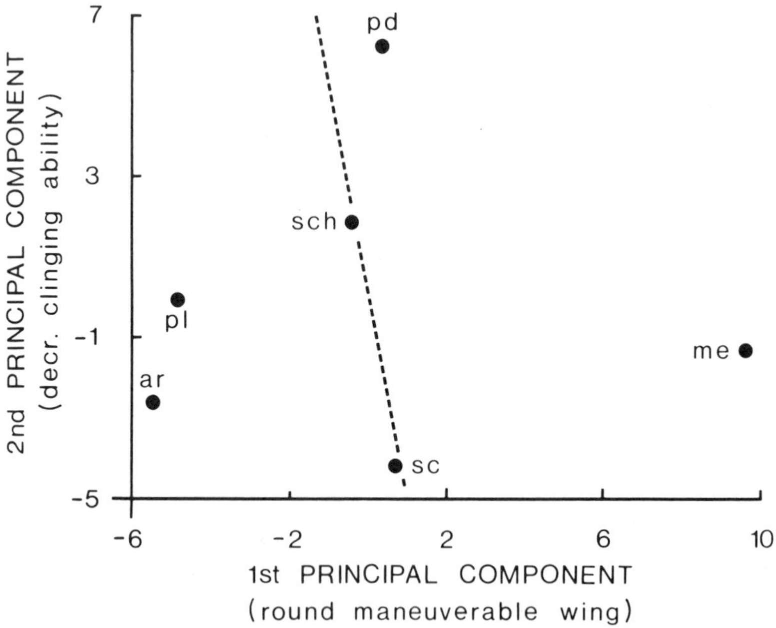

Fig. 6. PCA of total morphology (48 characters) of *Acrocephalus* warblers and its relation to vegetation height (broken line) (R^2 = 0.907). Species designations as in Fig. 7.

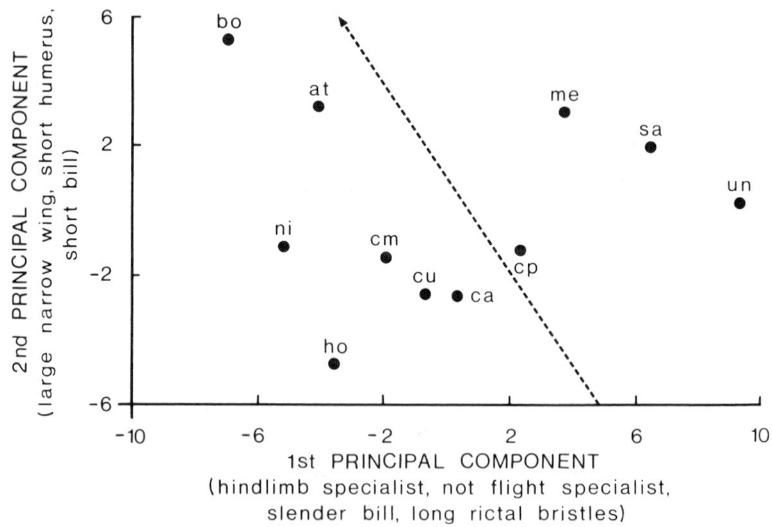

Fig. 7. PCA of total morphology (48 characters) of *Sylvia* warblers and its relation to vegetation height (R^2 = 0.740). Arrow points into direction of high vegetation. Species designations follow Fig. 3.

to measure but means very different things to different organisms (and it means different things in different vegetation types too). Vegetation height therefore may produce morphological responses that are dissimilar even within relatively close taxonomic units. Leisler and Winkler (1985) have shown that the genera *Acrocephalus* and *Sylvia* follow different strategies in solving the problems incurred by vegetation height. *Acrocephalus* warblers cover vertical distances with pedal locomotion, and they have a corresponding development of the hind limb (Fig. 6). *Sylvia* warblers use their wings for the same purpose, which results in adaptations mainly in the flight apparatus (Fig. 7). It should be noted that the correlation between morphology and vegetation height is good in both cases [the method of Albrecht (1979) was used]; the axis of major morphological variation paralleled habitat variation only in *Acrocephalus,* whereas in *Sylvia* habitat variation cut across two major axes of morphological variation (see Leisler and Winkler, 1985, for more details).

These results demonstrate that vegetation height as a parameter is different for different species or groups of species and that other more relevant habitat characteristics should also be studied. In our case the important feature of reed stems is that they provide unimpeded pathways over the full range of vegetation height but do not leave much room for wing extension. In the shrub habitats utilized by *Sylvia* warblers, the number and width of gaps within the vegetation are the main features which control access to the various vegetation stories (cf. Heydemann, 1956). Pedal locomotion is obstructed by the spatial configuration of twigs and leaves. Vegetation height, while presenting itself to the unreserved observer as one single dimension of the Hutchinsonian niche, may indeed be composed of several orthogonal dimensions which can only be disentangled after careful ecological, morphological, and behavioral studies.

Morphological thinking, always devoted to the study of individual adaptations, should help overcome the concept of organisms as "black boxes" (Bock, 1977). Inquiry into the mechanisms at work within them may inspire the selection of heuristically decisive habitat features. Kilham (1970), Jackson (1970), and other workers found a remarkable segregation of the sexes in the Downy Woodpecker (*Picoides pubescens*) in relation to feeding-site selection. Examination of museum specimens showed that there was a highly significant difference not only in bill size (males have the longer bill) but also in tail length (males have the shorter tail). The middle toes of the males tended to be longer too (H. Winkler, unpublished results). These structures are thought to be related to locomotion (e.g., Winkler and Bock, 1976). In larger woodpecker species such as *P. major,* differences in the locomotory apparatus are even more pronounced (Winkler, 1979b). These results suggest that such characters should be measured in addition to those related to the feeding apparatus commonly measured in ecological studies (e.g., Eckhardt, 1979) and that both types of characters should be measured in a more standardized manner. The morphologically relevant hab-

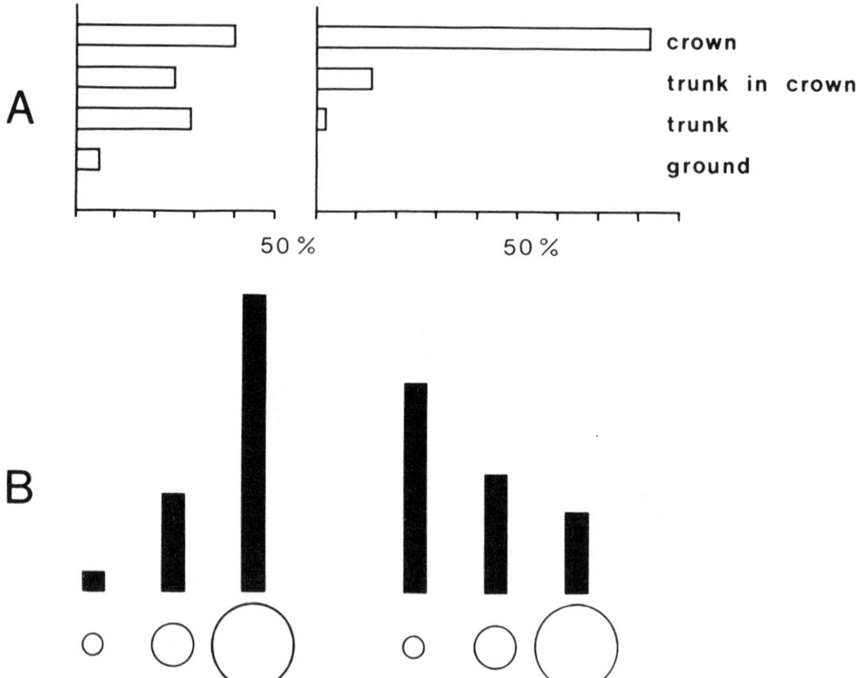

Fig. 8. Sexual differences in the habitat selection of Downy Woodpeckers. Data for female birds appears on the left, male birds on the right. (A) Selection of strata; female birds (n = 52), male birds (n = 41); $p < 0.0005$. (B) Selection of limb diameters; small circles denote diameters smaller than, middle circles diameters equal to, and large circles diameters greater than diameter of the woodpecker's body; female birds (n = 42), male birds (n = 45); $p < 0.0005$.

itat feature for these woodpeckers is the diameter of the substrate relative to its body. This relative diameter determines grip and hence the employment of the toes and is more relevant than the absolute diameter of the substrate, which is the usual feature measured (Jackson, 1970; Willson, 1970) according to ecological reasoning (resource partitioning).

With this idea in mind, the habitat exploitation of Downy Woodpeckers during feeding was measured (H. Winkler unpublished data). Males preferred the upper stories in the forest; females preferred them much less (Fig. 8). These data and additional observations strongly suggested that height selection is not the main reason for this difference but that choice of substrates with different diameters plays an important role. The females preferred thicker substrates (Fig. 8), a difference even more pronounced than the difference in exploitation of strata. Jackson (1970), although finding similar differences, did not arrive at the same conclusion (but see Jenkins, 1979, for *Picoides nuttallii*). A closer look reveals that there are even more morphologically and ecologically relevant differences. Males much more often than females exploit the underside of horizontal structures. Consequently, ideas inspired by morphological findings can help define the correct resource dimensions.

The role of habitat-character choice in the testing of ecological hypotheses may further be exemplified by another look at Fig. 8. Using one measure (A) would in one sense make the female the generalist, but using another (B) would label the male the generalist. Workers using only one or the other of such alternative measures would arrive at very different conclusions when, say, discussing the role of dominance in the width of habitat niches (Morse, 1974). The morphologist may start with a knowledge of the relevant adaptations which, when recognized, could help the ecologist to further refine the definition of habitat characters.

In other cases different habitats may appear to vary widely when viewed superficially but in fact possess certain features in common. One well-known example is the utilization of reed by many forest birds in the winter. Chiffchaffs (*Phylloscopus collybita*), Blue Tits, Winter Wrens (*Troglodytes troglodytes*), and even Lesser Spotted Woodpeckers possess adaptations which allow them to exploit reed, although with quite different modes of locomotion. Their specific morphological features evolved in forests of superficially quite different structure. The European Blackbird (*Turdus merula*) has a flight apparatus and a flight pattern which allow this bird to change flight direction rapidly and to dodge around corners quickly (Stork, 1968). The relevant habitat character for this bird, then, is the organization of the aerial space. Here one can see the basic similarities between the original forest habitats and the suburbs and city parks where they are now found. Hence, instances in which birds occupy new habitats either by range extension or by introduction may serve well for testing hypotheses about the correct evaluation of niche dimensions as discerned by ecological, morphological, and behavioral studies (Morse, 1980).

REFERENCES

Albers, P. H. (1978). Habitat selection by breeding red-winged blackbirds. *Wilson Bull.* **90,** 619–634.

Albrecht, G. H. (1979). The study of biological versus statistical variation in multivariate morphometrics: The descriptive use of multiple regression analysis. *Syst. Zool.* **28,** 338–344.

Bairlein, F. (1981). Ökosystemanalyse der Rastplätze von Zugvögeln. *Ökol. Vögel* **3,** 7–137.

Baker, M. C. (1979). Morphological correlates of habitat selection in a community of shore birds (Charadriiformes). *Oikos* **33,** 121–126.

Banks, R. C. (1970). On ecotypic variation in birds. *Evolution (Lawrence, Kans.)* **24,** 829–831.

Berthold, P. (1973). Relationships between migratory restlessness and migration distance in six *Sylvia* species. *Ibis* **115,** 594–599.

Berthold, P. (1979). Beziehungen zwischen Zugunruhe und Zug bei der Sperbergrasmücke *Sylvia nisoria:* Eine ökophysiologische Untersuchung. *Vogelwarte* **30,** 77–84.

Bezzel, E. (1964). Zur Ökologie der Brutmauser bei Enten. *Anz. Ornithol. Ges. Bayern* **7,** 43–79.

Bock, W. J. (1977). Toward an ecological morphology. *Vogelwarte* **29,** 127–135.

Catchpole, C. K. (1978). Interspecific territorialism and competition in *Acrocephalus* warblers as revealed by playback experiments in areas of sympatry and allopatry. *Anim. Behav.* **26,** 1072–1080.

Catchpole, C. K., Leisler, B., and Winkler, H. (1985). The polygyny in the great reed warbler *Acrocephalus arundinaceus:* A possible case of deception. *Behav. Ecol. Sociobiol.* (in press).

Cody, M. L. (1978). Habitat selection and interspecific territoriality among the sylviid warblers of England and Sweden. *Ecol. Monogr.* **48,** 351–396.

Cody, M. L. (1981). Habitat selection in birds: The role of vegetation structure, competitors, and productivity. *BioScience* **31,** 107–113.

Cody, M. L., and Walter, H. (1976). Habitat selection and interspecific interactions among Mediterranean sylviid warblers. *Oikos* **27,** 210–238.

Cox, G. W. (1968). The role of competition in the evolution of migration. *Evolution (Lawrence, Kans.)* **22,** 180–192.

Curio, E. (1977). Some aspects of individual variation in birds. *Vogelwarte* **29,** 111–120.

Curry-Lindahl, K. (1981). "Bird Migration in Africa," Vol. 1, Academic Press, London and New York.

Eckhardt, R. C. (1979). The adaptive syndromes of two guilds of insectivorous birds in the Colorado Rocky Mountains. *Ecol. Monogr.* **49,** 129–149.

Ficken, M. S., and Ficken, R. W. (1967). Age-specific differences in the breeding behavior of the American redstart. *Wilson Bull.* **79,** 188–199.

Fretwell, S. (1969). Ecotypic variation in the non-breeding season in migratory populations: A study of tarsal length in some fringillidae. *Evolution (Lawrence, Kans.)* **23,** 406–420.

Gaston, A. J. (1974). Adaptation in the genus *Phylloscopus. Ibis* **116,** 432–450.

Glutz v. Blotzheim, U. N., Bauer, K., and Bezzel, E. (1973). "Handbuch der Vögel Mitteleuropas," *Vol. 5. Akad. Verlagsges., Frankfurt on Main.*

Greenberg, R. (1979). Body size, breeding habitat, and winter exploitation systems in *Dendroica. Auk* **96,** 756–766.

Harrington, B. A. (1982). Morphometric variation and habitat use of semipalmated sandpipers during a migratory stopover. *J. Field Ornithol.* **53,** 258–262.

Haukioja, E. (1971). Flightlessness in some moulting passerines in Northern Europe. *Ornis. Fenn.* **48,** 101–116.

Heydemann, B. (1956). Die Biotopstruktur als Raumwiderstand und Raumfülle für die Tierwelt. *Verh. Dtsch. Zool. Ges.* **1956,** 332–347.

Hildén, O. (1965). Habitat selection in birds. *Ann. Zool. Fenn.* **2,** 53–75.

Hoesch, W., and Niethammer, G. (1940). Die Vogelwelt Deutsch-Südwestafrikas. *J. Ornithol.* **88,** Suppl., 1–404.

Impekoven, M. (1962). Die Jugendentwicklung des Teichrohrsängers (*Acrocephalus scirpaceus*) *Rev. Suisse Zool.* **69,** 77–191.

Jackson, J. A. (1970). A quantitative study of the foraging ecology of downy woodpeckers. *Ecology* **51,** 318–323.

James, F. C. (1971). Ordinations of habitat relationships among breeding birds. *Wilson Bull.* **83,** 215–236.

Jenkins, J. M. (1979). Foraging behavior of male and female Nuttall woodpeckers. *Auk* **96,** 418–420.

Jenni, L. (1983). Habitatnutzung, Nahrungserwerb und Nahrung von Mittel-und Buntspecht (*Den-*

drocopus medius und *D. major*) sowie Bemerkungen zur Verbreitungsgeschichte des Mittelspechts. *Ornithol. Beob.* **80,** 29–57.

Karr, J. R., and James, F. C. (1975). Eco-morphological configurations and convergent evolution of species and communities. *In* "Ecology and Evolution of Communities" (M. L. Cody, ed.), pp. 258–291. Harvard Univ. Press (Belknap), Cambridge, Massachusetts.

Kilham, L. (1970). Feeding behavior of downy woodpeckers. I. Preference for paper birches and sexual differences. *Auk* **87,** 544–556.

Leisler, B. (1977a). Die ökologische Bedeutung der Lokomotion mitteleuropäischer Schwirle (*Locustella*). *Egretta* **20,** 1–25.

Leisler, B. (1977b). Ökomorphologische Aspekte von Speziation und adaptiver Radiation bei Vögeln. *Vogelwarte* **29,** 136–153.

Leisler, B. (1981). Die ökologische Einnischung der mitteleuropäischen Rohrsänger (*Acrocephalus,* Sylviinae). I. Habitattrennung. *Vogelwarte* **31,** 45–74.

Leisler, B., and Thaler, E. (1982). Differences in morphology and foraging behaviour in the goldcrest *Regulus regulus* and firecrest *R. ignicapillus. Ann. Zool. Fenn.* **19,** 277–284.

Leisler, B., and Winkler, H. (1985). Ecomorphology. *Curr. Ornithol.* **2,** 155–186.

Lennerstedt, I. (1974). Pads and papillae on the feet of nine passerine species. *Ornis Scand.* **5,** 103–111.

Lennerstedt, I. (1975). A functional study of papillae and pads in the foot of passerines, parrots, and owls. *Zool. Scripta* **4,** 111–123.

Löhrl, H. (1970). Unterschiedliche Bruthöhlenansprüche von Meisenarten und Kleinbern als Beitrag zum Nischenproblem. *Verh. Deutsch. Zool. Ges.* **64,** 314–317.

Löhrl, H. (1977). Nistökologische und ethologische Anpassungserscheinungen bei Höhlenbrütern. *Vogelwarte* **29,** 92–101.

Lundberg, A., Alaltalo, R. V., Carlson, A., and Ulfstrand, S. (1981). Biometry, habitat distribution and breeding success in the pied flycatcher *Ficedula hypoleuca. Ornis Scand.* **12,** 68–79.

Morrison, D. F. (1976). "Multivariate Statistical Methods," 2nd ed. McGraw-Hill, New York.

Morrison, M. L. (1982). The structure of western warbler assemblages: Ecomorphological analysis of the black-throated gray and hermit warblers. *Auk* **99,** 503–513.

Morse, D. H. (1974). Niche breadth as a function of social dominance. *Am. Nat.* **108,** 818–830.

Morse, D. H. (1980). "Behavioral Mechanisms in Ecology." Harvard Univ. Press, Cambridge, Massachusetts.

Partridge, L. (1974). Habitat selection in titmice. *Nature* **247,** 573–574.

Partridge, L. (1979). Differences in behaviour between blue and coal tits reared under identical conditions. *Anim. Behav.* **27,** 120–125.

Parz-Gollner, R. (1983). Einflüsse auf die Habitatwahl und die Verteilung des Kampfläufers, *Philomachus pugnax,* im Seewinkel, Burgenland. Thesis, Univ. Vienna, Vienna.

Perrins, C. M. (1979). "British Tits." Collins, London.

Puttick, G. M. (1981). Sex-related differences in foraging behaviour of curlew sandpipers. *Ornis Scand.* **12,** 13–17.

Riehm, H. (1970). Ökologie und Verhalten der Schwanzmeise (*Aegithalos caudatus*). *Zool. Jahrb. Syst.* **97,** 338–400.

Rüggeberg, T. (1960). Zur funktionellen Anatomie der hinteren Extremität einiger mitteleuropäischer Singvogelarten. *Z. Wiss. Zool. Abt. A.* **164,** 1–118.

Selander, R. (1966). Sexual dimorphism and differential niche utilization in birds. *Condor* **68,** 113–151.

Stork, H. j. (1968). Morphologische Untersuchungen an Drosseln. Eine Analyse von Anpassungsstrukturen im Körperbau von sechs europäischen Arten der Gattung *Turdus* L. *Z. Wiss. Zool. Abt. A.* **178,** 72–185.

Stresemann, E. and Stresemann, V. (1966). Die Mauser der Vögel. *J. Ornithol.* **107,** Suppl., 1–445.

Svensson, S. E. (1978). Territorial exclusion of *Acrocephalus schoenobaenus* by *A. scirpaceus* in reed beds. *Oikos* **30,** 467–474.

Thorpe, R. S. (1976). Biometric analysis of geographic variation and racial affinities. *Biol. Rev.* **51,** 407–452.

Tiainen, J. (1982). Ecological significance of morphometric variation in three sympatric *Phylloscopus* warblers. *Ann. Zool. Fenn.* **19,** 285–295.

Van Valen, L. (1965). Morphological variation and width of ecological niche. *Am. Nat.* **99,** 377–390.

Willson, M. F. (1970). Foraging behavior of some winter birds of decidious woods. *Condor* **72,** 169–174.

Winkler, H. (1973). Nahrungserwerb und Konkurrenz des Blutspechts, *Picoides (Dendrocopos) syriacus. Oecologia* **12,** 193–208.

Winkler, H. (1979a). Foraging ecology of Strickland's woodpecker in Arizona. *Wilson Bull.* **91,** 244–254.

Winkler, H. (1979b). Bemerkungen zum Maurenspecht, *Picoides major numidus. J. Ornithol.* **120,** 290–298.

Winkler, H. (1983). The ecology of cormorants. *In* ''Limnology of Parakrama Samudra—Sri Lanka.'' (F. Schiemer, ed.), pp. 193–199. Junk, The Hague.

Winkler, H., and Bock, J. W. (1976). Analyse der Kräfteverhältnisse bei Klettervögeln. *J. Ornithol.* **117,** 397–418.

Zbinden, N., and Blondel, J. (1981). Zu Raumnutzung, Territorialität und Legebeginn mediterraner Grasmücken (*Sylvia melanocephala, S. undata, S. cantillans, S. hortensis*) in Südfrankreich. *Ornithol. Beob.* **78,** 217–231.

Chapter 15

Habitat Selection: Behavioral Aspects

PETER H. KLOPFER
Department of Zoology
Duke University
Durham, North Carolina

and

JÖRG U. GANZHORN[1]
Abt. Verhaltensphysiologie
Universität Tübingen
Tübingen, Federal Republic of Germany

[1]Present address: Gluckstrasse 1, D-7032 Sindelfingen, Federal Republic of Germany.

HABITAT SELECTION IN BIRDS

ISBN 0-12-178080-5

I. INTRODUCTION

The formal articulation of the notion that animals themselves select a particular habitat is generally attributed to David Lack (see Lack, 1954, for overview). Irrespective of this or other particular claims for priority, art historians could affirm a more ancient origin of the idea. Examinations of medieval and pre-Renaissance paintings reveal their artists to have often been endowed not only with keen appreciation of the appearances of different species and the poses normal to them but also the association of specific organisms with particular features of the landscape. Fanciful renderings aside, peacocks do not appear in drawings of moors nor moorhens in wheatfields.

Our earth is a patchwork of many different habitats—obvious enough to travelers in high-altitude aircraft—but apparent, too, though on a different scale, to the farmer who has never left his homestead.

> Almost any physiographic atlas provides illustrations of the patchwork pattern of our planet: here there is a yellow patch of low desert, there a Rorschach blot of montane forest, and over there broad streaks of upland prairie. Even as the scale is increased, the patches remain evident. On a large map the North Carolina Piedmont may appear uniform, but a drive along a Piedmont road dispels the illusion: a north-facing slope is capped with cool, dense beech trees; on the facing slope, toward the bottom, is a stand of tulip-poplar trees; above these oaks and hickories. But a few yards farther on, on an exposed flat, stand torches of loblolly or a cultivated field. We could increase the scale still further; for even at the levels of single particles of sand, patchiness in the physical substrate remains compellingly evident [Klopfer, 1969, pp. 3–4].

One might well imagine that a Creator who produced some finite number of species might have distributed them uniformly or randomly across all the different patches. Yet, as attested by early bestiaries and, later, nineteenth-century naturalist's documents (see Chapter 1, this volume), it was apparent to all observers that this was not their distribution. Not that a Creator could not be assumed sufficiently prescient to associate particular species with particular sets of conditions, but any degree of interspecific competition would assuredly produce such segregation. If species differ (as, by definition, they do), it is likely they will not be equally efficient in the business of surviving in a particular habitat. Given finite resources relative to reproductive capacity and stability of conditions within a reasonable duration of time, species will segregate themselves, each to be found in a particular place. This is but a restatement of the familiar Darwinian paradigm. It establishes the most general ultimate explanation for particular associations of species with particular habitats though leaves unanswered the question as to mechanisms: just why is one species more "efficient" in grasslands than another? And what gets it to or keeps it in the proper place? If a "choice" is actually made by an animal, rather than natural selection eliminating those individuals which made inappropriate choices, how is the correct habitat recognized? We review current views on this question in the following section.

Significance of Habitat Choice

A word need first be said on the significance of the issue of habitat selection. First, let us consider its significance to the animal itself. The selection, or choice, of the proper habitat, i.e., that area which provides optimum conditions for survival and reproduction, presumably provides a far more secure prospect for a long life than would a random choice. Ultimately, individuals exercising wise choices must fare better than those that do not—or so goes the conventional wisdom. Consistent, species-specific preferences also reduce interspecific competition, providing yet another stabilizing and advantageous influence. Where two species do show overlap in habitat preferences, as do the Willow Warbler, (*Phylloscopus trochilus*) and Chiffchaff (*P. collybita*), they may occupy mutually exclusive territories. These two species use the same sort of trees, forage at almost the same height, and almost exclusively exploit the outer third of branches (Saether, 1983; Cody, 1978). Although a powerful discriminant analysis shows differences in their use of the habitat, most vegetation parameters such as cover or tree species diversity in the areas used by the two species are almost identical. This broad overlap in their habitat niche is associated with a spatial (horizontal) segregation of the two species which seems to be maintained more by interspecific aggression than by minor habitat differences (Saether, 1983). Interspecific territoriality–associated preferences for the same food and vegetation parameters is common among Sylviid warblers (Cody, 1978, 1979; see also Chapter 3, this volume). Warblers which forage at different heights can occupy the same habitat and form stable assemblages. But two species that have similar foraging heights will exclude each other and form interspecific territories.

It appears that song convergence is the major mechanism, though not the only one, for the establishment and maintenance of interspecific territories among warblers. In Sweden the species *Sylvia nisora, S. borin,* and *S. communis* interact strongly with each other over territories. Here *S. nisora* sings longer songs while interacting with S. borin, whereas the songs are shorter when it interacts with *S. communis* (Cody, 1978, 1979).

There is evidence that related species with differing habitat preferences are each more efficient within their own habitat. Coal Tits (*Parus ater*) feed more often by closely examining and gently probing, while the Blue Tit (*P. caeruleus*) hammers, hacks, or tears at food items. These behavior patterns are adapted to the natural food supply in conifers, preferred by Coal Tits, and in broad-leaved trees, the habitat of the Blue Tit. The same behavioral differences are also seen under experimental conditions when both species were exposed to the same habitat (Partridge, 1976).

But, species are not inflexibly tied to a specific type of habitat. As population densities rise, normally less prized areas may be chosen (review in Davies, 1978), and habitat and weather may also influence foraging tactics. The Swainson's Hawk (*Buteo swainsoni*) does not hunt in fields with dense plant cover but does use such fields after the harvest (Bechard, 1982), and Mockingbirds (*Mimus polyglottos*) use shrubs and aerial hawking techniques more often in areas with tall grass than they do in short grass plots (Roth, 1979). Changes in foraging behavior such as selection of different tree species, foraging substrate, or height from one area to the next can be explained by a shift in vegetation structure and availability of resources (Franzreb, 1983; Landres and MacMahon, 1983). At high temperatures, plovers (*Charadrius* spp.) take large prey but ignore small prey items, which are frequently taken at lower temperatures (Piencowski, 1983), and the Bananaquit (*Coereba flaveola*), a nectarivore, becomes insectivorous when blossoms are rare due to drought (Faaborg, 1982a).

In these examples behavior is also influenced by the availability of resources, presumably in response to selection to maximize energy intake and to minimize energy loss as demonstrated by tits, which use less energy-consuming feeding postures at low ambient temperatures (Alatalo, 1983). All of this is consistent with the conventional view of habitat choice as an evolutionarily stabilizing process. However, it is a view now challenged, for example by the findings of Wiens and Rotenberry (1981). Their challenge is considered in Chapters 6 and 7 (this volume), but it must be borne in mind lest our conclusions on behavioral aspects of habitat selection be accorded more importance than they perhaps deserve.

A second matter of significance to the issue of habitat selection concerns the biologist studying the phenomenon. Perhaps no other area of research is more likely to educate one as thoroughly on the perils of a human perception–oriented approach. Just as an anthropologist studying the symbols of an alien culture must try to shed the preconceptions he brings from his culture, so the lesson here is the need to ''see'' the world from the bird's eye view. Von Uexkull (1921) made this point many decades past, but it is a lesson that must apparently be relearned with each generation. Study habitat choice, and it will be taught you! No other species sees the world as do we, and without insight into how another's perceptions are formed, our correlations are likely to be spurious. Chipping Sparrows (*Spizella pusilla*) may select woods on the basis of foliage shapes (Klopfer, 1963), but for reed warblers (*Acrocephalus* spp.) an assessment of moisture may be equally important (Leisler, 1981). Hummingbirds can be much more readily trained to go to different blossoms on successive foraging bouts than to return to the same ones, while for titmice (*Parus* spp.) and their foraging sites the reverse is true (Smith and Sweatman, 1974; Cole *et al.*, 1982). How can they be so perverse? Don't they ''see'' things as do we? Obviously, no. And in this important lesson lies the second significant fact about habitat selection.

II. BEHAVIORAL CIRCUMVENTION OF MORPHOLOICAL AND PHYSIOLOGICAL CONSTRAINTS

A. Nature of the Constraints

Physiological constraints in relation to habitat selection are dealt with separately in Chapter 13, (this volume). We nonetheless review certain aspects here, in order to examine behavioral strategies to circumvent such constraints.

"Constraints" may be regarded as a measure of the degree to which morphological or physiological adaptations fall short of a postulated optimum. Operationally, the measure can be based on energetic values (note Bock, 1977). The constraints may vary from those that produce gross differences to minor, seemingly trivial ones, those that differentiate species or sexes or merely individuals. An example of the former is provided by the webbed feet of penguins, which, though allowing efficient locomotion in water, are far more costly to use on land (Pinshow *et al.,* 1977). Indeed, where penguins are compelled to spend much time on land, as in some zoos, a high incidence of leg and foot injuries or deformities result, evidencing the remorseless nature of some morphological constraints. The latter are illustrated by sex-dependent differences in size and the concommitant differences in the optimum size of food or perches.

B. Interactions of Morphology and Behavior

Most descriptions of habitat choice on the basis of anatomical features entail minor morphological variations: specific differences in wing length, leg structure, or bill shape, for example (Leisler, 1977a; Chapter 14, this volume; Bock, 1970). Among the "classic" cases are the Galapagos finches (Bowman, 1961; Lack, 1947) whose bill shapes appear to be associated with preferences for particular foods, hence habitats. Hespenheide (1966) has shown that, while finches of different species can all utilize the same seeds, they do differ in the efficiency with which the differently sized seeds can be hulled. The more easily extracted kernels, when other considerations (e.g., nutritive value and taste) are equal, are in fact preferred (and see Pulliam and Enders, 1971).

Some bill shapes or sizes may allow for more flexibility than others, at the cost of efficiency with any particular seed type. Thus, Boag and Grant (1981) showed that the large-billed *Geospiza majors* were better able to survive dry periods in which the smaller (but not the larger) seeds became scarce, than were smaller-billed competitors, which were more tightly constrained, presumably, by their anatomy.

Doubts on morphologic similarities as indicators of dietary similarities, however, have been raised anew by Strong (1983), who cites work by Schluter that indicates, contra Lack, that *Geospiza fulginosa* and *G. difficilis* are, respectively, granivorous and carnivorous and hence not food competitors despite similarities in bill shape. However, it is not necessarily the case that trophic considerations are the cause—they could be the consequence of the selection of a particular habitat for an (originally) entirely different reason, such as suitability for nesting. For instance, nonmigrating Rooks (*Corvus frugilegus*) of some Scottish Highland valleys leave their rookeries in the fall to feed and roost in areas up to 45 km away, when food is more abundant than in the vicinity of their rookeries. During the second half of the winter, however, they return to their rookeries daily. These visits are not due to feeding conditions having improved nearby but are necessary to allow them to defend their nest sites, which were chosen in the fall (McKilligan, 1980).

Pine Warblers (*Dendroica pinus*) and Blue-gray Gnatcatchers (*Pulioptila caerulea*) forage on bark on one island but use foliage on a neighboring island in the Bahamas. Since food abundance and distribution were found to be similar on these two islands, Emlen (1981) explained this difference in foraging strategy by postulating different substrate preferences, which are presumed to have evolved under a different competitive situation in the past and are maintained now even while the original circumstances have changed.

Levin (1970) emphasized that not all species are resource limited but that some species can dominate several resources, and an equilibrium might be reached by the action of a predator. Finally, historical geography, man's influence, and the predictability of unfavorable conditions can all be influential, too, as is indicated by Karr's explanation of differences in the avian communities of two tropical regions in Panama and Liberia (Karr, 1976). If a particular beak structure were indeed the *conditio sine qua non* for a choice of a particular habitat, phenomena such as the rapid adaptative radiation of the Hawaiian honeycreepers would be hard to explain. Equally mysterious then would be the greater phenotypic variability in birds of the West Indies (Faaborg, 1982b) and the increased sexual dimorphism of territorial Chaffinchs (*Fringilla coelebs*) with decreasing interspecific competition. This last result was noted in a comparison of two regions with a similar level of habitat heterogeneity and food sources but differing levels of competition (Ebenman and Nilsson, 1982). Furthermore, Pulliam (1983) showed that the structure of an avian community in southeast Arizona could be as well described by a random aggregation model as by a community matrix theory. The theory was supported no better during periods of food shortage than during periods of abundance. Though Pulliam admits to shortcomings in his measures, if he is correct it does cast doubt on the traditional Hutchinson–MacArthur notion of habitat–trophic segregation in avian communities.

Even though morphologic differences between species may vary with the competitive situation and niche breadth (Grant *et al.*, 1976; Selander and Kaufman, 1973; Van Valen, 1965; MacArthur *et al.*, 1972), behavioral traits have also developed that can altogether circumvent the effects of morphology.

Locustella warblers in Central Europe settle in dense ground vegetation for which rounded wings are generally best suited. These birds have pointed wings but apparently surmount this handicap by disdaining wings for legs when they move (Leisler, 1977b, in Leisler, 1977a; see Chapter 14, this volume). An altogether new behavior pattern has circumvented the constraint.

Egyptian Vultures (*Neophron percnopterus*) also provide an intriguing instance of behavior overcoming structure. They are unable to pierce ostrich eggs with their curved bills—adapted for tearing flesh—so they bombard the eggs with stones thrown from the bill, thereby cracking the shells and gaining access to the contents. Common Crows (*Corvus brachyrhynchos*) may use stones to hammer acorns, and Scrub Jays hammer acorns on stones; thrushes hammer snails on stones, too. A comparable tool use as a means of surmounting structural constraints is illustrated also by the Galapagos Woodpecker Finch (*Geospiza pallida*), which uses twigs or cactus spines to retrieve insects from crevices. These and other examples are reviewed by Beck (1980).

C. Interactions of Physiology and Behavior

Physiological constraints are most readily noted through their effect on the distribution of particular species in respect to abiotic factors (e.g., Kendeigh *et al.*, 1977). Indeed, Whittaker *et al.* (1973) have used physical and chemical features of the environment to define habitats and their boundaries. Even where the "conglomerate of physical and biotic factors" (Partridge, 1978, p. 351) are used to define habitats, these are ultimately dependent on and clearly correlated with abiotic factors, which thus become directly relevant to problems of habitat selection and the operation of physiological contraints.

Extreme environmental conditions obviously have the most pronounced effect upon habitat choices. Strong winds drive Bald Eagles (*Haliaetus leucocephalus*) away from their usual feeding areas and make them roost in protected areas (Steenhof *et al.*, 1980). Cold and rain during the breeding season can chill young and even drown the brood of hole-nesting species such as the Blue Tit (*Parus caeruleus*), and high water levels can destroy the nests of birds breeding near the shoreline, as in the summer of 1981 in Finland with the nests of Common Gulls (*Larus canus*) (Hildén *et al.*, 1982). On the other hand, a heat wave killed up to 90% of the chicks of the Western Gull (*Larus occidentalis*) in California (Salzman, 1982), and heat is considered to be one possible factor explaining the absence of breeding gulls on many tropical islands. To avoid this constraint, Western Gulls often establish their nesting territories in a line along the water. This enables them to make quick trips to the water without leaving the eggs or young unattended for too long and exposing them to the heat or the risk of predation (Hand *et al.*, 1981).

The Rufus Hummingbird (*Selasphorus rufus*) is another interesting example of behavioral plasticity, relative to thermal constraints, for it alters its nest site during the course of the breeding season so as to compensate for changes in radiation (Horvath, 1964). Other hummingbirds (e.g., the Broadtail *S. platycercus*) characteristically choose a nest site under overhanging twigs with reduced heat loss in a cold environment (Calder, 1973). There is also the secondary effect of reduced predation under dense limbs (Glück, 1979). Species that prefer more open habitats, such as gulls, may achieve comparable results—i.e., control of radiative heat gain and loss—by a particular spatial orientation or the positioning of feathers so as to minimize (or maximize) exposure to the sun (Lustick *et al.*, 1978; Hennemann, 1982; Bryant, 1983).

The effects of the climate need not be direct, of course, for even though a prolonged drought may be tolerable physiologically to a particular species, if its food disappears it must do the same. For example, poor rainfall generates low production of grass seeds in southeastern Arizona. Since these seeds are the major food for overwintering sparrows, there is a strong correlation between summer rainfall and the population density of sparrows overwintering in this area (Dunning and Brown, 1982; Pulliam and Parker, 1979). Far more migrating sparrows, however, pass through this area than actually will settle down and remain for the winter. Since the food is not renewed during the winter, the sparrows must assess the population density of sparrows, as well as the food supply, and judge whether it will last for the winter. The birds probably measure these variables indirectly through the time they are involved in aggressive encounters and the time they need to search until they find another seed. An increase in these components of the time budget over a given threshold may indicate that either the population density is too high or the food supply is scarce. In both cases the birds are better off migrating further to the south (Pulliam and Parker, 1979).

Frugivorous birds are often much more affected by dry conditions than are insectivorous birds since the former are feeding on primary products. The food supply of insectivores, however, is removed from primary production and buffered by one or two trophic levels. Even under dry conditions, there is still some insect production (Janzen and Schoener, 1968, in Faaborg, 1982a). Therefore, the food for insectivores may be more reliable and predictable than the supply of seeds and fruits. In normal years fruit production in the West Indies is high enough to support high populations of frugivores. After drought conditions, however, only a small proportion of the original population of frugivores can be maintained. Insectivores, especially gleaning insectivores, are not as much affected by drought, and they show only minor population fluctuations. North American birds migrating south cannot know what the actual food supply on these islands might be. But, insectivores can count on stable resources, whereas frugivores cannot. Therefore, the latter are better off staying on the mainland where their potential feeding area is not restricted by surrounding water. This difference in reliability and predictability of the food supply can be used to explain the composition of overwintering bird assemblages found in the West Indies. Relatively many more small gleaning insectivores immigrate to and over-winter in the West Indies than would be expected from the total number of migrating North American species. Frugivores and granivores, many species of which migrate to the south, fly to the islands only in small numbers. Obviously, frugivorous birds cannot take the risk to fly all the way to an island if they do not know how much food there will be. Insectivores, however, can count on a predictable resource and are therefore less food limited in their migration than are frugivores (Faaborg and Terborgh, 1980; Faaborg, 1982a).

III. NATURE OF HABITAT CHOICE

A. Perceptual Mechanisms

All of the foregoing comments on the means by which anatomical or physiological limitations may be reduced or overcome in order to exploit particular habitats still beg the fundamental question: if birds do choose a potential habitat in which to settle, and the evidence indicates that they do so (see Klopfer, 1969), what characteristics does the bird use to identify the appropriate habitat? This question has been addressed by many researchers: Hildén (1965), Klopfer and Hailman (1965), Lack (1937), Lack and Venables (1939), and Partridge (1978). Recent studies suggest the answer to be more complex than originally assumed. These earlier views suggested either the identification of specific *Merkmale* or releasers or recognition of a general *Gestalt*. Now it appears that habitat choice may consist of a sequence of choices, the various criteria being hierarchically ordered. Wiens and Rotenberry (1981) showed that shrub–steppe species were correlated with gross features of habitat physiognomy at the continental level, but at the regional level the correlation was with the coverage by various shrubs. If one extrapolates their example, an ordered sequence of characteristics is suggested. Burger and Gochfeld (1982) made a similar proposal in their study of the Cattle Egret (*Bubulcus ibis*). These birds first seek a region in which potential "partners" are to be found. From there, they select first a herd of cattle and ultimately an individual in that herd, which they then accompany. They may switch individuals after some minutes, but they remain with the herd for hours and in the grazing area of the herd for days.

If habitat selection does proceed in a stepwise fashion, with the evaluation of different criteria at different stages, with differing temporal components at each stage, one could reconcile the differences often reported in correlations of particular species with particular features of their habitat (note Nudds, 1982; Faaborg,

1976; Wiens and Rotenberry, 1979). This is consistent with current neurophysiological models of perception, although it is only in the acoustic modality that relevent neurophysiological data pertaining specifically to birds are available (Kroodsma and Miller, 1982). Once a habitat has been selected, habitat use can still be varied during the course of the breeding season, in response, for example, to locally deteriorating conditions (Rice *et al.*, 1983).

B. Sensory Basis of Perception and Choice

A complication for human observers, of course, is that birds may have sensory capacities different from our own, which can obscure from us the basis for some choices. Infrasound, UV, polarized light, and electromagnetic fields are all apparently perceived by some birds (Burkhardt, 1982; Schmidt-Koenig, 1979), though such cues have not yet been implicated as important in habitat choice. However, Schichte and Schmidt-Koenig's (1971) demonstration that homing pigeons wearing heavily frosted lenses (which prevent image formation) still can find their way home must stand as an injunction for open-mindedness. The importance of cues not directly dependent on vegetation may be especially great for migratory species whose selection of a breeding site often must occur when that site has a very different appearance from that which it will obtain during the crucial weeks of reproduction. A New England hardwood forest in March is not, to our eyes, at all like its June successor. Not all birds are so prescient, of course. For instance, European Robins (*Erithacus rubecula*) are unable to find their normal habitat after dark; before sunrise they may often be espied in inappropriate places, only to move after dawn breaks (Bairlein, 1981). Vision here is clearly important, and the visual appearance of the habitat, dominant. For some birds, leaf shapes may be decisive, at least once they are within the general area of their home (e.g., Partridge, 1974; Klopfer, 1963), though the role of leaf shape could still be subordinate to other cues. In the cases of temperate zone migrants whose choice is made before leaves even emerge, other factors must dominate. What these are can only be surmised. The anticipatory capabilities of birds, however, must not be underestimated. Recall Lack's (1954) demonstration that clutch sizes are adjusted to the food supply at a time long before the nestling's insect food has emerged. There is additional evidence that the learning processes birds depend on are attuned to the relevant biological cues. An especially important case is provided by nectar-feeding hummingbirds, which can be more readily trained to switch from one food source to another than to make repetitive visits to the same source. In nature, of course, a blossom once visited will not immediately replace its nectar (Cole *et al.*, 1982).

When abiotic or other factors create a fluctuating food supply, changes in habitat preference may occur (Saari, 1977). Migratory birds, of course, are faced with making altogether different choices from one season to the next (Bairlein,

1981; see Chapter 16, this volume). This obviously adds another element to an already complex perceptual task. The assumption commonly made was that migrants are more plastic in their ability to use different habitats or foods than are residents, an assumption that was used to "explain" the greater avifaunal diversity of the tropics (Klopfer and MacArthur, 1961; Morse, 1971). However, field studies have since revealed no simple correlation: some migrants are consistent in their foraging techniques or the portion of the tree they inhabit, others are not (Keast and Morton, 1980; Greenberg, 1979). Greenberg (1983) noted that captive *Dendroica* warblers differed specifically in their readiness to approach and feed in novel microhabits. He suggests that neophobia may produce an apparent stereotopy in habitat (or food) choice that is not predictable from other anatomic, behavioral, or competitive parameters. The degree of neophobia, in turn, may be fixed by the probability of predation. The foraging plasticity of insular species, noted by Yeaton and Cody (1974) and Terborgh and Faaborg (1973), Greenberg argues, could well be due to the reduced predation pressure of islands. Group foragers presumably are less at risk than solitary foragers (e.g., Bay-breasted versus Chestnut-sided Warblers) and so are less neophobic.

C. Learning

How do birds know how to select the right habitat? Juveniles are often less specific in their preferences than older birds (Bairlein, 1981), with feedback from their experiences in different locations apparently influencing later choices (Partridge, 1979). For example, *Delichion urbica,* a multibrooded swallow, will stay *in situ* if its first brood fledges successfully. If not, it will relocate elsewhere (Hund and Prinzinger, 1979). *Ficedula hypoleuca* similarly, as well as the Mountain Bluebird (*Sialia currocoides*), is more likely to reuse a particular site in subsequent years after a successful outcome than after failure (Herlugson, 1981; Winkel, 1982). In these cases, breeding success is what the bird presumably considers in evaluating and selecting the habitat.

The vital importance of a successful brood underscores the supposition that the cues used in choosing a habitat are those that mark the availability of the resources the young require (Trivelpiece *et al.*, 1980). This also makes credible failures to breed in appropriate habitats but in unusual circumstances, as in Skuas in years of food shortage (Andersson, 1981). Habitat choice is here more dependent on prior events than upon immediate conditions.

The most spectacular instances of choices predicated upon prior experience are in nest parasites (brood parasites), where the bird deposits her eggs in the nest of another species. Host specificity is often very marked and, apparently, conditioned by the prefledging experiences of the young (Nicolai, 1964). In other species, imprinted attachments to the site of hatching are not the exclusive nor primary means for the development of habitat preferences. For many species, the habitat in which the juvenile period is passed becomes the bird's later preference choice (Brewer and Harrison, 1975; Catchpole, 1972; Hildén, 1965; Löhrl, 1959).

For many small migratory passerines, the breeding site may be habitable for but a few weeks of each year. Even if alternative habitats might be more suitable, temporal constraints may limit explorations or trials of new sites (Levins, 1968). The importance of a bird being able to precisely know where and when food is to be found is supported by the demonstration that, in fact, they do possess this information (Smith and Sweatman, 1974). In colonial species there exists a potential for sharing individually acquired information. The result is that in such species as gulls, which normally show a pronounced *Ortstreue* based on the selection of particular features of their habitat, may, under unfavorable conditions, give a higher priority to group allegiance and even relocate from their normal site (McNicholl, 1975). Populations of high and increasing density may also preclude settlement in the preferred area, and then, too, peripheral sites are chosen (note Kluijver, 1951; Fretwell and Lucas, 1970; Fretwell, 1972).

In many cases, a bird appears not to select a habitat per se, but rather an area in which conspecifics or individuals of another particular species are to be found (Buckley and Buckley, 1980; Nuechterlein, 1981). Male Spotted Sandpipers (*Arctitis macularia*) settle those areas inhabited by females, which, in turn, select those sites which in previous years produced viable broods. Here traditions have replaced the formal selection of particular habitat markers (L. W. Oring, personal communication).

The study of the role of social factors in the selection of habitats is not a field crowded with eager investigators. There has been, however, a considerable body of work published on observational learning and social facilitation, particularly as it relates to food choice. Much of this has been reviewed by Davis (1973). In general, birds may come to approach or avoid particular feeding sites (and, one may presume, breeding sites also) through observational encounters with conspecifics. Thus, traditional preferences may come to be established which have no immediate functional significance.

Careful parametric studies of observational learning, or the social transfer of information, have been conducted with rodents, in particular by Galef and his associates (see Galef and Wigmore, 1983). They were able to demonstrate that observer rats can learn from "demonstrators" the location at which each of three different foods was available. If extrapolation to birds is permitted, and the learning abilities of birds have never seemed to me to be inferior to those of rats, one is provided with a mechanism whereby the outcome of the experiences made by adults can be transmitted to their (or other's) offspring; this could assure continuity and consistency in habitat preferences.

There are further instances where periodic climatic catastrophes, as commonly occur in steppe biotopes, so decimate populations that resettlement of different species is scarcely influenced by the usual competitive processes. Habitat choice may then be more a random or haphazard affair than the choice of an optimum which has been forced by selective or competitive pressures (e.g., Cody, 1974; Diamond, 1978; and see especially Vuilleumier and Simberloff, 1980, who argue that short-term ecological factors and historical factors are indissolubly intertwined and causally inseparable). Optimal choices cannot always be made, a fact often overlooked in our inevitable tendency to emulate Pangloss.

IV. CONCLUSIONS

It was but two decades ago that it appeared as if little more could be added to our understanding of habitat selection. Most species, it was believed, were preferentially associated with a particular habitat, that habitat in which they could optimally function and reproduce. Only under conditions of exceptional population density or more intense than usual competition would individuals move from their preferred to a less preferred (and suboptimal) habitat. The cues used to identify the appropriate habitat were believed (in the case of birds) to be primarily visual and associated with the form or density or distribution of the dominant vegetation. In a few species, the availability of special nesting sites (as cliffside ledges for marine birds) or of one particular type of food plant might be the critical element.

Habitats were generally thought to be recognized on the basis of prior experience. In the case of migratory species, imprinting-like processes were believed important, while associative learning was considered more important for nidiculous birds. Extrapolations from work on habitat selection with mammals (e.g., Wecker, 1963) suggested that learning effects were not infinitely plastic: some associations or preferences were more readily assimilated than others. But that, except for the details pertaining to individual species (on which a great deal more has been said in the two decades past), was the end of the chapter on habitat selection.

What has changed? First is the growing recognition that the fixity of the correlation between a given species and its habitat can vary considerably according to the level of resolution. Wiens and Rotenberry (1981) noted that, in shrub–steppe environments, bird species exhibited different patterns of habitat correlations on two spatial scales. "In particular, the characteristic shrub–steppe species showed strong correlations with the features of habitat physiognomy in the continental analysis, but in the regional study such associations were generally lacking, and these birds instead were correlated with coverages of various shrub species. This suggests that at a large scale, between-habitat level of analysis these birds may respond to some element of general habitat configuration, but their within-habitat responses may be more strongly associated with details of habitat floristics'' (Wiens and Rotenberry, 1981, p. 21). If true, and if general, it indicates that the traditional behavioral studies of habitat selection in which subjects are asked to choose between various simple (simulated) habitats (see Klopfer, 1969) have presented us with a grossly oversimplified and misleading picture. The recognition process more likely follows a branching pattern, not unlike the flowchart of many computer searches. There is no reason to believe that whatever genetic or learning processes determine a choice at one particular node will necessarily determine the choices at other nodes.

A second factor that is contributing to the rewriting of the chapter is the new information on the role of particular sensory inputs at specific developmental periods on subsequent perceptual organization. Of course, the literature on imprinting (see Hess, 1973; Bateson, 1972; Oyama, 1979; Gottlieb, 1971) has all along intimated as much, but the experimental demonstration of direct neural field influence is more recent. (Consider the results of Mower *et al.*, 1983, for example, who have shown that but 6 hr of exposure to light triggered the onset of a particular development which otherwise would remain labile for at least a few weeks more). This intrusion of sensory physiology and development into a problem traditionally the domain of behavioral ecology is long overdue and should help resolve a number of paradoxes involving habitat choice.

Finally, the interactive nature of the processes that produce preferences is becoming more evident than ever before. Some decades past, a book on migratory paths could include the notion of traditions in its title (Hochbaum, 1955). The implication was that preferences are genetically fixed or result from imitative or traditional behavior; Klopfer (1973) posited comparable processes for the development of food preferences. Sociobiological studies suggest different emphasis. Despite the controversial and, in the views of many, questionable conclusions on cultural evolution presented by Lumsden and Wilson (1981), the sociobiological notion that minor genetic changes, through feedback from cultural factors, lead to major behavioral changes, appears sound. The older epigenetic models of Schneirla (1956) and Lehrmann (1953) here find a modern guise. More to the point, the process of developing a preference for a particular habitat cannot be viewed as based merely on the expression of ''genetic preferences'' (whatever those are) or particular experiences. Actually, few behavioral ecologists have ever been so naive as to embrace so crass a dichotomy, but, unfortunately, their experimental designs have ignored the niceties of distinction and isolation (*mea culpa,* Klopfer, 1969). The ontogenetic development of habitat preferences is no simpler a process than any other ontogeny and will not be revealed through simple-minded experimental designs predicated on ''either/or'' models.

REFERENCES

Alatalo, R. V. (1983). Effects of temperature on foraging behavior of small forest birds wintering in northern Finland. *Ornis Fenn.* **59,** 1–12.

Andersson, M. (1981). Reproductive tactics in the Long-tailed Skua *Stercorarius longicaudus*. *Oikos* **37,** 287–294.

Bairlein, F. (1981). Ökosystemanalyse der Rastplätze von Zugvögeln: Beschreibung und Deutung der Verteilungsmuster von ziehenden Kleinvögeln in verschiedenen Biotopen der Stationen des "Mettnau-Reit-Illmitz-Programmes." *Ökologie der Vögel* **3,** 7–137.

Bateson, P. P. G. (1972). Single units and sensation: A neuron doctrine for perceptual psychology. *Perception* **1,** 371.

Bechard, M. J. (1982). Effect of vegetation cover on foraging site selection by Swainson's Hawk. *Condor* **84,** 153–159.

Beck, B. B. (1980). "Animal Tool Behavior." Garland STPM Press, New York.

Boag, P. T., and Grant, P. R. (1981). Intense natural selection in a population of Darwin's Finches (*Geospizinae*) in the Galapagos (Ecuador). *Science* **214,** 82–85.

Bock, W. J. (1970). Microevolutionary sequences as a fundamental concept in macroevolutionary models. *Evolution* **24,** 704–722.

Bock, W. J. (1977). Toward an ecological morphology. *Vogelwarte* **29,** Sonderheft, 127–135.

Bowman, R. J. (1961). Morphological differentiation and adaptation in the Galapagos Finches. *Univ. Calif. Publ. Zool.* **58,** 1–302.

Brewer, R., and Harrison, K. G. (1975). The time of habitat selection by birds. *Ibis* **117,** 521–522.

Bryant, D. M. (1983). Heat stress in tropical birds: Behavioral thermoregulation during flight. *Ibis* **125,** 313–323.

Buckley, F. G., and Buckley, P. A. (1980). Habitat selection and marine birds. *In* "Behavior of Marine Animals" (J. Burger, B. Olla, and H. E. Winn, eds.), Vol. 4, pp. 69–112. Plenum, New York.

Burger, J., and Gochfeld, M. (1982). Host selection as an adaptation to host-dependent foraging success in the Cattle Egret (*Bubulcus ibis*). *Behaviour* **79,** 212–229.

Burkhardt, D. (1982). Birds, berries, and UV: A note on some consequences of UV vision in birds. *Naturwissenschaften* **69,** 153–157.

Calder, W. A. (1973). Microhabitat selection during nesting of hummingbirds in the Rocky Mountains. *Ecology* **54,** 127–134.

Catchpole, C. K. (1972). A comparative study of territory in the Reed Warbler (*Acrocephalus scirpaceus*) and the Sedge Warbler (*A. schoenobaenus*). *J. Zool.* **166,** 213–231.

Cody, M. L. (1974). "Competition and the Structure of Bird Communities." Princeton Univ. Press, Princeton, New Jersey.

Cody, M. L. (1978). Habitat selection and interspecific territoriality among the sylviid warblers of England and Sweden. *Ecol. Monogr.* **48,** 351–396.

Cody, M. L. (1979). Resource Allocation Patterns in Palaerartic Warblers (*Sylviidae*). *Fortschr. Zool.* **25,** 223–234.

Cole, S., Hainsworth, F. F., Kamil, A. C., Mercier, T., and Wolf, L. L. (1982). Spatial learning as an adaptation in hummingbirds. *Science* **217,** 655–657.

Davies, N. B. (1978). Ecological questions about territorial behaviour. *In* "Behavioural Ecology—An Evolutionary Approach" (J. R. Krebs and N. B. Davis, eds.), pp. 317–350. Blackwell, Oxford.

Davis, J. M. (1973). Imitation: A review and critique. *In* "Perspectives in Ethology" (P. P. G. Bateson and P. H. Klopfer, eds.), Vol. 1, pp. 43–72. Plenum Press, New York.

Diamond, J. M. (1978). Niche shifts and the rediscovery of inter-specific competition. *Am. Sci.* **66,** 322–331.

Dunning, J. B., Jr., and Brown, J. H. (1982). Summer rainfall and winter sparrow densities: A test of the food limitation hypothesis. *Auk* **99,** 123–129.

Ebenman, B., and Nilsson, S. G. (1982). Components of niche width in a territorial bird species: Habitat utilization in males and females of the Chaffinch (*Fringilla coelebs*) on islands and mainland of Sweden. *Am. Nat.* **119,** 331–334.

Emlen, J. T. (1981). Divergence in the foraging response of birds on 2 Bahama Islands. *Ecology* **62,** 289–295.

Faaborg, J. (1976). Habitat selection and territorial behavior of the small grebes of North Dakota. *Wilson Bull.* **88,** 390–399.

Faaborg, J. (1982a). Avian population fluctuations during drought conditions in Puerto Rico. *Wilson Bull.* **94,** 20–30.

Faaborg, J. (1982b). Trophic and size structure of West Indian bird communities. *Proc. Natl. Acad. Sci. U.S.A.* **79,** 1563–1567.

Faaborg, J., and Terborgh, J. E. (1980). Patterns of migration in the West Indies. *In* "Migrant Birds in the Neotropics: Ecology, Behavior, Distribution and Conservation" (A. Keast and E. S. Morton, eds.), pp. 157–163. Random House (Smithson. Inst. Press), New York.

Franzreb, K. E. (1983). A comparison of avian foraging behavior in unlogged and logged mixed coniferous forest. *Wilson Bull.* **95,** 60–76.

Fretwell, S. D. (1972). Populations in a Seasonal Environment. Princeton Univ. Press, Princeton, New Jersey.

Fretwell, S. D., and Lucas, H. L. (1970). On territorial behaviour and other factors influencing habitat distribution in birds. *Acta Biotheor.* **19,** 16–36.

Galef, B. G., and Wigmore, S. W. (1983). Transfer of information concerning distant foods: A laboratory investigation of the "information centre" hypothesis. *Anim. Behav.* **31,** 748–758.

Glück, E. (1979). Abhängigkeit des Bruterfolgs von der Lichtmenge am Nestandort. *J. Ornithol.* **120,** 215–220.

Gottlieb, G. (1971). "Development of Species Identification in Birds." Univ. Chicago Press, Chicago, Illinois.

Grant, P. R., Grant, B. R., Smith, J. N. M., Abbott, I. J., and Abbott, L. K. (1976). Darwin's finches: Population variation and natural selection. *Proc. Natl. Acad. Sci. U.S.A.* **73,** 257–261.

Greenberg, R. (1979). Body size, breeding habitat and winter exploitation systems in *Dendroica. Auk* **96,** 756–766.

Greenberg, R. (1983). The role of neophobia in determining the degree of foraging specialization in some migrant warblers. *Am. Nat.* **122,** 444–453.

Hand, J. L., Hunt, G. L., Jr., and Warner, M. (1981). Thermal stress and predation: Influences on the structure of a gull colony and possibly on breeding distributions. *Condor* **83,** 193–203.

Hennemann, W. W. (1982). Energetics and spread-winged behavior of Anhingas (*A. anhinga*) in Florida, U.S.A. *Condor* **84,** 91–96.

Herlugson, C. J. (1981). Nest site selection in Mountain Bluebirds (*Sialia currucoides*). *Condor* **83,** 252–255.

Hespenheide, H. A. (1966). The selection of seed size by finches. *Wilson Bull.* **78,** 191–197.

Hess, E. H. (1973). "Imprinting." Van Nostrand Reinhold, New York.

Hildén, O. (1965). Habitat selection in birds. *Ann. Zool. Fenn.* **2,** 53–75.

Hildén, O., Järvinen, A., Lehtonen, L., and Soikkeli, M. (1982). Breeding success of Finnish birds in the bad summer weather of 1981. *Ornis Fenn.* **59,** 20–31.

Hochbaum, A. (1955). "The Travels and Traditions of Waterfowl." Univ. of Minnesota Press, Minneapolis.

Horvath, O. (1964). Seasonal differences in Rufous Hummingbird nest height and their relation to nest climate. *Ecology* **45,** 235–241.

Hund, K. and Prinzinger, R. (1979). Studies on site-tenacity, pair bond and nesting survival rate of House Martins, *Delichon urbica,* in southwest Germany. *Vogelwarte* **30,** 107–117.

Janzen, D. H. and Schoener, T. W. (1968). Differences in insect abundance and diversity between wetter and drier sites during a tropical dry season. *Ecology* **49,** 96–110.

Karr, J. R. (1976). Within and between-habitat diversity in African and neotropical lowland habitats. *Ecol. Monogr.* **46,** 457–481.

Keast, A. and Morton, E. S. (1980). "Migrant Birds in the Neotropics: Ecology, Behavior, Distribution, and Conservation." Random House (Smithsonian Inst. Press), New York.

Kendeigh, S. C., Dol'nik, V. R., and Gavrilov, V. M. (1977). Avian energetics. *In* "Granivorous Birds in Ecosystems" (J. Pinowski and S. C. Kendeigh, eds.), pp. 127–204. Cambridge Univ. Press, London and New York.

Klopfer, P. H. (1963). Behavioral aspects of habitat selection: The role of early experience. *Wilson Bull.* **75,** 15–22.

Klopfer, P. H. (1969). "Habitats and Territories: A Study of the Use of Space by Animals." Basic Books, New York.

Klopfer, P. H. (1973). "Behavioral Aspects of Ecology." Prentice-Hall, Englewood Cliffs, New Jersey.

Klopfer, P. H. and Hailman, J. P. (1965). Habitat selection in birds. *In* "Advances in the Study of Behavior" (D. S. Lehrman, R. A. Hinde, and E. Shaw, eds.), pp. 279–303. Academic Press, New York.

Klopfer, P. H., and MacArthur, R. H. (1961). On the causes of tropical species diversity: Niche overlap. *Am. Nat.* **95,** 223–226.

Kluijver, H. N. (1951). The population ecology of the Great Tit, *Parus m. major* L. *Ardea* **39,** 1–135.

Kroodsma, D. E., and Miller, E. H. (1982). "Acoustic Communication in Birds." Academic Press, New York.

Lack, D. (1937). The psychological factor in bird distribution. *Br. Birds* **31,** 130–136.

Lack, D. L. (1947). "Darwin's Finches." Cambridge Univ. Press, London and New York.

Lack, D. (1954). "The Natural Regulation of Animal Numbers." Oxford Univ. Press (Clarendon), London and New York.

Lack, D., and Venables, L. S. V. (1939). The habitat distribution of British woodland birds. *J. Anim. Ecol.* **8,** 39–71.

Landres, P. B., and MacMahon, J. A. (1983). Community organization of arboreal birds in some oak woodlands of western North America. *Ecol. Monogr.* **53,** 183–208.

Lehrman, D. S. (1953). A critique of Lorenz's theory of instinctive behavior. *Q. Rev. Biol.* **28,** 337–363.

Leisler, B. (1977a). Ökomorphologische Aspekte von Speziation und adaptiver Radiation bei Vögeln. *Vogelwarte* **29,** Sonderheft 136–153.

Leisler, B. (1977b). Die ökologische Bedeutung der Lokomotion mitteleuropäischer Schwirle (*Locustella*). *Egretta* **20,** 1–25.

Leisler, B. (1981). Die ökologische Einnischung der mitteleuropäischen Rohrsanger (*Acrocephalus, Sylviinae*) I. Habitattrennung. *Vogelwarte* **31,** 45–74.

Levin, S. A. (1970). Community equilibria and stability, and an extension of the competitive exclusion principle. *Am. Nat.* **104,** 413–423.

Levins, R. (1968). "Evolution in Changing Environments." Princeton Univ. Press, Princeton, New Jersey.

Löhrl, H. (1959). Zur Frage des Zeitpunktes einer Prägung auf die Heimatregion beim Halsbandschnäpper (*Ficedula albicollis*). *J. Ornithol.* **100,** 132–140.

Lumsden, D., and Wilson, E. O. (1981). "Genes, Mind and Culture." Harvard Univ. Press, Cambridge, Massachusetts.

Lustick, S., Battersby, B., and Kelty, M. (1978). Behavioral thermoregulation: Orientation toward the sun in Herring Gulls. *Science* **200,** 81–82.

MacArthur, R. H., Diamond, J. M., and Karr, J. (1972). Density compensation in island faunas. *Ecology* **53,** 330–342.

McKilligan, N. G. (1980). The winter exodus of the Rook (*Corvus fruglgus*) from a Scottish, U.K., highland valley. *Bird Study* **27,** 93–100.

McNicholl, M. K. (1975). Larid site tenacity and group adherence in relation to habitat. *Auk* **92,** 98–104.

Morse, D. H. (1971). The insectivorous bird as an adaptive strategy. *Ann. Rev. Ecol. Syst.* **2,** 177–200.

Mower, G. D., Christen, W. G., and Caplan, C. I. (1983). Very brief visual experience eliminates plasticity in the cat visual cortex. *Science* **221,** 178–9.

Nicolai, J. (1964). Der Brutparasitismus der *Viduinae* als ethologisches Problem. *Z. Tierpsychol.* **21,** 129–204.

Nudds, T. D. (1982). Ecological separation of grebes and coots: Interference competition or microhabitat selection? *Wilson Bull.* **94,** 505–514.

Nuechterlein, G. L. (1981). "Information parasitism" in mixed colonies of western grebes and Forster's terns. *Anim. Behav.* **29,** 985–989.

Oyama, S. (1979). The concept of the sensitive period in developmental structure. *Merritt-Palmer Q.* **25,** 83.

Partridge, L. (1974). Habitat selection in titmice. *Nature* **247,** 573–574.

Partridge, L. (1976). Field and laboratory observations on the foraging and feeding techniques of Blue Tits (*Parus caeruleus*) and Coal Tits (*P. ater*) in relation to their habitats. *Anim. Behav.* **24,** 534–544.

Partridge, L. (1978). Habitat selection. *In* "Behavioural Ecology" (J. R. Krebs and N. B. Davis, eds.), pp. 351–376. Blackwell, Oxford.

Partridge, L. (1979). Differences in behaviour between Blue and Coal Tits reared under identical conditions. *Anim. Behav.* **27,** 120–125.

Piencowski, M. W. (1983). Changes in foraging patterns of plovers in relation to environmental factors. *Anim. Behav.* **31,** 244–264.

Pinshow, B., Fedak, M. A., and Schmidt-Nielsen, K. (1977). Terrestial locomotion in penguins: It costs more to waddle. *Science* **195,** 592–594.

Pulliam, H. R. (1983). Ecological community theory and the coexistence of sparrows. *Ecology* **64,** 45–52.

Pulliam, H. R., and Enders, F. (1971). The feeding ecology of five sympatric finch species. *Ecology* **52,** 557–566.

Pulliam, H. R., and Parker, T. H. (1979). Population regulation of sparrows. *Fortschr. Zool.* **25,** 137–147.

Rice, J., Ohmart, R. D., and Anderson, B. W. (1983). Habitat selection attributes of an avian community: A discriminant analysis investigation. *Ecol. Monogr.* **53,** 263–290.

Roth, R. R. (1979). Foraging behavior of Mockingbirds: The effect of too much grass. *Auk* **96,** 421–422.

Saari, L. (1977). Change of habitat preference during the summer in certain passerines. *Ornis Fenn.* **54,** 154–159.

Saether, B-E. (1983). Habitat selection, foraging niches and horizontal spacing of Willow Warbler *Phylloscopus trochilus* and Chiffchaff *P. collybita* in an area of sympatry. *Ibis* **125,** 24–32.

Salzman, A. G. (1982). The selective importance of heat stress in gull nest location. *Ecology* **63,** 742–751.

Schlichte, H. J., and Schmidt-Koenig, K. (1971). Zum Heimfindevermögen der Brieftaube bei erschwerter optischer Wahrnehmung. *Naturwissenschaften* **58,** 329–330.

Schmidt-Koenig, K. (1979). "Avian Orientation and Navigation." Academic Press, London and New York.

Schneirla, T. C. (1956). Interrelationships of the "innate" and the "acquired" in instinctive behavior. *In* "L'Instinct dans le Comportement des Animaux et de l'Homme." (P.-P. Grasse, ed.), pp. 387–452. Masson et Cie, Paris.

Selander, K., and Kaufman, D. W. (1973). Genetic variability and strategies of adaption in animals. *Proc. Natl. Acad. Sci. U.S.A.* **70,** 1875–1877.

Smith, J. N. M., and Sweatman, H. P. (1974). Food searching behaviour of titmice in patchy environments. *Ecology* **55,** 1216–1232.

Steenhof, K., Berlinger, S. S., and Frederickson, L. H. (1980). Habitat use by wintering Bald Eagles (*Haliaeetus leucocephalus*) in South Dakota, U.S.A. *J. Wildl. Manage.* **44,** 798–805.

Strong, D. R., Jr. (1983). Natural variability and the manifold mechanisms of ecological communities. *Am. Nat.* **122,** 636–660.

Terborgh, J., and Faaborg, J. (1973). Turnover and ecological release of the avifauna of Mona Island, Puerto Rico. *Auk* **90,** 759–779.

Trivelpiece, W., Butler, R. G., and Volkman, N. J. (1980). Feeding territories of Brown Skuas (*Catharacta lonnbergi*) *Auk* **97,** 669–676.

Van Valen, L. (1965). Morphological variation and width of ecological niche. *Am. Nat.* **99,** 377–390.

Von Uexkull, J. (1921). Umwelt und Innenwelt der Tiere. Springer-Verlag, Berlin and New York.

Vuilleumier, F., and Simberloff, D. S. (1980). Ecology versus history as determinants of patchy and insular distributions in high Andean birds. *In* "Evolutionary Biology" (M. K. Hecht, W. C. Steere, and B. Wallace, eds.), Vol. 12, pp. 235–379. Plenum, New York.

Wecker, S. C. (1963). The role of early experience in habitat selection by the Prairie Deermouse *Peromyscus maniculatus bairdi. Ecol. Monogr.* **33,** 307–325.

Whittaker, R. H., Levin, S. A., and Root, R. B. (1973). Niche, habitat, and ecotope. *Am. Nat.* **107,** 321–338.

Wiens, J. A. and Rotenberry, J. T. (1979). Diet niche relationships among North American grassland and shrub–steppe birds. *Oecologia* **42,** 253–292.

Wiens, J. A. and Rotenberry, J. T. (1981). Habitat associations and community structure of birds in shrubsteppe environments. *Ecol. Monogr.* **51,** 21–41.

Winkel, W. (1982). Zum Ortstreue-Verhalten des Trauerschnäppers (*Ficedula hypoleuca*) im westlichen Randbereich seines mitteleuropäischen Verbreitungsgebietes. *J. Ornithol.* **123,** 155–174.

Yeaton, R. I., and Cody, M. L. (1974). Competitive release in island song sparrow populations. *Theor. Popul. Biol.* **5,** 42–58.

Chapter 16

Habitat Selection by Nonbreeding, Migratory Land Birds

RICHARD L. HUTTO

Department of Zoology
University of Montana
Missoula, Montana

I. HABITAT SELECTION: SOME DEFINITIONS AND CONCEPTS

A. A Working Definition of Habitat

I offer a working definition of the term "habitat" so that we begin from a common point concerning what it is that a bird selects for or against (see Par-

HABITAT SELECTION IN BIRDS

ISBN 0-12-178080-5

tridge, 1978, for a similar perspective). For the purposes of this report, a habitat will be defined as a spatially contiguous vegetation type that appears more or less homogeneous throughout and is physiognomically distinctive from other such types. Granted, large groups of animal or plant species have not been shown to have entirely coincident distributions in space (Whittaker, 1975), as the super-organism concept would have it, so the point at which one habitat type ends and another begins is sometimes quite arbitrary. Nonetheless, there is a tendency for the physically dominant plant species to alter the environment to such an extent that important subsets of species are coincident in their distributions, and, in general, habitat types can be said to correspond well with biome types (Whittaker, 1975) or plant community types (Ricklefs, 1979). These might include categories such as grassland, pine–oak woodland, desert, deciduous forest, and so forth. More refined categories are possible if they are spatially and physiognomically distinct (e.g., shortgrass prairie, tallgrass prairie, desert bajada, desert riparian, etc.). Birds that differ in, say, the height at which they forage within an oak woodland illustrate within-habitat (microhabitat) differences in foraging locations, rather than habitat differences, according to this definition.

B. Habitat Selection: Proximate versus Ultimate Causation

Questions regarding habitat selection in birds can generally be thought of as belonging to one of two types: (1) Why does a species occur in the range of habitats that it does? and (2) What cues do individuals use as indications that they are in an appropriate habitat or not? The two really represent a division of the concept into ultimate and proximate causation. The former asks what are the evolutionary costs and benefits of using each of a possible range of habitat types, and the latter asks what is the mechanism by which the "choice" of habitat is achieved. In this chapter, I wish to concentrate on the first question as it relates to habitat selection in nonbreeding migrants and will consider whether the mechanisms involve direct or indirect assessments of those evolutionarily important costs and benefits by individual birds.

C. Habitat Selection: Intrinsic and Extrinsic Factors Associated with Ultimate Causation

Although habitat selection, in the evolutionary sense, involves a weighing of the costs and benefits associated with use of each available habitat type, it must be made clear that the costs and benefits may be determined by different factors at different levels of organization. These include what I call *intrinsic* (within-habitat) evolutionary constraints on suitability (how much, say, food or predator protection is afforded by the habitat) and *extrinsic* (extra-habitat) evolutionary constraints on suitability (habitat accessibility or weather patterns that might

TABLE I

Relative Importance of Intrinsic and Extrinsic Costs and Benefits of Habitat Use as a Function of Geographic Scale

Geographic scale or distance between habitats	Types of factors contributing to costs and benefits of habitat use	
	Intrinsic	Extrinsic
Broad scale, distant	Unimportant	Important
Local scale, close	Important	Unimportant

influence habitat use during migration, for example). The overall costs and benefits of using a given habitat will be the sum of those associated with the intrinsic and extrinsic constraints. Note that extrinsic factors could override a ranking of habitats determined solely from intrinsic criteria when the habitats become more greatly separated in space, because, at some point, the absolute costs and benefits associated with extrinsic factors become much greater than the costs and benefits associated with intrinsic factors (Table I). In the next section, I suggest how both intrinsic and extrinsic constraints might interact to determine the patterns of habitat use in nonbreeding migratory birds.

D. Mechanisms of Habitat Selection

The very phrase ''habitat *selection*'' carries with it the unfortunate connotation that organisms consciously choose among alternative habitats. In no instance do I mean to imply such here. Even though the chapter title includes the phrase ''habitat selection,'' I prefer ''habitat *use*'' because it does not connote a conscious choice; it merely indicates the actual distribution of individuals, through whatever mechanism (see Morse, 1980, for a similar, but not identical, distinction).

Consciousness aside, the ultimate range of habitats used by a species is likely to be a combined result of the following processes, or mechanisms: (1) individuals may cue into a geographical area (Hildén, 1965, p. 65) and thereby select (or avoid) habitat types, simply as a geographical or historical consequence of the fact that they have not been given an equal opportunity to discover each of a variety of habitats; (2) the habitat types (or geographic locations) currently used or avoided by an individual may reflect innate instructions passed on by ancestors who made relatively successful evolutionary ''choices'' in the past; (3) individuals may actively choose among habitats based on their own previous experience; or (4) individuals may ''explore'' possibilities and settle according to a ranking based on such exploration.

The extent to which exploratory ranking is possible depends on the geographic scale of resolution. Location decisions that are based on ultimate cost–benefit criteria, such as the economics of foraging or differential susceptibility to predation, are almost certainly important at local (microhabitat) scales but become less likely at levels that we feel comfortable calling the level of "habitat selection," i.e., choices among distinct vegetation types separated by several kilometers or more. In most of the latter cases, habitat use is believed to reflect innate responses to some proximate environmental cues that are themselves well correlated with the presence of ultimately important needs (Klopfer, 1976; Partridge, 1974), and the possibility of exploratory choice has been played down at this level. Habitat types separated by several hundred kilometers or more lie at the far end of the spectrum, where it seems certain that exploratory ranking of alternative locations is not possible and where habitat use undoubtedly reflects innate responses to proximate orientation cues that place the bird in an appropriate location (Emlen, 1975).

Since it is unlikely that the broad-scale cues alone will ensure that all necessary requisites for survival and reproduction are present, it is useful to think of the process of habitat selection as a hierarchical series of decisions that an animal must make (see also Hildén, 1965; Gass and Montgomerie, 1981; Tinbergen, 1981; Cody, 1981). At the most general level, a geographic location is chosen. Then a particular habitat is chosen. Finally, the precise location within the habitat (microhabitat) is chosen, based on cues that probably differ from those used at the more general levels. The proximate cues used at the microhabitat level will be close to the ultimate factors that determine a bird's success, if not the ultimate factors themselves. Cues at the broad-scale level may therefore be necessary but insufficient for habitat use to occur.

Using my own research on migratory wood warblers as a focus, I now review the evidence that local-scale nonbreeding microhabitat selection is based on some sort of exploratory comparison of intrinsic ultimate factors, and I discuss the implications in terms of more broad-scale patterns of habitat use. Next, I review the evidence that broad-scale habitat selection might also be based, in part, on an exploratory assessment of intrinsic ultimate factors. Finally, I discuss how extrinsic ultimate factors act to limit the range of possible choices and thus place constraints on a migratory species' overall pattern of habitat use during the nonbreeding season(s).

II. INTRINSIC FACTORS AND MICROHABITAT USE

A. Theoretical Expectations

The factors that contribute to a bird's choice about a precise location within which to feed, roost, nest, sing, etc., are characteristics that occur within a habitat—intrinsic factors such as food type or abundance, perch characteristics, branch configurations, and so forth. For nonbreeding birds, food acquisition must figure prominently in decisions about use of space, since getting enough food to store fat, or merely survive, could be considered the single most important constraint during these periods. I have therefore chosen to illustrate the role of intrinsic factors by concentrating on the relationship between food availability and a bird's use of space within a given habitat.

On a theoretical basis, we expect food availability to influence a bird's use of space. In fact, a major subset of optimal foraging theory involves patch choice (Pyke *et al.*, 1977), and habitat selection might be considered a logical extension of patch choice (Morse, 1980). So long as food is limited in supply there will be selective pressure on individuals to use space in the most profitable (Royama, 1970) manner possible. The key here is that this is true even if food is not limit*ing*—it only need be limit*ed* in supply (Hespenheide 1973, p. 218). No matter how much food there is, as long as it is limited in supply, there will be better and worse (relatively good and bad) places to settle, and selection will favor discrimination among sites on that basis alone. Admittedly, selective pressure will vary directly with the degree to which food is limited, but there will always be some pressure related to the economics of foraging that will tend to lead toward a close match between resource production and its utilization (MacArthur, 1969, 1970). At the other extreme, it could be argued that foraging constraints (competition for food) are unimportant except under extreme conditions (ecological crunches) and that the positions of bird territories in space may very well be products of random, stochastic processes (checkerboard model described by Wiens and Dyer, 1975; Rotenberry and Wiens, 1980; and Wiens, 1981). The answer may very well depend on the scale of consideration (Wiens, 1981). For example, local-scale patterns might be influenced markedly by relative differences in food availability among sites, but large-scale (habitat) patterns may not be so influenced, because of an inability of individuals to track differences in food availability on broad, geographic scales.

B. Empirical Evidence

As an example of a close match between resource availability and use of space, consider the combined foraging-height profile of three wood warbler species that I observed in Grand Teton National Park during May 1975 (Fig. 1). The relative availability of food at each height was estimated from the product of relative insect density (as determined from daily counts of four sets of five sticky boards that were hung for 4 days at 0.5-m height intervals) times the vegetation density at each height interval (see Hutto, 1980, for further explanation). The insectivorous bird "impact" histogram was calculated as the product of each species' height distribution times its relative abundance, summed over all spe-

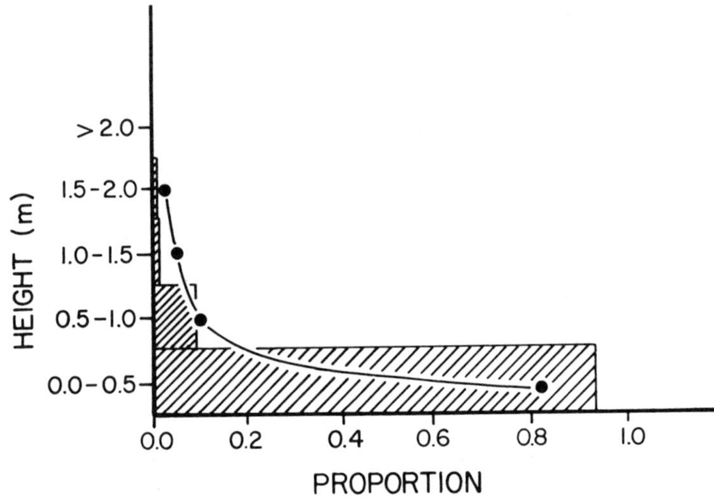

Fig. 1. Relationship between the combined foraging height distribution of three warbler species (*Dendroica petechia, D. coronata,* and *Geothlypis trichas*) (hatched box, $n = 115$) that were observed from 25–31 May 1975 in Grand Teton National Park, Wyoming, and the adjusted insect density (●——●, $n = 7282$) present at each height (see text).

cies. The most striking aspect of the birds' use of space in this instance was the atypically low foraging heights of both the Yellow Warbler (*Dendroica petechia*) and Yellow-rumped Warbler (*Dendroica coronata*); at other times of the year it is common to find either species foraging from 4–8 m, in the canopy of more forested habitats (see Fig. 3 in Hutto, 1981). The uncommon observation recorded here can only be explained, I believe, on the basis of food distribution at that time.

There is much stronger evidence that food is nonrandomly distributed in space on a local level and that the use of space by birds (microhabitat use) corresponds strikingly well with the distribution of available food. An especially clear example comes from the work of Tinbergen (1981), who demonstrated that female Starlings (*Sturnus vulgaris*) foraging within pastureland spent most of their time in areas that yielded the highest food intake rates, which coincided with locations of highest prey (leatherjacket) density. Tinbergen's study represents but one of a host of studies that have produced similar results (e.g., Goss-Custard, 1970; Smith and Dawkins, 1971; Smith and Sweatman, 1974; Cody and Walter, 1976; Gill and Wolf, 1977; Davies, 1977; Greenwood and Harvey, 1978; Zach and Falls, 1979; Gradwohl and Greenberg, 1980; Thomas, 1980; Holmes and Robinson, 1981). These references include breeding and nonbreeding season studies—situations in which absolute food densities and additional constraints on the birds' time are extremely varied. The results are even more remarkable when one considers that attributes of food other than its density, such as patch distribution or degree of isolation from other patches, also influence the probability of patch use (Gradwohl and Greenberg, 1980). Moreover, additional intrinsic factors, such as predator avoidance (Pulliam and Mills, 1977; Grubb and Greenwald, 1982), territorial defense, or other constraints, may also influence the probability of patch use and should therefore decrease the predictability of a bird's use of space that is based on food availability alone. For example, Tryon and MacLean (1980) found that Lapland Longspurs (*Calcarius lapponicus*) used the space immediately surrounding nest sites more than would be expected solely on the basis of food availability.

The point I wish to stress is that the evolutionarily important costs and benefits associated with use of an area are largely, if not exclusively, a result of intrinsic factors at this level, and the factors to which the birds appear to be responding are close to, if not identical with, the ultimate factors determining their evolutionary success. Food is one such factor, and its power of influence has been emphasized here.

The implications are profound. If birds choose locations based largely on food availability, then large-scale habitat distribution patterns may be determined to some extent by factors operating at this level. For example, it is possible that an individual of one species could influence the food distribution or abundance enough (as has been described by Davies and Houston, 1981; and Tinbergen, 1981) so that an individual of another species might not choose to settle nearby, thereby influencing community-level patterns, such as species diversity.

It is also possible that decisions to settle in a given habitat are made in response to more proximate cues and that the most these smaller-scale foraging decisions can do is rearrange territories or home ranges slightly or cause individual modifications in foraging behavior in response to differences in vegetation structure (Holmes *et al.*, 1978) or to differences in food abundance among patches (Tinbergen, 1981). Whether the intrinsic factors (e.g., food availability) that are ultimately responsible for local-scale patterns of microhabitat use might also influence the patterns of space use on a broader scale (habitat use) is, as yet, unknown. The next section approaches this question.

III. INTRINSIC FACTORS AND CHOICE AMONG HABITAT TYPES

A. Theoretical Expectations

The idea that the breadth of habitat types occupied by a species is directly related to its population size was perhaps originally published by Mayr (1926) and further developed by Svärdson (1949) and Hildén (1965). A more thorough theoretical treatment of this problem of how population size might affect the intrinsic quality of a habitat and thereby influence patterns of habitat use was presented by Fretwell and Lucas (1970) and Fretwell (1972). In their model, they described how habitats could be ranked on the basis of a suitability derived from intrinsic costs and benefits. The absolute suitability of a given habitat is then assumed to decrease with increasing bird population size, because of changes in costs and benefits due to changes in levels of competition or predation. The relevance to us here is that, according to the model, the final decision a bird makes about which habitat to settle in must be based on an exploratory mechanism whereby the individual assesses the costs and benefits of possible choices in some direct or indirect manner. The alternative extreme is that birds are innately programmed to settle in a particular habitat type and are inflexible in this behavior; intrinsic factors would, of course, still be involved in determining the evolutionary success of individuals in a given habitat, but the birds would merely be programmed to settle in the habitat that, on average, is the best place to settle.

In a compromising fashion, both processes could operate if an innate preference to settle in a particular habitat type were contingent upon additional microhabitat needs being met or if the threshold of acceptance for a habitat changed with "motivation" level of the bird (Svärdson, 1949). Here, I wish to explore the possibility that assessment of intrinsic factors influences habitat settlement patterns, just as the process apparently influences decisions about microhabitat selection (last section).

B. Empirical Evidence

A wealth of examples of flexibility in habitat use in relation to population size was provided by both Svärdson (1949) and Hildén (1965). For example, Hildén described how the Northern Lapwing (*Vanellus vanellus*) occupied only the most productive coastal habitat at the end of the nineteenth century but then began to occupy suboptimal habitats (less productive, small patches of meadow and tilled land and peatlands of forest areas) as its population size increased. In all, Svärdson and Hildén referenced more than two dozen examples of direct relationships between population density and habitat breadth. Such findings would be difficult to explain if birds were inflexibly programmed to "choose" a habitat on the basis of the absolute level of some cue(s). The most compelling argument against an inflexible mode of habitat selection, however, comes from examples of niche shifts in habitat use in the presence or absence of competitors (Lack, 1971; Diamond, 1978)—particularly from short-term removal experiments in which one species has been observed to invade a previously unoccupied habitat after removal of another species (Davis, 1973).

To what intrinsic factor(s) might birds be responding during such shifts in habitat use? Such a factor may reflect an important basis for habitat selection in general. Cody (1978) has suggested, for example, that birds might use cues that are directly related to the presence or absence of competitors, such as vocalizations. In addition to such direct cues, field studies of a correlative nature suggest, once again, that food availability might also be a critically important variable. During the nonbreeding season in western Mexico, resident insectivorous birds of the lowland evergreen forests fail to exploit the edge and second-growth situations for reasons apparently associated with an innate "mental conservatism" (Willis, 1966; Terborgh and Weske, 1969; Lack, 1971), which may have evolved in response to high predation rates in edge situations (Gates and Gysel, 1978). Food availability is four times greater in edge than interior habitats, and migratory insectivorous bird density is, correspondingly, five times greater in edge than interior habitats (Hutto, 1980). In speculation, perhaps the migrants can afford the "luxury" of high food levels in edge situations because, for some reason, they are less susceptible to predation in edge habitats than are residents.

In a similar fashion, during the 1975 spring migration period in Grand Teton National Park, Wyoming, the combined density of three wood warbler species was 12 times greater in the grass willow (*Salix spp.*) habitat than in the adjacent Lodgepole Pine (*Pinus contorta*) habitat (Fig. 2). This corresponded with an estimate of food density which was 13 times greater in the grass willow habitat (Fig. 2). The preference for grass willow habitat was particularly interesting in

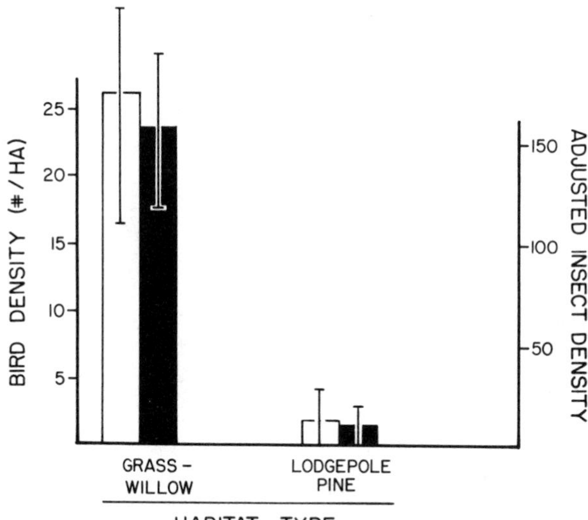

Fig. 2. Correspondence between the average (±SD) combined density of three warbler species (*Dendroica petechia, D. coronata,* and *Geothlypis trichas*) (□) and adjusted insect density (■) in each of two habitat types in Grand Teton National Park, Wyoming.

the case of the Yellow-rumped Warbler which, in midsummer, nests in the lodgepole habitat and is not seen in the grass willow habitat. Clearly, with nesting constraints removed, food availability appears to be an intrinsic factor of major importance in determining patterns of habitat use in this case.

Burger *et al.* (1977), Davies (1977), Connors *et al.*, (1981), and Sabo and Holmes (1983) have also demonstrated a close relationship between food availability and switches in habitat use by birds. Presumably, birds spend time exploring various habitats to keep track of the relative profitability of each. Consistent with this hypothesis is the observation by Gass and Montgomerie (1981) that (exploration?) time away from a territory by nonbreeding hummingbirds (*Amazilia rutila*) is inversely related to food productivity on their territories at any given time (Fig. 3). Habitat selection based on exploratory assessment of food availability has also been hypothesized to be important in fishes (Werner *et al.*, 1981) and mammals (Lewis, 1980).

As habitats become more widely separated from one another, we would expect single individuals to be unable to assess the relative costs and benefits of using each and that nonexploratory mechanisms of habitat selection would come into play exclusively. However, as long as individuals explore adjacent habitats, the logical consequence will be large-scale shifts in use to match food productivity (or the availability of some other intrinsic factor) over a wide range of habitats through a kind of domino effect.

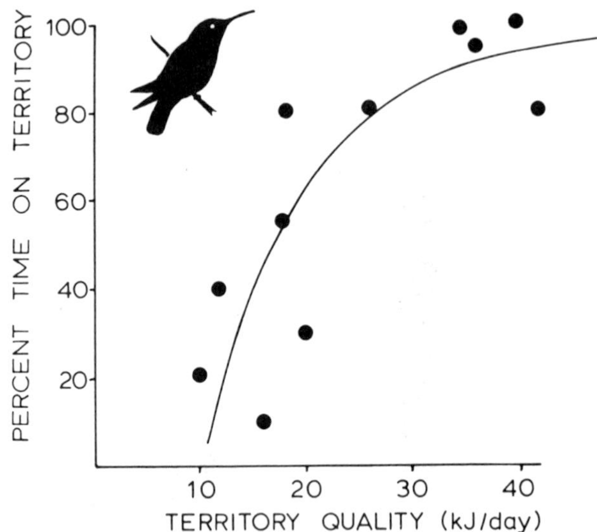

Fig. 3. The proportion of time *Amazilia rutila* individuals spend on their territories is directly related to the availability of food therein ($r_s = 0.86$). Time away from their territories when food is scarce is possibly spent in exploratory activities (from Gass and Montgomerie, 1981, by permission of Garland Publishing, Inc., New York, New York).

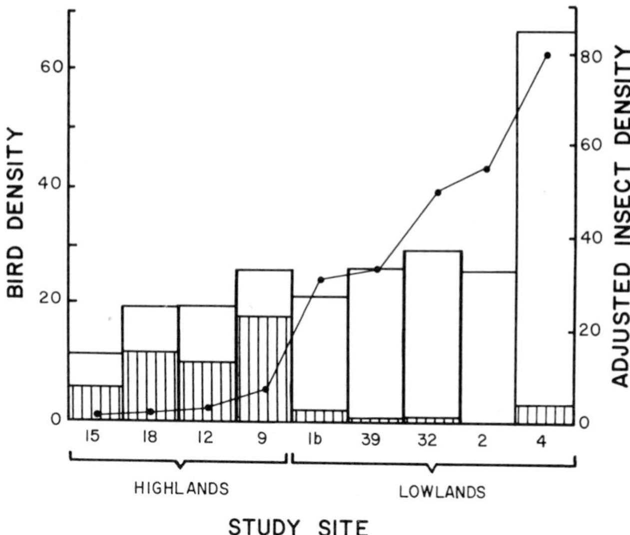

Fig. 4. The correspondence between insectivorous bird density (number of foliage-gleaning insectivores/ha) (migrants, open box; residents, hatched box) and adjusted insect density (●——●) (see text) over eight west Mexican habitat types (By permission of Smithsonian Institute Press from "Migrant Birds in the Neotropics: Ecology, Distribution, Behavior, and Conservation" (A. Keast and E. C. Morton, eds.), p. 148, Fig. 6 (by R. L. Hutto), © 1980, Smithsonian Institution, Washington, D.C.

Is there any evidence that birds collectively assess the relative suitability of a number of more widely spaced habitats and settle accordingly? Although the precise mechanism remains unknown, I have recorded strong correlations between a measure of food availability and bird density among widely distributed habitats on two occasions. Such correlations would not be expected if food were not at least ultimately involved in habitat settlement patterns. In winter, the density of west Mexican insectivorous birds belonging to the small, foliage-gleaning guild was significantly correlated ($r = 0.79, p < 0.05$) with a measure of food availability over eight habitat types that varied in elevation from sea level to almost 3000 m elevation (fig. 4). In contrast with the permanent resident species, the more eurytopic migratory species seem to have been largely responsible for the close match between food availability and bird density ($r = 0.93$ *versus* $r = -0.71$ for migrants and residents, respectively); migrants may be in a better position to explore alternative habitats and incorporate an assessment of food availability into their settling response.

Similarly, the small, insectivorous foliage-gleaning species that migrate through southeastern Arizona during the spring and fall seasons reveal a shift in their use of the available habitats from one season to the next (Hutto, 1985). During the northward migration, most species occupy the low elevation desert

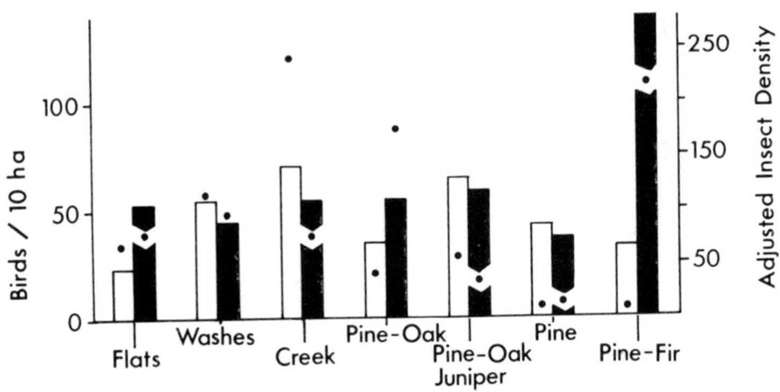

Fig. 5. The correspondence between insectivorous bird density and adjusted insect density (see text) over seven habitat types during two seasons in southeastern Arizona (spring season, open histograms; fall season, hatched histograms; insects, circles).

habitats, while during the southward migration there is a dramatic increase in migratory bird density in the highest elevation pine–fir forests (Fig. 5). Within a season there is a good correlation between bird density and a measure of food availability such that the seasonal shifts in habitat occupancy correspond well with the seasonal shifts in food availability among habitats. Food availability thus appears, again, to be an important intrinsic factor that is ultimately responsible for the habitat distribution pattern of nonbreeding birds during migration (see Balda *et al.*, 1975, and Graber and Graber, 1983, for similar conclusions). An alternative hypothesis that might be invoked to explain such a correlation is that populations of both the predators (birds) and their prey (arthropods) are independent but affected similarly by a third variable, such as weather. This is most improbable, however, because the birds are in southeastern Arizona for only a matter of days in comparison with the arthropods, which carry out their entire life cycles in that location.

The mechanism by which birds assess the relative food availability among habitats remains unknown as mentioned earlier but must be a result of their cuing in on either some proximate factor that is well correlated with food availability between seasons (such as vegetation density) or on food availability itself, as has been suggested on occasion in the past (Hildén, 1965; Root, 1967; Verner, 1975; Pearson, 1977; Cody, 1981; Meents *et al.*, 1982). In either case, it is worth pointing out that whatever the cues, they are much closer to the ultimately important intrinsic factors that determine success than those that have been suggested to be important during the breeding season, such as percentage of canopy cover or tree density (James, 1971; Whitmore, 1975; Clark *et al.*, 1983), because the latter cues simply do not change seasonally in this case, while the habitat distribution of birds does. In addition, shifts in habitat use from summer to winter, which are quite common among migratory land birds (Lack, 1971), imply either a shift in proximate cues used by individual birds or use of cues that are much closer to those of ultimate importance than those implied as important through breeding season studies. This idea of a facultative settlement pattern which is conditional upon food availability has been expressed by several authors in the past (Serventy, 1971; Ward, 1971; Fretwell, 1972; Sinclair, 1978; Pulliam and Parker, 1979; Smith, 1982; Nudds, 1983) and, in my view, deserves more attention in the future.

IV. GEOGRAPHICAL AND HISTORICAL CONSTRAINTS ON PATTERNS OF NONBREEDING HABITAT USE

The overwhelming majority of northern temperate migratory land bird species move more or less directly northward and southward during migration (Cooke, 1904; Wallace and Mahan, 1975). Consequently, the longitudinal separation of species and populations that exists during the breeding season persists during the winter season (see also Barlow, 1980; Fitzpatrick, 1980, 1982, for examples involving the Vireonidae and Tyrannidae). In the case of New World wood warblers, let us define 40 "Eastern" and 15 "Western" species and subspecies as those whose entire geographic ranges [as determined from the AOU checklist (1959)] are restricted to one side of the Rocky Mountains or the other and for which the geographic centers of their ranges lie north of the Mexican border during the summer. Using this definition and the specific or subspecific identity of populations as a convenient marker, one can draw the summer and winter ranges of each "species" and calculate the geographic overlap between all possible pairs. During the breeding season, Western–Western species pairs share 23.7% (*SD* = 23.9) of their combined geographic ranges, on average, and Eastern–Eastern species pairs share 24.1% (*SD* = 25.5) of their combined geographic ranges, on average. Eastern–Western species pairs share no area in common, by definition. To illustrate the maintenance of an East–West separation in winter, note that during the nonbreeding season, Western–Western species pairs share 38.5% (*SD* = 27.2) of their combined geographic ranges, on average, while Eastern–Eastern species pairs share 16.2% (*SD* = 18.6) of their combined ranges, on average. The Eastern and Western species remain largely separate in winter, as only 4.4% (*SD* = 7.1) of the combined ranges of Eastern–Western species pairs is held in common, on average. Each of the last three mean values is statistically distinct from the other two (t -tests, $p<0.001$). Thus, the Eastern species and subspecies remain geographically distinct from the Western species and subspecies in both summer and winter; a more random migratory pattern to and from the wintering grounds would produce average proportions of joint overlap that are statistically indistinguishable for all three groupings. The Western warblers winter almost entirely within a narrow strip of west Mexican

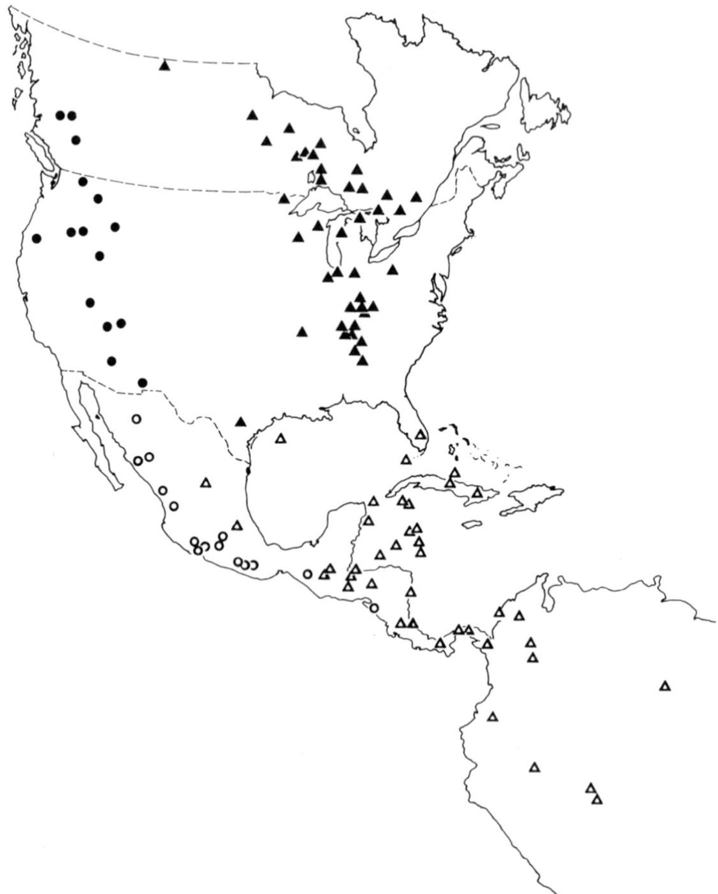

Fig. 6. The geographic centers of distribution are illustrated for Western (circles) and Eastern (triangles) migratory wood warblers during the breeding (closed symbols) and nonbreeding (open symbols) seasons. Only those species whose breeding range is centered north of the Mexican border are illustrated. The Western species and subspecies include (from north to south, closed circles) *Dendroica townsendi, Seiurus noveboracensis, Wilsonia pusilla, Oporornis tolmiei, Vermivora celata, Geothlypis trichas, D. petechia, V. ruficapilla, D. occidentalis, D. coronata, D. nigrescens, V. virginiae, Icteria virens, V. luciae,* and *D. graciae.* The Eastern species and subspecies include (from north to south, closed triangles) *V. celata, D. striata, D. coronata, V. peregrina, W. pusilla, O. agilis, D. palmarum, S. noveboracensis, D. magnolia, D. castanea, D. tigrina, W. canadensis, O. philadelphia, D. virens, D. caerulescens, D. petechia, V. ruficapilla, D. fusca, D. pensylvanica, S. aurocapillus, D. kirtlandii, Setophaga ruticilla, Parula americana, V. chrysoptera, G. trichas, Mniotilta varia, V. pinus, D. pinus, D. cerulea, D. discolor, S. motacilla, Helmitheros vermivorus, D. dominica, I. virens, Protonotaria citrea, O. formosa, W. citrina, V. bachmanii, Limnothlypis swainsonii,* and *D. chrysoparia.*

mainland from southern Sonora south to Guatemala, while the Eastern warblers generally winter in geographically separate areas in the Bahamas, West Indies, eastern Mexico, Central America, and northern South America (Fig. 6).

There are minor exceptions to this generalization. For example, the "Eastern" Black-and-White Warbler (*Mniotilta varia*) and American Redstart (*Setophaga ruticilla*) both winter, in part, in western Mexico (see Chapter 1, this volume); however, each represents a species whose breeding range bends westward across the Rocky Mountains in the north, and the individuals that winter in western Mexico may very well be "Western" individuals of these primarily Eastern species. On the other hand, there is at least one recorded case in which an Eastern individual [Nashville Warbler (*Vermivora ruficapilla*) banded in Oklahoma] was recovered in Nayarit, Mexico, in winter (Shackford, 1983), so the generalization is not without exception.

A further point of interest is suggested from the previous analysis. In the case of most migratory European passerines (Lack, 1971) and Eastern North American wood warblers (MacArthur, 1958; Chipley, 1980; Keast, 1980, this analysis) there is an increase in geographical isolation in winter compared with summer. In the case of sylviid warblers, the degree of geographic overlap among the various species is unchanged from summer to winter (M. Cody, personal communication). The situation is clearly different for Western North American wood warblers, and they may be unique among passerine bird species in this respect. The trend is unequivocally toward an increase in geographical overlap among species within the relatively tiny area in winter compared with summer. The relevance of all this to the question of habitat selection is that the range of habitats used during the nonbreeding season may be influenced, in part, by the geographic area occupied during the breeding season.

Why do the Western wood warblers "choose" not to occupy a greater geographic range in winter and thereby use a wider range of habitat types than they presently do? Before attempting to answer this question, I must address another issue that may be in the minds of some readers. Some might claim that the question of where to winter, or what geographic range to occupy, is entirely a product of a northern temperate bias, and if viewed from a more southerly perspective the question would be moot, because the species simply occur in those areas where they originated; the question should be "Why do they occupy the breeding range that they do?" However, even if the taxa in question were derived from tropical ancestors that were once permanent residents (Mengel, 1964), the eventual use of more seasonal and increasingly widespread northern habitats following glacial retreats would have allowed increased population sizes, which would have eventually placed pressure on individuals to expand into new geographic locations or habitats in winter. Hence, discussion of the influence of the geographic locations of breeding season habitats on the geographic locations and variety of habitat types used during the winter by a species is quite valid.

Back to the question of why the Western wood warblers occupy such a restricted winter range relative to their Eastern counterparts—it could be that both intrinsic and extrinsic (distance) factors make the cost of using other habitats outweigh any potential benefits, such that ancestors that have tried other options have been selected against. Alternatively, it could be that extrinsic factors alone have played the important role here, i.e., historically, Eastern habitats have always been avoided by Western migrants because of geographic constraints associated with restricted migratory routes. Certainly, the latter would be the more likely explanation for why Western migrants do not use, say, African scrub habitats, and such extrinsic constraints may play an important role within the hemisphere as well.

In addition, many of the habitats in western Mexico (desert scrub, riparian, oak woodlands, pine–oak woodlands, pine forests, pine–fir forests) are nearly contiguous with the same types farther north into the western United States and Canada (Mengel, 1964; Szafer, 1975). Moreover, these more xeric west Mexican habitats differ vegetatively from the other west Mexican habitats more than the others differ among themselves (the other habitat types are primarily variations of tropical evergreen or tropical deciduous forest types) (Szafer, 1975). This similarity between, and contiguity of, breeding and wintering habitats of the western North American migrants is unmatched by Palaearctic or eastern North American migratory systems and may have allowed many of the Western migrant species to become specialized toward year-round use of the more xeric Western habitat types. For Western migrants to winter in geographic locations (and habitats) farther removed from their present wintering grounds in western Mexico would mean dealing with the additional costs associated with using habitat types that are very dissimilar to their breeding habitats and encountering additional ecologically similar species farther eastward. Those additional costs are apparently outweighed by the advantages of using similar habitat types year-round and avoiding ecologically similar Eastern species, despite the build up of relatively high migrant densities that occur in most west Mexican habitats in winter (Hutto, 1980). In contrast, Eastern North American migrants winter in forests that, although once extensive and coterminous with their breeding forests (Braun, 1950; Graham, 1973; Howard, 1973), are now disjunct and relatively dissimilar to the more northern forests. Restriction of close competitors to different wintering zones (allopatry) might be expected to have evolved in this case, because the habitat differences that have developed among disjunct zones have provided a more geographically varied environmental backdrop against which the competitive differences among species have been expressed, creating the geographic replacement patterns commonly observed among the Eastern migratory species. MacArthur (1958), for example, found only one of his five warbler species (*Dendroica virens*) in Costa Rica when he traveled south to study the birds during winter; the other four species were wintering primarily in the West Indies (*D. tigrina*), southeastern United States and Mexico (*D. coronata*), South America (*D. fusca*), and Colombia (*D. castanea*).

Whether the above generalization is correct in detail, the example illustrates

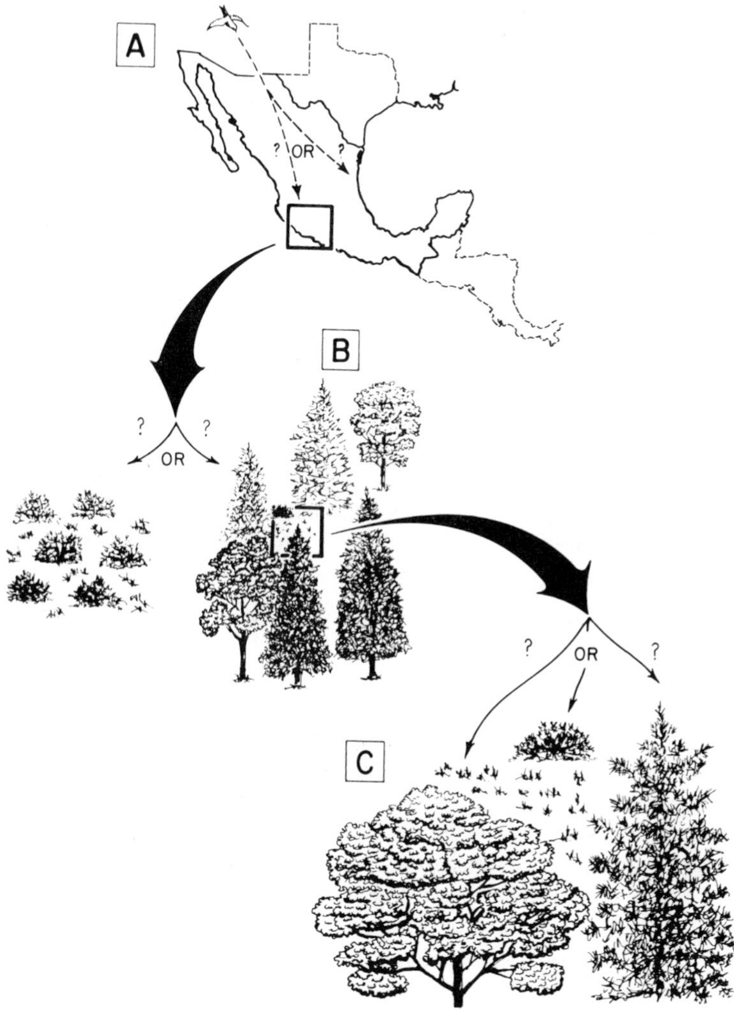

Fig. 7. A diagrammatic illustration of the hierarchical decision-making process involved in "choice" of nonbreeding habitats by a migratory bird. The process at level (A) is probably inflexible and innate, while the process at levels (B) and (C) may very well involve exploratory assessment of intrinsic costs and benefits associated with each option.

how extrinsic factors might operate to constrain the patterns of habitat use during the nonbreeding season.

V. SUMMARY

Habitat selection in nonbreeding, migratory land birds is seen as a hierarchical decision-making process (Fig. 7). At the highest (broadest) level (Fig. 7A), most land birds are probably inflexibly programmed to head along a certain migratory route to a wintering area. Thus, even though the intrinsic factors that determine the suitability of a habitat for nonbreeding season use may have influenced the evolutionary choice of the currently used migratory routes and wintering areas, once established, some habitats that lie outside the major migratory route or wintering areas may be unused simply because of the high extrinsic costs associated with getting *to* them, rather than the high intrinsic costs associated with their use. Similarly, the use of some habitats (particularly while in route) may be largely determined by the extrinsic benefits of using a particular (wind-aided, more direct, etc.) route, rather than being a result of the habitat's intrinsic quality.

The next level in the decision-making process occurs when the bird views the situation from a more restricted geographic perspective (Fig. 7B). At this level, where birds can actually explore alternatives, the intrinsic costs and benefits associated with the use of each habitat may be assessed directly or indirectly by the birds and each used accordingly. There is strong evidence that food availability is one intrinsic factor that plays an important role in the birds' assessment of habitat quality during the nonbreeding season and that use of direct or indirect cues enables birds to settle such that their densities match food resource production quite well.

At the lowest (most local) level (Fig. 7C), the same process involving an assessment of intrinsic factors (especially food availability) also appears to influence the use of space within a habitat (microhabitat use). This implies that decisions involving use of space on a local level may be partly responsible for the patterns of space use that we observe at a higher level—the level of habitat selection. Thus, patterns of habitat use during the nonbreeding season are simultaneously based on extrinsic factors that predispose birds to use a given habitat or not and on the intrinsic quality of the habitat itself, which appears to be determined, in part, by food availability.

ACKNOWLEDGMENTS

I received financial support for the research described herein from a number of sources, including the University of California through Regents' Research Grants, the University of Wyoming through the Jackson Hole Biological Research Station, the American Museum of Natural History through the Frank M. Chapman Memorial Fund, and the University of Montana through the Research Advisory Council. To each of these sources, I wish to express my gratitude. I am also grateful to Joe Wright and Joe McAuliffe for discussion of the results, to Martin Cody for comments on the manuscript, to DeWayne Williams for preparation of the figures, and to Mary Harris and Myra Shulman for their field assistance.

REFERENCES

Balda, R. P., McKnight, B. C., and Johnson, C. D. (1975). Flammulated Owl migration in the southwestern United States. *Wilson Bull.* **87,** 520–533.

Barlow, J. C. (1980). Patterns of ecological interactions among migrant and resident vireos on the wintering grounds. *In* "Migrant Birds in the Neotropics: Ecology, Behavior, Distribution, and Conservation" (A. Keast and E. S. Morton, eds.), pp. 79–107. Random House (Smithsonian Inst. Press), New York.

Braun, E. L. (1950). "Deciduous Forests of Eastern North America." Blakiston Co., Pennsylvania.

Burger, J., Howe, M. A., Hahn, D. C., and Chase, J. (1977). Effects of tide cycles on habitat selection and habitat partitioning by migrating shorebirds. *Auk* **94,** 743–758.

Chipley, R. M. (1980). Nonbreeding ecology of the Blackburnian Warbler. *In* "Migrant Birds in the Neotropics: Ecology, Behavior, Distribution, and Conservation" (A. Keast and E. S. Morton, eds.), pp. 309–317. Random House (Smithsonian Inst. Press), New York.

Clark, K., Euler, D., and Armstrong, E. (1983). Habitat associations of breeding birds in cottage and natural areas of central Ontario. *Wilson Bull.* **95,** 77–96.

Cody, M. L. (1978). Habitat selection and interspecific territoriality among the sylviid warblers of England and Sweden. *Ecol. Monogr.* **48,** 351–396.

Cody, M. L. (1981). Habitat selection in birds: The roles of vegetation structure, competitors, and productivity. *BioScience* **31,** 107–113.

Cody, M. L., and Walter, H. (1976). Habitat selection and interspecific interactions among Mediterranean sylviid warblers. *Oikos* **27,** 210–238.

Connors, P. G., Myers, J. P., Connors, C. S. W., and Pitelka, F. A. (1981). Interhabitat movements by Sanderlings in relation to foraging profitability and the tidal cycle. *Auk* **98,** 49–64.

Cooke, W. W. (1904). Distribution and migration of North American warblers. *U.S. Dept. of Agric. Biol. Survey* **(Bull. 18),** 1–142.

Davies, N. B. (1977). Prey selection and social behaviour in wagtails (Aves: Motacillidae). *J. Anim. Ecol.* **46,** 3757.

Davies, N. B., and Houston, A-. I. (1981). Owners and satellites: The economics of territory defence in the Pied Wagtail, *Motacilla alba*. *J. Anim. Ecol.* **50,** 157–180.

Davis, J. (1973). Habitat preferences and competition of wintering juncos and Golden-crowned Sparrows. *Ecology* **54,** 174–180.

Diamond, J. M. (1978). Niche shifts and the re-discovery of interspecific competition. *Am. Sci.* **66,** 322–331.

Emlen, S. T. (1975). Migration: Orientation and navigation. *In* "Avian Biology," Vol. 5. (D. S. Farner and J. R. King, eds.), pp. 129–219. Academic Press, New York.

Fitzpatrick, J. W. (1980). Wintering of North American tyrant flycatchers in the Neotropics. *In* "Migrant Birds in the Neotropics: Ecology, Behavior, Distribution, and Conservation" (A. Keast and E. S. Morton, eds.), pp. 67–78. Random House (Smithsonian Inst. Press), New York.

Fitzpatrick, J. W. (1982). Northern birds at home in the tropics. *Nat. Hist.* **91,** 40–47.

Fretwell, S. W. (1972). Populations in a seasonal environment. Monographs in Population Biology, Vol. 5, pp. 1–217. Princeton Univ. Press, Princeton, New Jersey.

Fretwell, S. W., and Lucas, H. L. (1970). On territorial behavior and other factors influencing habitat distribution in birds. *Acta Biotheŏ.* **19,** 16–36.

Gass, C. L., and Montgomerie, R. D. (1981). Hummingbird foraging behavior: Decision-making and energy regulation. *In* "Foraging Behavior: Ecological, Ethological, and Psychological Approaches." (A. C. Kamil and T. D. Sargent, eds.), pp. 159–194. Garland STPM Press, New York.

Gates, J. E., and Gysel, L. W. (1978). Avian nest dispersion and fledging success in field-forest ecotones. *Ecology* **59,** 871–883.

Gill, F. B., and Wolf, L. L. (1977). Nonrandom foraging by sunbirds in a patchy environment. *Ecology* **58,** 1284–1296.

Goss-Custard, J. D. (1970). The responses of Redshank (*Tringa totanus* [L.]) to spatial variations in the density of their prey. *J. Anim. Ecol.* **39,** 91–113.

Graber, J. W., and Graber, R. R. (1983). The feeding rates of warblers in spring. *Condor* **85,** 139–150.

Gradwohl, J., and Greenberg, R. (1980). The formation of antwren flocks on Barro Colorado Island, Panama. *Auk* **97,** 385–395.

Graham, A. (1973). History of the arborescent temperate element in the northern Latin American biota. *In* "Vegetation and Vegetational History of Northern Latin America" (A. Graham, ed.), pp. 301–314. Elsevier, Amsterdam.

Greenwood, P. J., and Harvey, P. H. (1978). Foraging and territory utilization of Blackbirds (*Turdus merula*) and Song Thrushes (*Turdus philomelos*). *Anim. Behav.* **26,** 1222–1236.

Grubb, T. C., Jr., and Greenwald, L. (1982). Sparrows and a brushpile: Foraging responses to different combinations of predation risk and energy cost. *Anim. Behav.* **30,** 637–640.

Hespenheide, H. A. (1973). Ecological inferences from morphological data. *Ann. Rev. Ecol. Syst.* **4,** 213–229.

Hildén, O. (1965). Habitat selection in birds: A review. *Ann. Zool. Fenn.* **2,** 53–75.

Holmes, R. T., and Robinson, S. K. (1981). Tree species preferences of foraging insectivorous birds in a northern hardwoods forest. *Oecologia* **48,** 31–35.

Holmes, R. T., Sherry, T. W., and Bennett, S. E. (1978). Diurnal and individual variability in the foraging behavior of American Redstarts (*Steophaga ruticilla*). *Oecologia* **36,** 141–149.

Howard, R. A. (1973). The vegetation of the Antilles. *In* "Vegetation and Vegetational History of Northern Latin America" (A. Graham, ed.), pp. 1–38. Elsevier, Amsterdam.

Hutto, R. L. (1980). Winter habitat distribution of migratory land birds in western Mexico, with special reference to small, foliage-gleaning insectivores. *In* "Migrant Birds in the Neotropics: Ecology, Behavior, Distribution, and Conservation." (A. Keast and E. S. Morton, eds.), pp. 181–203. Random House (Smithsonian Inst. Press), New York.

Hutto, R. L. (1981). Seasonal variation in the foraging behavior of some migratory Western wood warblers. *Auk* **98,** 765–777.

Hutto, R. L. (1985). Seasonal changes in the habitat distribution of transient insectivorous birds in southeastern Arizona: Competition mediated? *Auk,* **102,** 120–132.

James, F. C. (1971). Ordinations of habitat relationships among breeding birds. *Wilson Bull.* **83,** 215–236.

Keast, A. (1980). Spatial relationships between migratory parulid warblers and their ecological counterparts in the Neotropics. *In* "Migrant Birds in the Neotropics: Ecology, Behavior, Distribution, and Conservation" (A. Keast and E. S. Morton, eds.), pp. 109–130. Random House (Smithsonian Inst. Press), New York.

Klopfer, P. H. (1967). Behavioral stereotypy in birds. *Wilson Bull.* **79,** 290–300.

Lewis, A. R. (1980). Patch use by gray squirrels and optimal foraging. *Ecology* **61,** 1371–1379.

Lack, D. (1971). "Ecological Isolation in Birds." Harvard Univ. Press, Cambridge, Massachusetts.

MacArthur, R. H. (1958). Population ecology of some warblers of Northeastern coniferous forests. *Ecology* **39,** 599–619.

MacArthur, R. H. (1969). Species packing and what interspecies competition minimizes. *Proc. Nat. Acad. Sci. U.S.A.* **64,** 1369–1371.

MacArthur, R. H. (1970). Species packing and competitive equilibrium for many species. *Theor. Popul. Biol.* **1,** 1–11.

Mayr, E. (1926). Die ausbreitung des girlitz (*Serinus canaria serinus* L.). Ein beitrag zur tiergeographie. *J. Ornithol.* **74,** 571–671.

Meents, J. K., Anderson, B. W., and Ohmart, R. D. (1982). Vegetation relationships and food of Sage Sparrows wintering in honey mesquite habitat. *Wilson Bull.* **94,** 129–138.

Mengel, R. M. (1964). The probable history of species formation in some northern wood warblers (Parulidae). *Living Bird* **3,** 9–43.

Morse, D. H. (1980). "Behavioral Mechanisms in Ecology." Harvard Univ. Press, Cambridge, Massachusetts.

Nudds, T. D. (1983). Niche dynamics and organization of waterfowl guilds in variable environments. *Ecology* **64,** 319–330.

Partridge, L. (1974). Habitat selection in titmice. *Nature* **247,** 573–574.

Partridge, L. (1978). Habitat selection. *In* "Behavioural Ecology, an Evolutionary Approach" (J. R. Krebs and N. B. Davies, eds.), pp. 351–376. Sinauer, Sunderland, Massachusetts.

Pearson, D. L. (1977). Ecological relationships of small antbirds in Amazonian bird communities. *Auk* **94,** 283–292.

Pulliam, H. R., and Mills, G. S. (1977). The use of space by wintering sparrows. *Ecology* **58,** 1393–1399.

Pulliam, H. R., and Parker, T. H. (1979). Population regulation of sparrows. *Fortschr. Zool.* **25,** 137—147.

Pyke, G. H., Pulliam, H. R., and Charnov, E. L. (1977). Optimal foraging: A selective review of theory and tests. *Q. Rev. Biol.* **52,** 137–154.

Ricklefs, R. E. (1979). "Ecology." Chiron Press, Concord, Massachusetts.

Root, R. B. (1967). The niche exploitation pattern of the Blue-gray Gnatcatcher. *Ecol. Monogr.* **37,** 317–350.

Rotenberry, J. T., and Wiens, J. A. (1980). Temporal variation in habitat structure and shrubsteppe bird dynamics. *Oecologia* **47,** 1–9.

Royama, T. (1970). Factors governing the hunting behavior and selection of food by the Great Tit (*Parus major* L.). *J. Anim. Ecol.* **39,** 619–668.

Sabo, S. R., and Holmes, R. T. (1983). Foraging niches and the structure of forest bird communities in contrasting montane habitats. *Condor* **85,** 121–138.

Serventy, D. L. (1971). Biology of desert birds. *In* "Avian Biology," Vol. 1. (D. S. Farner and J. R. King, eds.), pp. 287–339. Academic Press, New York.

Shackford, J. S. (1983). Nashville Warbler banded in Oklahoma recovered in Nayarit, Mexico. *Bull. Okla. Ornithol. Soc.* **16,** 22–23.

Sinclair, A. R. E. (1978). Factors affecting the food supply and breeding season of resident birds and movements of Palaearctic migrants in a tropical African savannah. *Ibis* **120,** 480–497.

Smith, J. N. M., and Dawkins, R. (1971). The hunting behaviour of individual Great Tits in relation to spatial variations in their food density. *Anim. Behav.* **19,** 695–706.

Smith, J. N. M., and Sweatman, H. P. A. (1974). Food-searching behavior of titmice in patchy environments. *Ecology* **55,** 1216–1232.

Smith, K. G. (1982). Drought-induced changes in avian community structure along a montane sere. *Ecology* **63,** 952–961.

Svärdson, G. (1949). Competition and habitat selection in birds. *Oikos* **1,** 157–174.

Szafer, W. (1975). "General Plant Geography." PWN-Polish Scientific Publishers, Warszaw.

Terborgh, J., and Weske, J. S. (1969). Colonization of secondary habitats by Peruvian birds. *Ecology* **50,** 765–782.

Thomas, D. G. (1980). Foraging of honeyeaters in an area of Tasmanian sclerophyll forest. *Emu* **80,** 55–58.

Tinbergen, J. M. (1981). Foraging decisions in Starlings (*Sturnus vulgaris* L.). *Ardea* **69,** 1–67.

Tryon, P. R., and MacLean, S. F. (1980). Use of space by Lapland Longspurs breeding in arctic Alaska. *Auk* **97,** 509–520.

Verner, J. (1975). Avian behavior and habitat management. *U.S.D.A. For. Serv., Gen. Tech. Rep.* **W0–1,** 39–58.

Wallace, G. J., and Mahan, H. D. (1975). "An Introduction to Ornithology." Macmillan, New York.

Ward, P. (1971). The migration patterns of *Quelea quelea* in Africa. *Ibis* **113,** 275–297.

Werner, E. E., Mittelbach, G. G., and Hall, D. J. (1981). The role of foraging profitability and experience in habitat use by the bluegill sunfish. *Ecology* **62,** 116–125.

Whitmore, R. C. (1975). Habitat ordination of passerine birds of the Virgin River Valley, southwestern Utah. *Wilson Bull.* **87,** 65–74.

Whittaker, R. H. (1975). "Communities and Ecosystems." Macmillan, New York.

Wiens, J. A. (1981). Scale problems in avian censusing. Stud. Avian Biol. **6,** 513–521.

Wiens, J. A., and Dyer, M. I. (1975). Rangeland avifaunas: Their composition, energetics, and role in the ecosystem. *U.S.D.A. For. Serv., Gen. Tech. Rep.* **W0–1,** 146–182.

Willis, E. O. (1966). The role of migrant birds at swarms of army ants. *Living Bird* **5,** 187–231.

Zach, R., and Falls, J. B. (1979). Foraging and territoriality of male Ovenbirds (Aves: Parulidae) in a heterogeneous habitat. *J. Anim. Ecol.* **48,** 33–52.

Chapter 17

Habitat Selection in Island versus Mainland Birds

JACQUES BLONDEL

Centre National de la Recherche Scientifique
Centre d'Etudes Phytosociologiques et Ecologiques
Montpellier, France

I. THEORETICAL BACKGROUND

Habitat selection on islands is a central point in any theory of island biogeography; the ability of a species to select a habitat and eventually adapt itself to new environmental conditions in insular impoverished biotas has an influence on such important features as species–area relationships, community organization and

HABITAT SELECTION IN BIRDS

ISBN 0-12-178080-5

dynamics, resource partitioning, predator–prey interactions, and so on. Two sets of factors are involved in a probability of colonization. The first is related to intrinsic properties of the species, i.e., demography, competitive ability, and flexibility to cope with new habitat characteristics once a propagule has landed on an island. The second set concerns the island itself and includes such features as its size, the diversity of habitats and available resources, the composition, and the degree of saturation of the established communities. The fate of a candidate for colonization results from the combined effects of these two sets of factors, which have been integrated in different but not necessarily conflicting ways by MacArthur and Wilson (1963, 1967) and somewhat revised by MacArthur (1972), Lack (1969, 1976), Diamond (1969, 1970a, 1970b, 1975), Simberloff (1974), Carlquist (1974), and Abbott (1978), among others. Critical comments and alternative views are given for instance by Sauer (1969), Lynch and Johnson (1974), Simberloff (1976), Simberloff and Abele (1976), Connor and Simberloff (1978), Abbott (1978, 1980), Gilbert (1980), and Williamson (1981).

Island impoverishment is not a simple phenomenon, but it is of central importance as far as habitat selection is concerned. A basic statement is that species impoverishment on islands leads to new species assemblages with, as a consequence, modifications in interspecific interactions. It is hypothesized that interspecific competition should be reduced on species-poor islands, allowing each species to expand in the ecological space which would be occupied on the mainland by competitors. Making the assumption that the distribution of resources is the same on the island and in a mainland area of similar size, each island population has a greater share of these resources than does its mainland counterpart. Whether the mechanism is behavioral adjustment or an evolutionary adaptation, the result is an expansion of the species niche and habitat utilization. As a consequence of Darwin's work, competition has been held for a long time as a focal point in shaping community attributes, such as mutually exclusive spatial distributions, characteristic species/genus ratios, character displacement, and number of coexisting species (see Case and Sidell, 1983). Following the developments in the theory of species packing and community organization (Levins, 1968; MacArthur, 1970), there was in the last 15 years a large amount of work focusing on the study of ecological relationships of species belonging to assemblages with overlapping resource requirements (see, for instance, Vandermeer, 1972; Cody, 1974). Most studies on population biology on islands are primarily based upon this competition theory and its corollaries.

MacArthur and Wilson's paradigm of island biogeography does not consider competition per se but is largely based on the assumption that the number of species and community organization on islands are a result of dynamic population phenomena, leading through processes of immigration and extinction to a turnover of species. But most of the subsequent works on island biogeography, especially those of Diamond and co-workers, are based on the premise that interspecific competition is the key factor in explaining equilibrium between immigration and extinction. It must be recognized that alternative hypotheses have often been overlooked or even rejected without a careful examination of their validity (see Simberloff, 1976; Diamond, 1978; Abbott, 1980; Gilbert, 1980; Williamson, 1981). Habitat selection and the number of species on islands do not depend only on the number of potential or actual competitors. Lack's theory emphasizes the match between the species occurring on the islands and the diversity of the habitats, the number of species being determined ecologically by environmental factors. He argued that the number of species found on an island reflects such features as the climate, habitat sizes, and floristic composition. The importance of other factors, such as the structure of prey communities for insectivorous birds and predation, have recently been pointed out.

For the purpose of this chapter I shall make four points which will be the background for hypotheses to be examined. These points are acknowledged as proven facts largely supported by observations, even if there are divergences in their interpretation. The problem is to know their causality and their consequences on life history parameters and evolutionary processes, especially as far as habitat selection is concerned: (1) Islands have fewer species of land birds than areas of the same size on the nearby continent, and there is a relation between size and/or remoteness of an island and its species richness. (2) Certain species of land birds occupy a wider range of habitats on islands than on the mainland. Moreover, population densities are often modified, the trend usually to increase. (3) Some species have modified their niche to a large and unexpected extent, others have not. (4) Because of different selective pressures, the morphology of island populations is sometimes modified by comparison with that of conspecific populations on the mainland.

Each of these four points obviously has something to do with habitat selection, which I shall envisage in a broad sense, involving different kinds of adaptation of a population to an area, that is to its space and resources. This will be done mainly from Mediterranean bird faunas, and the question will be examined at three levels: the level of whole bird faunas, of communities, and, finally, of populations. Emphasis will be given to comparisons of similar habitats, although there are few examples of such comparisons in the literature; exceptions are the works of Crowell (1962), Yeaton (1974), Yeaton and Cody (1974), Nilsson (1977), Blondel (1979, 1981), and others.

II. BIRD FAUNAS OF MEDITERRANEAN ISLANDS

Because of the many islands of different sizes, degrees of isolation, and physiographic diversities, the Mediterranean region is suited to produce "natural experiments" on habitat selection. There are in this region 335 species of breed-

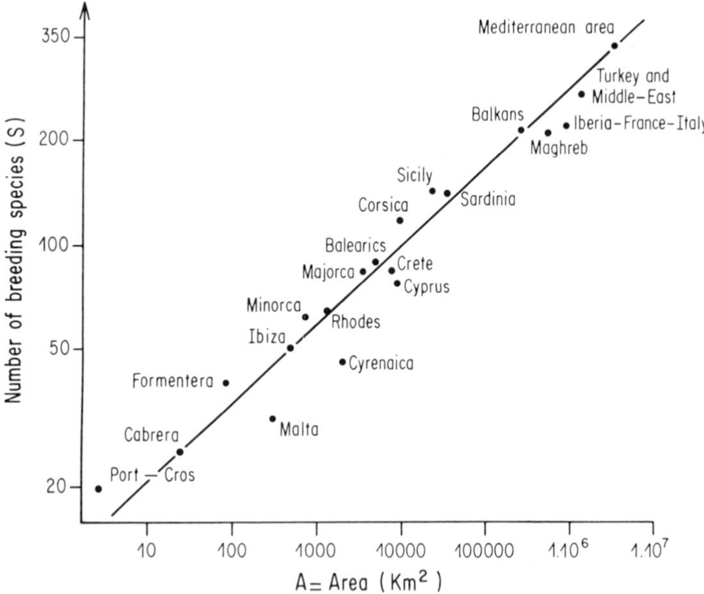

Fig. 1. Log–log relationship between species richness and area for Mediterranean islands and continental regions. $S = 12.9\ A^{0.22}$; $r = 0.97$.

ing birds on an area of 2,970,000 km^2, as compared to 419 (Voous, 1960) on the 10,100,000 km^2 on the whole of Europe. Thus, the ratio of richness to area (S/A 10^{-5}) is 4.15 for Europe but rises to 11.3 for the Mediterranean area (Blondel, 1982). I have divided the Mediterranean region into 4 continental sectors and 10 insular sectors (including such biological islands as the Cyrenaica, which is a Mediterranean enclave surrounded by deserts). A log–log regression of the richness against area for all breeding species shows a highly significant correlation (Fig. 1). Such a result just confirms this well-known relationship, which is observed in virtually all studies. Its biological meaning is not fully understood and has recently been questioned and discussed by several authors (see reviews by Connor and McCoy, 1979; Gilbert, 1980; and Williamson, 1981). At this point the crucial questions are: How do species numbers become adjusted to island areas; How are available space and habitats shared by the potential colonists; Are successful colonists a random sample of the species pool of the nearby continent and, if not, why and how do they differ? On account of the distribution of these 335 species breeding in the Mediterranean region, Blondel (1982) has defined for each of the 14 geographic areas an index of geographic amplitude (IGA), which is the average value of the total number of sectors occupied by each species breeding in this sector. As shown in Fig. 2, there is a highly significant inverse relationship between the IGA and the size of the sector.

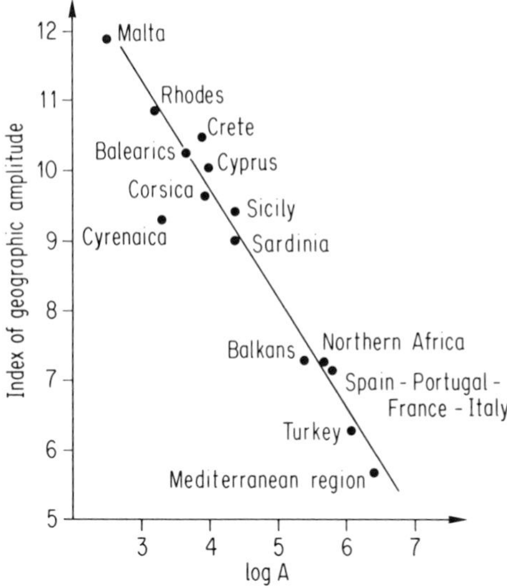

Fig. 2. Inverse relationship between the index of geographic amplitude *IGA* (IGA = 15.89−1.52A) and area for island and continental sectors of the Mediterranean region ($r = -0.99$). High values of *IGA* mean that the bird fauna is composed of ubiquitous species. See text.

Mainland bird faunas have on the average a smaller IGA (7.01) than the insular ones (IGA = 10.03). This means that there are on the average more localized and specialized species on the mainland than on islands, which suggests that widespread and abundant species on the continent have better chances to colonize islands than localized and rare species. Thus, the smaller an island, the poorer its bird fauna and the more ubiquitous its constituent elements. For instance, there are only 28 species of land birds breeding in Malta (Sultana *et al.*, 1975), and all of them are widely distributed everywhere in the Mediterranean. Since there are only three endemic species in the Mediterranean islands—*Sylvia sarda* in the western islands, *Sitta whiteheadi* in Corsica, and *Sylvia melanothorax* (superspecies with *S. melanocephala*) in Cyprus—these localized insular species represent only a negligible fraction of the fauna. These figures support the view that insular communities are not random samples of the species pool of the nearby continent. But it must be recognized that since large islands have a higher diversity of habitats than small ones, the more diverse the available habitats, the greater the probability of colonization by different species. Simberloff (1978) points out that the biota of a larger island is a larger sample of the total number of species of the continental source; hence there is a higher probability that rare species will be present by chance on large islands. For this author, chance colonization is sufficient to explain the bird species composition on islands. But Diamond (1974) and Lack (1976) gave evidence that island species are drawn from the most versatile bird species of the continental source, which is supported by observations at the community level in the Mediterranean Islands, as will be shown later.

Four main causes may explain the absence of a given species from an island: (1) Some localized, specialized, and highly sedentary species on the mainland have rates of dispersal too low to have any chance of successful immigration on islands. Indeed, population sizes of potential immigrant species on the mainland are varying, and so the probability of immigration varies in parallel. (2) Not only will very rare species on the mainland be poor colonists, but weak dispersal (island too far or too small) may be typical even in species which are widespread on the mainland. (3) Then successful immigration may be precluded by a lack of suitable habitats or resources on the island. (4) Or failure to establish an island population may be attributable to the already existing communities, which are saturated and resist the invasion of intruders through some mechanism of competition. The difficulty in deciding which of these possibilities account for absences on islands is due to our lack of experimental studies involving manipulations of populations and ecological factors. These questions will be examined in the next section, with Corsica taken as an example.

III. BIRD FAUNA OF CORSICA

Corsica is a large island of 8,700 km^2. It is 2700 m in elevation and 80 km off the Italian coast to its east. It has not been connected to the mainland since the beginning of the Pliocene. Apart from a large lowland plain in its eastern part, Corsica is a mountain in the sea with a succession of vegetational belts from thermomediterranean phytosociological series near sea level up to oromediterranean series (Gamisans, 1975). Its richness in bird species is on an average 37% lower than that of three similar (in respect to area, altitude, and physiographic diversity) regions in southern France (Table I). Only 109 species regularly breed there (Blondel and Frochot, 1976; Thibault, 1983) plus 16 occasional species (see Appendix), the reproduction of which has been recorded at least once but no more than three times. By comparison with the land-bird faunas of these similar continental regions, species impoverishment in Corsica is not evenly distributed among the different families. Some groups are hardly impoverished, for instance Apodidae, Caprimulgidae, and Hirundinidae. These birds are aerial feeders, and many of them are anthropophilous or semianthropophilous species, like the *Passer* sparrows (3 species, both on Corsica and on the mainland) and many Fringillidae and Carduelidae which live in open habitats (9 species on Corsica, 10 on the mainland). On the other hand, some groups are heavily impoverished. This is the case for Galliformes, Strigiformes, Picidae, Alaudidae, Motacillidae, Laniidae, Sylviidae, Paridae, and Emberizidae.

What kind of explanation if any can be invoked for such an impoverishment?

TABLE I

Comparison of the Number of Breeding Bird Species in Corsica and in Three Homologous Continental Areas in Southern France[a]

Region	Area (km^2)	Maximal altitude (m)	Number of species
Corsica	8,700	2,700	109[b]
Alpes maritimes and Var	10,320	2,800	170
Bouches du Rhône and Vaucluse	8,850	1,912	173
Pyrénées orientales and Aude	10,480	2,800	172

[a] From Blondel and Frochot, (1976).

[b] This number does not include the 16 occasional species listed in the Appendix.

A. Corsica Is outside the Geographical Range of the Species

This zoogeographical explanation can be put forward objectively for species which are absent from northern Africa. This is the case for 12 species which are at the southern fringe of their distributional range on the European mainland; this explanation is especially likely for species such as *Tetrao tetrix, Tetrastes bonasia,* or *Aegolius funereus* which are primarily boreal species of the large mixed forests of central and northern Europe (see Appendix).

B. Weak Dispersal

Although difficulties of overwater colonization may account for absences in remote islands, especially in the tropics, and for weak dispersers, as stressed by Abbott (1980), such an explanation is hardly conceivable in the Mediterranean. This region is crossed each spring and autumn by millions of birds migrating between Eurasia and Africa (Blondel, 1969; Moreau, 1972). Many of these species halt on the Mediterranean islands but have no local breeding populations, although there are suitable habitats for them. Difficulty of dispersal cannot be put forward for such species. The most obvious cases of migrants which pass over Corsica in large numbers but have not established themselves as breeding birds are listed in the Appendix. Many of them have breeding populations on both sides of the Mediterranean Sea (see Appendix), which rules out the possibility that Mediterranean islands are outside their geographical range. Detailed studies of song characteristics made by Chappuis (1976) have shown that even for sedentary species, such as the Treecreeper (*Certhia brachydactyla*), dispersal of propagules over the Mediterranean exists to such an extent that the sea cannot be considered a barrier for colonization, although we have no data on the absolute rates of dispersal of the species. Simpson (1940) made a distinction between "island hopping," where the distances to be covered are within the average dispersal ranges of the species, and "sweepstake dispersion," which requires exceptional and improbable events. There are many examples of sweepstake dispersion in remote islands and archipelagos where the occurrence of stragglers is evidence that many species are potential colonists (Lack, 1969; Diamond, 1975). Sweepstake dispersion followed by successful colonization must be a very rare event in remote islands, and this allows the evolution of endemic bird faunas on such archipelagos as the Galapagos and the Hawaiian Islands, where evolutionary processes leading to speciation were faster than immigration rates. In the Tristan de Cunha archipelago, which is among the most remote in the world (2800 km off Africa and 3200 km off South America) only five land bird species breed, but Elliott (1957) has seen no fewer than nine additional species of passerines within 2 years; none succeeded in becoming established. But the geographical situation of the Mediterranean islands is completely different; clearly, dispersal abilities of most of the Palearctic birds as well as the proximity of these islands from the nearby continents are of the island-hopping type. Thus, weak dispersal can be ruled out as an explanation of island impoverishment in this region. Island-hopping dispersal is probably not a general rule for any island close to a continent. Abbott (1980) reported that few of the species missing in Tasmania and Kangaroo Island have been recorded on these islands, and Willis (1974) and Diamond (1975) gave examples of tropical birds which are unable to cross even very narrow stretches of water. This may be because dispersal abilities are much lower for highly sedentary tropical birds firmly tied to forest habitats than for temperate birds.

C. Lack of Suitable Habitats

The problem of habitat availability and suitability has often been put forward to explain insular impoverishment, especially by Lack (1969, 1976). But curiously, studies on habitat availability and diversity are very scarce if one excepts those of Watson (1964) and Johnson (1975). Watson found a strong correlation between habitat diversity and area in the islands of the Aegean Sea. Johnson found that, on isolated mountaintops in the Great Basin of the Sierra Nevada and the Rockies, the best correlation with bird species numbers is not area per se but an index of habitat diversity and the minimum interisland distance. Since habitat diversity is difficult to measure, most authors implicitly assume that this parameter is correlated with area and altitude (Abbott, 1978), which seems reasonable since altitude is accompanied by a succession of life zones, which enhances both structural and floristic habitat diversity. For instance, Diamond (1972) has shown in New Guinea that four congeners divide an altitudinal gradient with virtually no overlap between the adjacent numbers of the replacing species.

An important factor affecting the occurrence of species, especially large ones, is not only availability of habitats but also the size of continuous habitats. Diamond (1975), Moore and Hooper (1975), Galli et al. (1976), Blondel (1980), and Soulé and Wilcox (1980) have shown that species with large territories or home ranges will be precluded from colonizing islands which have only small patches of suitable habitat, because they would be too vulnerable to stochastic extinction. Some absences from Corsica could be explained by this factor, especially absences of large species since, for the same number of individuals, a population of a large bird needs a larger area than a population of a small bird (Appendix). Some of the unsuccessful attempts of colonization listed in the Appendix, for instance *Egretta garzetta, Ardea cinerea, Sterna hirundo, Asio otus,* and *Dryocopus martius* could fall in this category. But even small species, such as *Parus cristatus, Parus palustris,* or *Pyrrhula pyrrhula,* are very scattered on the mainland and need large blocks of homogeneous habitat to build up viable populations.

One consequence of habitat size is that on small islands a degree of habitat variation which could support two species in larger continental areas supports only one. This is thought to be a niche-widening mechanism in very small patches of habitat. Lack (1971) gave many examples of this process, which promotes the population size of a species and hence its probability of survival at the expense of species richness. Finally, it must be pointed out that species-specific requirements are generally guessed via rather crude methods. It may be that habitats which look at first sight suitable for a given species on the mainland are actually different on the island and that this species is not flexible enough to cope with this new situation (Morse, 1971). For instance, there is no leaf warbler of the genus *Phylloscopus* breeding in Corsica although two species, *P. collybita* and *P. bonelli,* breed commonly on both sides of the Mediterranean Sea. These species need very low bushes and grass to build nests, but such microhabitat features are lacking in most parts of Corsica, because of severe overgrazing by pigs and cattle. This could be the explanation for the absence of these species.

In some circumstances a paucity of species is due to a lack of suitable colonists rather than a lack of suitable habitats for them. Australia and New Zealand now support more land birds than they did before man succeeded in introducing so many European species (Abbott and Grant, 1976).

D. Competition

It follows from the MacArthur–Wilson theory of dynamic biogeography that the number of species on islands remains constant even though species composition changes; turnover of species is one of the basic statements of the theory. At equilibrium, faunas must be saturated, which means that no more species can be added without causing the extinction of species already present. This is the concept of "closed communities" developed by these authors (MacArthur and

Wilson, 1967). But the existence of species turnover resulting from the equilibrium theory has been disputed especially by Simberloff (see review and references in Gilbert, 1980).

The importance of competition has been emphasized in another way by Lack (1969, 1976) who stressed that islands have a reduced diversity of habitats and food resources. As a consequence, island avifaunas are saturated with fewer species than those of the mainland and resist invasion through competition by enlarging their habitat niche. There is in the literature very little direct evidence of competition, because experimentation is almost impossible with birds; however, the many cases of character displacement, ecological shifts, and niche broadening suggest that actual or potential competition plays a role in determining the colonizing abilities of species.

So far as the Corsican bird fauna is concerned, the problem of competition and habitat selection will be discussed at the community level in the next section. Some absences from the Corsican list of breeding species can be discussed here, although in most cases it is very difficult to know the exact effect of competition, space, food, or both. Thus, competition is not independent of the problem of availability and size of habitats. Some kind of direct or potential competition may account for the absence from Corsica of as many as 42 species (Appendix), that is to say more than half of the missing species. For instance, food resources for raptors and owls may be too scarce and habitats too small and patchy to allow coexistence of more than 1 or 2 species, since there are only 6 species of micromammals in Corsica (versus 15 on the mainland, P. Orsini, personal communication), and such important prey species as *Clethrionomys glareolus, Pitymys duodecimcostatus,* and *Microtus arvalis* are absent from the island.

Competition for food and space may account for some particular patterns of distribution in the Mediterranean area as a whole. For instance, partridges of the genus *Alectoris* have speciated in the Mediterranean region (Watson, 1962) and are represented there by four species, *A. graeca, A. rufa, A. chukar,* and *A. barbara*. Not a single island has more than one species, with *A. rufa* in Corsica, *A. barbara* in Sardinia, A. *graeca* in Sicily, and *A. chukar* in Crete. Such patterns obviously show geographical replacement of these potential competitors, because there is neither the space nor the resources for more than one species on each island. Thus, nonoverlap in geographical range is a means of avoiding competition. Such a mechanism is further supported by the geographical replacement, with only very narrow marginal overlap of *A. graeca* and *A. chukar* in European Turkey (Watson, 1964). Such geographical replacements have the same meaning as the displacement patterns described by Diamond (1972) for the partitioning of altitudinal gradients in New Guinea by close relatives.

A somewhat counterexample is that of the large vultures. There are five species of vultures in the Mediterranean region, *Neophron percnopterus,*

Gypaetus barbatus, Torgos tracheliotus, Aegypius monachus, and *Gyps fulvus.* As many as three or four species occur together in the larger islands (e.g., *G. barbatus, Ae. Monachus,* and *Gyps fulvus* in Sardinia, and these three plus *N. percnopterus* in Crete.) But ecological isolation between these species, which belong to different genera, is probably more pronounced than in the partridges, especially between such species as *N. percnopterus* and *G. barbatus,* whose diets are quite different. A detailed analysis of habitat and niche partitioning between these sympatric vultures would be extremely interesting.

Taxonomic diversity may be a clue to the competition hypothesis on islands. It could be argued that there are fewer close relatives sharing the same set of resources on an island than on the mainland, because more severe competition would preclude coexistence between close congeners. I examined the species/genus ratio in the Mediterranean islands but found no significant difference in taxonomic diversity between the islands and the mainland. Simberloff (1970) and Abbott (1975a) argued that islands have a greater proportion of congeners than expected on the basis of a random selection from the continental pool. This would be because congeners have more similar ecological requirements than noncongeners and they could coexist better than unrelated species on a same set of resources, because neither could exclude the others in case of food shortage (Abbott, 1975a). Figures from the Mediterranean islands show that in some cases the proportion of congeners is higher than on the mainland and in other cases, not.

For most of the species missing from Corsica, especially for passerines, direct evidence of competition is lacking and a reference to the concept of "diffuse competition" (MacArthur, 1972) is far from satisfactory, because it is almost impossible to test. Diffuse competition is a mechanism whereby the species of a species set indirectly compete with one another, thus preventing potential intruders from colonization. Cases of what can be interpreted as opposition to colonization by established communities in Corsica are suggested by many instances of birds which attempted to breed but either failed or bred only once or a very few times. This could be the case for some of the vagrants or occasional breeders listed in the Appendix (*Coracias garrulus, Picus viridis, Picoides minor,* and all the passerines). One case has been especially well documented by Chappuis (1976) for treecreepers although later challenged by Thibault (1983). There are in Europe two very similar species of treecreepers, *Certhia brachydactyla* and *C. familiaris.* The latter is particularly a mountainous species where the two species occur together on the mainland. But in Corsica there is only *C. familiaris,* which is widespread everywhere and occurs in a variety of forest habitats, exhibiting a large niche shift on the island. Chappuis found in 1969, in a coastal locality of Corsica, several pairs of *C. brachydactyla,* but this small party failed to survive on the island and quickly disappeared. From sonogram examination of the song of these birds, which were territorial, it appeared to Chappuis that this propagule came from nothern Africa. This is probably one of the best examples of resistance to invasion by a local population.

To summarize, three possible broad "explanations" may account for the absences of species from Corsica (Appendix). The most important by far is the competition hypothesis which can account for 42 (59%) of the absences. An interesting fact is that the most impoverished groups are the forest birds. It will be shown in the next section that it is precisely these species which present the most striking changes in habitat shifts, density compensation, demography, and morphological variation. To go further into the investigation of habitat selection patterns in relation to the competition hypothesis, one needs to go deeper into the study of community organization and structure.

IV. HABITAT SELECTION IN MATORRALS AND FOREST COMMUNITIES OF CORSICA: A "NATURAL EXPERIMENT"

This section will deal with between-habitat selection patterns. The main question is: Is the insular impoverishment in species numbers accompanied by a proportional reduction in the total number of birds, or is there some kind of disharmony due, for instance, to different trophic or taxonomical relationships? In other words, are the island patterns of community organization only a simplified facsimile (i.e., density stasis of Williamson, 1981) of homologous and more complex continental communities?

There are actually many papers in the literature which clearly show that insular communities present important changes in habitat utilization (Lack, 1942, 1969, 1976; Crowell, 1962; Grant, 1966a, 1966b, 1968; MacArthur *et al.*, 1966, 1972; Keast, 1968; Diamond, 1970a, 1970b, 1973; Yeaton, 1974; Yeaton and Cody, 1974, Cox and Ricklefs, 1977; Blondel, 1979, 1981; Abbott, 1980), although there is debate on their causes, and even the existence of such changes was disputed by Simberloff (1978, 1980) and Haila and Järvinen (1983).

I shall briefly summarize here some of the results obtained in Corsica, at the community level, through the study of ecological succession (see Ferry *et al.*, 1976; Blondel, 1979, 1981; Martin, 1982). Seven study sites were chosen in the French Mediterranean region of Provence and in Corsica. These sites were selected according to an increasing complexity of structure of the vegetation, and they form two habitat gradients which match each other and allow comparisons between bird communities. Detailed measurements of habitat characteristics for the two gradients were made according to the procedure of Blondel and Cuvillier (1977). These measurements include the total height of vegetation, the number of strata, the percentage of cover of the vegetation in each stratum, and both the horizontal and the vertical diversity of vegetation (Table II, but see Blondel, 1981, 381–383 for more details). Data show that these two habitat gradients

TABLE II

Some Characteristics of Habitats and Bird Communities in Seven Stages of Homologous Habitat Gradients in Provence (Southern France) and in Corsica[a]

Characteristics	Stages							Total	Mean
	1	2	3	4	5	6	7		
Height of vegetation (m)									
Provence	0.20	0.50	1	4	6	10	20		
Corsica	0.30	0.80	1	2	6	12	25		
Number of layers									
Provence	1	2	3	5	6	7	8		
Corsica	2	3	4	4	6	7	8		
Complexity index of vegetation									
Provence	46	94	174	354	466	442	577		
Corsica	94	168	240	318	350	526	571		
Number of breeding bird species									
Provence	9	10	14	11	16	24	23	48	
Corsica	22	29	24	31	23	24	18	42	
Overall density (breeding pairs/10 ha)									
Provence	16.7	8.7	17.1	22.6	35.6	40.0	61.3		28.9
Corsica	18.5	29.0	29.5	40.7	49.3	51.6	63.9		40.4
Average density per species									
Provence	1.86	0.87	1.22	2.05	2.23	1.67	2.67		0.60
Corsica	0.84	1.00	1.23	1.31	2.14	2.15	3.55		0.96
Mean habitat–niche breadth ($e^{H'}$)									
Provence	1.83	2.53	2.99	3.12	2.89	2.59	1.98		1.92
Corsica	3.63	3.68	4.10	3.75	4.10	3.81	3.81		3.24
Mean body weight of the species (gm)									
Provence	53.8	27.2	35.2	43.5	41.9	35.1	29.0		38.0
Corsica	35.1	35.5	19.0	23.5	20.0	28.0	19.8		25.8
Total biomass (gm)									
Provence	1797	476	1205	1967	2988	2827	2942		2029
Corsica	1298	2063	1119	1910	1971	2886	2528		1967

[a] Mean niche breadth calculated with $e^{H'}$ on the densities along the two gradients (see more details in Blondel, 1981, pp. 381–383).

match each other reasonably well. I tried to have as precise a recognition of habitats as possible, because such important features as habitat structure are too often overlooked in island biogeography studies. Table II gives the main results. There are 48 species on the mainland gradient and 42 on the insular one. Twenty-five species are common to both gradients. Thus, the island communities are not heavily impoverished, and the large number of species shared in common will make it easier to determine a measure of habitat and niche shifts. How the species distribute themselves among the habitats gives a clue to whether habitat selection differs in the island. An analysis of the data leads to the following conclusions.

1. Although the species pool is smaller in Corsica than in Provence, there are on the average more species per habitat on the island than on the mainland, especially in the grassland and matorral, but not in the forest. Each species occupies on an average 2.25 habitats in Provence and as many as 4.05 in Corsica. In other words 32% (19% in habitat 1 to 50% in habitat 6) of the mainland species pool occupy each mainland habitat, compared to 58% (43% in habitat 7 to 74% in habitat 4) of the species pool on the island. Thus, there is a tremendous expansion in habitat utilization by the island populations.

2. It follows tautologically from (1) but with a species-specific quantification that habitat–niche breadth (calculated with $e^{H'}$) is much higher on the island (Table II).

3. In all cases total densities are higher in the insular sites. At the gradient-wide level there is on an average 40.4 pairs/10 ha versus only 28.9 pairs/10 ha on the mainland. This much higher density for a smaller richness is a widespread phenomenon called ''density compensation'' (MacArthur *et al.*, 1972), recorded by many authors. Note that the average density per species does not follow the same trend in all stages of the gradient. It is definitely higher in Corsica than on the mainland only in habitats 6 and 7, that is, in forest. However, at the gradient-wide level there are on an average 0.96 breeding pairs/species/10 ha in Corsica to be compared to only 0.60 on the mainland.

4. Total biomass per habitat is of the same order of magnitude in the two regions, especially in the last stages of the gradients. Thus, productivity is nearly similar despite a very different distribution of organisms. Since for the same biomass there are 28% more breeding pairs on the island, the average size of species must be smaller. Hence, the hypothesis that large species should be handicapped for colonizing insular habitats and that there is some kind of selection for small size on islands is supported. At the scale of the two gradients, Table II shows that, except for habitat 2, island species are lighter than mainland ones. At the level of the Corsican land-bird fauna as a whole, some examples can be given which illustrate this selection of small species among the candidates to colonize. Among the Strigiformes the largest species which breeds in Corsica is

TABLE III

Composition of Bird Communities of Old Stands of *Quercus ilex* in Provence and in Corsica[a]

Provence		Corsica	
Species	Density	Species	Density
Parus caeruleus	11.5	*Parus caeruleus*	14.1
Fringilla coelebs	7.7	*Regulus ignicapillus*	9.0
Certhia brachydactyla	6.8	*Sylvia atricapilla*	6.7
Erithacus rubecula	6.1	*Fringilla coelebs*	6.4
Sylvia atricapilla	5.5	*Erithacus rubecula*	5.7
Troglodytes troglodytes	5.0	*Troglodytes troglodytes*	5.0
Turdus merula	3.8	*Parus major*	4.7
Parus major	3.2	*Parus ater*	4.1
Sitta europaea	2.1	*Muscicapa striata*	1.8
Picoides major	1.8	*Certhia familiaris*	1.6
Parus cristatus	1.8	*Turdus merula*	1.1
Regulus ignicapillus	1.5	*Picoides major*	1.1
Luscinia megarhynchos	1.4	*Aegithalos caudatus*	1.1
Phylloscopus bonelli	1.0	*Garrulus glandarius*	1.0
Garrulus glandarius	0.6	*Carduelis citrinella*	0.4
Picus viridis	0.3	*Buteo buteo*	0.1
Picoides minor	0.2	*Cuculus canorus*	0.1
Parus ater	0.2	*Columba palumbus*	0.1
Turdus viscivorus	0.2		
Cuculus canorus	0.1		
Oriolus oriolus	0.1		
Columba palumbus	0.1		
Accipiter nisus	0.1		
Total density	60.9		64.1
Average density per species	2.6		3.6

[a] Seventh stage of the gradients; data are densities/10 ha. From Blondel (1979).

Tyto alba (330 gm), but *Strix aluco* (500 gm) and *Bubo bubo* (2000 gm) are missing. Still more instructive is the case of Picidae: the only two species which regularly breed in Corsica are *Jynx torquilla* (35 gm) and *Picoides major* (76 gm); *Picus viridis* (160 gm) and *Dryocopus martius* (280 gm) are missing, although there are suitable habitats for them.

5. Another trend appears from an examination of the specific densities. Table III shows that species missing from Corsica are those which are less abundant on the mainland. For instance, nine of the most abundant species in the mainland old forest are found in its insular counterpart. But most of the rarest species on

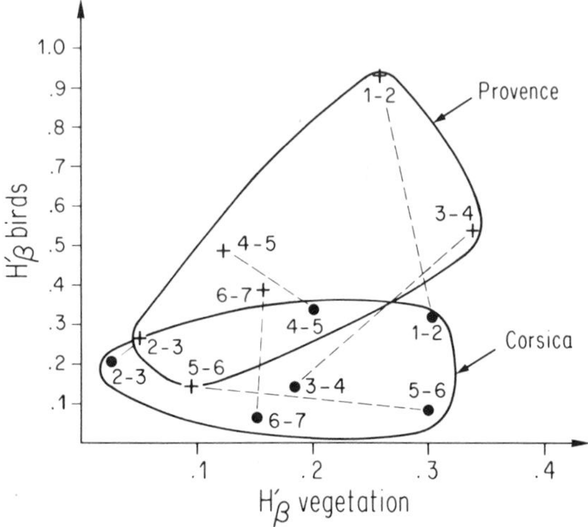

Fig. 3. Influence of between-habitat differences in the structure of vegetation on between-habitat differences in bird communities. For a same difference in habitat structure, between-habitat diversities in bird communities are much higher on the mainland than in Corsica. This is a consequence of habitat–niche expansion on the island.

the mainland are absent from Corsica. Examples are *Parus cristatus, Luscinia megarhynchos, Phylloscopus bonelli, Picus viridis,* and *Picoides minor*. Species which occur in both mainland and island gradients have on an average 4.15 breeding pairs/10 ha, whereas the mainland densities of species which do not occur in Corsica have densities of only 2.67 breeding pairs.

6. Measurements of habitat parameters and community parameters allow one to examine to what extent changes in species numbers and densities at the within-habitat level are related to changes at the between-habitat level. From these measurements it is possible to measure the β-component of diversity, that is, the amount of change between one stage of the habitat gradient and the next. Thus, it is possible in this way to plot bird communities' β-diversity against habitat β-diversity (Fig. 3). Although differences in habitat structure do not match equally well for all pairs of mainland versus island habitats, especially for pairs 3–4, 4–5, and 5–6 (see Fig. 3), mainland β-diversity scores in the birds are in all cases above the insular scores. This means that for the same value of habitat change, insular communities differ less in structure and composition than mainland ones. Hence, there is a much better predictive power in relations such as bird species diversity to habitat structure diversity on the mainland than on the island.

7. Finally, some of the above points, namely (1), (2), and (6), stipulate that species packing is achieved in a much different way in Corsica than on the mainland, since, despite a reduced richness at the pool-wide level, there are on the average more species in each insular habitat than in its mainland counterpart.

It is possible to position the species in relation to one another by calculating the extent of their coexistence in space, that is, an index of habitat–niche overlap which gives insight to the amount of potential interactions between each species and the whole bird community. I measured habitat–niche overlap at the whole gradients scale using Cody's (1974) procedure: $\alpha = p_{ij}/[(p_{ii} + p_{ij})(p_{jj} + p_{ij})]^{0.5}$, where p_{ij} is the proportion of individuals of species i and j found in the same habitats, p_{ii} and p_{jj} are the respective proportions of individuals of the species i and j found in habitats from which the other species is absent. Alpha = 0 when the two species never co-occur in the same habitat, and $\alpha = 1$ when there is total co-occurrence. A matrix habitat–niche overlap has been computerized for the two gradients. For the 48 mainland and 42 island species, the average habitat–niche overlap between each species pair was calculated. Species with high mean α values have a great network of potential interactions with others, while those with low mean α values must be more specialized in habitat selection. Thus, the higher the amount of overlap, the greater number of potential interspecific interactions, and the looser the species' habitat selection pattern. For species which occur on both gradients, mean α values are much higher on the island. As shown by Blondel (1981), niche overlap values are higher in more complex vegetation. Such measures of ecological overlap are not competition coefficients of the Lokta–Volterra equations, because they say nothing about morphological similarities, foraging techniques, behavior, and so on (Ricklefs and Cox, 1977). They are only indices to describe some aspects of community structure. Nevertheless, they indicate that, despite the species impoverishment on the island, competition within and between populations there could be potentially more severe than on the mainland.

To summarize, communities are impoverished, but the total number of organisms is not lower on the island because of density compensation and habitat-use expansion. Thus, each population occupies more space on the island. If niche is a function and not just a compartment of the ecosystem, then the concept of unfilled niches on islands (Slud, 1976) is false, since island populations can expand and perform a wider variety of functions than on the mainland. Since there are, on average, fewer species but more individuals in a wider range of habitats and species of smaller size but increased density, it is hypothesized that such patterns have repercussions on habitat and resource utilization. This will be examined in the next section.

V. NICHE SHIFTS IN CORSICAN BIRDS

I shall now consider habitat selection in congeneric species, several of which can coexist in the same patch of uniform habitat. The question is, are there ecological and behavioral shifts in the mode of utilization of ecological space by insular populations? By ecological space, I mean microhabitat structure of the vegetation and food supply, as well as other features such as foraging techniques, nest sites, and so on. Here again and for the same reasons as those invoked at the between-habitat level, current theory hypothesizes a competitive release (Cody, 1974) that would lead to an enlarging of the resources exploited by island populations. A larger spectrum of occupied habitats and changes in feeding habits, involving a greater versatility or shifts in feeding behavior, have been noticed by many authors (Crowell, 1962; Grant, 1965, 1966a,b; MacArthur *et al.*, 1966, 1972; Selander, 1966; Keast, 1968, 1970; Diamond, 1970a,b; Abbott, 1973; Yeaton and Cody, 1974; Lack, 1976; see also Chapter 3, this volume). Shifts in microhabitat selection patterns suggest an ecological adjustment to a new competitive environment. But as will be shown later, evolutionary shifts can give rise to morphological modifications. I shall examine this problem for two groups of closely related species or guilds, because such species are likely to exhibit interesting differences in the way they partition the same resource gradients, such as habitats, vertical feeding ranges, and food.

A. Warblers of the Genus *Sylvia*

Along with the genera *Alectoris, Monticola, Sitta,* and others, warblers of the genus *Sylvia* are one of the few groups of birds which have speciated in Mediterranean matorral (Blondel, 1982, 1985). Out of the 18 species of *Sylvia* breeding in the Palaearctic region, 9 (*S. hortensis. S. rüppelli, S. melanocephala, S. melanothorax, S. mystacea, S. cantillans, S. conspicillata, S. undata,* and *S. sarda*) are particularly adapted to Mediterranean matorral, and 2 are insular endemics, *S. sarda* on the large western Mediterranean islands and their satellites and *S. melanothorax* in Cyprus. In the western part of the Mediterranean, as many as 7 species are sympatric at a regional scale. Because of the extreme difficulty in collecting reliable life history parameters and behavior in these small buoyant birds, which live secretly in thick and evergreen vegetation, there is little information on the isolating mechanisms of these species, which must be quite subtle for such close relatives. This problem has already been studied by Lack (1971), Berthold and Berthold (1973), Cody and Walter (1976), Blondel (1969, 1979, 1981), Zbinden and Blondel (1981), and Martin and Thibault (1983). In this section I shall deal at some length with the ecological isolation of these birds, using the above references and a large amount of personal unpublished data, especially as far as differences in habitat selection patterns between mainland and Corsica are concerned.

Distributional patterns and densities of the species in the two homologous gradients of vegetation discussed in the previous section are given in Table IV. There are six species on the mainland and five in Corsica. *Sylvia conspicillata* is

TABLE IV

Distribution and Densities (Breeding Pairs/10 ha) of Warblers of the Genus *Sylvia* in Homologous Habitat Gradients in Provence and in Corsica[a]

	Density at stage								
Distribution	1	2	3	4	5	6	7	$e^{H'}$	g
Provence									
S. conspicillata		.65						1.00	2.00
S. undata		1.05	4.44	1.53		.24		2.78	3.17
S. melanocephala			.38	6.95	.35	2.27		2.29	4.45
S. cantillans			1.82	4.18	8.30	2.97		3.44	4.72
S. hortensis					.27	.49		1.92	5.64
S. atricapilla					6.11	6.95	5.55	2.99	5.97
Total		1.70	6.64	12.66	15.03	12.92	5.55	Mean 2.40	4.87
Percentage *Sylvia*		18.9	38.8	55.9	42.2	32.3	9.0		
Corsica									
S. sarda	.86	3.34	3.88	2.99				3.60	2.81
S. undata		.38	1.40	1.46	.07			2.87	3.37
S. melanocephala		.29	1.80	3.56	.17	.06		2.62	3.64
S. cantillans			.99	4.24	1.98	.08		2.69	4.16
S. atricapilla		.29	.29	1.22	7.15	6.84	6.66	3.89	5.78
Total	.86	4.30	8.36	13.47	9.37	6.98	6.66	Mean 3.13	4.48
Percentage *Sylvia*	4.6	14.8	28.3	33.0	19.0	13.5	10.4		

[a] $e^{H'}$ = niche breadth; g = barycenter of the species' distributions ($g = \Sigma id_i/\Sigma d_i$ where i is the stage number of the gradient and d_i the density of the species).

a scarce but regular breeder in Corsica, where it occurs in some coastal localities. *Sylvia hortensis* is only an occasional breeder in Corsica, since there are no more than five records of breeding (Thibault, 1983). Problems of coexistence and niche segregation in these very closely related species are particularly interesting to study, because they appear to present very subtle species-specific differences in habitat utilization.

From Table IV it appears that

1. The guild can include as many as five species in the same habitat, in Corsica as well as on the mainland, which is a maximum since *Sylvia hortensis* does not occur on the Corsican gradient nor *S. sarda* on the mainland. Thus, there is no insular impoverishment for this genus. The same situation was observed by Cody and Walter (1976) in Sardinia, where the ranking of the species according to habitat characteristics is nearly the same as that described in Table IV, except that *S. conspicillata* was present in the Sardinian sites but not in mine.

2. Densities at the gradient level (seven habitats combined) are slightly lower on Corsica than on the mainland. As expected, the ratio of *Sylvia* densities to total bird densities is maximum in the bushy habitats in the middle of the

gradient, but this ratio is higher on the mainland than on Corsica. This is a result of habitat expansion of forest birds in lower matorral in Corsica, as was shown in the previous section. There is no density compensation for these Mediterranean warblers except for *S. atricapilla,* which is more broadly distributed and prefers forest habitats. Highest densities of warblers occur in habitats 4, 5, and 6 in Provence but in habitats 3, 4, and 5 in Corsica. Hence, there is a shift in habitat use toward a lower vegetation on the island. This is obvious for three out of the four species which occur on both gradients, namely *S. melanocephala, S. cantillans,* and *S. atricapilla,* since the centers of their distributions (see g in Table IV) have lower values in Corsica than in Provence. This shift may be due to interspecific competition caused by the expansion of forest birds into the Corsican matorral. There is a significant expansion of habitats occupied by the *Sylvia* guild as a whole in Corsica in comparison to the mainland, since $e^{H'}$ averages 3.13 in Corsica versus 2.40 in Provence (Table IV; χ^2 test, $p<0.05$). But this is due only to the high values of $e^{H'}$ for *S. sarda* and *S. atricapilla.* The former is absent from the mainland, and the latter is not a bird specialized for matorral, since its preferred habitats are the high coppices and forests of the last stage of the gradient. Thus, it is interesting to notice that *S. atricapilla* shows the same patterns of habitat expansion and density compensation as do the other forest species. For the three species which occur in common on both gradients, $e^{H'}$ is slightly but not significantly higher in Corsica for *S. undata* and *S. melanocephala* but significantly lower in *S. cantillans.* These results show that these Mediterranean warblers do not present a definite syndrome of insularity in Corsica, and this contrasts sharply with the situation for forest birds, such as the titmice discussed later.

A more detailed analysis of habitat selection and niche partitioning was performed using the vertical foraging range and nest location in the vegetation (J. Blondel, unpublished observations). As to vertical foraging range, Fig. 4 shows the pattern of segregation in habitat 4 on the gradient, where three species co-occur in Provence and five in Corsica. Figures were drawn by multiplying the percentage of foraging in each stratum *i* of vegetation by the density of species *i* in the habitat and then weighing the product by the ratio of the density of this species to the total warbler density $(s_i \cdot d_i/100d_i)$ $(d_i/\Sigma d)$. These data give the relative impact of each *Sylvia* species in the habitat and show that:

1. The two species which co-occur only in Corsica, *S. sarda* and *S. atricapilla,* have the widest range in foraging heights. *Sylvia atricapilla* occupies a wider range of habitats on the island as already shown, and this is associated with a wider range of foraging heights. Thus, both habitat niche and foraging niche are broader on the island. *Sylvia sarda* is an insular endemic and a very versatile bird for both habitat choice (see Table IV) and foraging height. It feeds largely on the ground and in the lowest parts of the vegetation, especially in

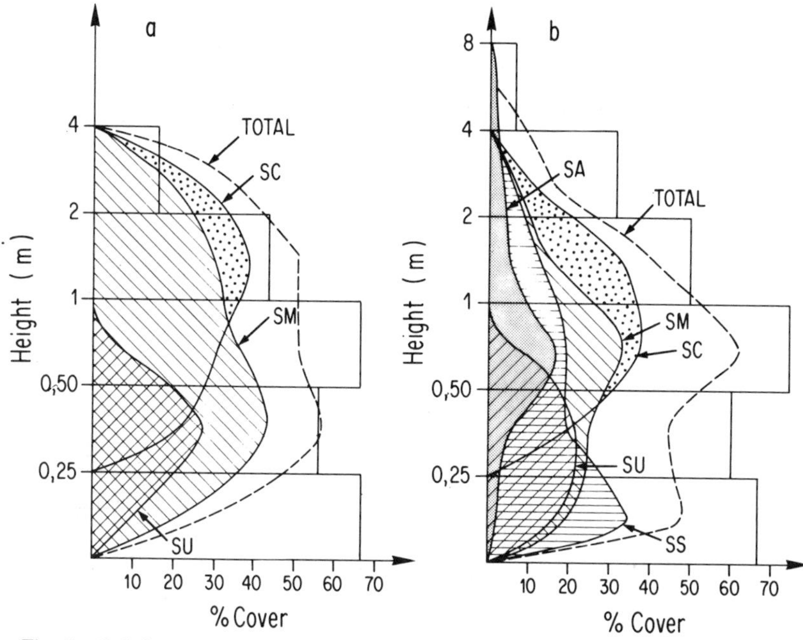

Fig. 4. Relation between the foraging-height distributions of the *Sylvia* warblers and vegetation profiles (percentage of cover, histograms) in two similar sites in Provence (a) and in Corsica (b). Notice that there are more coexisting species in the insular site and that there is no definite shift in Corsica for the three species which occur in the two regions. *S. atricapilla* (SA), *S. melanocephala* (SM), *S. cantillans* (SC), *S. undata* (SU), *S. sarda* (SS).

Cistus monspelliensis, but can also be found up to the top of trees (Berthold and Berthold, 1973).

2. The three remaining species which occur in Provence and in Corsica show nearly the same patterns of habitat utilization. *Sylvia undata* forages mainly in the stratum 0.25–0.50 m; *S. melanocephala* forages higher in shrubby vegetation, while the migrant *S. cantillans* searches for its food mainly in the canopy of the highest bushes and in trees, especially in the Holm Oak (*Quercus ilex*), where it takes advantage of the spring flush of insects (Blondel, 1969). The three sedentary species take a large part of their diet on the ground. From the figure it is obvious that there is no within-habitat displacement in Corsica for *S. undata* and *S. cantillans,* since these species have exactly the same profiles in Provence and Corsica. In particular *S. undata* is not displaced by *S. sarda* in Corsica, although these two species are very closely related. But populations of *S. undata* are much lower on the island, which may be an indication of competition between the two. *Sylvia melanocephala* forages slightly higher in Corsica than in Provence, perhaps as a result of some competition with *S. undata* and *S. sarda,*

but the overall pattern is not very different from that observed on the mainland. *Sylvia melanocephala* and *S. sarda* are the two most versatile species, and intraspecific variance was found to be especially high for all parameters: habitat utilization, foraging height, extension of breeding season, and diet (Blondel, 1969, Zbinden and Blondel, 1981, Cody and Walter, 1976).

For the seven species studied there is a significant correlation between measures of habitat utilization and both foraging heights and the nest heights in the vegetation. Thus, as for *Sylvia atricapilla* mentioned previously, foraging height reflects relative habitat preference. Species which occupy low habitats search for their food on the ground and in the lower part of the vegetation, while species which prefer taller habitats forage higher. Berthold and Berthold (1973) and Cody and Walter (1976) came to the same conclusion.

For each parameter, I measured niche overlap using the procedure of Cody (1974) and defined overall niche overlap by the product of the three α's ($\alpha_{habitat} \times \alpha_{foraging\ height} \times \alpha_{nest}$), assuming that the probabilities of a joint utilization of resources by species pairs is the product of the separate probabilities. Although such a procedure is empirical and only a rough approximation of resource utilization, the product of the α's were assembled in community matrices, from which dendrograms of niche overlap relations were drawn (Fig. 5). The species are ordered in this dendrogram in the same way as their ranking on the habitat gradient (Table IV), which is not surprising, since habitat utilization is one of the three parameters employed and since there are parallels between habitat use and foraging height. Thus, the sequence of the species among the gradients of vegetation (Table IV: see also p. 214 in Cody and Walter, 1976) is also the sequence of mean foraging heights. These sequences are *Sylvia conspicillata, S. undata, S. melanocephala, S. cantillans, S. hortensis, S. atricapilla* for Provence and *S. sarda, S. undata, S. melanocephala, S. cantillans, S.*

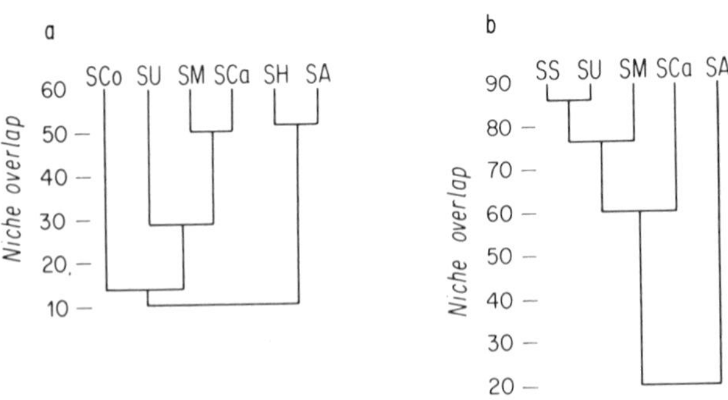

Fig. 5. Dendrogram of ecological relations among the warbler guild in Provence (a) and in Corsica (b). Data calculated from alpha products for habitat, foraging height, and nest location. Symbols as in Fig. 4.

atricapilla for Corsica. The figures show that overall niche overlap is much higher on the island.

Despite the fact that overall niche overlap is higher in Corsica, I found no clear-cut differences in resource utilization between Corsica and continental populations, except that *Sylvia melanocephala* is ecologically closer to *S. cantillans* in Provence than in Corsica. But I did not find any evidence of close interactions and interspecific aggression between *S. melanocephala* and *S. undata* either in Provence or in Corsica, which contrasts to the findings of Cody and Walter (1976; cf. also Chapter 3, this volume). These authors found that in Sardinia the proportion of space monopolized by a single species is far greater than expected on the grounds of a random dispersion of species' territories with respect to each other, given the interspecific differences in habitat use. My own experience suggests that the distribution of territories is more related to microhabitat heterogeneity than to interspecific interaction.

It appears that in Mediterranean *Sylvia* species there is a considerable amount of ecological overlap which is nearly the same on the mainland as on the island. It is suggested by Blondel (1982) that Mediterranean warblers speciated in recent times, during the Pleistocene, in separate patches of matorral which were ice free throughout the glacial periods. Severe degradation of the vegetation due to man's activities over more than 6000 years brought about a dramatic extension and homogenization of matorral with a closer spatial contact between these species, as a consequence. The species were ecologically isolated in such a way that their coexistence is now possible without the need of character displacement or convergence. I suggest that the circumstances in which these species evolved, i.e., in more or less isolated patches of vegetation within the Mediterranean basin, explain why insular populations do not show any indication of ecological shifts in comparison with the mainland. In this respect, the obvious changes in habitat selection in *S. atricapilla* are illuminating because it is the only warbler of the Corsican gradient which did not evolve in Mediterranean matorral.

B. Tits of the Genus *Parus*

I shall very briefly summarize here some features of the comparative biology of tits on the mainland and in Corsica which are relevant to the theme of this chapter.

As shown in Table V, the distributional patterns of tits on the two homologous habitat gradients show striking differences that contrast sharply with what is observed in the warblers. (1) Four species breed on the mainland and three in Corsica, where *Parus cristatus* has never bred (although suitable habitats are not lacking); (2) There is a very large broadening of the habitat niche in the three insular populations; (3) In all cases but one, average densities of tits per habitat are much higher on Corsica than on the mainland. At the gradient-wide level,

TABLE V

Distribution and Densities (Breeding Pairs/10 ha) of Tits of the Genus *Parus* in Homologous Habitat Gradients in Provence and in Corsica[a]

Distribution	Density at stage								
	1	2	3	4	5	6	7	$e^{H'}$	g
Provence									
P. major					2.16	3.08	3.20	2.96	6.10
P. caeruleus							11.55	1.00	7.00
P. ater							.18	1.00	7.00
P. cristatus							1.76	1.00	7.00
Total					2.16	3.08	16.67	Mean 1.49	6.78
Percentage *Parus*					6.0	8.0	26.0		
Corsica									
P. major		1.60	1.70	2.55	3.59	2.62	4.67	5.59	5.07
P. caeruleus			.52	.20	3.33	7.86	14.15	2.92	6.34
P. ater					1.25	2.10	4.10	2.68	6.38
Total		1.60	2.22	2.75	8.17	12.58	22.92	Mean 3.73	5.93
Percentage *Parus*		5.5	7.5	6.8	16.6	24.4	35.9		

[a] $e^{H'}$ = niche breadth; g = barycenter of the species' distributions ($g = \Sigma id_i/\Sigma d_i$ where i is the stage number of the gradient and d_i the density of the species).

there are more than twice as many birds on the island as on the mainland, hence a larger amount of density compensation.

The distributional patterns in tits are completely different from those of the warblers (see Table IV), since the warblers, with comparable mainland–island diversity, exhibit no obvious habitat–niche broadening except in *Sylvia atricapilla* and no increase in specific densities on the island. I have unfortunately no precise data on within-habitat partitioning of these tits in Provence and Corsica, but circumstantial evidence suggests that there are large differences in foraging habitats and diet, at least for the Blue Tit (*Parus caeruleus*), which has been more extensively studied than the others. It was shown by Michelland (1982) that (1) the range of prey items of this species is wider on the island that on the mainland, and (2) there is a sexual dimorphism in prey selection, with males taking, on average, larger prey that females, a difference noted on the island only. Furthermore, spiders make up a much larger proportion of the diet on the island; more casual observations indicate that there is a substantial broadening of the vertical foraging range on the island. For example, Blue Tits were seen to forage actively on the ground and in the low *Cistus* bushes, as well as in the canopies of large evergreen oaks (*Quercus ilex*) in Corsica, whereas on the mainland this species forages mostly on the leaves and twigs of broadleaved Downy Oaks (*Q. lanuginosa*).

But the question is raised whether these shifts are evolutionary adaptations to enlarge the niche of the birds on the island or whether they are ecological adjustments to changes in insular arthropod abundances; the causality of niche expansion would be different in the two cases. Some reports, for instance those of Drewry (1970), Janzen (1973a,b), and Allan *et al.*, (1973), have shown that arthropod communities on Caribbean islands are much impoverished and comprise an unusually high proportion of predators such as spiders. Although I have no precise data on the food supplies of titmice on Corsica, the same could be true there. A reduced food supply and a release in interspecific competition leads to a broadening of habitat and foraging-height utilization. Changes in foraging behavior and diet of Blue Tits on the island, as well as the sexual dimorphism in prey selection, supports the view that the niche shift is in this case an evolutionary adaptation to minimize intraspecific competition in these crowded populations.

Changes in diet may be accompanied by a change in bill structure and body size. Although an expanded diet is often related to increases in bill and body size, especially in differentiated island populations (Cody, 1974), Corsican Blue Tits are on average 12% lighter than mainland relatives, in opposition to the general trend. On the other hand, selection for small size could be an adaptation to a reduced productivity of prey populations on the island and to a high population density in the insular stable environment. Thus, small size could be advantageous despite a reduced number of competitors (Schoener, 1970). Actually, preliminary observations suggest that the variance in morphological characters is larger on the island than on the mainland, which may explain a broadening of the island niches. Anyway, morphological differences between insular and continental tit populations are pronounced, with subspecific endemics on Corsica of both Blue Tits (*Parus caeruleus ogliastrae*) and Coal Tits (*Parus ater sardus;* Vaurie, 1959).

In summary, the situation of the tits sharply contrasts with that of the *Sylvia* warblers. For the latter there was no clear-cut indication of within-habitat displacement nor changes in resource partitioning between the mainland and the island, whereas for the former there are strong changes in habitat utilization. Such changes in forest species on the island must be related to a much greater impoverishment in species numbers in the forests than in the matorral. Actually, there are on the average more species in the island matorral than in the mainland equivalent, whereas the opposite is true for old forests: 18 species in the Corsican forest against 23 in the mainland forest (Table III). Thus, for the same foliage density in the vertical plane, there are fewer passerine species on the island than on the mainland, but many more individuals per species since the overall densities of the communities are nearly the same. Thus, niche expansion could be interpreted as a result of relaxation of interspecific competition and increase in intraspecific competition.

VI. HABITAT SELECTION AND EVOLUTIONARY CHANGES

Evolutionary changes in insular populations may include any shift in morphological, behavioral, or physiological characters. Since an island population is founded with a reduced genome relative to that of the mainland population it comes from, differences can be ascribed either to founder principle or to local selection. Actually, the distinction is probably not a clear-cut one, because the local effects of selection remodel the genome in the same direction whatever its initial divergence from the genome of the source population. The problem is to discover to what extent evolutionary changes are adaptations to insular environments. Morphological shifts may be increases or decreases in such characteristics as bill shape (length, width, depth), tarsus, wing length, and body size. But there is no definite theory on their causality. Such shifts have obvious ecological significance, since these morphological characteristics are utilized in foraging activities and hence for microhabitat selection (Ricklefs and Cox, 1977). Interspecific ratios of bill size or body size are related to overlap in prey size (Storer, 1966; Hespenheide, 1971), and there seems to be a strong relationship between wing shape and leg length and the feeding sites exploited by birds (Karr and James, 1975).

As far as morphology is concerned, three main hypotheses can account for evolutionary changes on islands: (1) Modifications arise through new patterns of competition (character displacement). (2) Changes evolve due to differences in habitat structure and food resources. (3) Chance accounts for the shifts.

In so far as morphological shifts are ascribed to new interspecific interactions among different sets of species on islands, changes in numbers of species and competitors on islands could abruptly change character values from those optimal on the mainland in the context of other sets of species. Such new values could be quickly reached through genetic variations and isolation (Cody, 1974). Perhaps the most frequent shift in island populations of birds is a trend to increased bill length on islands. Yeaton (1974) reports that the Scrub Jay (*Aphelocoma caerulescens*), in the absence of its mainland competitor the California Thrasher (*Taxostoma redivivum*), has a longer bill on Santa Cruz Island (32.5 mm versus 26.4 mm on the mainland), which presumably enables the jay to forage in the microhabitats which would otherwise be utilized by the thrasher. Abbott (1977) has shown that, in the absence of competitors, single species in Tasmania tend toward increased lengths of bill, tarsus, and wing. Other examples are given by Keast (1968) for Tasmanian land birds, by Johnson (1972) for the birds of the Channel Islands off California, and by Higuchi (1976) for the tit *Parus varius* on islets off the coast of Japan. Generally speaking, most herbivores, omnivores, and insectivores show larger body sizes on islands (Cody, 1974). Absence of large predators and large competing herbivores together with the simplicity of island communities are thought to be strong selective pressures for large size (Arnold, 1979). But morphological increase is not a general rule for insectivorous and herbivorous species as evidenced by the Corsican Blue Tit, which is definitely smaller than the mainland conspecfic. For herbivorous species Lack (1970) gives evidence that island subspecies of ducks are smaller than their mainland relatives.

There are in fact many cases of decreased size of morphological characters on islands. Examples are raptor species that feed upon larger prey. They decrease in body size when they colonize habitats from which their smaller competitors are absent (Cody, 1974). A simple explanation for smaller size on islands is that limited food resources give the advantage to smaller individuals, especially in crowded populations showing a strong amount of density compensation. In this context, selection for small size also allows populations to be larger and, hence, less vulnerable to stochastic extinction.

Perhaps more important than absolute differences in morphological measurements is their variability. Increased morphological variability allows the population to be more versatile in habitat and resource utilization, which in turn are prerequisites for niche enlargement on islands: a jack-of-all-trades is a master of none! Changes and increased variability in bill length on islands have been related to a reduced number of competitors and to an increased ecological versatility. Indeed, Van Valen (1965) found a higher variability in morphological measurements for six species in the Atlantic islands (but only 6 out of 14 tests performed were significant). The trend of increased morphological variability on islands is not ubiquitous, however, since subsequent workers failed to confirm it on other islands (see Abbott, 1980). Hypothetically, a longer bill is more efficient for dealing with the wider range of prey items available on islands and left unused by competitors (Lack, 1947; Grant, 1965; Keast, 1968, 1970; Yeaton, 1974; Sheppard *et al.*, 1968). Differences in tarsus length between island and mainland populations have been related by Grant (1965, 1971) to a larger spectrum of microhabitats utilized. The competition hypothesis is further supported by the fact that congeneric species on islands sometimes show exaggerated differences in bill morphology (Schoener, 1965; Grant, 1966a). It is surprising that very few attempts have been made to relate morphological shifts on islands to intraspecific competition, which must be quite severe in crowded populations exhibiting a large amount of density compensation. If intraspecific pressures are likely to be stronger on islands, broadening the range of habitats occupied and resources utilized, then morphological shifts may be accompanied by a greater variability. Sexual dimorphism in prey selection such as that in Corsican Blue Tits (see Section V,B) could have a morphological basis and promote niche expansion at the population level (Selander, 1966).

An interesting question is whether morphological variation is achieved by species independently or is a community-wide phenomenon of coevolution leading to particular coadapted assemblages of species such as those described by

Diamond (1975) in his "assembly rules." Community-wide morphological coadaptation could be produced either by a given set of species, which adjust their morphology through natural selection until the new community reaches a viable configuration, or by species' assorting nonrandomly through interactive mechanisms of immigration–extinction processes until species sets are compatible (Case and Sidell, 1983). In both cases some kind of competition is at work, either through character displacement or competitive exclusion. In this view, size assortments in communities must arise from a greater persistence of some species combinations in which compatible morphologies provide a resistance to invasion.

Another possible source of evolutionary changes on islands emphasizes differences in habitat structure and the distribution of prey size and its variance. For instance Abbott *et al.* (1977) noticed that the bill length of the largest beaked *Geospiza* species breeding on islands in the Galapagos correlates with the abundance of large hard seeds and fruits available. There is little support for this food size hypothesis because of few comparative measurements of food supply between the island and the nearby mainland. In Tasmania the Boobook Owl (*Ninox novaeseelandiae*), which eats vertebrates, has a wing 20 mm shorter than in mainland Victoria, but the Masked Owl (*Tyto novaehollandiae*), which eats insects, has a wing 20 mm longer than on the mainland (Mees, 1964). This could be related to the fact that island insects are proportionately larger than those on the mainland (Janzen, 1973a). Another example is *Parus varius*, which is larger on the island as a result of its adaptation to eat large nuts during the winter (Higuchi, 1976), contrasted with the Blue Tit, which is smaller in Corsica than on the mainland. To determine the causality of shifts to either smaller or larger island sizes we need to know much more about microhabitat characteristics and food resources on each particular island.

Finally, morphological changes may be due to chance only. Strong *et al.* (1979) analyzed Galapagos finches to determine if the body and bill sizes of sympatric species were more different than a random assemblage of species might predict. Using Monte Carlo protocols, they found a close resemblance between expected and observed morphological ratios and concluded that random colonization explains the size structure of Galapagos finch communities better than competition. But Case and Sidell (1983) reanalyzed the Galapagos finch data using values of limiting size similarity set by ecological interactions; they concluded that there is a community-wide character displacement, a conclusion reached also by Hendrickson (1981) and Grant and Schluter (1983).

The evidence of morphological shifts on islands confronts the problem of turnover rates, a keystone of the MacArthur–Wilson equilibrium theory. Jones and Diamond (1976) believe that turnover on islands must always occur, although it may have been increased by human impact on habitats. But shifts in morphology and habitat selection patterns, obvious for so many species, clearly indicate that many populations remain established for a very long time without turnover (see Gilbert, 1980). Clearly, species turnover does exist, and for certain categories of species on small temperate islands, the rates can be high (Morse, 1977; Väisänen and Järvinen, 1977); turnover may be much less apparent on large islands where there is often a high degree of subspecific endemism.

VII. TOWARD A SYNTHESIS

I return now to the fundamental question: are habitat selection patterns modified on islands? The answer is definitely yes, but the mechanisms and the biological significance of such changes are far from clear. Most studies interpret the differences in insular avifaunas via interspecific competition and assume that the availability of resources and habitats does not differ between islands and mainland. MacArthur and Wilson (1967) wrote that increased population densities on islands are "the most convincing evidence of interspecific competition." In studies where population densities are taken into account (Abbott, 1975b; Cox and Ricklefs, 1977; Crowell, 1962; Diamond, 1970a; Grant, 1966b; MacArthur *et al.*, 1972; Morse, 1977; Nilsson, 1977; Yeaton, 1974; Yeaton and Cody, 1974; Emlen, 1978; Keast, 1970; Case *et al.*, 1979; Blondel, 1979, 1981), increases in population density are generally interpreted as resulting from changes in competitive interactions on islands. The term "density compensation," coined by MacArthur *et al.* (1972), implicitly assumes competition as a causal agent for such shifts. Cases of "excess density compensation" have been discussed by MacArthur *et al.* (1972), Case (1975), Diamond (1975), and Case *et al.* (1979). Diamond describes this phenomenon for "supertramp" bird species in species-poor islands subject to high variability in habitats. These supertramp species have broad and unspecialized habitat preferences, high reproductive potential and high dispersal ability. According to Diamond (1975) and Case *et al.* (1979), supertramps are competitively excluded from species-rich islands by *K*-selected species which harvest resources to levels lower than those tolerable by supertramps.

Thus, it is indisputable that competition plays an important role in shaping insular communities, and many observations and field studies support this view (Connell, 1983; Schoener, 1983). But alternative explanations must still be considered. One difficulty is with the definition of competition, used indiscriminately for the process and the results and for direct interactions, as well as for ecological shifts and evolutionary changes. The changes in phenotype which occur where two related species coexist "result in the avoidance of competition and do not suggest that competition is a major force in the day-to-day working of island ecosystems" (Williamson, 1981). These difficulties are partly overcome by the term "apparent competition," introduced by Holt (1977), which could perhaps better be replaced by the term "potential competition." Partly because of these difficulties, there has been a trend to minimize the role of competition and to emphasize the role of chance (see, e.g., *The American Naturalist,* November 1983). But this standpoint is as extreme as that which considers competition the only agent of community organization. Comprehensive reviews by Connell (1983) and Schoener (1983) show that some kind of competition is revealed by most of the tests that use controlled field experiments. The MacArthur–Wilson model has been and remains extremely useful to direct field work and generates testable hypotheses, but now it must be acknowledged that this is an oversimplified and somewhat controversial model (Williamson, 1981). MacArthur and Wilson were quite aware of the limits of their model when they wrote in the preface of their book, "We do not seriously believe that the particular formulations advanced in the chapters to follow will fit for very long the exacting results of future empirical investigations." Islands and their biotas are not interchangeable functional units; they differ from one another in resources, habitat diversity, proportion of predators, location in relation to land masses, colonizing abilities of the species of the continental pool, and so on. Each set of parameters does not work in the same manner for islands so far studied in North America, Central America, Australia, and Europe.

The following cautionary views are offered with respect to habitat selection studies on islands.

1. Although they closely interplay, interspecific and intraspecific competition act in a somewhat opposite way for many specific life history parameters and therefore have different ecological and evolutionary consequences. As was pointed out a long time ago by Svärdson (1949), intraspecific competition tends to broaden the spectrum of habitat and resource utilization, whereas interspecific competition tends to contract it. In this context it is surprising that so many authors invoke only interspecific competition to account for habitat and niche shifts on islands; on islands where the land-bird fauna is impoverished, intraspecific competition can be expected to enlarge habitat occupancy and niche breadth, especially if food utilization is likewise expanded. I explain the sexual dimorphism of prey selection in the crowded Blue Tit populations on Corsica by reference to intraspecific pressures.

2. Many studies examine only species numbers and not population densities and demographic parameters. Some species, uncommon on the mainland, are poor colonists, not because of dispersal failure or intraisland competition or otherwise unsuitable island environments, but simply because their initial propagules fail to reach populations of a viable size. Examples of species missing from Corsica because of demographic failure may be *Pica pica* and *Sylvia hortensis*.

3. Much emphasis has been laid purposely on differences between conspecific

populations of islands and the mainland, but shifts in habitat selection, resource utilization, and abundances are not a general rule for all species (namely, the Mediterranean warblers); changes in densities and species-specific attributes are not a universal prerequisite for survival on islands. As rightly pointed out by Cody (1974, 1983), many field studies, including those in Corsica discussed in this chapter, show that overall densities can be attributed to density increases of just a few species. These are presumably smaller-sized generalists predisposed for niche expansion.

4. Detailed studies on habitat suitability and diversity in islands by comparison with adjacent mainland areas are lacking in most cases. One consequence of an island's being small is that its habitats may be less diverse and certainly of reduced area than those of a larger island or of mainlands. Clearly, the broad habitat descriptions such as those given by Diamond (1975) for the Bismark archipelago or by Watson (1964) in the islands of the Aegean Sea are too crude to justify species-specific explanations of absences from islands (see Terborgh *et al.*, 1978; Abbott, 1980). The extent to which islands and the nearby mainland really match one another in habitats and ecological resources is poorly understood. Because numbers of species of plants and insects are reduced on islands, resources available for birds may be likewise reduced, but this is no more than an unchecked assumption. Fewer insect species do not necessarily mean a less abundant food supply, although Janzen's (1973a, 1973b) work suggests this interpretation. Insect biomass was found by Yeaton (1974) to be 20% greater on Santa Cruz Island than in a chaparral site in California; island and mainland resources must be quantified by exact comparative measurements.

More specialized species may be excluded from islands because their specific food is absent: the absence of the Tawny Owl (*Strix aluco*) from Ireland can be attributed to the absence of voles there (Williamson, 1981). Power (1972) showed plant species numbers affect the numbers of bird species on the islands off California, and Bowman (1961) suggested that differences in the floristic composition of the different Galapagos Islands could account for absence or presence of certain Darwin's Finches, a serious alternative to the competition paradigm. This explanation may apply to many highly specialized birds in the tropics (Abbott, 1980); the distribution of species-specific requirements may explain better than anything else the presence or absence of bird species.

VIII. CONCLUDING REMARKS

At the present state of the art it is difficult to decide which mechanisms are responsible for changes in habitat selection patterns on islands. Therefore, a synthesis is premature before sound evaluation is made of the respective roles of such ecological processes as inter- and intraspecific competition, resources and habitat availabilities, microstructural changes in vegetation, the role of chance, and so on. Island avifaunas are often strongly modified in comparison with those of homologous continental areas. But besides the bird comunities many other features such as climate, predators, and resources are also modified. Habitat selection patterns are just one component of a highly integrated set of adjustments and adaptations which allow the species to cope with new environmental conditions. There are undoubtedly large shifts in habitat utilization by land birds on islands at both between- and within-habitat levels, and in many of these cases interspecific competition plays a role. But it is difficult to experiment with birds, and most of the evidence for competition between birds is indirect. So-called natural experiments are not true experiments involving artificial modifications of ecological factors and removal or addition of populations of birds or their prey. Indeed, the correct procedure in scientific investigation, the adoption or refutation of a null hypothesis, is extremely difficult to follow in this field (Grant and Abbott, 1980).

Habitat selection patterns on islands are interpreted with reference to the following factors: (1) Increased densities. There are often unusually high densities in confined areas, and density compensation processes can account for these. Extinction probabilities are a function of population size, and thus there is a selective advantage to increased densities on islands. Since the size of each population is likely a decreasing function, on average, of the total species number, here again is an advantage for island communities in having low numbers of coexisting species. (2) Selection for small size. Since in habitats of the same area larger birds have larger territories and hence lower densities than smaller birds (Schoener, 1968), they are more vulnerable to extinction and are underrepresented on islands. Larger species are selectively disfavored in the process and also disfavored through evolutionary processes leading to smaller size of birds on islands (MacArthur *et al.*, 1972). (3) Carnivores are often large, and there are often proportionately fewer predators on islands than on the mainland. This in turn allows their prey species to be more abundant and to enlarge their spectrum of resources utilization. Connell (1975) summarized evidence that predation rather than competition has the larger effect on community structure. Predation is undoubtedly a powerful pressure for an optimal spacing pattern of breeding passerines (Andersson and Wiklund, 1978; see also Chapter 3, this volume), and reduced predation on islands could have effects on niche expansion and demography of passerine birds. (4) Selection for sedentarity. High sedentarity is reflected by the degree of subspecific differentiation, which can be high. More than half of the species of the old forests of Corsica discussed in this chapter are subspecific endemics (Ferry *et al.*, 1976). (5) Selection for lower fecundity. Island populations have a lower fecundity than conspecific populations on the mainland (Cody, 1971; Crowell and Rothstein, 1981; Blondel and Isenmann, 1979). Along with a higher adult survival, this leads to a slowing down of population turnover and an increased stability of the populations, a selective advantage in isolated biotas.

Such sets of adaptations are highly integrated into an insular syndrome that must make island communities more stable than continental ones; yet island populations are particularly vulnerable to all forms of introductions (Elton, 1958; Lawlor and Maynard-Smith, 1976). This and the high endemism recorded on many islands suggest that established populations show a very low turnover, in contrast to the most important point of the MacArthur–Wilson theory. Mixtures of species on islands exhibit disharmonic patterns both at levels of taxonomical and of trophical composition by comparison with those on the mainlands, and this is all the more true on remote oceanic islands where species can be derived from a variety of biogeographical regions.

APPENDIX

Small and medium-sized land birds which are widespread in Mediterranean Europe and Northern Africa are missing as breeding birds in Corsica. Long-distance migrants are indicated by asterisks in the tabulation below.

*Falco naumanni**	*Sylvia hortensis**
*Coracias garrulus**	*Phylloscopus collybita*
Alcedo atthis	*Phylloscopus bonelli**
Picus viridis	*Oenanthe hispanica**
Picoides minor	*Phoenicurus ochruros*
Melanocorypha calandra	*Phoenicurus phoenicurus**
Galerida cristata	*Sitta europaea*
*Riparia riparia**	*Certhia brachydactyla*
Motacilla alba	*Emberiza cia*
*Motacilla flava**	*Emberiza hortulana**
Lanius excubitor	*Pica pica*
*Lanius minor**	*Pyrrhocorax pyrrhocorax*
*Hippolais polyglotta**	*Corvus monedula*
*Sylvia communis**	

Species are listed in the tabulation below which attempted to breed in Corsica but failed to establish themselves as regular breeders. Proven cases of breeding at least once but no more than 5 times are indicated by asterisks (J. Blondel, unpublished data; Thibault, 1983).

Podiceps nigricollis	*Picoides minor*
Ixobrychus minutus	*Galerida cristata*
*Egretta garzetta**	*Tringa hypoleucos*
*Ardea cinerea**	*Hirundo daurica**

(*continued*)

(*Continued*)

*Falco naumanni**	*Motacilla flava**
*Anas clypeata**	*Motacilla alba**
Himantopus himantopus	*Sylvia hortensis**
Burhinus oedicnemus	*Sylvia communis**
*Scolopax rusticola**	*Phylloscopus bonelli**
*Sterna hirundo**	*Certhia brachydactyla*
Bubo bubo	*Pica pica*
*Asio otus**	*Corvus monedula*
Coracias garrulus	*Montifringilla nivalis**
Picus viridis	*Carduelis spinus**
Dryocopus martius	*Sturnus vulgaris**

Possible explanations for absences of bird species from Corsica (sea birds excluded) are tabulated below.

Corsica out of the geographical range	Lack of suitable habitats or habitats too small	Competition for food and/or space
Pernis apivorus	*Ixobrichus minutus*	*Neophron percnopterus*
Tetrao tetrix	*Egretta garzetta*	*Hieraaëtus fasciatus*
Tetrastes bonasia	*Nycticorax nycticorax*	*Circaetus gallicus*
Aegolius funereus	*Botaurus stellaris*	*Milvus milvus*
Dryocopus martius	*Otis tetrax*	*Circus pygargus*
Anthus trivialis	*Burhinus oedicnemus*	*Falco naumanni*
Prunella modularis	*Tringa totanus*	*Alectoris graeca*
Turdus philomelos	*Himantopus himantopus*	*Scolopax rusticola*
Sylvia borin	*Recurvirostra avosetta*	*Tringa hypoleucos*
Sylvia curruca	*Larus ridibundus*	*Columba oenas*
Phylloscopus trochilus	*Sterna hirundo*	*Bubo bubo*
Parus palustris	*Sterna albifrons*	*Strix aluco*
Total = 12	*Clamator glandarius*	*Asio otus*
	Alcedo atthis	*Coracias garrulus*
	Riparia riparia	*Picus viridis*
	Turdus torquatus	*Picoides minor*
	Total = 16	*Melanocorypha calandra*
		Galerida cristata
		Motacilla alba
		Motacilla flava
		Lanius minor
		Lanius excubitor
		Oenanthe hispanica
		Phoenicurus phoenicurus
		Phoenicurus ochruros
		Hippolais polyglotta
		Sylvia hortensis
		Sylvia communis
		Phylloscopus collybita

(Continued)

Corsica out of the geographical range	Lack of suitable habitats or habitats too small	Competition for food and/or space
		Phylloscopus bonelli
		Muscicapa hypoleuca
		Parus cristatus
		Sitta europaea
		Certhia brachydactyla
		Emberiza cia
		Emberiza hortulana
		Emberiza schoeniclus
		Pyrrhula pyrrhula
		Sturnus vulgaris
		Corvus monedula
		Pica pica
		Pyrrhocorax pyrrhocorax
		Total = 42

REFERENCES

Abbott, I. (1973). Birds of Bass Strait. Evolution and ecology of the avifaunas of some Bass Strait islands, and comparisons with those of Tasmania and Victoria. *Proc. R. Soc. Victoria* **85,** 197–223.

Abbott, I. (1975a). Coexistence of congeneric species in the avifaunas of Australian islands. *Aust. J. Zool.* **23,** 487–494.

Abbott, I. (1975b). Density and species diversity of bird populations in *Eucalyptus* forests in Victoria, Bass Strait islands and Tasmania. *Proc. R. Soc. Victoria* **87,** 187–196.

Abbott, I. (1977). The role of competition in determining morphological differences between Victorian and Tasmanian passerine birds. *Aust. J. Zool.* **25,** 429–447.

Abbott, I. (1978). Factors determining the number of land bird species on islands around South-Western Australia. *Oecologia* **33,** 221–233.

Abbott, I. (1980). Theories dealing with the ecology of land birds on islands. *Adv. Ecol. Res.* **11,** 329–371.

Abbott, I., and Grant, P. R. (1976). Non-equilibrium bird faunas on islands. *Am. Nat.* **110,** 507–528.

Abbott, I., Abbott, L. K., and Grant, P. R. (1977). Comparative ecology of Galapagos ground finches (*Geospiza* Gould): Evaluation of the importance of floristic diversity and interspecific competition. *Ecol. Monogr.* **47,** 151–184.

Allan, J. D., Barnhouse, L. W., Prestbye, R. A., and Strong, D. R. (1973). On foliage arthropod communities of Puerto Rico second growth vegetation. *Ecology* **54,** 628–632.

Andersson, M., and Wiklund, C. G. (1978). Clumping versus spacing out: Experiments on nest predation in Fieldfares (*Turdus pilaris*). *Anim. Behav.* **26,** 1207–1212.

Arnold, E. N. (1979). Indian Ocean giant tortoises: Their systematics and island adaptations. *Philos. Trans. R. Soc. London Ser. B* **286,** 127–145.

Berthold, P., and Berthold, H. (1973). Zur Biologie von *Sylvia sarda balearica* und *S. melanocephala. J. Ornithol.* **114,** 79–95.

Blondel, J. (1969). ''Synécologie des Passereaux résidents et migrateurs dans le Midi méditerranéen français.'' Centre Régional de Documentation Pédagogique, Marseille.

Blondel, J. (1979). "Biogéographie et Ecologie." Masson, Paris.

Blondel, J. (1980). L'influence du morcellement des paysages sur la structure des communautés. *Acta Oecologica/Oecol. Gener.* **1,** 91–100.

Blondel, J. (1981). Structure and dynamics of bird communities in Mediterranean habitats. *In* "Ecosystems of the World, Vol. 11: Mediterranean-type Shrublands" (F. di Castri, D. W. Goodall, and R. L. Specht, eds.), pp. 361–385. Elsevier, Amsterdam.

Blondel, J. (1982). Caractérisation et mise en place des avifaunes dans le Bassin méditerranéen. *Ecologia Mediterranea* **8,** 253–272.

Blondel, J., and Cuvillier, R. (1977). Une méthode simple et rapide pour décrire les habitats d'oiseaux: Le stratiscope. *Oikos* **29,** 326–331.

Blondel, J., and Frochot, B. (1976). Caractères généraux de l'avifaune Corse. Effets de l'insularité et influence de l'homme sur son évolution. *Rev. Soc. Sci. Hist. Nat. Corse.* **96,** 63–74.

Blondel, J., and Isenmann, P. (1979). Insularité et démographie des mésanges du genre *Parus* (Aves). *C. R. Hebd. Séances Acad. Sci.* **289,** 161–164.

Bowman, R. I. (1961). Morphological differenciation and adaptation in the Galapagos finches. *Univ. Calif. Publ. Zool.* **58,** 1–302.

Carlquist, S. (1974). "Island Biology." Columbia Univ. Press, New York.

Case, T. J. (1975). Species numbers, density compensation, and the colonizing abilities of lizards in the gulf of California. *Ecology* **56,** 3–18.

Case, T. J., and Sidell, R. (1983). Pattern and chance in the structure of model and natural communities. *Evolution* **37,** 832–849.

Case, T. J., Gilpin, M. E., and Diamond, J. M. (1979). Overexploitation, interspecific competition and excess density compensation in insular faunas. *Am. Nat.* **113,** 843–854.

Chappuis, C. L. (1976). Origine et évolution des vocalisations de certains oiseaux de Corse et des Baléares. *Alauda* **44,** 475–495.

Cody, M. L. (1971). Ecological aspects of reproduction in birds. *In* "Avian Biology," Vol. 1 (D. Farner and J. King, eds.), pp. 462–503. Academic Press, New York.

Cody, M. L. (1974). "Competition and the Structure of Bird Communities." Princeton Univ. Press, Princeton, New Jersey.

Cody, M. L. (1983). Bird diversity and density in South African forests. *Oecologia* **59,** 201–215.

Cody, M. L., and Walter, H. (1976). Habitat selection and interspecific interactions among Mediterranean sylviid warblers. *Oikos* **27,** 210–238.

Connell, J. H. (1975). Some mechanisms producing structure in natural communities: A model and evidence from field experiments. *In* "Ecology and Evolution of Communities" (M. L. Cody and J. M. Diamond, eds.), pp. 460–490. Harvard Univ. Press, Cambridge, Massachusetts.

Connell, J. H. (1983). On the relative prevalence and relative importance of interspecific competition: Evidence from field experiments. *Am. Nat.* **122,** 661–698.

Connor, E. F., and McCoy, E. D. (1979). The statistics and biology of the species–area relationship. *Am. Nat* **113,** 791–833.

Connor, E. F., and Simberloff, D. (1978). Species number and compositional similarity of the Galapagos flora and avifauna. *Ecol. Monogr.* **48,** 219–248.

Cox, G. W., and Ricklefs, R. E. (1977). Species diversity and ecological release in Caribbean land bird faunas. *Oikos* **28,** 113–122.

Crowell, K. (1962). Reduced interspecific competition among the birds of Bermuda. *Ecology* **43,** 75–88.

Crowell, K., and Rothstein, S. I. (1981). Clutch sizes and breeding strategies among Bermudan and North American passerines. *Ibis* **123,** 42–50.

Diamond, J. M. (1969). Avifaunal equilibria and species turnover on the Channel Islands of California. *Proc. Nat. Acad. Sci. U.S.A.* **64,** 57–63.

Diamond, J. M. (1970a). Ecological consequences of island colonization by southwest Pacific birds. I. Types of niche shifts. *Proc. Nat. Acad. Sci U.S.A.* **67,** 529–536.

Diamond, J. M. (1970b). Ecological consequences of island colonization by southwest Pacific birds. II. The effects of species diversity on total population density. *Proc. Nat. Acad. Sci. U.S.A.* **67,** 1715–1721.

Diamond, J. M. (1972). Biogeographic kinetics: Estimation of relaxation times for avifaunas of southwest Pacific islands. *Proc. Nat. Acad. Sci. U.S.A.* **69,** 3199–3203.

Diamond, J. M. (1973). Distributional ecology of New Guinean birds. *Science* **179,** 759–769.

Diamond, J. M. (1974). Colonisation of exploded volcanic islands by birds: The supertramp strategy. *Science* **184,** 803–805.

Diamond, J. M. (1975). Assembly of species communities. *In* "Ecology and Evolution of Communities" (M. L. Cody and J. M. Diamond, eds.), pp. 342–444. Harvard Univ. Press, Cambridge, Massachusetts.

Diamond, J. M. (1978). Niche shifts and the rediscovery of interspecific competition. *Am. Sci.* **66,** 322–331.

Drewry, C. (1970). A list of insects form El Verde, Puerto Rico. *In* "A Tropical Rain Forest" (H. T. Odum, ed.), pp. E 129–E 150. U.S. Atom. Energ. Comm., Washington D.C.

Elliott, H. F. (1957). A contribution to the ornithology of the Tristan da Cunha group. *Ibis* **99,** 545–586.

Elton, C. S. (1958). "The ecology of invasions by animals and plants." Methuen, London.

Emlen, J. T. (1978). Density anomalies and regulation mechanisms in land bird populations on the Florida peninsula. *Am. Nat.* **112,** 265–286.

Ferry, C., Blondel, J., and Frochot, B. (1976). Plant successional stage and avifaunal structure on an island. *Proc. Int. Ornithol. Congr. 16th Canberra,* Australia, pp. 643–653.

Galli, A. E., Leck, C. F., and Forman, R. T. (1976). Avian distribution patterns within sized forests islands in central New Jersey. *Auk* **93,** 356–365.

Gamisans, J. (1975). "La végétation des montagnes Corses." Thesis, Univ. Aix-Marseille III, Marseille.

Gilbert, F. S. (1980). The equilibrium theory of island biogeography: Fact or fiction? *J. Biogeogr.* **7,** 209–235.

Grant, P. R. (1965). The adaptive significance of some size trends in island birds. *Evolution* **19,** 355–367.

Grant, P. R. (1966a). Ecological compatibility of bird species on islands. *Am. Nat.* **100,** 451–462.

Grant, P. R. (1966b). The density of land birds on the Tres Marias Islands in Mexico. I. Numbers and biomass. *Can. J. Zool.* **44,** 391–400.

Grant, P. R. (1968). Bill size, body size, and the ecological adaptations of bird species to competitive situations on islands. *Syst. Zool.* **17,** 319–333.

Grant, P. R. (1971). Variation in the tarsus length of birds in island and mainland regions. *Evolution* **25,** 599–614.

Grant, P. R., and Abbott, I. (1980). Interspecific competition, island biogeography and null hypothesis. *Evolution* **34,** 332–341.

Grant, P. R., and Schluter, D. (1983). Interspecific competition inferred from patterns of guild structure. *In* "Ecological Communities: Conceptual Issues and the Evidence" (D. R. Strong, D. Simberloff, L. G. Abele, and A. B. Thistle, eds.). Princeton Univ. Press, Princeton, New Jersey (in press).

Haila, Y., and Järvinen, O. (1983). Landbird communities on a Finnish island: Species impoverishment and abundance patterns. *Oikos* **41,** 255–273.

Hendrickson, J. A., Jr. (1981). Community-wide character displacement reexamined. *Evolution* **35,** 794–810.

Hespenheide, H. A. (1971). Food preference and the extent of overlap in some insectivorous birds, with special reference to the Tyrannidae. *Ibis* **113,** 59–72.

Higuchi, H. (1976). Comparative study on the breeding of Mainland and Island Subspecies of the Varied Tit, *Parus varius*. *Tori* **25,** 11–20.

Holt, R. D. (1977). Predation, apparent competition, and the structure of prey communities. *Theor. Popul. Biol.* **12,** 197–229.

Janzen, D. H. (1973a). Sweep samples of tropical foliage insects: Description of study sites, with data on species abundances and size distributions. *Ecology* **54,** 659–680.

Janzen, D. H., (1973b). Sweep samples of tropical foliage insects: Effects of seasons, vegetation types, elevation, time of day, and insularity. *Ecology* **54,** 681–702.

Johnson, N. K. (1972). Origin and differenciation of the avifauna of the Channel islands, California. *Condor* **74,** 295–315.

Johnson, N. K. (1975). Controls of numbers of bird species on montane islands in the Great Basin. *Evolution* **29,** 545–574.

Jones, H. L., and Diamond, J. M. (1976). Short-time-base studies of turnover in breeding bird populations on the California Channel Islands. *Condor* **78,** 526–549.

Karr, J. R., and James, F. C. (1975). Eco-morphological configurations and convergent evolution in species and communities. *In* ''Ecology and Evolution of Communities'' (M. L. Cody and J. M. Diamond, eds.), pp. 258–291. Harvard Univ. Press, Cambridge, Massachusetts.

Keast, A. (1968). Competitive interactions and the evolution of ecological niches as illustrated by the Australian honeyeater genus *Melithrepus* (Meliphagidae). *Evolution* **22,** 762–784.

Keast, A. (1970). Adaptive evolution and shifts in niche occupation in island birds. *Biotropica* **2,** 61–75.

Lack, D. (1942). Ecological features of the bird faunas of British small islands. *J. Anim. Ecol.* **11,** 9–36.

Lack, D. (1947). ''Darwin's Finches.'' Cambridge Univ. Press, London and New York.

Lack, D. (1969). The numbers of bird species on islands. *Bird Study* **16,** 193–209.

Lack, D. (1970). The endemic ducks of remote islands. *Wildfowl* **21,** 5–10.

Lack, D. (1971). ''Ecological Isolation in Birds.'' Blackwell, Oxford.

Lack, D. (1976). ''Island Biology, Illustrated by the Land Birds of Jamaica.'' Blackwell, Oxford.

Lawlor, L. R., and Maynard-Smith, M. (1976). The coevolution and stability of competing species. *Am. Nat.* **110,** 79–99.

Levins, R. (1968). ''Evolution in Changing Environments.'' Princeton Univ. Press, Princeton, New Jersey.

Lynch, J. F., and Johnson, N. K. (1974). Turnover and equilibrium in insular avifaunas, with special reference to the California Channel Islands. *Condor* **76,** 370–384.

MacArthur, R. H. (1970). Species packing and competitive equilibrium for many species. *Theor. Popul. Biol.* **1,** 1–11.

MacArthur, R. H. (1972). ''Geographical Ecology.'' Harper and Row, New York.

MacArthur, R. H., and Wilson, E. O. (1963). An equilibrium theory of insular zoogeography. *Evolution* **17,** 373–387.

MacArthur, R. H., and Wilson, E. O. (1967). ''The theory of island biogeography.'' Princeton Univ. Press, Princeton, New Jersey.

MacArthur, R. H., Recher, H., and Cody, M. L. (1966). On the relation between habitat selection and species diversity. *Am. Nat.* **100,** 319–332.

MacArthur, R. J., Diamond, J. M., and Karr, J. R. (1972). Density compensation in island faunas. *Ecology* **53,** 330–342.

Martin, J.-L. (1982). L'infiltration des oiseaux forestiers dans les milieux buissonnants de Corse. *Rev. Ecol. (Terre et Vie)* **36,** 187–210.

Martin, J.-L., and Thibault, J.-Cl. (1983). Les oiseaux de la Réserve naturelle de Scandola (Corse). Inventaire et structure des peuplements terrestres. *Bull. Ecol.* **14,** 279–296.

Mees, G. F. (1964). A revision of the Australian Owls (Strigidae and Tytonidae). *Zool. Verh.* **65,** 3–62.

Michelland, D. (1982). Survie en milieu insulaire: Quelle stratégie? Le cas des Mésanges en Corse. *Rev. Ecol. (Terre et Vie)* **36,** 187–210.

Moore, N. W., and Hooper, M. D. (1975). On the number of bird species in British woods. *Biol. Conserv.* **8,** 239–250.

Moreau, R. E. (1972). "The Palaearctic–African Bird Migrations Systems." Academic Press, London and New York.

Morse, D. H. (1971). The foraging of warblers on small, isolated islands. *Ecology* **52,** 216–228.

Morse, D. H. (1977). The occupation of small islands by passerine birds. *Condor* **79,** 399–412.

Nilsson, S. G. (1977). Density compensation and competition among birds breeding on small islands in a south Swedish lake. *Oikos* **28,** 170–176.

Power, D. M. (1972). Numbers of bird species on the California islands. *Evolution* **26,** 451–463.

Ricklefs, R. E., and Cox, G. W. (1977). Morphological similarity and ecological overlap among passerine birds on St. Kitts, British West Indies. *Oikos* **29,** 60–66.

Sauer, J. D. (1969). Oceanic islands and biogeographic theory. *Geogr. Rev.* **59,** 582–593.

Schoener, T. W. (1965). The evolution of bill size differences among sympatric species of birds. *Evolution* **19,** 189–213.

Schoener, T. W. (1968). Sizes of feeding territories among birds. *Ecology* **49,** 123–141.

Schoener, T. W. (1970). Size patterns in West Indian highlands of New Guinea. *Publ. Nuttall Ornithol. Club,* Cambridge, Massachusetts.

Schoener, T. W. (1983). Field experiments on interspecific competition. *Am. Nat.* **122,** 240–285.

Selander, R. (1966). Sexual dimorphism and differential niche utilization in birds. *Condor* **68,** 113–151.

Sheppard, D. H., Klopfer, P., and Oelke, H. (1968). Habitat selection: Differences in stereotypy between insular and continental birds. *Wilson Bull.* **80,** 452–457.

Simberloff, D. (1970). Taxonomic diversity of island biotas. *Evolution* **24,** 23–47.

Simberloff, D. (1974). Equilibrium theory of island biogeography and ecology. *Annu. Rev. Ecol. Syst.* **5,** 161–182.

Simberloff, D. (1976). Experimental zoogeography of islands: Effects of island size. *Ecology* **57,** 629–648.

Simberloff, D. (1978). Using island biogeographic distributions to determine if colonization is stochastic. *Am. Nat.* **112,** 713–726.

Simberloff, D. (1980). A succession of paradigms in ecology: Essentialism to materialism and probabilism. *Synthèse* **43,** 3–39.

Simberloff, D., and Abele, J. G. (1976). Island biogeography theory and conservation practice. *Science* **191,** 285–286.

Simpson, G. G. (1940). Mammals and land bridges. *J. Wash. Acad. Sci.* **30,** 137–163.

Slud, P. (1976). Geographic and climatic relationships of avifaunas with special reference to comparative distribution in the Neotropics. *Smithson. Contrib. Zool.* **212,** 1–149.

Soulé, M. E., and Wilcox, B. A. (1980). "Conservation Biology." Sinauer, Sunderland, Massachusetts.

Storer, R. W. (1966). Sexual dimorphism and food habits in three North American *Accipiter. Auk* **83,** 423–436.

Strong, D. R., Szyska, L. A., and Simberloff, D. (1979). Tests of community-wide character displacement against null hypothesis. *Evolution* **33,** 897–913.

Sultana, J., Gauci, C., and Beaman, M. (1975). "A Guide to the Birds of Malta." Malta Ornithol. Soc., Malta.

Svärdson, G. (1949). Competition and habitat selection in birds. *Oikos* **1,** 157–174.

Terborgh, J., Faaborg, J., and Brockmann, H. J. (1978). Island colonization by Lesser Antillean birds. *Auk* **95,** 59–72.

Thibault, J.-Cl. (1983). "Les oiseaux de la Corse." Parc Naturel Régional de la Corse, Ajaccio.

Väisänen, R. A., and Järvinen, O. (1977). Dynamics of protected bird communities in a Finnish archipelago. *J. Anim. Ecol.* **46,** 891–908.

Vandermeer, J. H. (1972). Niche theory. *Annu. Rev. Ecol. Syst.* **3,** 107–132.

Van Valen, L. (1965). Morphological variation and width of ecological niche. *Am. Nat.* **99,** 377–399.

Vaurie, C. (1959). "The birds of the Palearctic Fauna, Order Passeriformes." Witherby, London.

Voous, K. H. (1960). "Atlas of European Birds." Nelson, Amsterdam.

Watson, G. E. (1962). Three sibling species of *Alectoris* Partridges. *Ibis* **104,** 353–367.

Watson, G. E. (1964). Ecology and evolution of Passerine birds on the islands of the Aegean sea. Thesis, Yale University, New Haven, Connecticut.

Williamson, M. (1981). "Island populations." Oxford Univ. Press, Oxford.

Willis, E. O. (1974). Population and local extinctions of birds on Barro Colorado Island, Panama. *Ecol. Monogr.* **44,** 153–169.

Yeaton, R. I. (1974). An ecological analysis of chaparral and pine forest bird communities on Santa Cruz Island and mainland California. *Ecology* **55,** 959–973.

Yeaton, R. I., and Cody, M. L. (1974). Competitive release in island Song Sparrow populations. *Theor. Popul. Biol.* **5,** 42–58.

Zbinden, N., and Blondel, J. (1981). Zu Raumnutzung, Territorialität und Legebeginn Mediterraner Grasmücken (*Sylvia melanocephala, S. undata, S. cantillans, S. hortensis*) in Südfrankreich. *Der Ornithol. Beob.* **78,** 217–231.

Chapter 18

Some Theoretical Aspects of Habitat Selection

MICHAEL L. ROSENZWEIG
Department of Ecology and Evolutionary Biology
University of Arizona
Tucson, Arizona

I. INTRODUCTION

I can remember Robert MacArthur proclaiming often in conversation that any good and experienced bird watcher could walk into a habitat and tell what birds would be found living there. But for MacArthur this was not sufficient. He wanted to be able to systematize that knowledge so that it could be communicated efficiently to inexperienced students. Otherwise the patterns would go on being sensed without ever being expressed, quantified, and interpreted. I think that the great progress made in achieving these goals is cogently attested to in the preceding chapters of this volume. For example, Chapter 3 summarizes the habitat selection patterns of sylviine warblers. These species generally have very broad habitat ranges with a great deal of interspecific overlap. Yet they do in fact

HABITAT SELECTION IN BIRDS

ISBN 0-12-178080-5

pay considerable attention to the structure of vegetation in choosing a place to live (Cody, 1978). Some are also influenced in their habitat choice by the densities of certain congeners (Cody, 1978; Cody and Walter, 1976).

But why do we see the habitat selection patterns we do? This is the underlying question that theory addresses. Can we understand what we see? How have the patterns been produced by evolutionary processes? Building theories to help us understand the observations must now be high on the agenda of priorities, because without comprehension there is no real prediction, merely extrapolation. Since this second step is founded on the first step of pattern elucidation, it is logical to put it at the end of the book. When a successor volume is compiled in a decade or two, it ought to have one or more theoretical chapters much closer to its beginning, followed by a distinguished series of contributions which test those theories. But that is another step, and we have hardly begun to ready our feet in order to take it.

I shall approach theories of habitat selection with a bias. I shall assume that individual organisms select particular habitats because they thereby enhance their fitness. However, the point of elucidating theories is to see them tested, and to expect them to be tested is to expect them to be found wanting.

Nevertheless, if all habitat selection theory does is to explain how habitat selection can be produced by natural selection, this contribution would be nothing more than a guiding arrow on a nature trail. However, it asks how the variety of answers to certain questions can influence the selective regime so as to affect whether there is habitat selection and what sort of habitat selection patterns we are likely to observe. It concentrates on graphical theories, because it is my experience that although these are harder to achieve, they are more general, are easier to understand, and offer more insight than do theories built on sets of differential equations.

Before we can turn to the questions, there is one point of definition we need to address. There is a laxity about the way ecologists use the words generalist and specialist. Sometimes we use them to describe behavior, sometimes to describe abilities. In this chapter I shall try to be consistent and use them only in the context of abilities. Thus, generalist means an individual able to use all habitats equally well; specialists are more adept at using some subset of patches. Note that these abilities must be measured with all else except patch type held constant. Now we need a pair of words to describe behavior. ''Selector'' and ''opportunist'' seem suitable and will be used. An opportunist uses habitats in their encountered proportions.

This is not merely an academic nicety, this distinction between behavior and ability. It is possible for an individual to be an opportunistic specialist or a selecting generalist. It is even possible that an individual will choose to avoid using the habitats for which it is specialized.

II. HABITAT SELECTION AND OPTIMAL FORAGING THEORY

Habitat selection theory is really a branch of optimal foraging theory. This has been clear since the seminal work of MacArthur and Pianka (1966). The approach imagines that habitats occur as patches and that organisms make choices about how to allocate their time among those patches. Theoreticians in this field often begin by assuming that there is a fixed environment with a given patch structure, then, ask whether an individual should be selective. For example, one can model the fitness of a patch user as

$$\overline{\ln W} = \Sigma\, t_i \ln W_i / \Sigma\, t_i \tag{1}$$

where i is a foraging activity type, $\ln W_i$ is net fitness while engaging in activity i, and t_i is the amount of time spent engaging in i during a foraging period. Note that the total time is not fixed in this model: the proportion of time spent foraging will vary according to how relatively rewarding foraging is. The activity types may include foraging in a patch or searching for a patch (if there is a searching cost).

Once such a model is established, it is differentiated with respect to one of its t_i values. For example, if there are two patch types, we may evaluate $\partial \ln W / \partial t_1$ to determine whether increasing the time spent foraging in patch 1 will increase fitness.

The result is a criterion for habitat selection. In the case of the two-patch universe, the criterion is typical: decrease use of patch 1 if and only if:

$$\ln W_2 / \ln W_1 > 1 + t_3/t_2 \tag{2}$$

where t_3 is the time spent looking for two patches, one of each type (Rosenzweig, 1974).

As simple as it is, Eq. (2) already presents many of the important theoretical issues in bold relief. It says that habitat selection against a patch is to be practiced only if one's fitness while using it is less than while using another patch (the ratio must be larger than 1). Moreover, it says the ratio of fitnesses must exceed 1 by an amount t_3/t_2, which is the ratio of the search time to the amount of time the individual gets to forage in it after it is found. Thus, anything that promotes a difference in the reward structure of patch types promotes habitat selection. So does anything which makes the better patches richer or larger (so that more time can be spent in one after it is found). In the limit, when an individual can spend its whole life in a patch (so that t_3 is zero and t_2, virtually infinite), profit from habitat selection requires only some differences between patches.

Another point implicit in Eq. (2) is more subtle. Notice that it does not contain

t_1. Thus, if it is a good strategy to reduce t_1 a bit, it is good to reduce it altogether to zero. This leads to the principle of "bang-bang control." There is no partial preference. Individuals should either use a patch type whenever it is encountered or ignore it entirely.

Equation (2) was derived using a set of ludicrous assumptions analogous, for example, to the set of assumptions used in deriving the Hardy–Weinberg distribution of basic population genetics. It assumes a constant environment except while the individual is foraging. (Foraging must reduce patch resources or else why would a patch ever be abandoned?) Yet all patches of a given type are uniform when they are first encountered; thus, patch recovery (resource renewal) must occur instantaneously after the patch is left behind. It assumes the animal has perfect knowledge of the reward and cost structure of its environment. That is, the animal identifies patches and traveling costs among them, knows their food value, and realizes the costs of searching for and harvesting this food. It also implicitly assumes that the animal does not interact with other members of either its own or other species. Equation (2) should therefore not be taken too seriously except as a foil against which to compare other more reasonable sets of assumptions or as a starting model on which to build other more interesting ones.

One of the first offshoots of the basic habitat selection model is "quitting time" theory (Charnov, 1976). The question here is: when should an individual leave a patch in which it is foraging? The first answer obtained, termed the "Marginal Value Theorem," is that an individual should forage in a patch until the yield from its foraging activity there drops to a rate equal to its average profit rate (over all habitat patches). Then it quits the patch and looks for another. This is not a contradiction, since the average includes not only foraging but traveling between patches [t_3 in Eq. (2)]. Oaten (1977; Stewart-Oaten, 1982) has demonstrated that the precise value of the quitting time is quite sensitive to the assumptions a model makes about the individual's knowledge of its environment. Nevertheless, the question asked by quitting time theorists seems sound. Moreover, if patches are variable in type, quitting time theory and habitat selection theory merge. Using the Marginal Value Theorem, for instance, we can say that a patch should not be used at all if its initial rate of profit return is below the average rate. And if a variety of patch types is used, each should be used until its resources have declined to produce the same rate of return, the average. We shall need this result in the following discussion.

Quitting time tests have been successfully performed in both the laboratory and the field (e.g., Cowie, 1977; Howell and Hartl, 1980). Baharav and Rosenzweig (1985) observed that Dorcas gazelles do remain in a patch in direct proportion to its usefulness, but their criterion of quality changes from energy–biomass content during the rainy season to something else (presumably the moisture content of the forage eaten) during the dry season.

Pyke's (1981) tests led to a successful alternative quitting time theory for a special situation. Australian honeyeaters (Meliphagidae, see Chapter 12, this volume) foraging on inflorescences of *Lambertia formosa* face a patch organization which is precise. Thus, they can completely drain it in one visit. A honeyeater can tell in a trice whether a patch has been depleted by sampling just one flower. Its quitting time should therefore be bimodal and it is: some flower patches are left quickly, others explored completely. Similarly, the visits of nectar-seeking beeflies to mustang clover (*Linanthus*) flowers in the Sierra Nevada show the same striking bimodality, presumably for the same reason (M. L. Cody, personal communication).

Another issue which is concerned with an individual's knowledge of patch structure has been recently addressed by W. Mitchell (personal communication). Mitchell points out that some species in certain environments are "myopic": they know about the location and quality of only one patch at a time. Given other species or other environments, however, individuals may be able to sense the location of many patches simultaneously; Mitchell terms such animals "periscopic."

A periscopic animal may not rank habitats strictly on the basis of how rapidly it accrues energy or fitness in them. For example, if a poor patch is near a set of rich ones and can be visited with little extra time and energy expended, it may be included in the foraging pathway. On the other hand, myopic animals have to visit a patch to tell what's in it, so they may find it useful to forage in it, once they have arrived there, even if it is a bit poorer than their optimum patch type. In sum, myopic animals should discriminate less between fairly similar patch types and more between rather different ones than periscopic animals.

W. Mitchell (personal communication) has tested this prediction in free living hummingbirds using artificial feeders whose sucrose concentration and identifying marks he controls. Sure enough, when the feeders were identifiable as rich or poor by sight (the periscopic case), poor feeders were used more often if they were near rich ones. Moreover, when the rich (1.0 *M* sucrose) and poor (0.8 *M*) food sources were similar, persicopic birds were selective, but myopic birds were not. And when the rich (1.0 *M*) and poor (0.5 *M*) were more different in quality, the periscopic birds were very slightly less selective than were myopic birds.

III. DENSITY-DEPENDENT HABITAT SELECTION IN ONE SPECIES

Since Eq. (2) is so sensitive to rewards and patch richness, it must be sensitive to user densities. A large population is likely to reduce the values of ln W_i and t_i for every patch type it uses. The consequences of this fact for habitat selection are critical (Svärdson, 1949; Morisita, 1950, 1969; Fretwell and Lucas, 1970); quite simply, it can turn a selector into an opportunist.

Fretwell's (1972) theoretical scheme is elegant and the easiest to explain. Fretwell imagines that individuals accumulate in a uniform patch. He then graphs their fitness against their density. Of course, as density increases, their fitness declines. First, it declines to the point at which they are just replacing themselves. Then, as they grow denser, their fitness becomes so low that their population declines. This is nothing more than the simplest sort of observation one makes to any beginning course in ecology: in any environment, a successful species has a carrying capacity.

Now Fretwell imagines more than one patch type yielding positive carrying capacities. Each has its own curve of declining fitness with increasing density. He also assumes there is no cost to reaching a patch. He draws a set of curves indicating how density and fitness are related in each patch (Fig. 1a). From these and a model like Eq. (2), it is clear that when a species is rare, it should select only the best of its patches, but as its density grows, it should spread out to equalize the return generated by all used patch types. This means that better patches support more individuals than do poorer ones. The resulting distribution is termed an "ideal-free distribution."

With somewhat different assumptions, organisms show an alternative ideal-despotic distribution. An ideal-despotic distribution is different, because it includes the effects of contest competition. These effects may actually result in lower densities in the better patches, as dominant or "alpha" individuals establish themselves in and protect their grip on a disproportionately high share of the best places to the exclusion of subordinate individuals. Yet even an ideal-despotic distribution produces the most interesting result: species abandon habitat selection and become more opportunistic as their density rises.

Fretwell's theory offers a practical message for field workers. If one wants to compare the utility or quality of habitats, compare their utilization patterns when

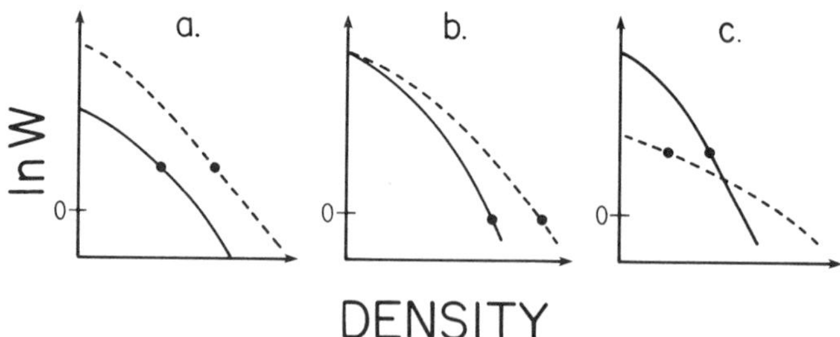

Fig. 1. Fitness as a function of density in two patch types. In (a) one patch is superior at every density. In (b) physiological limits render the patches effectively identical at low densities. In (c) one patch is better at low densities and poorer at high ones. In each graph, an example of an ideal-free distribution is given by the points on the lines. (After Holt, 1985.)

a species is rare. Habitat ranking will be much more transparent when the species is very common. This lesson will be reinforced in situations where there are more than one species. But there is one situation in which it will not work.

R. D. Holt has pointed out (personal communication; 1985) that patches may not differ in their ability to support a species until after the density of the occupants has risen to or beyond some threshold (Fig. 1b). This situation could be produced, for example, if there were physiological and/or physical limits to foraging and reproduction potential which are reached at low densities in all habitat types. Under such circumstances, there will be no habitat selection at low densities, but at high densities we can expect ideal-free distributions apportioning species according to the patches' different capacities to support life.

Finally, Holt (1985) notes the possibility that the effects of density could be so profound as to reverse the rank order of patch types (Fig. 1c) beyond some density threshold. We include this prospect without further comment. It could happen if resource standing crop and renewal rates were inversely related, a not too unlikely prospect. Bear in mind that it was an assumption like that of Fig. 1c that led MacArthur (1962) to the very influential idea of r and K selection.

The theory of density-dependent habitat selection has been tested quite often. A few prominent examples are demonstrated in Whitham (1978) for aphids, Bovbjerg (1970) for the crayfish *Orconectes immunis,* and Holmes (1961) for the rat tapeworm. Since this is after all a volume on avian habitat selection, I should point out that the ideas of both Fretwell and Svärdson are founded on a thorough knowledge of birds; the theory works for them too (Pimm *et al.*, 1985) as one can see from the example of Blue-throated Hummingbirds (*Lampornis clemenciae*).

Blue-throated Hummingbirds living in Cave Creek Canyon, Cochise County, Arizona, live mostly on sucrose solution in the springtime. Pimm *et al.* (1985) controlled these solutions and manipulated the Blue-throated densities. Habitat patches were created which contained either 1.2 *M* or 0.35 *M* sucrose. The time birds spent actually feeding from these solutions was recorded at various bird densities. The result was that the birds spent most of their time on the richer solution when they were uncommon, but as they increased, they divided their time more and more equally between rich and poor patches (Fig. 2a). (I shall take up this experiment again when population interactions are discussed later.)

We have developed a general technique for examining density dependence of habitat selection in species which are difficult to observe directly (Rosenzweig and Abramsky, 1984). It might apply, for example, to mist-netted birds in a dense vegetation or understorey. The technique uses censuses from grids or sets of points encompassing as much variation as possible with respect to a variable which is hypothesized to be important in determining habitat quality. The habitat within a grid should be as uniform as possible, the variation being between grids. In an ideal situation, density perturbations are now performed, and the censuses are repeated again and again. In a real situation, one might take advantage of

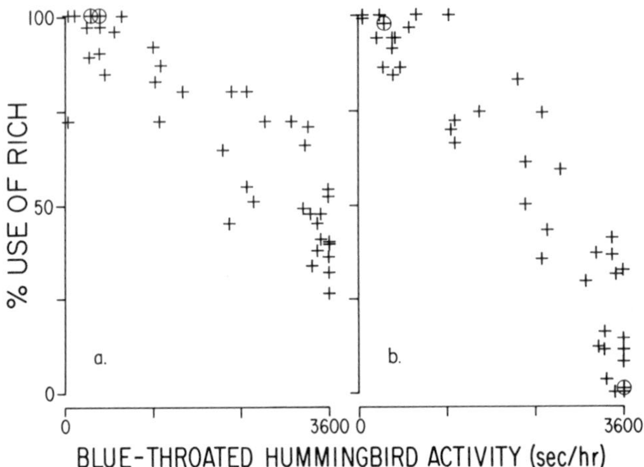

Fig. 2. (a) The more time Blue-throated Hummingbirds spend feeding in and defending patches with feeders, the smaller the proportion of time they spend in richer patches. (b) Rivoli's Hummingbird responds the same way, except more strongly. Greater blue-throated densities actually force Rivoli's to avoid the rich patches. (Adapted from Pimm *et al.*, 1985.)

some natural perturbations and also establish several sets of grids at similar sites to produce enough data for the test.

Simpson's index is now calculated for each set of censuses at a site,

$$y = \sum^{m} n_i^2/N^2 \tag{3}$$

where n_i is the number of individuals living in grid i and $N = \sum n_i$ is the total census on the m grids of a site. Simpson's index is itself density dependent. However, assuming that the individuals are multinomially distributed in the grids, it is possible to transform Simpson's index into a variable whose relationship to density is actually an aid to discerning habitat selection.

Making the assumption that individuals are multinomially distributed among the grids is like imagining that each grid is a bin and each individual is dropped into a bin with a certain chance, P_i. This allows us to use some standard machinery of probability, the most important of which is that in a multinomial distribution, it is possible to calculate the expectation of the numerators of Simpson's Index:

$$\epsilon(n_i^2) = N(N-1)P_i^2 + NP_i \tag{4}$$

where $\epsilon\ (n_i^2)$ is the expectation of n_i^2 and P_i is the true probability of an individual living in grid i (David and Barton, 1962). From Eq. (4) we have deduced the expectation of a linear transform of Simpson's index (Rosenzweig and Abramsky, 1984):

$$\epsilon(mNy-N-m+1) = (m \sum^{m} P_i^2-1)\,(N-1) \tag{5}$$

The critical part of Eq. (5) is the term ΣP_i^2. This is the true strategy of selectivity adopted by an individual. If, for example, it accepts all grids equally, then $P_i = 1/m$ and $\Sigma P_i^2=1/m$. In this case, the expectation of Eq. (5) is zero for all N. If there is selectivity, however, so that some P_i's are larger than others, then $\Sigma P_i^2>1/m$. In this case, the expectation of Eq. (5) will rise with N.

Now if we let

$$y' = mNy - N - m + 1 \tag{6}$$

$$x' = N - 1 \tag{7}$$

we can plot our census data and determine true selectivity. Figure 3a shows y' and x' for various values of ΣP_i^2 if $m=4$. Notice that all selectivities are straight lines with slope = $(m \Sigma P_i^2-1)$. All selectivities also transect the origin. Hyperdispersed data will appear as negative numbers.

We have been able to apply this technique to small nocturnal rodents of the Negev desert, Israel (Rosenzweig and Abramsky, 1984). Figure 3b is an example of data from Allenby's gerbil. Such data are easily analyzed by standard polynomial regression. Notice in Fig. 3b that somewhat selective behavior over low densities is replaced quite suddenly with rather unselective distributions at densities greater than 100 animals per site. This strongly supports the notion of density-dependent habitat selection as proposed by Fretwell and the others. The very rapid shift to a nearly random behavior seems to indicate bang-bang control in this case.

IV. COMPETITION AND HABITAT SELECTION

Early optimal foraging theorists in ecology were surely motivated by a desire to understand the community structure of competing species. Theorists were certainly aware that field and laboratory data implicated competition as a major influence on habitat choice. Some of the most recent of these data solidify this conclusion. For example, Cody (1983) showed that the relative habitat–niche breadths of "warbler-type" foliage-gleaning insectivores are inversely related to diversity in South African montane forests. Yet not until Lawlor and Maynard Smith (1976) could one point to a single theoretical paper which treated the relationship between competition and optimal habitat selection.

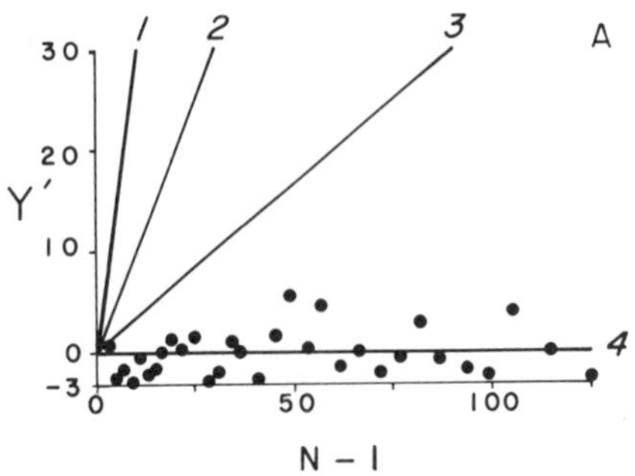

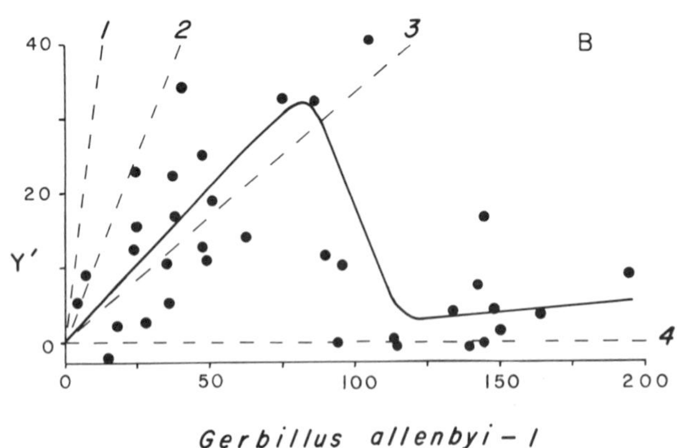

Fig. 3. (a) The transformed selectivity variable, y′, behaves linearly for any constant strategy of selectivity. As examples, the integer strategies are depicted: 1, 2, 3, and 4 equally common patch types (out of four patch types). A Monte Carlo simulation of four equally common patch types is plotted. (b) The data on patch distribution of *Gerbillus allenbyi,* whose total census is the variable *A*. This gerbil is selective until it reaches a density of 80–100, then it becomes distributed almost randomly. [Both (a) and (b) from Rosenzweig and Abramsky, 1984; copyright 1984 by Michael Rosenzweig.]

The difficulty was that, approached traditionally, the problem confronts the researcher with a sheer, ice-covered mountain of unwritable, unsolvable systems of partial nonlinear differential equations. The temptation to simplify such systems until one can at least solve them numerically is very hard to resist. Even Lawlor and Maynard Smith, who did resist, were constrained to examine their system near an assumed equilibrium point. What they found, however, was fascinating. Given a two-patch, two-species system in which each species' specialty was unique, each species emerged total habitat selectors. Species *A* used only its specialty, patch *a,* and *B* used only *b*. The competitive interaction

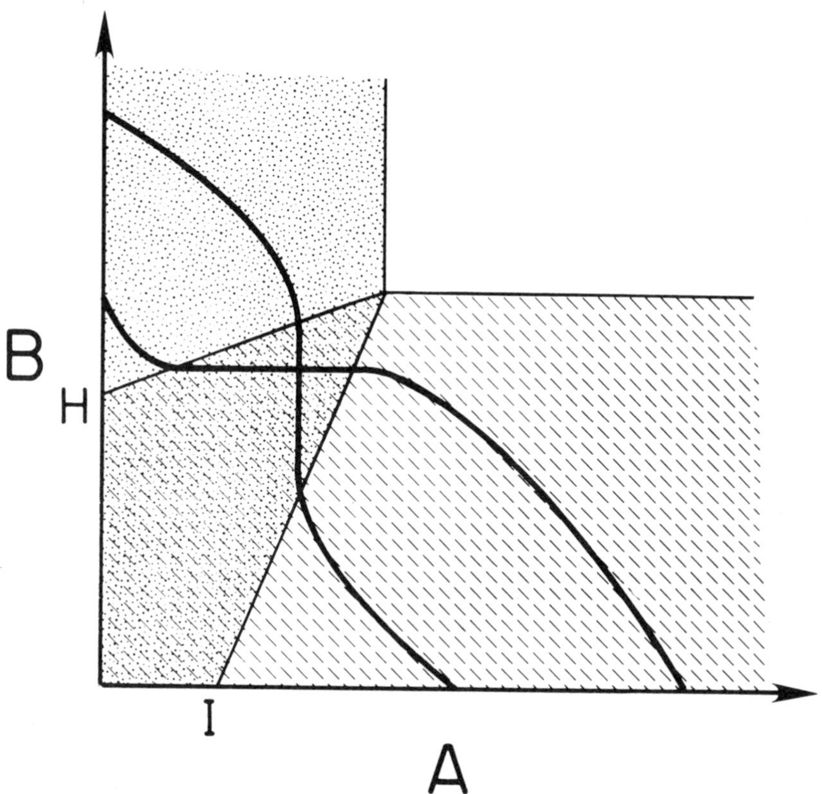

Fig. 4. Isolegs of two competing species whose habitat preferences are distinct. The axes are the species densities. *A* selects only habitat *a* in the stippled region: *B* selects only *b* in the shaded one. *A*'s isoleg is marked with I at its intercept; *B*'s is marked with H. Each species uses both habitats on the other side of its isoleg. If the isolegs cross (as in this case), they probably lose their positive slope and become perpendicular to their axes. Sample isoclines appear as thick lines.

between species had restored the habitat selection which the intraspecific competition removed.

Our mountain of equations becomes a pasture if we look for the patterns revealed by mapping behaviors onto graphs whose axes are population densities. This technique is called "isoleg" analysis after the Greek words for equal (*isos*) and choice (*lego*). An isoleg is a line in a state space of animal densities such that some aspect of a species' habitat selection is constant at every point on this line.

The first system analyzed with isolegs was the same as that of Lawlor and Maynard-Smith: two species, *A* and *B*, each have two acceptable habitat types, *a* and *b*, but are distinct and different habitat selectors at low densities, *A* in *a*, *B* in *b*. The isoleg I chose to follow was the threshhold isoleg: the set of points such that any increment in the density of species *i* would force its individuals to begin using the second patch type (Fig. 4). Both the *A* and *B* threshhold isolegs turned out to have positive slopes (Rosenzweig, 1981); in other words, the interaction between species *A* and *B* does indeed strengthen the tendency for each species to be selective just as Lawlor and Maynard-Smith had said. Whether they were separated completely by habitat, however, was a more complex question, which I shall treat later.

As yet there have been only three isoleg analyses made, and none show a pair of positively sloped isolegs. Luckily, before we even began our tests of the theory, Stuart Pimm made me realize the need for an isoleg model to cover the case of interference competition.

Suppose two species share an ability to use patch *a* better, but *A* is dominant, *B* subordinate. *B* can also use *b* successfully; *A* may or may not use *b* with profit, but if it does, it does not dominate *B* therein. The most important deduction from this model is that *B* has a negatively sloped isoleg which divides the graph into a region where it selects *a* preferentially and one in which it does not. The reason is that as *A*s are added, they depress $\ln W_{B,a}$, since the *A*s also select *a*. Given an inequality like Eq. (2), *B* finds its ratio of fitnesses, $\log W_b/\log W_a$, depressed, its ratio of $t_{3,B}/t_{B,a}$ increased, and habitat selection a less and less profitable choice.

Suppose we follow along the *A* axis in Fig. 5a until we have added enough *A*s to depress the profitability of *a* patches so that Eq. (2) becomes an equation. That will be the *A* intercept of *B*'s isoleg. But if $t_{3,B}$ is zero (i.e., the habitat selection is cost free) and we add even one more *A*, *b* will have to become the superior habitat for *B* and will need to be selected at low *B* density. Somehow there is another *B* isoleg and we have crossed it too! This second *B* isoleg separates the graph into a region where *B* selects only *b* and a region where it does not.

If $t_{3,B}>0$, then these two isolegs' intercepts are separated along the *A* axis, because, although Eq. (2) is reduced first to an equation, then reversed, $\ln W_{B,a}$ remains greater than $\ln W_{B,b}$ at first. In other words, both habitats must be accepted opportunistically. The larger $t_{3,B}$, the greater the separation and the larger the region of *B*'s opportunism.

The second isoleg can touch the first only on the *A* axis (if at all), as you can probably now deduce for yourself. As we trace it up, we expect a positive slope instead of a negative one. To see this, imagine a point in the region just to the right of the second isoleg. Adding more *B*s (i.e., going up) should reduce *b*'s attractiveness and eventually produce another isoleg point.

There may also be an *A* isoleg. If there is, it should be rather steeply positive, because *B* has such a small effect on *A* in *a* that it probably also has a small effect on *A* in *b* (Pimm *et al.*, 1985). All these features are summarized in Fig. 5a.

Pimm *et al.* (1985) tested the interference isoleg system in two pairs of hummingbird species living mostly on sucrose solutions in Cave Creek Canyon,

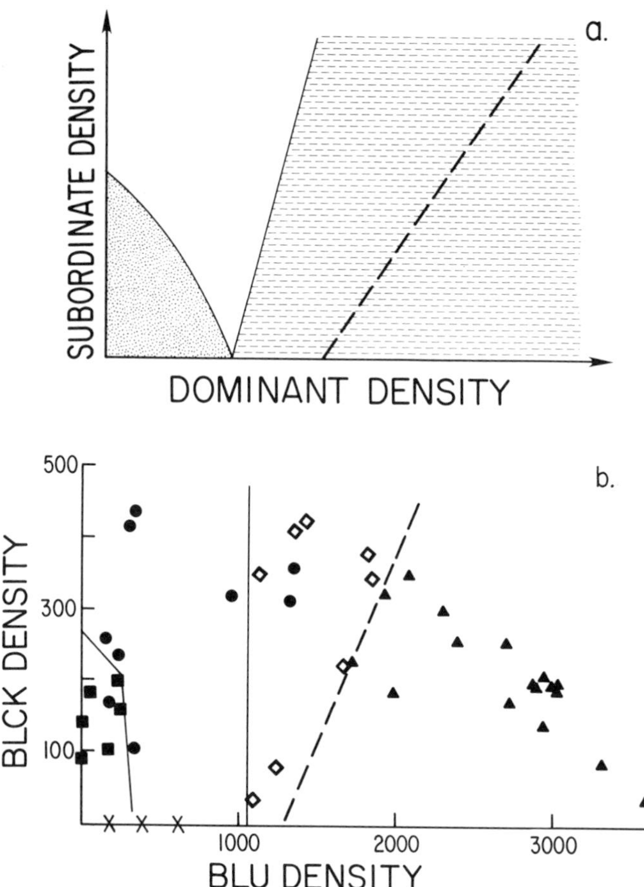

Fig. 5. (a) Isolegs of a two-patch, cost-free, shared preference system with interference. The subordinate selects the better in the stippled region and the poorer in the shaded region. The subordinate is opportunistic in the clear portion. The dominant may have an isoleg too (dashed line). If so, it selects the better patch to the left of its isoleg. (b) Behavior and isolegs of Blue-throated and Black-chinned Hummingbirds in an experimental system. Code: solid squares, both species select rich; dots: blue-throats select rich, and black-chinneds are opportunistic; diamonds: blue-throats select rich, and black-chinneds select poor; triangles: blue-throats are opportunistic, and black-chinneds select poor.

Cochise County, Arizona. Habitat patches were created which contained either 1.2 *M* sucrose or 0.35 *M* sucrose. In both pairs, the Blue-throated Hummingbird was a territorialist and dominant over Black-chinned Hummingbirds (*Archilochus alexandri*) and Rivoli's Hummingbirds (*Eugenes fulgens*). Of course the habitats with 1.2 *M* sucrose were better for *all* species. We explored the state space, both by temporary artificial perturbations and by taking advantage of natural variations in density. The result was that both pairs conformed to the isoleg theory. Figure 5b shows the isolegs of the interaction with Black-chinned Hummingbirds (*Archilochus alexandri*). Figure 2b shows how the foraging activity of Rivoli's Hummingbird declined from strongly selecting the rich patches to strongly selecting the poor ones as *Lampornis* density increased.

A third set of isolegs has recently been obtained from a pair of gerbils, *Gerbillus allenbyi* and *G. pyramidum,* which live in sandy habitats in Israel (Rosenzweig and Abramsky, 1984). To obtain their isolegs, we used estimates of selectivity based on Simpson's index. In fact, using Eq. (5), we showed that a density independent estimate of $1/\sum P_i^2$ which behaves linearly is

$$Z = (N - 1)/(Ny - 1) \tag{8}$$

where Z stands for number of equally common habitats accepted.

Figure 6 shows the result of the analysis. The result was both encouraging and disconcerting. It was encouraging because both species' Z values showed clear patterns of response to the other species. In each case, the second species density was a significant predictor of the Z value of the other. It was disconcerting, because no one had previously predicted the pattern we got: both isolegs have a negative slope.

In response to the gerbil data, we produced a new model, which does lead to the pattern. Moreover, I think it leads to a better conceptual understanding of habitat selection theory.

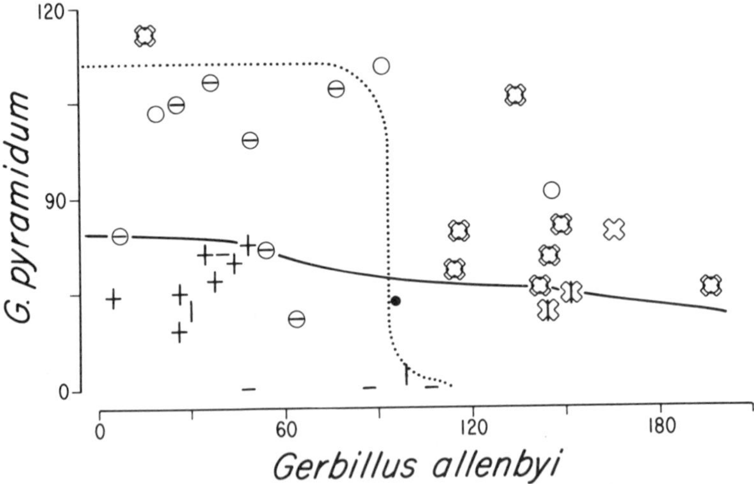

Fig. 6. Behavior and isolegs of two psammophilic gerbils. *Allenbyi:* selective (horizontal line), opportunistic (open cross); *Pyramidium:* selective (vertical line), opportunistic (open circle). At points where either species' behavior was indeterminate, there is no symbol for that species. Thus, the dot near the isolegs' intersection was a case of both species having intermediate behavior. Behavior of *G. pyramidum* is also not recorded at three points having none of these animals.

Assume that there are three habitat types and two species. Both A and B prefer b (at least when A and B are rare). But as A gets common, it begins to include a in its patch set; whereas when B gets common, it begins to include c in its patch set. The result is two negatively sloped isolegs.

The reason for the negative slopes is that when either species is common enough to include its secondary habitat type, it is facing competition from a species which is not using that habitat but is using the primary habitat type. So the more dense this second species is, the less valuable is the primary habitat and the sparser the first species is when it needs to include its secondary patch type in its repertory.

Previously, we had understood negatively sloped isolegs as products of interference competition (Rosenzweig, 1979; Pimm *et al.*, 1985). The dichotomy was exploitative or interference competition, the former yielding positively sloped isolegs, the latter negative. Now we can see that the real distinction is between shared preferences and distinct preferences. I did not need to postulate interference to explain the gerbil case. Indeed, it is difficult to understand how interference could ever explain it. But the gerbil situation became possible to understand without difficulty as soon as we allowed the species to share their preference in an environment with three patch types.

Similarly, we can understand both the negative and positive isoleg slopes in an interference case (Fig. 5a): the negatively sloped subordinate isoleg occurs in a region of the graph where the species share a preference. The positively sloped isolegs occur where there are so many dominants that preferences of the species are distinct.

How does the assumption of discrete patch types influence the results of theory? We may address this question in two ways. First, we may allow for variation in the resource level within each patch type; this is easily imagined since in fact each patch is depleted while it is foraged and then requires time to reach again an equilibrium level of resources. Second, we may imagine that there is so much variation that the exercise of identifying patch types is pointless; there is instead a continuum of patch quality.

Joel Brown and I (Brown and Rosenzweig, 1985) have considered the first situation. We imagined that each individual must forage in a number of patches to prosper. Each visited patch is foraged down to a quitting level of resources. (The Marginal Value Theorem or any other quitting rule can be in force.) The patch then begins recuperating and is reencountered by some forager in an amount of time which fits some unspecified random variable. The quitting rule is the same for any patch type, and it is this rule which generates any habitat avoidance (the converse of which we call habitat selection).

Using such a model and assuming two patch types and two distinct-preference foraging species, we determined the following. When only one species is present, suppose the equilibrium resource level in patch a is below the quitting

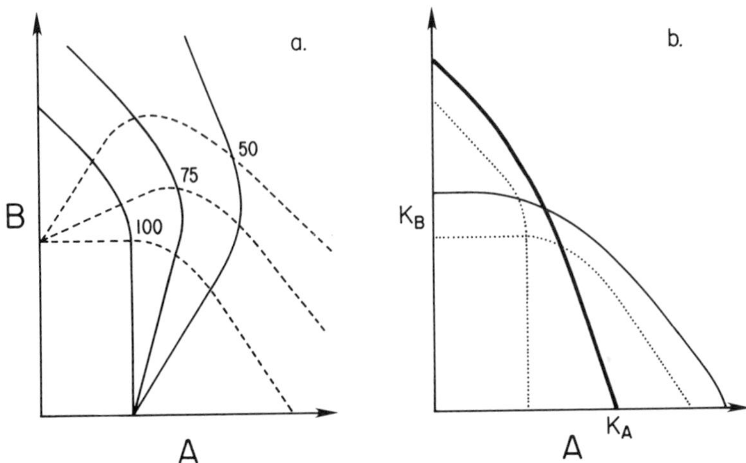

Fig. 7. (a) Isolegs of distinct preference species whose densities are A and B, when patches are semicontinuous (see text). Solid lines are A's isolegs; dashed lines are B's. Numbers are the percentage of the better habitat type in the foraging of a species and thus indicate the partial preference. (b) An example of isoclines on the isolegs of Fig. 6a. Isolegs of 100% are dotted lines.

level. This species will select b alone. As the species grows in density, the average recuperating patch b will suffer a revisit in a shorter time and thus contain fewer resources. The result is that the average yield is reduced and so is the quitting level. At some high enough density, the quitting level is low enough so that the a patches are also accepted. The result duplicates the single species predictions, even to bang-bang control.

Now let us add the second species. We assumed the second, when rare, to prefer the a patches. Thus, they impose on them a distribution of resource density so that only some are as high as equilibrium. When species one has reduced the yield of the b's so that quitting time has reached the equilibrium level in a, species one will be at exactly the same density as before two was added, but only some a's will be acceptable. There is, in other words, no change in the density of one at which it encounters its isoleg. But now there remains a partial preference of one for the b patches. As one's density increases, its quitting time declines, so its partial preference does too. The result for both species is typified by the two fan-shaped parts of Fig. 7a. Isolegs here tend to retain their positive slope (like the purely discrete patch case), but the isoleg which indicates the border of 100% habitat selection has become a line perpendicular to its species' density axis.

After the isolegs cross, however, their pattern changes. Let us focus on A's 100% isoleg. This begins, as I have said, by being vertical. But once it crosses B's 100% isoleg, it enters a region where B is using both habitat types. Thus, A's quitting time in its preferred patch must be lowered, and A's isoleg, along with it. As A's isoleg ventures past more and more of B's partial preference isolegs, B's effect on A's preferred habitat is increased, causing A's 100% isoleg to curve more away from the vertical. Finally, the last of B's isolegs is crossed, and A's isoleg follows a constant slope. The same argument can be repeated for the other isolegs of both species, resulting in the rest of Fig. 7a.

Brew (1982) has modeled and analyzed the second case, the case of a continuum of patch qualities. He has allowed for any number of species. His results are astonishingly similar in some ways to the case of discrete or semidiscrete patch types, but they also exhibit a crucial difference which makes the continuous case even more extreme in its predictions about habitat selection. This surprised me, because I had expected the continuous case to result in less extreme habitat selection.

Brew uses the Q function (MacArthur, 1970), which is the sum of the squared differences between available and actual resource production at all resource qualities. But it is easy to imagine that the resource qualities are really patch qualities in a habitat continuum (Joel Brown, personal communication). In fact, Joel Brown points out that even in a thoroughly homogeneous environment with only one resource type supplied at a constant rate, there will be a heterogeneous distribution of resource densities over space; individuals will forage each spot down to the quitting level and return to the spot in variable amounts of time. Pulliam (1983) has already hypothesized that such variation coupled with variation in species' minimum seed density requirements may account for coexistence of various sparrow species.

Like MacArthur, Brew seeks to minimize the Q function; i.e., minimize the community's waste in utilizing the spectrum of habitat types. But, unlike MacArthur, he allows individuals to adjust their behaviors to achieve the minimization. He does not show whether such minimization is in fact optimal for the individuals comprising the community. But he argues that since the minimization results in all habitat types supporting the identical per capita growth rate, i.e., zero, it is quite likely that the individuals' distribution to minimize Q will be selected for. This argument certainly seems reasonable, at least for exploitative consumers without territoriality.

When Q is minimized in the presence of several species that can coexist, Brew demonstrates that each species occupies a range of the habitat types exclusively. There is no overlap at all. Moreover, even when species densities are not at equilibrium, there is no overlap. The position of the dividing lines between the species will be different, but there will still be sharp dividing lines. If one assumes discrete patch types, this sort of result is obtained only under restricted circumstances (Rosenzweig, 1979, 1981; Pimm and Rosenzweig, 1981).

The isolegs that result from Brew's model are not exactly the same as in a discrete model. We cannot classify the behaviors into a few categories and

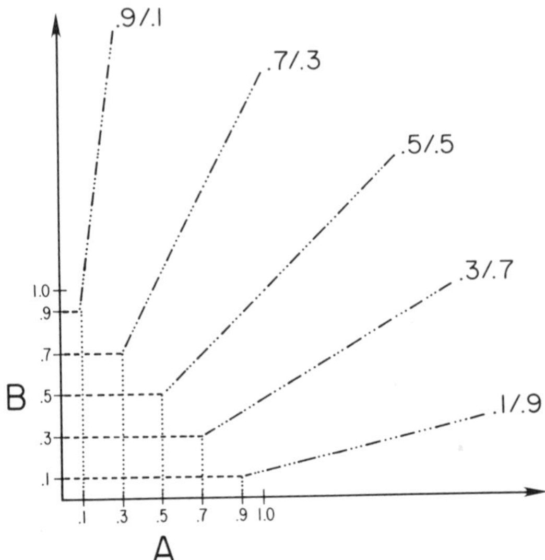

Fig. 8. Isolegs in a continuous spectrum of habitats. *A* and *B* are species densities. The numbers are the ratio of the range of habitats which *B*s use to that which *A*s use. Isoclines (not drawn) are concave.

delimit the regions of state space where they obtain. What we can do instead is very much like what was done to obtain the gerbil graph (Fig. 6). We can assess the range of habitat types used and ask how that is influenced by population densities.

Brew notes that if a specialist is alone and at low density, it will use a restricted range of habitats around its optimum. As its density increases, the range does too (see examples in Chapter 1, this volume). He also notes that addition of another species will reduce the range of the first (provided both are not so rare and different that their ranges are restricted to different regions of the spectrum). Hence, these isolegs too must have a positive slope. They are also constrained to total no more than 100% of the range (because species never should overlap in patch use.) Figure 8 is an example of a two-species case where the species have distinct preferences (they do better on opposite ends of the spectrum).

When testing such models as Brew's, we have to keep in mind the advice of Joel Brown. If the quality gradient is simply resource density and is produced not by inherent spatial variation but by random waiting times before patch revisitation, then given enough time, each patch will attain the whole spectrum of densities. We will therefore need to measure resource density every time we perform a census or observation on a patch (a formidable task in most circumstances, but exemplified in the work of Cody, 1981, 1983).

V. POPULATION CONSEQUENCES OF HABITAT SELECTION

Now that we have looked at how densities determine habitat selection, we can turn the question around. How do selectivities help determine population dynamics, population size, and even geographical range? Ultimately, we are asking the same suite of questions that has bedeviled community ecologists for a long time. Most of the answers offered by optimal foraging theory are tentative, incomplete, and untested. Nevertheless, I believe they are promising and could well represent the relief community ecology has been seeking for its severe case of frustration–depression.

I begin with a case that has been around a long time, but is not generally included when people consider the effects of habitat selection: the prey or victim refugium. Most ecologists are aware that refugia can promote global stability in predatory interactions (Rosenzweig and MacArthur, 1963). But they do not often recall that a necessary feature of this effect is density-dependent habitat selection: when victims are rare, they are assumed to stay in the refugia as much as possible. This was part of the original theory. If victims instead were to spread out at random, the refugia would lower predatory proficiency somewhat, but the interaction would be less stable globally unless the refugia were very, very abundant.

Holt (1985) has recently addressed the other side of the question: suppose it is the predator which is habitat selecting; what then is the effect on stability? Not at all surprisingly, the effect is reversed. If there are two habitats and in only one of them is the predator any good at obtaining food, selection of this habitat enhances the predator's ability to survive and reproduce. In general, any such enhancement tends to produce instability.

By assuming that more overlap means more competition, we can also address competitive dynamics. As I have shown elsewhere (Rosenzweig, 1981), coexistence can actually result from habitat selection; there is no doubt that it actually does. More interesting is how habitat selection transforms the shapes of isoclines and determines the values of competition coefficients.

In regions of state space where species are selecting entirely different patch types, their isoclines are perpendicular to their density axes. In other regions they are negatively sloped.

Given a two-species, two-patch-type, distinct preference case, Pimm and I (Pimm and Rosenzweig, 1981) showed that the isoclines are likely to intersect in the region where they are both perpendicular. Their shapes are also rather distinctive: they go from negative slope to perpendicular to negative slope. Brew (1982) has termed these shapes "strange but valid." One hopes their strangeness will make them early targets of field tests. Brew's own theories lead to concave isoclines (Brew, 1982).

The most unsettling prediction of habitat selection theory, however, has hardly been the isocline shapes. It is the prediction that equilibrium is likely to occur where there is no overlap in patch use. This conclusion has been reached also by Lawlor and Maynard-Smith and by Brew. It means that competitive alphas should be zero at community equilibria of exploitative competitors. And it gave rise to the idea that competitive dynamics could well be a ghost of competitions past, haunting only the fringes of state space to which interactions are rarely driven, and otherwise eliminated by optimal foraging.

The notion that competitive alphas may be constrained near zero in at least some mature interactions is not without empirical support in the literature of avian habitat selection. For example, the jays, orioles, *Basileuterus* warblers, and *Myiarchus* flycatchers of Chapter 1 (this volume) provide likely field examples of this phenomenon. Moreover, Cody (1981) has remarked: "On tropical mountainsides, both Terborgh (1971) . . . and Diamond (1972) . . . have shown that elevational gradients may be subdivided by three-to-six related bird species, with precise and non-overlapping turnovers between adjacent species at elevations that vary among species sets and . . . do not, in general, correspond to habitat ecotones."

The fact that alphas of zero are less likely for shared preference competitions (Rosenzweig, 1979) means that perturbation studies are still going to be useful. It may also explain why perturbation studies of putative competitors generally produce the result that if there is competition, it is based on a shared preference (usually the rich end of a habitat gradient). Classical exploitative competition should be hard to document. I still believe, and Brew's work only reinforces me in this, that the likeliest place to find classical exploitative competition is at the boundary where two parapatrically distributed species meet. Elevational turnover boundaries (e.g., Terborgh, 1971; Diamond, 1972) may be the most practical places at which to conduct the needed experiments.

Yet the model of Fig. 7a is an exception among purely exploitative models. It predicts that isoclines will cross in that region of the graph where both species have a partial preference. Here their alphas should be positive (Brown and Rosenzweig, 1985). Figure 7b presents one example of the appearance of isoclines in such a system.

Holt (1985) has begun to attack two related problems. How does habitat selection affect population size? And how does it affect a species' geographical range? Although the effects can be profound, they are in each case ambiguous. A single species which is optimally habitat selecting may experience an increase or decrease in either its population size or its range, compared to a species which is not.

VI. BUILDING A THEORY OF COMMUNITY ECOLOGY

There has been a growing feeling of despair among community ecologists. The limiting similarity model has been justifiably scrapped (Rummel and Roughgarden, 1983, are about as kind to it as any recent contribution.). The search to measure even one community's alpha matrix is becoming more and more difficult (Rosenzweig *et al.*, 1984; Rosenzweig *et al.*, 1985). Any species set we study is only an abstraction from a much larger interacting set whose properties are hard to judge from any subset (Schaffer, 1981). Virtually any qualitative dynamical property can be displayed by virtually any sort of interaction—given the right time lags and coefficients (May and Oster, 1976; May, 1976). Moreover, the promise that at least ultimately we could go and do perturbation experiments to clarify community structure is now haunted by the threat of the ghost of competitions past. And, finally, we are witness to the clashing views of Diamond (e.g., Diamond and Gilpin, 1982) and Simberloff (e.g., Connor and Simberloff, 1979), which has absorbed so much of the energies of community ecologists for the past 5 years. Should we take the advice of some and learn to do cloning experiments? Or should we become ecological reactionaries, deny the existence of pattern, and insist that theory has no proper role to play in ecology?

Let us instead persist! I suggest that habitat selection theory may well offer some hope and serve also as a guide for theories of community ecology.

Levins (1966) pointed out that some realistic theory is general and some precise. The most general makes simply a yes–no prediction or predicts the sign of a slope. This is surely suspect since it has half a chance of being found accurate even if wrong. On the other side, the most precise theories predict actual numbers; Slobodkin (1965) once warned against lusting after such precision. Ecological systems are simply too noisy and probably too fuzzy for us ever to predict exact numbers with accuracy (May, 1976).

What is left is a middle path, which is exemplified by habitat selection theory. A suite of qualitative predictions emerges. Each may be right half the time by chance, but the whole set of predictions is much less likely to hold up unless the theory is valid. For example, the theory which produced Fig. 7 is full of qualitative predictions: (1) Preferences should be partial. (2) Isolegs should be convex, have both positive and negative slopes but converge on an axis. (3) Isoclines should be nonlinear and articulate with isolegs in a restricted number of ways. Figure 4 affords another example. If we can produce other theories of community ecology which yield suites of qualitative features for testing, I think we are likely to make progress.

There is at least one other strategy of habitat selection theory which is worth mimicking. These theories begin by asking what individuals should do under the influence of natural selection. Community structure more or less accumulates as the effect of all these separate, simultaneous decisions. Roughgarden (1976) has taken much the same view.

Using evolutionary thought, we have seen how habitat selection can tend to disappear in a single species under the influence of density dependence, how it can be reinstated by competitors, and how positively sloped isolegs are generated by distinct habitat preferences, negative ones by shared preference. Elsewhere (Rosenzweig, 1979) I have used evolutionary thought to argue that specialization should intensify in distinct preference species with high positive density covariance but weaken if the covariance is highly negative. We could also use it to conjecture that communities of territorialists should be more habitat selective than trapliners, since the latter spend a considerable amount of time and energy traveling between patches [see Eq. (2)].

We may never be able to answer the traditional questions about community structure. But I am encouraged that habitat selection theory will help us answer related questions at least as interesting.

Why does exploitative competition seem so rare among habitat selectors (see Schoener, 1983)? Is it that exploitative habitat selectors tend to be parapatric and not co-occurring?

Why are active interspecific competitions so rare or elusive (Whittam and Siegel-Causey, 1981)? Is it that competitive alphas are often reduced to near zero by optimal habitat selection?

Under what circumstances should we expect competitive interactions to be asymmetrical (Hallett, 1982; Hallett, O'Connell and Honeycutt, 1983)? Does this occur when habitat preferences are shared? When should they be symmetrical? Are they symmetrical when habitat preferences are distinct? What might be the role of interference and territoriality in community structure? Do they appear only in shared preference systems?

Why should one congener be social, another in the same or similar habitat seem to lack sociality (Pulliam and Mills, 1977; Rosenzweig and Abramsky, 1984)? Does this have to do with the cost of predation in different habitats (Pulliam and Mills, 1977)? Or does it have to do with the permanence and predictability of a species' habitats?

What is the driving mechanism of the taxon cycle? And why is it that rare species tend to have restricted geographical distributions, while common ones are more widespread (Bock and Ricklefs, 1983; Brown, 1984)? Is the answer to both these questions that the righ end of the habitat spectrum is a relatively uncommon niche—however evolutionarily magnetic it is?

What reduces the effective dimensionality of natural systems so that they become somewhat predictable (Schaffer, 1984)? Is it that density-dependent habitat selection reduces the variance of fates among individuals? Why is allozymic variation in the pitcher plant mosquito related to temporal and not to spatial variation (Istock, 1985)? Is it that density-dependent habitat selection eliminates much of the selective differential among habitats in space but cannot eliminate it to as great an extent among habitats distributed in time?

If these questions intrigue you and you think the answers suggested in this chapter are at least worth developing and testing, then you too are optimistic about the potential impact of habitat selection theory on community ecology.

REFERENCES

Baharav, D., and Rosenzweig, M. L. (1985). Optimal foraging in Dorcas gazelles. *J. Arid Environ.* (in press).

Bock, C. E., and Ricklefs, R. E. (1983). Range size and local abundance of some North American songbirds: A positive correlation *Am. Nat.* **122,** 295–299.

Bovbjerg, R. V. (1970). Ecological isolation and competitive exclusion in two crayfish (*Orconectes virilis* and *Orconectes immunis*). *Ecology* **51,** 225–236.

Brew, J. S. (1982). Niche shift and the minimization of competition. *Theor. Pop. Biol.* **22,** 367–381.

Brown, James. (1984). On the relationship between abundance and distribution of species. *Am. Nat.* **124** (in press).

Brown, Joel, and Rosenzweig, M. L. (1985). Habitat selection in slowly regenerating environments. Unpublished manuscript.

Charnov, E. L. (1976). Optimal foraging. The marginal value theorem. *Theor. Pop. Biol.* **9,** 129–136.

Cody, M. L. (1978). Habitat selection and interspecific territoriality among the sylviid warblers of England and Sweden. *Ecol. Monogr.* **48,** 351–396.

Cody, M. L. (1981). Habitat selection in birds: The roles of vegetation structure, competitors, and productivity. *BioScience* **31,** 107–113.

Cody, M. L. (1983). Bird diversity and density in south african forests. *Oecologia* **59,** 201–215.

Cody, M. L., and Walter, H. (1976). Habitat selection and interspecific interactions among Mediterranean sylviid warblers. *Oikos* **27,** 210–238.

Connor, E. F., and Simberloff, D. (1979). The assembly of species communities: Chance or competition? *Ecology* **60,** 1132–1140.

Cowie, R. J. (1977). Optimal foraging in great tits (*Parus major*). *Nature* **268,** 137–139.

David, F. N., and Barton, D. E. (1962). "Combinatorial Chance." Griffin, London.

Diamond, J. M. (1972). "The Avifauna of the Eastern Highlands of New Guinea." Nuttall Ornithology Club, Cambridge, Massachusetts.

Diamond, J. M., and Gilpin, M. E. (1982). Examination of the "null" model of Connor and Simberloff for species co-occurences on islands. *Oecologia* **52,** 64–74.

Fretwell, S. D. (1972). "Populations in a Seasonal Environment." Princeton Univ. Press, Princeton, New Jersey.

Fretwell, S. D., and Lucas, H. L., Jr. (1970). On territorial behavior and other factors influencing habitat distribution in birds. I. Theoretical development. *Acta Biotheor.* **19,** 16–36.

Hallett, J. G. (1982). Habitat selection and the community matrix of a desert small-mammal fauna. *Ecology* **63,** 1400–1410.

Hallett, J. G., O'Connell, M. A., and Honeycutt, R. L. (1983). Competition and habitat selection: Test of a theory using small mammals. *Oikos* **40,** 175–181.

Holmes, J. C. (1961). Effects of concurrent infections on *Hymenolepsis diminuta* (Cestoda) and *Moniliformis dubius* (Acanthocephala) I. General effects and comparison with crowding. *J. Parasitol.* **47,** 209–216.

Holt, R. D. (1985). Population dynamics in two-patch environments: Some anomalous consequences of optimal habitat selection. *Theor. Pop. Biol.* (in press).

Howell, D. J., and Hartl, D. L. (1980). Optimal foraging in Glossophagine bats: When to give up. *Am. Nat.* **115,** 696–704.

Istock, C. A. (1985). Pattern and process in life history genetics. *In* "Ecology of Mosquitoes: Proceedings of a Workshop." (L. P. Lounibos, J. R. Rey, and J. H. Frank, eds.) Amer. Entomological Inst., Ann Arbor, Michigan. (In press).

Lawlor, L., and Maynard-Smith, J. (1976). The coevolution and stability of competing species. *Am. Nat.* **110,** 79–99.

Levins, R. (1966). Strategy of model building in population biology. *Am. Sci.* **54,** 421–431.

MacArthur, R. H. (1962). Some generalized theorems of natural selection. *Proc. Nat. Acad. Sci. U.S.A.* **48,** 1893–1897.

MacArthur, R. H. (1970). Species packing and competitive equilibrium for many species. *Theor. Pop. Biol.* **1,** 1–11.

MacArthur, R. H., and Pianka, E. R. (1966). On optimal use of a patchy environment. *Am. Nat.* **100,** 603–609.

May, R. M. (1976). "Theoretical Ecology: Principles and Application." Saunders, Philadelphia, Pennsylvania.

May, R. M. and Oster, G. F. (1976). Bifurcations and dynamic complexity in simple ecological models. *Am. Nat.* **110,** 573–599.

Morisita, M. (1950). Dispersal and population density of a water strider, *Gerris lacustris* L (in Japanese). *Contr. Physiol. Ecol.*, Kyoto Univ., No. 65.

Morisita, M. (1969). Measuring of habitat value by the "environmental density" method, *In* "Statistical Ecology, Vol. 1: Spatial Patterns and Statistical Distributions" (G. P. Patil, E. C. Pielou, and W. E. Waters, eds.), pp. 379–401. Penn. State Univ. Press., Univ. Park, Pennsylvania.

Oaten, A. (1977). Optimal foraging in patches: A case for stochasticity. *Theor. Pop. Biol.* **12,** 263–285.

Pimm, S. L., and Rosenzweig, M. L. (1981). Competitors and habitat use. *Oikos* **37,** 1–6.

Pimm, S. L., Rosenzweig, M. L., and Mitchell, W. (1985). Competition and food selection: Field tests of a theory. *Ecology* (in press).

Pulliam, H. R. (1983). Ecological community theory and the coexistence of sparrows. *Ecology* **64,** 45–52.

Pulliam, H. R., and Mills, G. S. (1977). The use of space by wintering sparrows. *Ecology* **58,** 1393–1399.

Pyke, Graham. (1981). Honeyeater foraging: A test of optimal foraging theory. *Anim. Behav.* **29,** 878–888.

Rosenzweig, M. L. (1974). On the evolution of habitat selection. *Proc. Int. Congr. Ecol. 1st,* 401–404.

Rosenzweig, M. L. (1979). Optimal habitat selection in two-species competitive systems. *Fortschr. Zool.* **25,** 283–293.

Rosenzweig, M. L. (1981). A theory of habitat selection. *Ecology* **62,** 327–335.

Rosenzweig, M. L., and Abramsky, Z. (1984). Detecting density-dependent habitat selection. Unpublished manuscript.

Rosenzweig, M. L., and MacArthur, R. H. (1963). Graphical representation and stability conditions of predator–prey interactions. *Am. Nat.* **97,** 209–223.

Rosenzweig, M. L., Abramsky, Z., and Brand, S. (1984). Estimating species interactions in heterogeneous environments. *Oikos* **43,** 329–340.

Rosenzweig, M. L., Abramsky, Z., Kotler, B., and Mitchell, W. (1985). Can interaction coefficients be determined from census data? *Oecologia* (in press).

Roughgarden, J. (1976). Resource partitioning among competing species—a coevolutionary approach. *Theor. Pop. Biol.* **5,** 163–186.

Rummel, J. D., and Roughgarden, J. (1983). Some differences between invasion-structured and coevolution-structured competitive communities: A preliminary theoretical analysis. *Oikos* **41,** 477–486.

Schaffer, W. M. (1981). Ecological abstraction: The consequences of reduced dimensionality in ecological models. *Ecol. Monogr.* **51,** 383–401.

Schaffer, W. M. (1984). Order and chaos in ecological systems. *Ecology* **66,** 93–106.

Schoener, T. W. (1983). Field experiments or interspecific competition. *Am. Nat.* **122,** 240–285.

Slobodkin, L. B. (1965). On the present incompleteness of mathematical ecology. *Am. Sci.* **53,** 347–357.

Stewart-Oaten, A. (1982). Minimax strategies for a predator–prey game. *Theor. Pop. Biol.* **22,** 410–424.

Svärdson, G. (1949). Competition and habitat selection in birds. *Oikos* **1,** 157–174.

Terborgh, J. (1971). Distribution on environmental gradients: Theory and a preliminary interpretation of distributional patterns in the avifauna of the Cordillera Vilcabamba, Peru. *Ecology* **52,** 23–40.

Whitham, T. G. (1978). Habitat selection by *Pemphigus* aphids in response to resource limitation and competition. *Ecology* **59,** 1164–1176.

Whittam, T. S., and Siegel-Causey, D. (1981). Species interactions and community structure in Alaskan seabird colonies. *Ecology* **62,** 1515–1524.

Index

D

E

G

H